Design of Experiments for Chemical, Pharmaceutical, Food, and Industrial Applications

Eugenia Gabriela Carrillo-Cedillo
Universidad Autónoma de Baja California, Mexico

José Antonio Rodríguez-Avila
Universidad Autónoma del Estado de Hidalgo, Mexico

Karina Cecilia Arredondo-Soto
Universidad Autónoma de Baja California, Mexico

José Manuel Cornejo-Bravo
Universidad Autónoma de Baja California, Mexico

A volume in the Advances in Chemical and Materials Engineering (ACME) Book Series

IGI Global
DISSEMINATOR OF KNOWLEDGE

Published in the United States of America by
IGI Global
Engineering Science Reference (an imprint of IGI Global)
701 E. Chocolate Avenue
Hershey PA, USA 17033
Tel: 717-533-8845
Fax: 717-533-8661
E-mail: cust@igi-global.com
Web site: http://www.igi-global.com

Library of Congress Cataloging-in-Publication Data

Names: Carrillo-Cedillo, Eugenia Gabriela, 1966- editor.
Title: Design of experiments for chemical, pharmaceutical, food, and industrial applications / Eugenia Gabriela Carrillo-Cedillo, Jose Antonio Rodriguez-Avila, Karina Cecilia Arredondo-Soto, Jose Manuel Cornejo-Bravo, editors.
Description: Hershey, PA : Engineering Science Reference, 2019. | Includes bibliographical references and index. | Summary: “”This book seeks to increase the use of design of experiments to optimize and improve analytical methods and productive processes in order to use less resources and time”--Provided by publisher”-- Provided by publisher.
Identifiers: LCCN 2019030353 (print) | LCCN 2019030354 (ebook) | ISBN 9781799815181 (h/c) | ISBN 9781799815198 (s/c) | ISBN 9781799815204 (ebook)
Subjects: LCSH: Experimental design.
Classification: LCC QA279 .D4655 2019 (print) | LCC QA279 (ebook) | DDC 607.2--dc23
LC record available at https://lccn.loc.gov/2019030353
LC ebook record available at https://lccn.loc.gov/2019030354

This book is published in the IGI Global book series Advances in Chemical and Materials Engineering (ACME) (ISSN: 2327-5448; eISSN: 2327-5456)

British Cataloguing in Publication Data
A Cataloguing in Publication record for this book is available from the British Library.

For electronic access to this publication, please contact: eresources@igi-global.com.

Advances in Chemical and Materials Engineering (ACME) Book Series

J. Paulo Davim
University of Aveiro, Portugal

ISSN:2327-5448
EISSN:2327-5456

Mission

The cross disciplinary approach of chemical and materials engineering is rapidly growing as it applies to the study of educational, scientific and industrial research activities by solving complex chemical problems using computational techniques and statistical methods.

The **Advances in Chemical and Materials Engineering (ACME) Book Series** provides research on the recent advances throughout computational and statistical methods of analysis and modeling. This series brings together collaboration between chemists, engineers, statisticians, and computer scientists and offers a wealth of knowledge and useful tools to academics, practitioners, and professionals through high quality publications.

Coverage

- Metallic Alloys
- Computational methods
- Statistical methods
- Chemical Engineering
- Polymers Engineering
- Fatigue and Creep
- Nanomaterials
- Ceramics and glasses
- Multifuncional and Smart Materials
- Electrochemical and Corrosion

IGI Global is currently accepting manuscripts for publication within this series. To submit a proposal for a volume in this series, please contact our Acquisition Editors at Acquisitions@igi-global.com or visit: http://www.igi-global.com/publish/.

The Advances in Chemical and Materials Engineering (ACME) Book Series (ISSN 2327-5448) is published by IGI Global, 701 E. Chocolate Avenue, Hershey, PA 17033-1240, USA, www.igi-global.com. This series is composed of titles available for purchase individually; each title is edited to be contextually exclusive from any other title within the series. For pricing and ordering information please visit http://www.igi-global.com/book-series/advances-chemical-materials-engineering/73687. Postmaster: Send all address changes to above address. ©

Titles in this Series

For a list of additional titles in this series, please visit:
https://www.igi-global.com/book-series/advances-chemical-materials-engineering/73687

Handbook of Research on Developments and Trends in Industrial and Materials Engineering
Prasanta Sahoo (Jadavpur University, India)
Engineering Science Reference • © 2020 • 370pp • H/C (ISBN: 9781799818311) • S $295.00

Nanocomposites for the Desulfurization of Fuels
Tawfik Abdo Saleh (King Fahd University of Petroleum and Minerals, Saudi Arabia)
Engineering Science Reference • © 2020 • 300pp • H/C (ISBN: 9781799821465) • US $225.00

Enhanced Heat Transfer Mechanism of Nanofluid MQL Cooling Grinding Emerging Research and Opprtunities
Changhe Li (Qingdao University of Technology, China) and Hafiz Muhammad Ali (University of Engineering and Technology Taxila, Pakistan)
Engineering Science Reference • © 2020 • 350pp • H/C (ISBN: 9781799815464) • US $235.00

Advanced Catalysis Processes in Petrochemicals and Petroleum Refining Emerging Research and Opportunities
Mohammed C. Al-Kinany (King Abdulaziz City for Science and Technology, Saudi Arabia) and Saud A. Aldrees (King Abdulaziz City for Science and Technology, Saudi Arabia)
Engineering Science Reference • © 2020 • 257pp • H/C (ISBN: 9781522580331) • US $195.00

Nanotechnology in Aerospace and Structural Mechanics
Noureddine Ramdani (Research and Development Institute of Industry and Defense Technologies, Algeria & Harbin Engineering University, China)
Engineering Science Reference • © 2019 • 356pp • H/C (ISBN: 9781522579212) • US $265.00

Polymer Nanocomposites for Advanced Engineering and Military Applications
Noureddine Ramdani (Research and Development Institute of Industry and Defense Technologies, Algeria & Harbin Engineering University, China)
Engineering Science Reference • © 2019 • 435pp • H/C (ISBN: 9781522578383) • US $225.00

Process Analysis, Design, and Intensification in Microfluidics and Chemical Engineering
Harrson Silva Santana (University of Campinas, Brazil) João Lameu da Silva Jr (Federal Institute of Education, Science, and Technology of South of Minas Gerais, Brazil) and Osvaldir Pereira Taranto (University of Campinas, Brazil)
Engineering Science Reference • © 2019 • 367pp • H/C (ISBN: 9781522571384) • US $215.00

IGI Global
DISSEMINATOR OF KNOWLEDGE

701 East Chocolate Avenue, Hershey, PA 17033, USA
Tel: 717-533-8845 x100 • Fax: 717-533-8661
E-Mail: cust@igi-global.com • www.igi-global.com

Table of Contents

Detailed Table of Contents

This chapter applies design of experiments to improve plating performance and better practices in an electrochemical process within a company making electronic components. Specifically, the electroplated process of a metal housing served as the object of study. This process consists of plating an aluminum housing with silver (Ag) to improve the electrical signal characteristics and properties. It includes multiple factors affecting the process, which are clearly seen in the diverse failures such as electric response, pollution by solid waste, among others. These directly impact production costs and delivery time. To minimize the mentioned failures, diverse critical factors were enlisted discovering that the principal problem is the homogeneous distribution of the final finish of the commented product. Particularly, the final finish is realized with silver so it directly affects the electric response as final quality test.

One of the great challenges of the environmental diagnoses of soils contaminated with hydrocarbons is the optimization of analytical determinations. For this reason, this chapter evaluates the extraction of hydrocarbons by the Soxhlet method through the design of experiments (DOE), variating three different solvents, three soils, and three extraction times. Soil was experimentally contaminated at different concentrations, and hydrocarbons totals relying on conditions organic matter, electrical conductivity, pH and textures, amount of sample, solvents, and NaSO4, were studied. The variables were evaluated by means of an analysis of the Taguchi design and a factorial design, with the results the significant and optimal parameters of the process were determined, which were solvent type and time (10 hours and dielectric constant of 9). Also, the model discards the soil properties. These results will save time and resources, and they reduce errors.

The design of experiments (DOE) is a useful tool to define the most significative variables of a process, and the optimal operation conditions, or reduce the noise caused by uncontrollable variables. An advantage of the employ of DOE is the reduction of time required for analysis and costs associated. The Factorial, surface response methodology, and Taguchi have been employed to analyze the variables involved in the removal of heavy metal ions coming from wastewater by adsorption employing low-cost adsorbents, which include agro-industrial waste and biomass. The most important variables associated to increase of adsorption capacity and evaluated by researchers include temperature, pH, adsorbent dose, initial concentration of metal, and particle size.

Biofuels emerge as an alternative to mitigate climate change. In this sense, four biofuels generations have been proposed to produce clean and renewable fuels. To achieve this, the development of these fuels requires an extensive and rigorous experimental work that will bring optimal results in short

Foreword

The book *Design of Experiments for Chemical, Pharmaceutical, Food, and Industrial Applications*, edited by Doctors Eugenia Gabriela Carrillo-Cedillo, Jose Antonio Rodríguez-Avila, Karina Cecilia Arredondo-Soto and José Manuel Cornejo-Bravo offers an important contribution to the emerging field of evaluation and optimization of different processes with a minimum of experimentation. The book contains 15 chapters provided by authors from four countries, and it is divided into five sections named Chemical, Pharmaceutical, Food, Industrial and Diverse applications.

The content provided in this book will help promote the use of Design of Experiments in more countries and in industrial sectors. This topic is an interesting alternative to evaluate different variables involve in a process. The topic is now included in many educational programs reason why this book contributes to understand the importance and applicability in educational institutions and in companies from different sectors.

All chapters include topics that represent a challenge in different industrial sectors. The readers will find new vision to solve a problem based on the cases of studies presented which were similar situations than their problems. For all these reasons, this book deserves to be in libraries and bookstores that deal with the evaluation and optimization of different processes employing Design of experiments.

Rafael Pardo Almudí
Universidad de Valladolid, Spain

Preface

INTRODUCTION

Statistics is the science that uses sets of numerical data to obtain, based on them, inferences based on the calculation of probabilities. Statistics are useful for a wide variety of factual sciences, from physics to social sciences, from health sciences to quality control. In addition, it is used in business areas or government institutions with the aim of describing the set of data obtained for decision making, or to make generalizations about the observed characteristics. The statistics are divided into two main areas: descriptive statistics (dedicated to the description, visualization and summary of data originated from the phenomena of study) and inferential statistics (dedicated to the generation of models, inferences and predictions associated with the phenomena in question taking into account the randomness of the observations). As a methodology of inferential statistics, Design of Experiments (DOE) is defined as a set of active techniques that manipulate a process to induce it to provide the information that is required to improve it through changes in its variables and its interaction or sequence of execution.

The statistical DOE is the most efficient way of testing processes. When performing controlled and process-related tests as well as establishing a method in the application sequence, we can obtain data that statistically can provide testable conclusions about the process or system. DOE is precisely the application of the scientific method to generate knowledge about a system or process. The experimental design presents a set of documented evidence so that the collected data can be analyzed through statistics to obtain valid and objective predictable patterns or responses that form an effective conclusion. DOE helps investigate the effects of input variables (factors) on an output variable (response) at the same time. These experiments consist of a series of runs, or tests, in which intentional changes are made to the input variables. Data is collected. The experimental design finds applications in industry, agriculture, marketing, medicine, ecology, behavioral sciences, pharmaceutical research, chemical process, food industry, among others; constituting an essential phase in the development of an experimental study. Therefore it is relevant to promote research on the emerging applications of DOE for process and products improvement.

This book was contributed by authors from four countries: Mexico, Portugal, Turkey and Spain. The general objective of this book is to increase the use of design of experiments to optimize, improve and optimize analytical methods and productive processes, in order to use less resources and time, the estimation of the effects is much more precise. The design of experiments allows the optimization of a reaction through the variation of multiple factors. Importantly, this allows the evaluation of a large number of reaction parameters in a relatively small number of experiments. It is a methodology that is routinely applied to industrial processes but is rarely applied to academic projects. The benefits by applying this related methods are:

- Evaluate more precisely the individual effects of each variable and the interaction between variables.
- It can be combined with other research methods Experimental research allows cause and effect to be determined.
- Reduce analysis times and costs.
- Improve the resolution.
- It provides researchers with a high level of control.

The target audience of this book is composed of professors, managers, employers, researchers and students working in the field of Advanced and Multivariate Statistics of any labor sector. More specifically, the target audience includes pharmaceutical, environmental, chemical and industrial researchers, manufacturing engineers, chemical engineers, industrial engineers, industrial designers, researchers, industry practitioners, research scientists, academics, and any employee that manages DOE at any level.

ORGANIZATION THE BOOK

This book is the result of the research and applications of different colleagues located in different countries Mexico, Spain, Turkey and Portugal. The design of experiments plays a very important role in the productive processes and in the investigations of the different areas such as chemistry, food and pharmacy. Currently it is necessary to optimize resources, use of reagents, raw materials, modify methods, and make time efficient.

Section 1: Chemical

Chapter 1: The purpose of this chapter is to apply Design of Experiments to improve (plating performance) and better practices in an electrochemical process within a company making electronic components. Specifically, the electroplated process of a Metal Housing served as object of study. This process consists on plating an aluminum housing with silver (Ag) in order to improve the electrical signal characteristics and properties. It includes multiple factors affecting the process, which are clearly seen in the diverse failures such as electric response, pollution by solid waste, among others. These impact directly in the production costs and delivery time. In order to minimize the mentioned failures, diverse critical factors were enlisted discovering that the principal problem is the homogeneous distribution of the final finish of the commented product. Particularly, the final finish is realized with silver so that it affects directly to the electric response as final quality test.

Chapter 2: The main objective of the present investigation was to evaluated the extraction of hydrocarbons by the Soxhlet method through the design of experiments (DOE), variating three different solvents, three soils and three extraction times. soil was experimentally contaminated at different concentrations and were studied hydrocarbons totals relying on conditions organic matter, electrical conductivity, pH and textures, amount of sample, solvents and NaSO4. The variables were evaluated by means of an analysis of the Taguchi design and a factorial design, with the results the significant and optimal parameters of the process were determined, which were, solvent type and time (10 hours and dielectric constant of 9), also model discards the soil properties, these results will save time, resources and they reduced errors.

Chapter 3: In this chapter, the factorial, Surface Response Methodology and Taguchi have been employed to analyze the variables involved in the removal of heavy metal ions coming from wastewater by adsorption employing low-cost adsorbents, which include agro-industrial waste and biomass. The most important variables associated to increase of adsorption capacity and evaluated by researchers include temperature, pH, adsorbent dose, initial concentration of metal and particle size.

Chapter 4: Several authors have studied and optimized the different biofuel production systems using Factorial Designs and Response Surface Design methods and statistical analysis with reliable results. This work reviews and classifies the results obtained by these investigations and demonstrates the scopes and limitations of the application of DoE. Biofuels emerge as an alternative to mitigate climate change. In this sense, four biofuels generations have been proposed to produce clean and renewable fuels. To achieve this, the development of these fuels requires an extensive and rigorous experimental work that will bring optimal results in short time periods. Hence, to accelerate the development of clean fuels, the Design of Experiments (DoE) methodologies are a useful tool to improve the operational conditions such as temperature, time, pressure and molar ratios.

Chapter 5: FTIR and Raman spectroscopy are complementary spectroscopic techniques that play an important role in the analysis of molecular structure and the determination of characteristic vibrational bands.Vibrational spectroscopy has wide range of applications including mainly in physics and biology.Especially in recent years its applications has gained tremendous speed in the field of biological macromolecules and biological systems, such as tissue, blood and cells. However, the vibrational spectra obtained from the biological systems contain a large number of data and information that make the interpretation difficult.To facilitate the analysis, multivariant analysis comprising the reduction of the dimension of spectrum data and classification of them by eliminating redundancy data, which are obtained from the spectra and does not have any role, becomes critical.In this chapter, the applications of Principal Component Analysis (PCA), Linear Discriminant Analysis (LDA) and their combination PCA-LDA, which are widely used among multivariant techniques, on biological systems will be disclosed.

Section 2: Pharmaceutical

Chapter 6: This chapter presents an overview of the methods of preparation of solid lipid nanoparticles (SLN) which are nanomaterials effective to improve the bioavailability of poorly soluble drugs, and also to achieve transdermal drug delivery. It includes a critical review of selected references about the optimization of the preparation of drug-loaded SLN using DOE. The chapter presents the study of the case of the optimization of the size of SLN containing capsaicin, an analgesic drug, pinpointing the importance of producing small size SLN for transdermal absorption.

Chapter 7: This chapter analyses de advantages and disadvantages of parallel design with respect to cross-over trials and cross-over trials coupled to design of experiments, used in nutritional clinical trials for foods, food ingredients, or bioactive compounds. The advantages of coupling cross-over trials with design of experiments to reduce the number of subjects and time employed is also discussed.

Section 3: Food

Chapter 8: In this chapter, the authors present the optimization of the hydrodynamic and chemical parameters of the FIA system in the determination of copper and manganese in wine samples by VIS spectrophotometry using DOE. The chapter illustrates that using factorial designs can reduce the number of experiments necessary to achieve successful quantification of copper and manganese in wines.

Chapter 9: In this chapter a general outline is presented on the general workflow; illustrated through examples and a current review of some of the applications of design of experiments in the context of food production help provide some directives and guidelines. The general steps are described, as well as some classifications and applications. A description of some of the more important principals needed to apply design of experiments are presented.

Chapter 10: This chapter evaluated the effect of temperature and time on the physicochemical properties and fatty acid composition of chia seed oil. The seeds, either whole or ground, were subjected to several treatments using a full factorial experimental design. The models can be used to find the areas of optimal response and obtain limit conditions in the application of thermal treatments to the chia seed.

Section 4: Industrial Applications

Chapter 11: This study applied the methodology of process validation in a medical device company; the engineering tests were used in the company's clean room, using an injection molding machine and tests of several parameters were used even without being validated to know which are the best run and the best parameters, for its daily use. This project addresses the validation of the blood filter process by injection molding. The Design of Experiments applied was a 2K factorial design with central points where the replicas consisted of five in total of the runs were 45 in total for the three dimensions, the pieces are shared for those three dimensions that are being evaluated.

Chapter 12: This chapter focuses on providing the reader different applications of DOE in industry, to highlight the importance and benefits of knowing and applying these techniques. It will present the application of this methodology in a general way and finally, it will discuss different case studies that use this methodology in industry.

Section 5: Diverse Applications

Chapter 13: In this chapter there are practical problem solving strategies used to obtain a product or chemical process with desirable characteristics in an efficient mode, focused on the use of full and fractional (2-level) designs. The information is presented as a tutorial and the main advantages and disadvantages are presented and discussed, emphasizing the effect of reduction of experimentation in the data quality.

Chapter 14: The objective of this research work is to compare the results that can be obtained with classical techniques against the results that alternative linear search techniques such as the Simultaneous Perturbation Stochastic Approaches (SPSA) algorithm can achieve. They start from the work reported by Gedi et al. 2015 to implement the SPSA algorithm. The experiments carried out allow us to affirm that for this case study, the SPSA is capable of equalizing, even improving the results reported by the authors.

Chapter 15: This chapter presents an overview of the challenges concerning engineering education. It also presents an overview of DOE with a brief historical analysis of the important developments in DOE being restricted to that of a group of techniques in design that boost the exploration of a region of design variables in one or more response variables. This chapter also presents a literature review regarding the utilization of DOE in engineering teaching and learning for general problem solving as well as for the optimization of product/process with special focus on engineering education. The main contribution of the chapter is to bring up ideas from those studies devoted to the use of DOE in teaching and learning engineering environments to better understand the deficit of the utilization of such type of approaches in academic projects/experiments regardless of the common utilization of DOE in statistics and quality literature.

Acknowledgment

The editors of the book want to acknowledge the chapter's authors for its insightful contributions. The book includes 15 chapters that amalgam the areas of Chemical, Pharmaceutical, Food, Industrial Aplicactions, and Diverse Applications, to give interesting perspectives to the versatile applications of Design of Experiments. It bridges the gap from basic sciences to technological applications. The chapter´s authors provide the readers with valuable information about different methods of DOE and how it can be applied to different fields and areas of interest.

This is an international collaborative effort with the participation of 13 academic institutions, from four countries (Mexico, Portugal, Turkey and Spain), which have enriched the discussion in the different fields of knowledge.

The edition of this book integrates the efforts of the Academic Bodies "Process and Products Innovation", and "Biopharmaceutics", from the Universidad Autónoma de Baja California, and the Academic Body "Analytical Chemistry and Physical Chemistry of Solutions and Surfaces", form the Universidad Autónoma del Estado de Hidalgo. The institutions provided their facilities and services to develop this book.

In addition, editors want to acknowledge to the National Council of Science and Technology (CONACYT) and PRODEP (Program for Strengthening of Academic Bodies), since these organizations have provided support to perform researches.

Editors want to acknowledge the editorial IGI Global Publishing for the promotion to integrate the chapters of the book, the technical support for edition, review process, and diffusion of the book.

Finally, editors want to recognize the effort of all who contributed to the development of this book with their valuable reviews of chapters, and to the researcher who contributed with the foreword.

Section 1
Chemical

Chapter 1
Design of Experiments in an Electrochemical Process

Guadalupe Hernández-Escobedo
https://orcid.org/0000-0002-7516-972X
Instituto Tecnológico de Tijuana, Tecnológico Nacional de México, Mexico

Jesus Ivan Alaniz-Muñoz
Instituto Tecnológico de Tijuana, Tecnológico Nacional de México, Mexico

Arturo Realyvásquez-Vargas
https://orcid.org/0000-0003-2825-2595
Instituto Tecnológico de Tijuana, Tecnológico Nacional de México, Mexico

Karina Cecilia Arredondo-Soto
https://orcid.org/0000-0002-8929-7319
Universidad Autónoma de Baja California, Mexico

ABSTRACT

This chapter applies design of experiments to improve plating performance and better practices in an electrochemical process within a company making electronic components. Specifically, the electroplated process of a metal housing served as the object of study. This process consists of plating an aluminum housing with silver (Ag) to improve the electrical signal characteristics and properties. It includes multiple factors affecting the process, which are clearly seen in the diverse failures such as electric response, pollution by solid waste, among others. These directly impact production costs and delivery time. To minimize the mentioned failures, diverse critical factors were enlisted discovering that the principal problem is the homogeneous distribution of the final finish of the commented product. Particularly, the final finish is realized with silver so it directly affects the electric response as final quality test.

DOI: 10.4018/978-1-7998-1518-1.ch001

INTRODUCTION

The Design of Experiments (DoE) is frequent in diverse areas of study. One of them is the industry, which is principally used in the design and improvement of products and processes. It is because the experimentation offers a close description on how the products could be used by certain individuals, and processes could be operated in determined environments. Particularly, the operation of the process served as a precursor for its improvement. The costs and the time employed in the improvement are two variables that limit it. However, the experimentation is one tool that is employed with this objective in mind. The principal idea is to use it until desired goals and/or indexes are achieved. This is considering the costs and time in order to reduce both using the experimentation. In other words, this means that experimentation is one form on improving the processes at low cost and using less time.

The EEC company is an industry that manufactures filters for electrical and radar devices, is autonomous and subsidiary of CE. The company started operations in 1987 and operates within the ''Maquiladora'' program. The design and manufacturing center is located in Tijuana, Mexico, while the sales, IT and customer service departments are located in California, United States. CEE is a company specifically dedicated to the manufacture of RF filters for High frequency and microwave. Over the years the product line has expanded to include ceramic filters, surface mounted filters and miniature filters. The company has 50,000 square feet assigned to assemblies, electrical tests, life tests, quality assurance, design engineering and technical service. It also has areas for machines, tools and electroplating finishing processes.

The production departments are organized to process engineering prototypes, issue small quantity orders and volume production lines. In addition, it has the ability to transform certain areas into clean rooms "class 100,000" for specific orders that require it. According to the company's sales catalog, Electronic Components has certifications in ISO 9000 and ISO 14000. This plant was certified in the functions of design, development, production, installation and customer service. On the other hand, clients vary in different sectors. Some examples of application of the final product are mobile and wireless telephones, GPS (global location systems), satellite receivers, radio frequency data processors and network communication systems.

It is important to note that within these production departments, there is a process in which attention is paid to the importance of the final product. According to the company's plating manual, electroplating is the chemical process where a layer of metal is deposited by applying electricity to a metal. An internal and / or external layer is deposited to said material and it is according to the material to be processed. In this area metals such as silver and nickel are deposited through an internal process in which the filters that are manufactured internally are placed. Components-based metals are aluminum, brass and steel to be coated with a specified thickness. These specifications are indicated on the traveling or work instructions sheets, following the QQN290 standards for nickel, QQS365 for silver and ASTM-B700 for silver. This electroplating process generates various properties and characteristics in the coated materials. To do this, an electrochemical process is followed that includes a series of steps to meet the manufacturing specifications and achieve the plating requested in the product. In this particular case it is an Aluminum Housing, which is presented in Figure 1. The products are plated to protect them from environmental corrosion and give them certain electrical properties. This is because they are metal parts of aluminum and a copper derivative called brass (copper and zinc alloy).

According to the plating manual, the department is made up of a process engineer, a production supervisor, a material prèparer and a plainer. Employees belonging to this department must have knowledge of the dangerous chemicals that are handled there. This is important for personal safety; Therefore,

Figure 1. The Aluminum Housing

employees use personal protective equipment such as masks for acid vapors, safety glasses, antacid apron, latex gloves, uniform and non-slip footwear. When there are risks in this department, the platinum department is located at the back of the plant. To make the correct plating deposit on the filters, a line of different chemicals is used that clean the material, then activate it and then deposit the required metal, which can be silver or nickel.

However, in the metal case there are problems in electrical response, defective material and rework in 50% of the final product. This causes low levels of quality and delays in the delivery time of the finished product. To solve these problems, a new platinum line was implemented in the company since previously the product was sent to platinum with an external supplier. This in order to reduce costs and help improve customer delivery. The process was designed for standard models with easy to meet specifications. A new process was subsequently requested using the same facilities to meet special specifications. At the start of production, the initial pieces presented other problems as follows:

1. Problems with electrical response: the geometry of the product is complex with areas of high and low density, complicating the normal electrodeposition process. The distribution of the final finish is of importance for the achievement of a favorable electrical response that meets the stipulated specifications, see figure 2.
2. Compliance with customer specifications: variable directly related to compliance with military standards and is dependent on the electrical response. It is required to meet the levels of electrical response and achieve internal product manufacturing requirements.
3. Improve quality levels: it is necessary to eliminate waste and thus take advantage of the consumption of silver, reducing the process time.

4. Cost of plating process: rework and waste increase the cost of manufacturing in the plating process. Therefore, it is important to eliminate or reduce it.

From the above, the objective of applying the DoE to improve an electrochemical process in a company of electronic components to increase the quality levels stipulated in said process was defined. This shows the effectiveness of this in the elimination or reduction of the problems mentioned above and that are related to the homogeneous distribution of silver in base metals. This experimentation helps reduce variations in said finish. This also leads to the improvement in the internal processes of the company and the fulfillment of goals with customers, thus generating economic benefit for the company. Establishing the new parameters through the DoE is intended to increase the quality level and thus improve the important parameters to improve the silver finish and quality levels. For these reasons the following hypotheses were defined:

- H_0 = The electrical response does not depend on the homogeneous distribution of silver in its internal and external areas
- H_1 = The electrical response does not depend on the homogeneous distribution of silver in its internal and external areas.

For this, the critical factors in the control of the distribution of silver were determined, which were: a) dependent variables: improvement of electrochemical process, homogeneous distribution of silver, reduction of platinum costs, desired electrical response, compliance with requirements of the client; b) independent variables: purity of silver, platinum standards, design of experiments, amperage, time, number of cycles, concentration of silver, geometry of piece to be plated; c) controlled variables: product family, operators, plating equipment, plating materials; and d) intervening variables: silver quality, staff turnover.

The rest of this chapter is structured as follows. The first part presents the bibliographic review, which refers to research related to the area of design of experiments. Also, reference is made to specialists in the field and the important concepts used are clarified. The following part sets out the development carried out in this project work describing the variables used, the systematized activity to carry out the research is proposed, the statistical treatment is decided and defined, also the required activities are presented. This continues explaining the results obtained to culminate with the conclusion and recommendations derived from them.

BACKGROUND

Design of Experiments

The Design of Experiments (DoE) is a technique that consists in carrying out a series of experiments in which deliberate changes in the variables of a process are induced. This allows us to observe and identify the causes of the changes in the output response. Montgomery (2005) explains that with this technique you can achieve, for example, improve the performance of a process and reduce its variability or production costs. Its application in industry includes fields such as chemistry (Ramírez, 2004), mechanics (Nataraj, 2005), materials (Ray, 2004), industrial engineering (Tong, 2004), or electronics (Wingbrant, 2003). Although DoE is defined as an effective tool to improve and optimize processes and products, its application is not common in industries (Tanco, 2006). Engineers use advanced statistical techniques limitedly in troubleshooting. This is because they do not have adequate statistical knowledge (Tanco 2008b). Most of the scientific articles published on DoE have focused on models, criteria and other topics that refer almost exclusively to the stage of choice of the model or design to carry out the experiment and data analysis.

Applications

Tanco (2008a) presented a case study conducted at a recognized company in the automotive sector, in which about 4,000 employees worked to produce a single car model. The process analyzed is the laser welding applied to join the sides of the car and its roof, which is carried out in a closed cabin by means of two robots. This process presented frequent quality problems due to the appearance of pores in both weld seams. Therefore, experimentation was used and in particular the Design of Experiments to improve the process. The project started with an average defect rate of 3% and concluded with an average defect rate of 0.09%. This represented a 97% reduction of the pores. After a while, a confirmatory analysis was carried out to verify that the results were maintained over time.

In addition, Padilla (2008) carried out a research project in a floricultural industrial field where it is common to do experiments in the different processes on which the final product depends. These tests were intended to discover the changes in the materials, methods or operating conditions of a process so that quality problems can be detected, resolved or minimized. It is common for these tests or experiments to be done on the fly (trial and error) and depending on experience and intuition. This results in only visible results with the passage of time because there is no adequate experimental plan that guarantees a response to the questions raised. The DoE is an effective way of testing the processes, since it provides the techniques and strategies necessary to effectively bring the processes to better operating conditions. In addition, it determines which tests and how they should be performed in order to obtain data that, when statistically analyzed, serve to obtain conclusions and thus improve the decision-making process in the process performance improvements. The objective of this research was to apply the aforementioned technique to correct and improve the planting processes and quality of ornamental plants. Materials such as chemical and organic substances, substrates and temperature variation were used so that through the observation, measurement and analysis of collected data the best decisions could be made regarding the variables of interest. The research presented practical applications of randomized experimental design and factorial experiments through which problems in production processes have been detected. Among these were the problems of seed germination due to problems with the substrate. Once corrected, germination efficiency increased by 15%. Also, at least an additional 20 grams of root mass was obtained in the cultures. This resulted from properly combining factors during planting. Thus, the overall productivity was increased by 8.2%, and the cost of the plant was reduced. This was 3.22 cents to 2.84 cents per hectare.

Also, Roy (2001) comments that all types of industries can benefit from DoE, including those of service. The application of this methodology can have maximum benefits when it is used in applied research and the design of new products. Montgomery (2005a) comments on this:

"Most of the industries were working for more than 75 years to improve the efficiency of their processes. For those cases in which they achieved a significant improvement in the efficiency of the processes there may not be much room for significant improvements. But in product design, the opportunities for

improvement are almost unlimited, because most of the costs of a product (production cost, warranties, repairs, re-work, after-sales service, etc.) are predetermined by the design stage (p.3-4)".

In addition, the DoE has been successfully applied in marketing (Starkey, 1997) and in finance (Celik, 2007). Along the same lines, Wu (2000) state that DoE can be used in industries to:

- Improve processes, either improving their efficiency, reliability or performance.
- Assist in troubleshooting.
- Learn about processes and their failures.
- Establish cause-effect relationships between the input (inputs) of a process and its outputs (output).
- Identify the factors that have the greatest impact and the least in the processes and / or products.
- Achieve a production of products that meet specifications that are robust to external noise.
- Establish a region (or window) of the process where some factors can operate, finding out the sensitivity to the change of some factors in the response.
- Set specifications and logical tolerances for products and processes.
- Obtain a polynomial equation that models the response behavior of a process in a region of variation of the factors.
- Verify if the solution adopted to improve a process really obtains the expected results.

Advantages of Designing Experiments One Factor at a Time

Funkenbush (2005) mentioned that the usual strategy in industries requires changing one factor at a time (OFAT). This is because it is one of the first lessons taught in formal studies, where experimentation is emphasized by changing one factor at a time. On the other hand, if several factors are changed at once, it would be inefficient to find out which variable is responsible for the differences generated. This is good advice, but it does not apply to the use of experiment design if proper designs are used. Also, Knapp (2003) explains that an experimenter who believes that only one factor must be modified at a time when experimenting, must be introduced into the concept of factorial designs. These designs involve experimenting in all combinations of the levels of each factor and are very simple to analyze (Box, 2005; Taguchi, 1986). For example, Knapp (2003) shows how Chevron Chemical Co., obtained negative results with the application of OFAT and, on the contrary, had savings of over US $ 50,000.00 in raw material costs due to the application of fractional factorial designs at two levels. Hence, factorial designs have the following advantages over an OFAT strategy:

- They require less resources for the amount of information obtained. This can be important for the industry since experiments tend to be expensive.
- The estimation of the effects of each of the factors in the response is accurate. Factorial designs make each observation double benefit. Using additional experiments to estimate an effect makes the estimate accurate.
- The interactions between the factors can be systematically estimated. In most cases when the OFAT is applied, it is not possible to estimate any interaction.
- It provides experimental information in a large area of the region under study. Because the experimentation region is smaller when one factor is modified at a time, the conclusions obtained from the analysis are limited.

- It prevents obtaining suboptimal responses. When experimenting with OFAT, because it is experienced in a small region, the direction of improvement can lead us to areas far from optimal.

To explain these five disadvantages of OFAT, a practical explanation is presented with an example (Anderson, 2005). The example shows the differences between a factorial design at two 2^3 levels and an equivalent OFAT design in terms of replication and parameter estimation. For example, in a factorial design, with factor A, the four experiments on the left side are used at their low level (1, 3, 5, 7) and the four experiments on the right side at their high level (2, 4, 6, 8). To obtain an equivalent OFAT design it is necessary to perform all 16 experiments to obtain the same parameter estimation. This shows that such a strategy requires additional experiments to obtain the same information. In addition, OFAT designs lack the ability to investigate interactions and therefore they experience in a limited region. It is now assumed that nine experiments are performed, according to an OFAT strategy, modifying factor A and keeping factor B constant at a medium level (0). The response obtained shows the maximum efficiency of 80% when the level of A is approximately 0.63. Therefore, A is set at its "optimum" and nine experiments are performed by modifying factor B. Subsequently, B is placed at 0.83, thus obtaining an efficiency of 82%. Therefore, this strategy helps to conclude that the maximum possible efficiency to be achieved is 82% and is obtained with the combination of (A, B) = (0.63, 0.83). However, the actual response surface can obtain efficiency values of the order of 95%. This value could have been obtained from the application of factorial designs sequentially. This method is usually called Response Surfaces (RSM).

On the other hand, Daniel (1973) identifies five different strategies for experiments that can be performed by modifying one factor at a time. Until now, reference has been made to the "classic" strategy since it is the simplest, known and used by industries. On the other hand, recently, Frey-Wang (2006) investigated that under certain circumstances an adaptation of the OFAT can function efficiently compared to factorial designs. This strategy considers interactions and does not move easily like the classical strategy. Despite the recent discovery, the strategy is difficult to apply. Therefore, Del Castillo (2007) recommends as a general principle that factorial designs are preferred over OFAT.

Evolution of Experiment Design

The DoE was first applied in agriculture and was carried out by Fisher (1935) in England. His experiences led him to publish his book "Design of experiments". These studies were focused on improving potato production, at the experimental agricultural station in Rothamsted, London. Since then, other researchers contributed to the development and application of the design of experiments in different fields. According to DeMast (2004), the contributions of Shainin and Taguchi created two new approaches to DoE. Both approaches offer additional aspects to the design of experiments and are therefore considered improvement methodologies. That is why Montgomery (2005a) considers that there were four stages in the development of the design of classical experiments. The first stage, initiated in the twenties of the last century, is characterized by the systematic introduction of scientific thinking, the application of factorial designs and the analysis of variance (ANOVA) in experimental scientific research. In the 1930s and 1940s, fractional factorial designs were introduced as a solution to the excess of experiments necessary to carry out full factorials. These designs consist of choosing an appropriate fraction of a complete factorial. In addition, they provide an effective and cheap way to study various factors at the expense of ignoring high-order interactions generally considered insignificant.

The second stage was initiated by Box and Wilson (1951). This is characterized by the development of what will later be called the response surface methodology (RSM). These authors noted that industrial experiments differed from experiments in agriculture in two aspects: a) immediacy, because the response can be observed quickly, without waiting as long as in agriculture; and b) sequentiality, the experimenter can perform few experiments and plan the following experiments based on the results. During the following years, the DoE and the RSM extended to the chemical industry and industrial processes, especially in the areas of research and development. At this stage, designs to estimate quadratic functions such as central composite designs (CCD) and Box-Behnken (BBD) designs arise.

The third stage began in the early 80s, with the emergence of the approaches of Taguchi and Shainin. These were presented as simple and efficient methodologies for experimentation. The first reaction was the criticism of these approaches attacking their statistical validity. In spite of this, mainly the works of Taguchi (1986) on robust parameter design (RPD) served to spread the interest and use of DoE in other areas, such as automation and the aerospace, electronics and semiconductor industries. As a consequence, various academics and statisticians began to appreciate ideas of the new approaches which generated a profound change in the classic approach of the DoE. For example, the reduction of variability began to be an area of research, as well as the effort to develop methodologies and guidelines to simplify the application of DoE. This created the foundation to start the fourth stage of the design of experiments in the 1990s.

The fourth stage was the democratization of statistics. This was thanks to the expansion of Six Sigma and the varied offer of statistical software that helped the DoE be applied in various types of industry (Montgomery, 2002). In recent years there has been an increase in publications on the DoE (Booker, 2005), which allowed the technique to be significantly developed. In addition, the automation of calculations and graphs through software allowed simplifying the technique to be used by engineers and scientists. Recent books on DoE, such as Funkenbush (2005) or Goupy-Creightom (2007) show how this methodology can be understood and applied by engineers. At this stage the concept of optimality arises which leads to the use of optimal designs. Statisticians, engineers and scientists also contributed to the development and application of DoE in industries throughout these almost 90 years of the classical approach. This caused that this approach is now consolidated and a valid and robust approach.

Approaches of Taguchi and Shainin

According to DeMast (2004), the contributions of Shainin and Taguchi created two new approaches in the DoE. Both approaches offer the design of experiments and can also be considered as methodologies to propose improvements. As a researcher in an electronic control laboratory in Japan, the engineer Geneichi Taguchi developed in the 40s a method for the design of experiments. Although their first Japanese publication dates back to the 1950s, this approach was introduced in the United States and Europe until the early 1980s. Tay and Butler (1999) mention that Taguchi promoted statistical techniques from a statistical perspective. of the moment Although Taguchi played an important role in popularizing this technique, and much of its success was due to it, it would be wrong to consider Taguchi's methods as a single approach to DoE. De Mast (2004) states that Taguchi developed a methodology for solving problems, which he called "Quality Engineering". The basic concepts about Taguchi methods can be summarized as follows:

- A quality product is one that minimizes losses to society throughout its useful life. The relationship between this loss and the technical characteristics is expressed in the loss function, which is proportional to the square of the deviations of the response over its desired value.
- Taguchi breaks down its "Quality Engineering" strategy in three phases: system design, parameter design and tolerance design. Systems design tries to innovate and investigate what are the factors and levels that should operate. The parameter design attempts to increase the performance of a process / product by adjusting the factor levels. Generally, this phase is known as the Taguchi Method and is the phase related to the technique. Finally, the tolerance design determines the control parameters of each factor, which are identified in the Taguchi Method.
- The experimentation objective must be changed to obtain results according to the specifications in order to achieve a desired value while minimizing variability.

According to Arvidsson and Gremyr (2008), because the DoE is a key tool for the design of parameters, Taguchi placed special emphasis on making it simple to apply. Basically, he simplified the application of the technique by incorporating a standard set of experimental designs (orthogonal matrices), a graphical tool to assign the factors to the experimental design (linear graphs), guides to interpret the results (cookbook), ways to transform the response to obtain reduction in variation (S / N ratios) and a formal method to study uncontrollable factors using the robust design technique (Roy, 2001). It also simplified the analysis of tolerances through the use of the technique (Pignatello, 1991). Taguchi's main contribution was the emphasis on reducing variation: "quality is something that cannot be characterized only by a measure of a desired characteristic. The variability of this characteristic must also be considered", (Robinson, 2004, p 81-101).

Therefore, he proposed the use of special designs in which the factors included in the experimentation could be classified in control and noise. The first are those factors that can be controllable; while noises are those difficult or expensive to control. The basic idea of parameter design is to identify the appropriate level of each control factor at which the system performance is robust to the presence of noise factors (Nair, 1992a). This with the aim of making the system robust to noise. This approach is known as "robust design." The methodology for problem solving created by engineer Dorian Shainin is called the Shainin system. In 1975, Shainin founded his own company and describes his method as the American approach to problem solving with the same objectives as the Taguchi approach (Shainin, 1988).

Limitations of Diverse Approaches

Goh (1993) states that until the Taguchi methods were disclosed in the United States, the classic DoE was considered a mathematical tool and was seen as a complement in the technical training of the engineer for the study of processes and products. Bhote (2000) explains that Taguchi and Shainin were the greatest critics of the classical approach. They argued that entrepreneurs, engineers and workers found DoE complicated to apply and ineffective, which made it a frustrating experience. Also, that this approach was preferred only by those individuals with mathematical or statistical inclination. Despite this, it is important to note that after a decade of strong opposition to new approaches, the classical approach began to detect the importance of several ideas proposed by Taguchi and Shainin. For example, there were several attempts to integrate the design of Taguchi parameters with the classical approach (Nair, 1992b). As a consequence, the use of the response surface methodology when a variation reduction is desired became an emerging research area (Quesada, 2004). Finally, the complexity of the technique, strongly criticized, was significantly diminished with the help of software that allows designing and analyzing the experiments. In addition, special emphasis was placed on presenting guides that included graphic tools to clarify and simplify it. Some examples of these simplifications are Pareto charts to see the magnitude of the effects or the use of charts to define the problem under study.

Montgomery (2005b) classifies the strategies used to carry out the experiments in companies in three categories. The first is the best guess ("best guess"), a strategy that consists of using prior knowledge about the process in question, modifying the variables involved and conducting the tests under conditions in which better results are expected. This strategy lacks scientific rigor since it is based solely on intuition and experience. The second strategy is that of one factor at a time ("OFAT"), Czitrom (1999) explains that it consists in modifying a single factor (variable) each time and distinguishing the effects that each factor has on the response separately. This strategy, although it is the most used in companies and according to the scientific method, is inefficient to find the best process conditions. In addition, it does not systematically detect the possible relationships (interactions) that may exist between several factors by driving only one at a time.

Thus, the statistical methodology par excellence to optimize experimentation is known as DoE, this being the third strategy. It is therefore necessary, as Deming (1982) states; more efficient to use a pre-established statistical methodology as a strategy for experimentation; which, in terms of results obtained and number of experiments, allows us to plan and analyze the experiments correctly. There is no knowledge that can contribute both to improving quality, productivity and competitiveness as well as statistical methods. DoE is defined by Lye (2005) simply as a method to systematically apply statistics to the experimentation process. Montgomery (2005b) defines it as performing a set of tests in which voluntary changes are made to the control parameters of a process or system, to observe and identify the reasons for the changes in the output variable or response of the process.

Methodologies of the Design of Experiments

According to Polished (2004) the statistical design of experiments is one of the techniques used in the industrial field. It is the planning of a set of experimental tests, whose objective is to obtain data that can be analyzed statistically and thus be able to obtain valid conclusions of a system or process under study in order to implement an improvement. The important thing about a statistical design of experiments is to know in advance what are the factors that you want to control and how many repetitions are appropriate to obtain a representative sample, so that a set of test conditions is obtained; It is a sequence of stages or activities that must be accomplished to achieve the objectives pursued. The statistical design of experiments consists of the following stages:

1. Planning. They are each and every one of the activities that are aimed at understanding the problem, designing and performing the appropriate experimental tests.
2. Analysis. The variance analysis model (ANOVA) or the statistical technique that best describes the behavior of the data is determined.
3. Interpretation. This part must go beyond the formal statistical analysis, analyze what has happened in the experiment in detail and contrast the initial hypotheses with the results obtained, observe the new learning process, verify the assumptions and choose another treatment to Efficientize the process.

4. Conclusions. It consists in deciding what measures will be implemented to generalize the result of the study and ensure that the improvements are maintained. In addition, it serves to present and disseminate the achievements to people who work directly in the process under study.

The DoE essentially deals with phenomena that are observable and repeatable, so care must be taken in planning and analyzing an experiment. This is due to the fact that being observable and repeatable experiments, these concepts by nature contradict each other. Anything observed is appreciated with variability; that is, nothing happens in the same way twice, even measurements of the same event vary. This is why the starting point for planning is to apply the basic principles of the statistical design of experiments, such as randomization, repetition and blocking, explained below:

- Randomization. It consists of making experimental runs in random order; This principle increases the possibility that the assumption of independence of errors will be fulfilled.
- Repeat. It is to run more than once the treatment or a set of given factors, which allows to distinguish which part of the total variability of the data is due to the random error and which to the factors. Care must be taken to confuse this concept with that of measuring several times the same item or product manufactured in the same combination of factors.
- Lock. It is to nullify or takes properly into account all the factors that can affect the response observed.

On the other hand, the objective is to study the effect of several factors on one or several responses or quality characteristics called factor design. In other words, this design seeks to study the relationship between the factors and their response in order to know what this relationship is like and generate knowledge that allows actions and decisions to be taken that improve the performance of the process being studied. Through this design you can determine a combination of levels of factors in which the process performance is optimal and that in current operating conditions and therefore can reduce or even eliminate quality problems in the output variable. Another advantage of this design is that the input variables or factors can be of qualitative type (machines, operator, type of material, etc.) and / or quantitative (temperature, humidity, speed, etc.). According to Montgomery (2005a), in the analysis of variances (ANOVA) variances ratios are used to test the hypothesis of equality of means, this means that it is a statistical technique that allows to identify differences of existence between the means of different groups.

This technique is common to use when two or more arithmetic means have to be compared with each other. It should be noted that the analysis of variance does not allow the Type I Error (probability of accepting that the means are different from each other, when they really are not) increase when comparing a set of treatment groups, as it would if the comparison were made two by two using a t-test for two treatments. The analysis of variance is used to determine the probability that the differences of the means between several groups are due merely to the sample error but not to the experimental random error. The logic of a contrast in the analysis of variance is simple if two independent calculations of the variance are compared (this depending on the experimental model that is applied) for the independent variance. One reflects the existing variability within the groups and the other represents the existing variability between the groups attributed to the effects of the treatment.

Silver Deposit Process

Electrodeposition or electroplating is an electrochemical plating process where metal cations contained in an aqueous electrolytic solution are deposited in a layer on a conductive object. The process uses the electric current to reduce on the cathode surface the metal cations contained in an aqueous solution. When reduced, the cations precipitate on the surface creating a coating, the thickness of which will depend on various factors. Mainly electrodeposition is used to adhere a metal layer with a desired property to a surface that otherwise lacks that property, for example: resistance to abrasion and wear, protection against corrosion, the need for lubrication, or qualities aesthetic Particularly, the silver-aluminum electrodeposition process used in the aforementioned company will be described below.

The piece of aluminum serving as a conductive body is immersed for ten to fifteen minutes in an alkaline solution containing sodium hydroxide to clean organic compounds, grease or oil. Then double rinse with deionized water. Subsequently, the piece is immersed for four to five minutes in an antioxidant solution, which contains nitric acid, ferric sulfate and ammonium in order to remove excess sodium hydroxide. A double rinse is performed with deionized water. Subsequently, the piece is immersed for three to fifteen minutes in caustic soda (sodium hydroxide) to perform a micro attack on the surface of the aluminum in order to condition the piece for nickel. A double rinse is performed with deionized water. Again, the piece is immersed for four to five minutes in an antioxidant solution to remove excess sodium hydroxide. Here a double rinse with deionized water is performed. Since the piece is conditioned, it is immersed for a minute in a zinc coating solution containing ferric hydrochloric acid and sodium hydroxide in order to give the piece a conditioning prior to immersion in nickel (non-electrolytic nickel plating). A double rinse is performed with deionized water. Subsequently, to neutralize the action of zinc plating, it is immersed (until complete dissolution of zinc plating) in a zinc plating solution that is formulated with a potassium compound. Subsequently, it is immersed for 30 seconds in the zinc coating solution. A double rinse is performed with deionized water. Then, the piece is immersed for 10 to 15 minutes in a solution of nickel (non-electrolytic nickel plating), which is composed of nickel sulfate and will give the addition of nickel to aluminum. At this point three laboratory tests are carried out to the solution:

- Ph level, which should be in the range of 4.7 to –5.1
- Metallic nickel, which must be from 5.7 to 6.3 g / l
- Sodium hypophosphite, maintaining the range of 20-50 g / l

A double rinse is performed with deionized water. Subsequently, the piece is immersed for eight to thirty-five minutes in a solution of nickel sulfamate that is composed of nickel sulfamate and that will give the piece a finish, shine and hardness. This solution has three laboratory tests:

- Ph level, which should be in the range of 3.6 to 4.4
- Nickel sulfamate, which should be 7 to 5 oz / gal
- Boric acid, maintaining the range of 2 to 5 oz / gal

A double rinse is performed with deionized water. Then, the aluminum-nickel piece is immersed for three to thirty minutes in an acidic copper solution which contains copper sulfate and sulfuric acid. There a layer of copper is deposited on the piece. The solution has two laboratory tests:

- Copper Sulfate (25-45 oz/gal)
- Sulfuric Acid (10-40 oz/gal)

A double rinse is performed with deionized water. Then, once the aluminum-nickel-copper piece is in place, it is immersed for one to five minutes in a silver solution that will give the piece an initial layer of silver. This solution contains potassium cyanide, silver cyanide and potassium. The solution has two laboratory tests:

- Silver cyanide (3.5-7.5 oz / gal).
- Potassium cyanide (14-18 oz / gal)

A double rinse is performed with deionized water. Immediately and depending on the finish that you want to give the piece can be:

- Opaque silver: Immerse the piece for two to ten minutes in an opaque silver solution, which is composed of silver cyanide, potassium and potassium cyanide. This will give the silver an opaque finish. This solution has two laboratory tests:
  - Silver and potassium cyanide (3.5-7.5 oz / gal)
  - Potassium cyanide (14-18 oz / gal)
- Shiny silver: Dip the piece for two to three minutes in a solution of shiny silver, which is composed of silver cyanide, potassium and potassium cyanide, in addition to brighteners, which will give the silver a shiny finish. This solution has two laboratory tests:
  - Silver and potassium cyanide (0.5-0.9 oz / gal)
  - Potassium cyanide (11-15 oz / gal)

A double rinse is performed with deionized water. The piece is dried in the oven at a temperature greater than 75 degrees Celsius.

METHODOLOGY

According to Montgomery (2002), the general guidelines for the design of experiments require that all participating parties have a clear idea of the objectives, procedures and methods of analysis that will be used in this task. Thus, the general scheme recommended for the procedure involves the following steps:

1. Identification and statement of the problem. This point may seem very obvious, but it is common that in practice it is not easy to realize that there is a problem that requires experimentation, and it is not easy to develop a clear statement, with which everyone agrees, of this problem. It is necessary to develop all ideas about the objectives of the experiment. Generally, it is important to request contributions from all the areas involved: engineering, quality assurance, manufacturing, marketing, administration, the client and the operation staff (who usually knows the process thoroughly and who too often is ignored). For this reason, a team approach is recommended to design experiments. In most cases it is convenient to make a list of the specific problems or questions that will be addressed in the experiment. A clear statement of the problem contributes substantially often to

achieve a better understanding of the phenomena under study and the final solution of the problem. It is also important to keep in mind the overall objective; For example, is it a new process or system (in which case the initial objective may be the characterization or screening of the factors) or is it a mature system that is known with reasonable depth and that has been previously characterized (in which case the objective may be optimization)? In an experiment there may be many possible objectives, including confirmation (does the system behave in the same way now as in the past?), Discovery (what happens if new materials, variables, operating conditions, etc. are explored. ?) and stability (under what conditions do the response variables of interest suffer serious degradation?). Obviously, the specific issues to be addressed in the experiment are directly related to the overall objectives. Often at this stage of the problem formulation many engineers and scientists realize that it is not possible for an extensive comprehensive experiment to answer the key questions and that a sequential approach in which a series of smaller experiments is used is a strategy more suitable.

2. Choice of factors, levels and ranges. When considering the factors that can influence the performance of a process or system, the experimenter usually discovers that these factors can be classified as potential design factors or as disturbing factors. The potential design factors are those that the experimenter may want to vary in the experiment. It is common to find that there are many potential design factors, so it is convenient to have some additional classification of them. Some useful classifications are design factors, factors that remain constant, and factors that are allowed to vary. The design factors are those that are actually selected to study them in the experiment. The factors that remain constant are variables that may have some effect on the response, but that for the purposes of the current experiment are not of interest, so they will remain fixed at a specific level. For example, in a chemical etching experiment in the semiconductor industry there may be an effect, which is unique, of the specific tool for plasma chemical etching that is used in the experiment. However, it would be very difficult to vary this factor in an experiment, so the experimenter can decide to perform all experimental runs on a particular chemical recorder (ideally "typical"). In this way, this factor remains constant. As an example of factors that are allowed to vary, the experimental units or the "materials" to which the design factors are applied are generally not homogeneous, however often this variability of a unit is ignored. another and randomization is relied on to compensate for any effect of the material or experimental unit. Many times we will work with the assumption that the effects of the factors that remain constant and the factors that are allowed to vary are relatively small.

On the other hand, disturbing factors can have considerable effects that must be taken into consideration, even though there is no interest in them in the context of the current experiment. Disturbing factors are usually classified as controllable, uncontrollable or noise factors. A controllable disturbing factor is one whose levels can be adjusted by the experimenter. For example, the experimenter can select different batches of raw material or several days of the week to conduct the experiment. The basic structure of block formation, discussed in the previous section, is usually useful for working with controllable disturbing factors. If a disturbing factor is not controllable in the experiment, but can be measured, the analysis procedure called covariance analysis can often be used to compensate for this effect. For example, the relative humidity in the process environment can affect the process performance, and if the humidity cannot be controlled, it can probably be measured and treated as a covariate. When a factor that varies naturally and cannot be controlled in the process can be controlled for the purpose of an experiment, it is often called a noise factor. In such situations, it is common that the objective is to find the settings of controllable design factors that minimize the variability transmitted by noise factors. Sometimes this is called the study of robustness of the process or the problem of robustness of the design.

Once the experimenter has selected the design factors, he must choose the ranges in which he will vary these factors, as well as the specific levels with which the runs will be performed. It should also be considered how these factors will be controlled at the desired values and how they will be measured. For example, in the liquid welding experiment, the engineer has defined 12 variables that can affect the occurrence of welding defects. The engineer will also have to make a decision as to the region of interest for each variable (that is, the range in which each factor will be varied) and as to the number of levels of each variable he will use. This requires knowledge of the process. This knowledge of the process is usually a combination of practical experience and theoretical knowledge. It is important to investigate all the factors that may be of importance and not be too influenced by past experience, particularly when one is in the initial stages of experimentation or when the process is not fully mature. When the objective of the experiment is the screening of the factors or characterization of the process, it is generally better to keep the number of factor levels reduced. In general, two levels work quite well in factor screening studies. Choosing the region of interest is also important. In screening factors, the region of interest should be relatively large; that is, the range in which the factors are varied should be wide. As more is known about the variables that are important and the levels that produce the best results, the region of interest will generally become narrower.

3. Selection of the response variable. To select the response variable, the experimenter must be certain that this variable actually provides useful information about the process under study. In most cases, the average or standard deviation (or both) of the measured characteristic will be the response variable. Multiple answers are no exception. The efficiency of measuring instruments (or measurement error) is also an important factor. If the efficiency of the measuring instruments is inadequate, the experimenter will only detect the relatively large effects of the factors or additional replications may be necessary. In some situations where the efficiency of the measuring instruments is poor, the experimenter may decide to measure each experimental unit several times and use the average of the repeated measurements as an observed response. It is often of decisive importance to identify the aspects related to the definition of the responses of interest and how they will be measured before carrying out the experiment. Sometimes experiments designed to study and improve the performance of measurement systems are used. It is reiterated how crucial it is to present all the views and information of the process in steps 1 to 3 above. This is referred to as pre-experiment planning. Coleman and Montgomery provide worksheets that can be useful in pre-experiment planning. In many situations, it is not possible for a single person to possess all the knowledge required to do this properly. Therefore, a broad recommendation is made for teamwork during the planning of the experiment. Most of the success will gravitate to how well the previous planning of the experiment has been done.
4. Choice of experimental design. If the pre-experiment planning activities are carried out properly, this step is relatively simple. The choice of design implies the consideration of the sample size (number of replicas), the selection of an order of runs suitable for experimental trials and the determination of whether block formation or other restrictions on randomization come into play. . This book reviews some of the most important types of experimental designs, and can ultimately be used as a catalog to select the most appropriate experimental design for a wide variety of problems. There are also several interactive statistical software packages that support this phase of experimental

design. The experimenter can enter information on the number of factors, levels and ranges, and these programs will present a selection of designs to the experimenter for consideration or recommend a particular design.

These programs also generally provide a worksheet (with the random order of runs) that will be used in conducting the experiment. When selecting the design, it is important to keep the experimental objectives in mind. In many engineering experiments it is known in advance that some of the factor levels will produce different values of the response. Consequently, the focus is on identifying what factors cause this difference in estimating the magnitude of the response change. In other situations there may be more interest in verifying uniformity. For example, two production conditions A and B can be compared, where A is the standard and B is an alternative with greater cost efficiency. The experimenter will then be interested in demonstrating that, for example, there is no difference in performance between the two conditions.

5. Performing the experiment. When the experiment is carried out, it is vital to carefully monitor the process in order to ensure that everything is being done according to the planning. Errors in the experimental procedure at this stage will generally destroy the experimental validity. Placing foreground planning is crucial to success. It is easy to underestimate the aspects of logistics and planning when running an experiment designed in a complex manufacturing or research and development environment. Coleman and Montgomery suggest that before carrying out the experiment, it is convenient on many occasions to conduct some pilot or test runs. These runs provide information about the consistency of the experimental material, a verification of the measurement system, an approximate idea of the experimental error and the opportunity to implement the global experimental technique. This also offers an opportunity to review, if necessary, the decisions made in steps 1 to 4.
6. Statistical analysis of the data. Statistical methods should be used to analyze the data so that the results and conclusions are objective and not appreciative. If the experiment has been designed correctly and if it has been carried out according to the design, the necessary statistical methods should not be complicated. There are several excellent software packages designed to assist in data analysis, and many of the programs used in step 4 to select the design have a direct interface for statistical analysis. It is often found that simple graphic methods play an important role in the analysis and interpretation of data. Because many of the questions that the experimenter wants to answer can be inserted in the hypothesis test framework, the procedures for testing hypotheses and estimating confidence intervals are very useful in analyzing data from a designed experiment. Most of the time, it is also very useful to present the results of several experiments in terms of an empirical model, that is, through an equation derived from the data that expresses the relationship between the response and the important design factors. The residual analysis and the verification of the adequacy of the model are also important analysis techniques. Remember that statistical methods cannot demonstrate that a factor (or factors) has a particular effect, they only provide general guidelines regarding the reliability and validity of the results. Applied correctly, statistical methods do not allow the experimental demonstration of anything, but they do serve to measure the possible error in a conclusion or assign a level of confidence to a statement. The main advantage of statistical methods is that they add objectivity to the decision-making process. Statistical

techniques, coupled with good engineering or knowledge of the process and common sense, will generally lead to strong conclusions.

7. Conclusions and recommendations. Once the data has been analyzed, the experimenter must draw practical conclusions about the results and recommend a course of action. Graphic methods are usually useful at this stage, in particular to present the results. Follow-up runs or confirmation tests should also be performed to validate the conclusions of the experiment. Throughout the entire process it is important to keep in mind that experimentation is an essential part of the learning process, in which tentative hypotheses about a system are formulated, experiments are conducted to investigate these hypotheses and new hypotheses are formulated based on results, and so on. This suggests that experimentation is iterative. It is usually a big mistake to design a single comprehensive and extensive experiment at the beginning of a study. A successful experiment requires knowing the important factors, the ranges in which these factors should be varied, the appropriate number of levels to be used and the appropriate units of measurement for these variables. In general, the precise answers to these questions are unknown, but you learn about them on the fly. As an experimental program progresses, it is common to abandon some input variables and incorporate others, modify the exploration region of some factors or incorporate new response variables. Therefore, experimentation is generally done sequentially and, as a general rule, no more than 25% of the resources available in the first experiment should be invested. This will ensure that there will be sufficient resources to carry out the confirmation runs and that the final objective of the experiment will ultimately be achieved.

RESULTS

It was found that there was a correlation between the insertion loss and the silver thickness. Additionally, there was a negative effect between the silver deposit and insertion loss. It is if there was a greater deposit of silver, the insertion loss was lower. Moreover, the copper use, silver purity, amperage and silver concentration were the factors with greater effect. This means that adding copper, maintaining the silver purity at 99.99%, amperage at 5 amperes, and silver concentration at 7.5 oz./gal were the ideal values of the factors to obtain a better silver distribution onto filter surface. Furthermore, the copper use and silver purity were the factors with greater effect in the electric response and insertion loss. It is adding copper and maintaining the silver purity at 99.99% were the ideal values of the factors to obtain an acceptable electric response. Consequently, there was a substantial increase in the levels of quality passing from 50% to 90%. The former was before the implementation of Design of Experiments, and the last one was after the mentioned implementation. Finally, the last phase was concentrated on obtaining conclusions and recommendations. It is important to remark that was necessary to clearly describe and understand the silver deposit process. This permitted to identify in detail the problem and to define the factors and additional information required in diverse phases of the mentioned technique. As a result, seven factors were evaluated and subsequently changed in order to increase the value of the electric response. Particularly, the amperage, the material of the box of metal, the silver cleanliness and the silver concentration were the factors that changed their levels. In general terms, the percentage of the quality increased 40%. Additionally, the Design of Experiments permitted to advise some changes in the electrochemical process so that some steps were reordered. Particularly, this technique permitted to find a better combination of factors affecting the process in study. It was seen during the experimentation and the statistic analysis of the obtained data. This admitted to uncover the new parameters of the factors considered subsequently enabling a better distribution of the silver finish. For this, it is suggested additional changes in the materials and equipment used in that process.

Problem Identification

The company's catalog defines an electric filter or electronic filter as an element that distinguishes a certain frequency or frequency range from an electrical signal that passes through it, being able to modify both its amplitude and its phase. A band filter allows the passage of a specific band, rejecting the passage of low and high frequencies within a range of 2000 to 12000 MHz. There, an attenuation is defined as the loss of signal from a filter, which is usually measured in decibels (dB). Attenuation is usually referred to as signal amplitude or signal strength. In telecommunication, it is called attenuation of a signal, be it acoustic, electrical or optical, to the loss of power suffered by it when passing through any transmission medium. Also, it defines an insertion loss as the signal loss caused when the filter is inserted into a circuit and measured in decibles (dB). A general definition is established as the amount of voltage allowed when the electric frequency is at its maximum potential. Telecom insertion losses are the loss of signal power due to the insertion of a device in a transmission line or optical fiber. In figure 2 both parameters are presented schematically.

These parameters are directly proportional to each other. Insertion losses are a measure of attenuation due to the insertion of a device in the signal path. They require a homogeneous surface finish so that the passage of the electrical signal flows in the same way without affecting the spectrum of signals defined in the parameters for which it was designed. On the other hand, silver, according to Marks (1995) is defined as a chemical element that has the maximum electrical conductivity and thermal conductivity of any metal, as well as being a metal with solution pressure or the tendency of the metal to dissolve in water, which is involved in electrochemical corrosion. Of the world's silver production, approximately 70% is used for industrial purposes and 30% for monetary purposes. In chemistry and electronic design, it is where their uses are mainly given. For these reasons, silver is the metal selected to electro deposit on the aluminum surface. Thus, in the process under study there was a loss of signal specifically in the BAND PASS model. This resulted in a 50% loss of quality levels and increased costs. Figure 3 shows this signal loss.

These parameters are directly proportional to each other. Insertion losses are a measure of attenuation due to the insertion of a device in the signal path. They require a homogeneous surface finish so that the passage of the electrical signal flows in the same way without affecting the spectrum of signals defined in the parameters for which it was designed. On the other hand, silver, according to Marks (1995) is defined as a chemical element that has the maximum electrical conductivity and thermal conductivity of any metal, as well as being a metal with solution pressure or the tendency of the metal to dissolve in water, which is involved in electrochemical corrosion. Of the world's silver production, approximately 70% is used for industrial purposes and 30% for monetary purposes. In chemistry and electronic design, it is where their uses are mainly given. For these reasons, silver is the metal selected to electro deposit on the aluminum surface. Thus, in the process under study there was a loss of signal specifically in the 3BD810-925 model. This resulted in a 50% loss of quality levels and increased costs. Figure 3 shows this signal loss.

The cause of the cost increase was attributed to the rework of the pieces as they were rejected and as a correction measure, silver was added to it. Therefore, the rejected filters were used to measure the

Figure 2. Frequency signal of the electronics filter

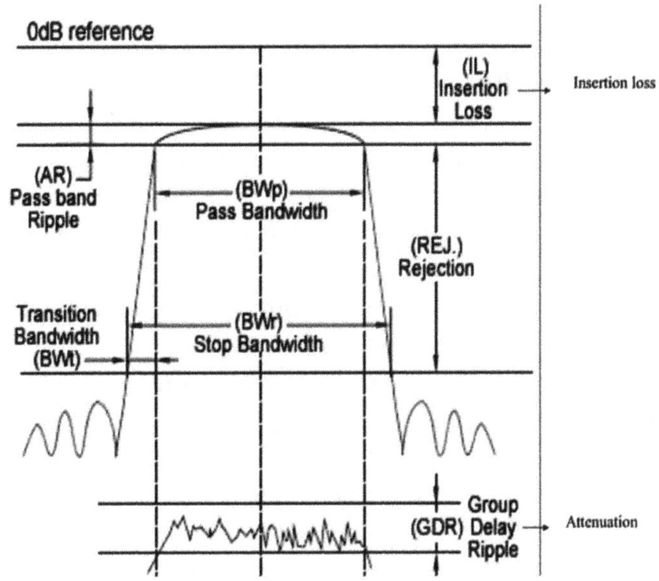

Figure 3. Signal loss

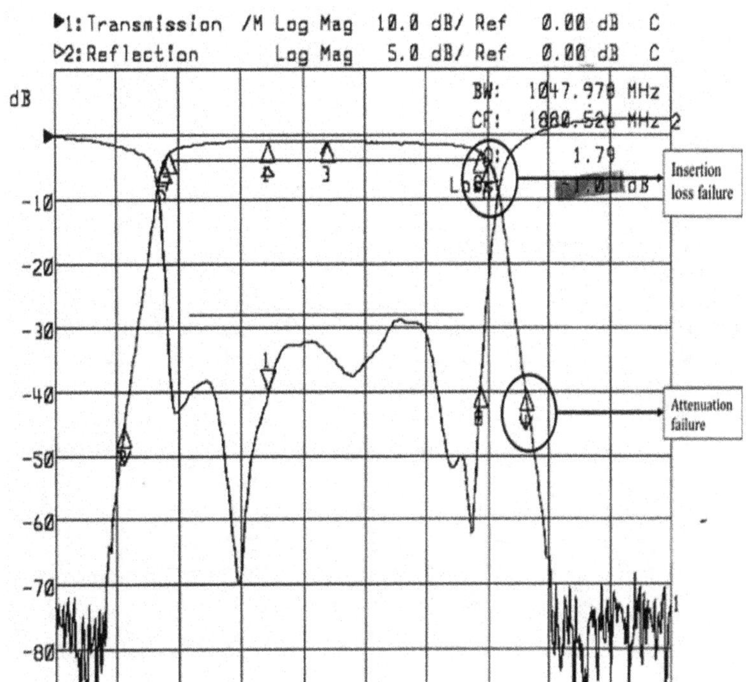

flatness of the outside and the cavities. This demonstrated an irregular shape in the distribution of silver, specifically within the cavities in micro-inches, as presented in Figure 4.

Choice of Factors and Levels

After doing several brainstorming sessions with experts in the process discussed here, seven factors affecting that process were selected. These factors were considered as those that can benefit the optimization of silver deposition. These factors are listed below, and their description is presented:

1. **Electric Current (Amperage):** Electric charge flow per unit of time that a material travels, the greater the electric charge, the greater the electrodeposition.
2. **Time (Minutes):** Physical magnitude with which we measure the duration of electrodeposition.
3. **Plating Standard:** The current standard without copper (Cu) was compared against a new standard with copper.
4. **Silver Purity:** 99.99% silver was compared against 99.75% silver in anodes.
5. **Type of Electrical Pulse:** A flow of continuous amperes from anode to cathode was compared against programmed amperage profiles and changes between current direction anode to cathode and vice versa.
6. **Rectifier Capacity:** 25 amperes and 50 amperes
7. **Silver Concentration:** 3.5 oz / gal and 7.5 oz / gal

Similarly, the levels were selected, taking the level (+) as the maximum permissible levels and the levels (-) those minimum permissible levels, allowing to be within the control levels, see table 1.

Figure 4. Silver thickness in product cavities

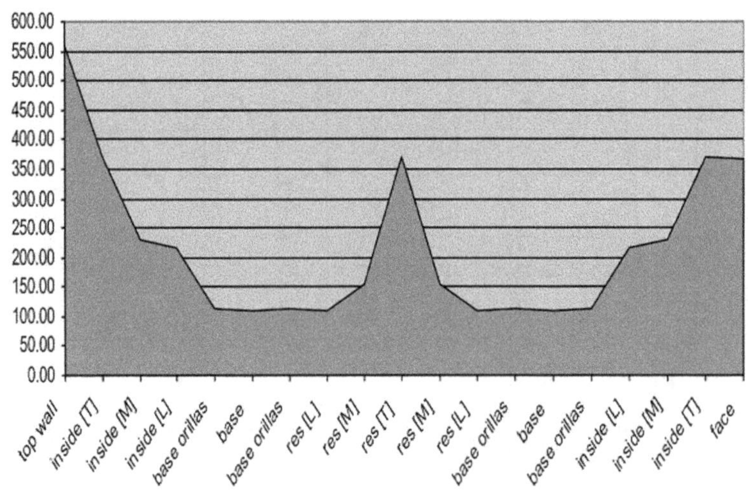

Selection of the Variable Response

The response variables were selected based on the hypothesis to be tested and the electrical specification required by the customer. Consequently, the selected variables were:

X_1: Silver thickness reading range (300 to 500 micro-inches)
X_2: Electrical response (insertion loss and attenuation)

Experimental Design Selection

The decision to use an orthogonal L8 Taguchi matrix design for seven variables and eight experiments was defined by the cost of experimentation, because the thickness of the silver represents an important cost for each run. The Taguchi method is a robust design that considers variation and it is not necessary to make too many runs to obtain significant results. Table 2 shows the matrix of variables and experiments.

Performing the Experiments

For a better control each of the experiments was defined by assigning the values of their levels. Table 3 shows the list of the eight experiments with their respective level values. Due to the implicit cost in each experiment, only one filter was used per experiment and the silver thickness measurements were averaged.

Figure 5 shows the positions where the thickness readings were taken.

Data Analysis

Appendix 1 shows the results of each experimental run performed on the platinum filters established in the design of experiments. The statistics of these results were generated with the Minitab software. Figure 6 shows the results of the regression analysis of these experiments. The results demonstrate a correlation between insertion loss and silver thickness. There is a negative effect: if there is a greater deposit of silver, the insertion loss is less. To solve the problem of failure in insertion loss it is necessary to maintain the homogeneous silver distribution.

Table 1. Factor description and levesl for the experimental design

Code	Factors	Levels	
		(+)	(-)
A	Amperage	5 A	2 A
B	Time	40 min	20 min
C	Plating standard	With Cu	No Cu
D	Silver purity	99.99%	99.75%
E	Type of electrical pulse	Direct	Inverted
F	Rectfier capacity	50 amps	25 amps
G	Silver concentration	7.5 oz/gal	3.5 oz/gal

Table 2. L8 matrix for 7 variables and 8 experiments

Experiment	Factors						
	A	B	C	D	E	F	G
1	(-1)	(-1)	(-1)	(-1)	(-1)	(-1)	(-1)
2	(-1)	(-1)	(-1)	(+1)	(+1)	(+1)	(+1)
3	(-1)	(+1)	(+1)	(-1)	(-1)	(+1)	(+1)
4	(-1)	(+1)	(+1)	(+1)	(+1)	(-1)	(-1)
5	(+1)	(-1)	(+1)	(-1)	(+1)	(-1)	(+1)
6	(+1)	(-1)	(+1)	(+1)	(-1)	(+1)	(-1)
7	(+1)	(+1)	(-1)	(-1)	(+1)	(+1)	(-1)
8	(+1)	(+1)	(-1)	(+1)	(-1)	(-1)	(+1)

The main response effects in silver thickness measurements and their relevant analysis are shown in Figure 7. It is shown that the use of copper, the purity of silver, the amperage and the concentration of silver are the factors with the greatest effect. The ratio shown in the graphs is that by adding copper, maintaining the purity of silver at 99.99%, the amperage at 5 amperes and the concentration of silver at 7.5 oz / gal, will result in a better distribution of silver on surface of the filters.

The main effects of electrical response on insertion loss are shown in Figure 8. It is shown that the use of copper and the purity of silver are only the factors with the greatest effect. The ratio shown in the graphs is that by adding copper and maintaining the purity of silver at 99.99%, an acceptable electrical response is obtained.

Conclusion and Recommendations

In conclusion, the employed technique was useful to uncover factors that affect the quality of the product in study. It agreed the active manipulation of some factors until some were discovered affecting the mentioned process. In addition, it simultaneously admitted that manipulation of diverse factors at the same time reducing the costs on it. With the establishment of the new parameters through the Design of experiments, the quality level was increased to 90% acceptance, which implies a reduction of waste costs of 40%. The objective was to analyze, through an experiment design, the important parameters to improve the distribution of the finish and improve the quality levels, see table 4.

Table 5 shows the previous parameters against the parameters established by the design of experiments.

Maintaining the purity of silver at 99.99% and adding copper before the final silver finish, proved to give better results, improving quality levels. The parameters of amperage and concentration of silver were modified since the graph of effects of silver thicknesses showed that they are factors with a positive effect, see figure 7.

FUTURE RESEARCH DIRECTIONS

In this way, the feasibility and efficiency of the application of the design of experiments focused on process improvement has been made clear, concluding with the following recommendations for the industry:

Table 3. Experiments and their levels

Run	Factors						
	A	B	C	D	E	F	G
1	2 amps	20 min	No Cu	97.30%	Inverted	20 amps	3.5 oz/gal
2	2 amps	20 min	With Cu	99.70%	Direct	50 amps	7.5 oz/gal
3	2 amps	40 min	No Cu	97.30%	Direct	50 amps	7.5 oz/gal
4	2 amps	40 min	With Cu	99.70%	Inverted	20 amps	3.5 oz/gal
5	5 amps	20 min	No Cu	99.70%	Inverted	50 amps	7.5 oz/gal
6	5 amps	20 min	With Cu	97.30%	Direct	20 amps	3.5 oz/gal
7	5 amps	40 min	No Cu	99.70%	Direct	20 amps	3.5 oz/gal
8	5 amps	40 min	With Cu	97.30%	Inverted	50 amps	7.5 oz/gal

- Carry out more projects where all the departments involved participate to propose ideas, doing so with a good organization to define activities and responsibilities.
- Propose various designs of experiments that consider additional factors that affect the process.

Figure 5. Positions where measurements were taken

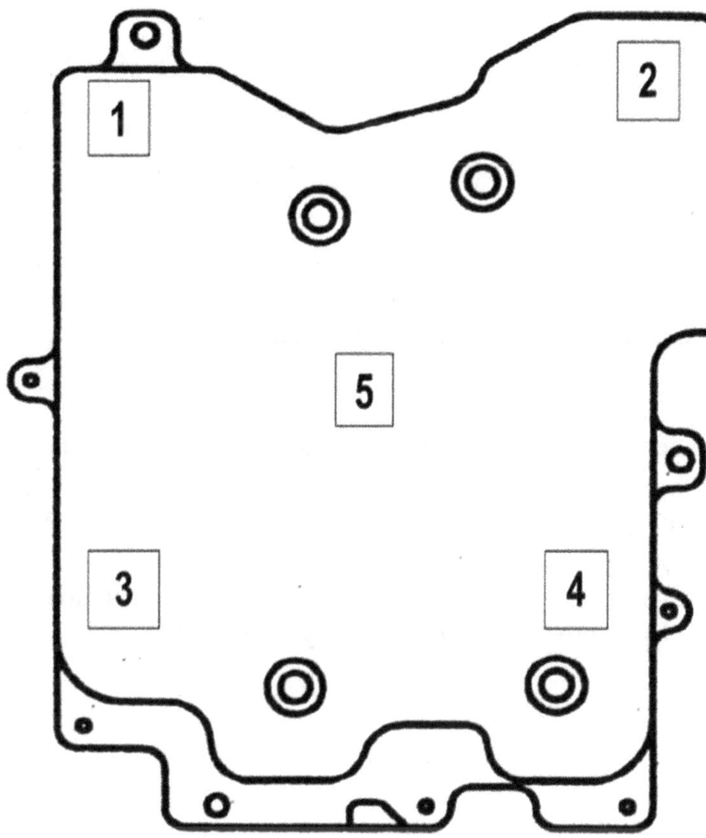

Figure 6. Regression analysis

x	y
760	281.1
764	172.2
759	359.6
770	18.06
762	258.3
772	104.4
758	281.2
764	104.9

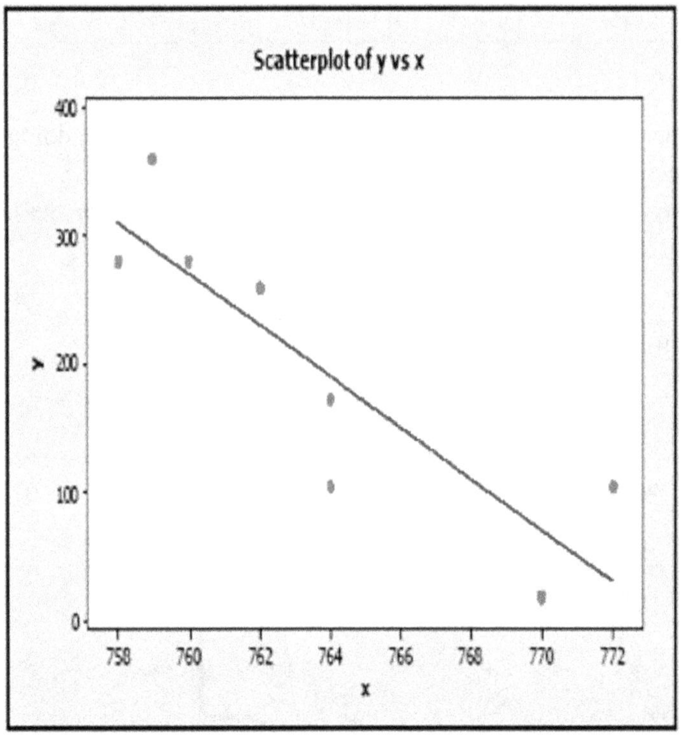

Scatterplot of y vs x

Regression Analysis: x versus y

The regression equation is
x = 771 - 0.0382 y

Predictor	Coef	SE Coef	T	P
Constant	771.163	1.962	392.96	0.000
y	-0.038174	0.008711	-4.38	0.005

S = 2.67149 R-Sq = 76.2% R-Sq(adj) = 72.2%

Analysis of Variance

Source	DF	SS	MS	F	P
Regression	1	137.05	137.05	19.20	0.005
Residual Error	6	42.82	7.14		
Total	7	179.88			

Unusual Observations

Obs	y	x	Fit	SE Fit	Residual	St Resid
6	104	772.000	767.178	1.245	4.822	2.04R

- Do research on the different electroplating processes, as well as the materials and their characteristics.
- Provide more training and training to staff for platinum operations, as well as experiment design courses.
- Constantly review that all activities that may affect electroplating operations are carried out in accordance with established procedures and constantly monitored.

Figure 7. Response to effects (silver measures)

A	B	C	D	E	F	G	Results
1	1	1	1	1	1	1	325
1	1	1	2	2	2	2	444.5
1	2	2	1	1	2	2	430
1	2	2	2	2	1	1	407.59
2	1	2	1	2	1	2	414.3
2	1	2	2	1	2	1	429.2
2	2	1	1	2	2	1	334.7
2	2	1	2	1	1	2	421.7

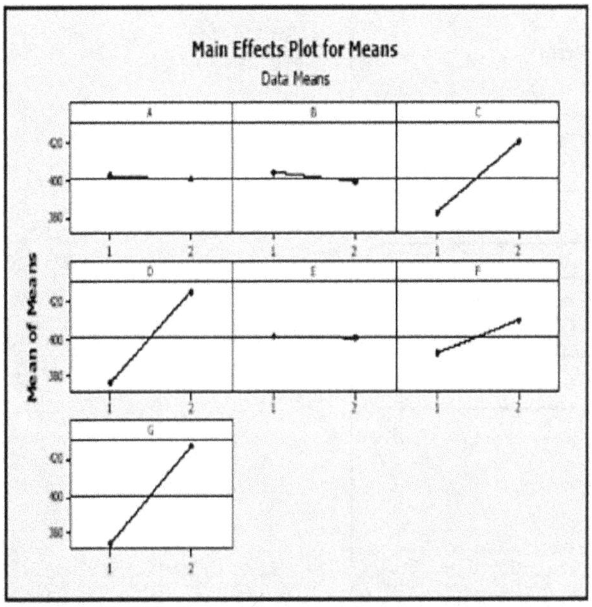

Taguchi Analysis: C8 versus A, B, C, D, E, F, G

Response Table for Signal to Noise Ratios
Nominal is best (10*Log10(Ybar**2/s**2))

Level	A	B	C	D	E	F	G
1	*	*	*	*	*	*	*
2	*	*	*	*	*	*	*
Delta	*	*	*	*	*	*	*
Rank	4	4	4	4	4	4	4

Response Table for Means

Level	A	B	C	D	E	F	G
1	401.8	403.3	381.5	378.0	401.5	392.1	374.1
2	400.0	398.5	420.3	425.7	400.3	409.6	427.6
Delta	1.8	4.8	38.8	49.7	1.2	17.5	53.5
Rank	6	5	3	2	7	4	1

Main Effects Plot for Means

CONCLUSION

It was found that there was a correlation between the insertion loss and the silver thickness. Additionally, there was a negative effect between the silver deposit and insertion loss. It is if there was a greater deposit of silver, the insertion loss was lower. Moreover, the copper use, silver purity, amperage and silver concentration were the factors with greater effect. This means that adding copper, maintaining the silver purity at 99.99%, amperage at 5 amperes, and silver concentration at 7.5 oz./gal were the ideal values of the factors to obtain a better silver distribution onto filter surface. Furthermore, the copper use and silver purity were the factors with greater effect in the electric response and insertion loss. It is adding copper and maintaining the silver purity at 99.99% were the ideal values of the factors to obtain an acceptable electric response. Consequently, there was a substantial increase in the levels of quality passing from 50% to 90%. The former was before the implementation of DoE, and the last one was after the mentioned implementation. Finally, the last phase was concentrated on obtaining conclusions and recommendations. It is important to remark that was necessary to clearly describe and understand the silver deposit process. This permitted to identify in detail the problem and to define the factors and ad-

Figure 8. Response to effects (insertion loss)

A	B	C	D	E	F	G	Results
1	1	1	1	1	1	1	760
1	1	1	2	2	2	2	764
1	2	2	1	1	2	2	759
1	2	2	2	2	1	1	770
2	1	2	1	2	1	2	762
2	1	2	2	1	2	1	772
2	2	1	1	2	2	1	758
2	2	1	2	1	1	2	764

Results for: Worksheet 1

Taguchi Analysis: C8 versus A, B, C, D, E, F, G

Response Table for Signal to Noise Ratios
Nominal is best (10*Log10(Ybar**2/s**2))

Level	A	B	C	D	E	F	G
1	*	*	*	*	*	*	*
2	*	*	*	*	*	*	*
Delta	*	*	*	*	*	*	*
Rank	4	4	4	4	4	4	4

Response Table for Means

Level	A	B	C	D	E	F	G
1	763.3	764.5	761.5	759.8	763.8	764.0	765.0
2	764.0	762.8	765.8	767.5	763.5	763.3	762.3
Delta	0.8	1.8	4.3	7.8	0.3	0.8	2.8
Rank	5.5	4	2	1	7	5.5	3

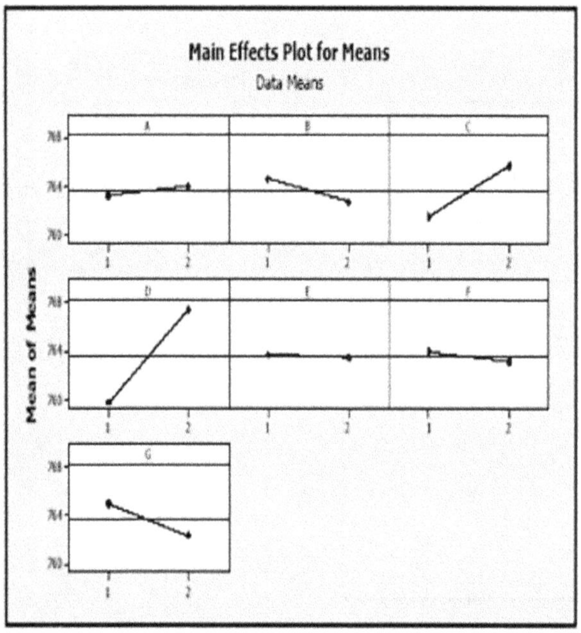

ditional information required in diverse phases of the mentioned technique. As a result, seven factors were evaluated and subsequently changed in order to increase the value of the electric response. Particularly, the amperage, the material of the box of metal, the silver cleanliness and the silver concentration were the factors that changed their levels. In general terms, the percentage of the quality increased 40%. Additionally, the Design of Experiments permitted to advise some changes in the electrochemical process so that some steps were reordered. Particularly, this technique permitted to find a better combination of factors affecting the process in study. It was seen during the experimentation and the statistic analysis of the obtained data. This admitted to uncover the new parameters of the factors considered subsequently enabling a better distribution of the silver finish. For this, it is suggested additional changes in the materials and equipment used in that process.

As a conclusion of this thesis project it was possible to verify that the design of experiments is an efficient tool for this specific case, where it is about finding based on certain input variables which are the ones that are most affecting the process. In this case there are many factors that can affect the electroplating process. The DoE is a useful tool, since it was possible to verify the best combination of

Table 4. Quality percentage before and after DoE implementation

Before	After
50%	90%

factors that affect the electrodeposition process. It could be seen that the data obtained in the design of experiments are reliable, because both in the experiment and in the analysis of the data it was shown that with the new parameters a better distribution of the silver finish can be obtained.

With the DoE, parameters supported by a statistical analysis were achieved, which is the DoE using the Taguchi Method. One of the important points that the Electronic Components company required was to have established the parameters for the silver electrodeposition process, and through the results obtained by the design of experiments, they were obtained and standardized in a work instruction for the plating process silver. The fundamental reason for experimental design as a tool for decision making in business is threefold. First, experiments require an active manipulation of the process. One way to know how the input affects the output is by combining it. Second, the experimental design employs the principle of average, which reduces noise. Finally, with a design approach, more than one aspect of the process can be investigated simultaneously, obtaining data with which it can be decided whether or not the project investment will be expensive.

ACKNOWLEDGMENT

We would like to acknowledge to the Tecnológico Nacional de México/Instituto Tecnológico de Tijuana, Universidad Autonóma de Baja California, PRODEP and CONACYT for partially funding the study.

Table 5. Parameters of the factors used before and after DoE implementation

Factors	Used Parameters	Established Parameters
A. Amperage	2 amps	5 amps
B. Time	20 min	20 min
C. Estándar	No Cu	With Cu
D. Pureza de plata	99.75%	99.99%
E. Tipo de pulso eléctrico	Direct	Direct
F. Capacidad de rectificador	25 amps	25 amps
G. Concentración de plata	3.5 oz/gal	7.5 oz/gal

REFERENCES

Anderson, M. J. (2005). Trimming the fat out of experimental methods. *OE Magazine*, *5*(6), 32–33.

Arvidsson, M., & Gremyr, I. (2008). Principles of robust design methodology. *Quality and Reliability Engineering International*, *9*(3), 185–202.

Bhote, K. R. (2000). *World class quality. Using design of experiments to make it happen* (2nd ed.). New York: Amacom.

Booker, B. W., & Lyth, D. M. (2005). Quality engineering from 1988 through 2005: Lessons from the past and trends for the future. *Quality Engineering*, *18*(1), 1–4. doi:10.1080/08982110500439452

Box, G. E. P., & Liu, P. Y. T. (1999). Statistics as a catalyst to learning by scientific method part 1. *Journal of Quality Technology*, *31*(1), 1–15. doi:10.1080/00224065.1999.11979889

Box & Wilson. (1951). On the experimental attainment of optimum conditions. *Journal of the Royal Statistical Society. Series B. Methodological*, (13), 1–45.

Celik, A. E., & Karatepe, Y. (2007). Evaluating and forecasting banking crisis through neural network models: An application for Turkish banking sector. *Expert Systems with Applications*, *33*(4), 809–815. doi:10.1016/j.eswa.2006.07.005

Czitrom, V. (1999). One factor at a time versus designed experiments. *The American Statistician*.

Daniel, C. (1973). One at a time plans. *Journal of the American Statistical Association*, *68*(342), 353–360. doi:10.1080/01621459.1973.10482433

Del Castillo, E. (2007). *Process optimization: A statistical approach. International Series in Operations Research & Management Science*. Springer. doi:10.1007/978-0-387-71435-6

DeMast, J. A. (2004). A methodological comparison of three strategies for quality improvement. *International Journal of Quality & Reliability Management*, *21*(2), 198–212. doi:10.1108/02656710410516989

Deming, W. E. (1982). Out of the crisis. Cambridge, MA: MIT Center for Advanced Engineering Study.

Fisher, R. A. (1935). *The design of experiments*. New York: John Wiley.

Frey, D. D., & Wang, H. (2006). Adaptive one-factor at a time experimentation and expected value of improvement. *Technometrics*, *48*(3), 418–431. doi:10.1198/004017006000000075

Funkenbush, P. D. (2005). *Practical guide to designed experiments. A unified modular approach*. Marcel Dekker.

Goh, T. N. (1993). Taguchi methods: Some technical, cultural, and pedagogical perspectives. *Quality and Reliability Engineering International*, *9*(3), 185–202. doi:10.1002/qre.4680090306

Goupy-Creightom. (2007). Introduction to design of experiments, with JMP examples (3rd ed.), SAS Press Series.

Knapp, G. (2003). How DoE saved a company 50 million dollars in raw materials costs. *DEO FAQ Alert*, *3*(2).

Lye, L. M. (2005). *Tools and toys for teaching design of experiments methodology. 33rd Annual General Conference of the Canadian Society for Civil Engineering*, Toronto, Canada.

Montgomery, D. C. (2002). Changing roles for the industrial statisticians. *Quality and Reliability Engineering International, 18*(5).

Montgomery, D. C. (2005a). Generation III Six Sigma. *Quality and Reliability Engineering International, 21*(6), 3–4. doi:10.1002/qre.751

Montgomery, D. C. (2005b). *Design and analysis of experiments.* John Wiley & Sons.

Nair, V. N., Abraham, B., MacKay, J., Box, G., Kacker, R. N., Lorenzen, T. J., ... Jeff Wu, C. F. (1992). Taguchi's parameter design: A panel discussion. *Technometrics, 31*(2), 127–161. doi:10.1080/00401706.1992.10484904

Nataraj, M. (2005). Optimizing diesel engine parameters for low emissions using Taguchi Method: Variation risk analysis approach –part 1. *Indian Journal of Engineering and Materials Sciences, 12*(3), 169–181.

Padilla, D. (2008). *Aplicación del Diseño de Experimentos a los ensayos realizados en la Unidad de Negocio PROPAGACION, El Quinche*. Escuela Superior Politécnica de Chimborazo, Facultad de Ciencias, Escuela de Física y Matemática.

Pignatello, J. (1991). Top ten triumphs and tragedies of Genechi Taguchi. *Quality Engineering, 4*(2), 211–225. doi:10.1080/08982119108918907

Pulido, H. (2004). *Analisis y diseño de experimentos.* McGraw-Hill, Interamericana.

Quesada, G. M., & Del Castillo, E. (2004). A dual response approach to the multivariate robust Parameter Design problem. *Technometrics, 46*(2), 176–187. doi:10.1198/004017004000000220

Ramirez, E., Recasens, F., Fernández, M., & Larrayoz, M. A. (2004). Sunflower oil hydrogenation on Pd/C in SC propane in a continuous recycle reactor. *AIChE Journal. American Institute of Chemical Engineers, 50*(7), 1545–1555. doi:10.1002/aic.10142

Ray, A. (2004). Development of coating colour formation for duplex board based on Plackett Burman statistical design. *Cellulose Chemistry and Technology, 38*(3-4), 265–275.

Robinson, T. J., Borror, C. M., & Myers, R. H. (2004). Robust parameter design: A review. *Quality and Reliability Engineering International, 20*(1), 81–101. doi:10.1002/qre.602

Roy, R. K. (2001). *Design of experiments using the Taguchi approach: 16 steps to product and process improvement.* Wiley.

Shainin, D., & Shainin, P. (1988). Better than Taguchi orthogonal tables. *Quality and Reliability Engineering International, 4*(2), 143–149. doi:10.1002/qre.4680040209

Starkey, M., Aughton, J., & Brewin, R. (1997). Extending process thinking: Design of experiments in sales and marketing. *The TQM Magazine, 9*(6), 434–439. doi:10.1108/09544789710187046

Taguchi, G. (1986). Introduction to quality engineering. New York: UNIPUB/Kraus International.

Tanco, M. (2006). *Is DoE really used in Industries? ENBIS 6, Wroclaw, Poland.*

Tanco, M., Ilzarbe, L., Viles, E., & Alvarez, M. J. (2008). Applying design of experiments to improve a laser welding process. *Proceedings of the Institution of Mechanical Engineers. Part B, Journal of Engineering Manufacture, 222*(8), 1035–1042. doi:10.1243/09544054JEM1155

Tanco, M., Viles, E., Ilzarbe, L., & Alvarez, M. J. (2008). Is design of experiments really used? A survey of Basque industries. *Journal of Engineering Design, 19*(5), 5. doi:10.1080/09544820701749124

Tay, K. M., & Butler, C. (1999). Methodologies for experimental design: A survey, comparison and future predictions. *Quality Engineering, 11*(3), 343–356. doi:10.1080/08982119908919250

Tong, K. W., Kwong, C. K., & Yu, K. M. (2004). Process optimisation of ranfer moulding for electronic packages using artificial neural networks and multiobjective optimisation techniques. *International Journal of Advanced Manufacturing Technology, 24*(9-10), 675–685. doi:10.100700170-003-1739-2

Wingbrant, H., Svenningstorp, H., Salomonsson, P., Tengström, P., Lundström, I., & Lloyd Spetz, A. (2003). Using a MISiCFET device as a cold start sensor. *Sensors and Actuators. B, Chemical, 93*(1-3), 295–303. doi:10.1016/S0925-4005(03)00227-2

Wu, C. F. J. (2000). *Experiments.* Wiley-Interscience.

ADDITIONAL READING

Bhote, K. R., & Bhote, A. K. (2000). *World Class Quality. Using Design of Experiments to make it happen* (2nd ed.). New York: Amacom.

Cervantes, M. J., & Engstro¨m, T. F. (2004). Factorial Design Applied to CFD. *Journal of Fluids Engineering-Transactions of the ASME, 126*(5), 791–798. doi:10.1115/1.1792277

Designation B700-08 (2011). *Standard Specification for Electrodeposited Coating of Silver for Engineering Use.* ASTM International.

Farhang-Mehr, A., & Azarm, S. (2005). Bayesian Meta-Modelling of Engineering Design Simulations: A Sequential Approach with Adaptation to irregularities in the response Behaviour. *International Journal for Numerical Methods in Engineering, 62*(15), 2104–2126. doi:10.1002/nme.1261

Hoadley, B. A., & Kettering, J. R. (1990). Communications between statisticians and engineers /physical scientist. *Technometrics, 32*(3), 243–247.

Ogle, T., & Hornberger, L. (2001). Reduction of Measurement Variation. Small Acoustic Chamber Screening of Hard Disk Drives. *Noise Control Engineering Journal, 49*(2), 103–107. doi:10.3397/1.2839645

QQ-S-365D (1979). *Federal Specification for Silver plating, electrodeposited: General Requirements.* U.S. Government Printing Office.

Qu-Wu. (2005). One-Factor at a time designs of resolution V. *Journal of Statistical Planning and Inference,* (131): 407–416.

Ruotolo, L., & Gubulin, J. C. (2005). A Factorial Design Study of the Variables Affecting the Electrochemical Reduction of Cr (VI) at Polyaniline Modified Elecodes. *Chemical Engineering Journal, 110*(1-3), 113–121. doi:10.1016/j.cej.2005.03.019

KEY TERMS AND DEFINITIONS

Amperage: The amount of energy applied to the materials at the time of electroplating. The unit of measurement is amps.

Current: Also called electrical intensity, is the load flow per unit of time that a material travels. It is due to a movement of the electrons inside the material.

DoE: Design of experiments.

Effect of a Factor: It is the change observed in the response variable due to a change in level in the factor.

Experiment: It is the moment of scientific research in which theories and hypotheses are put into practice to observe the results of them. The experiment, a word from Latin that means 'put to the test', is precisely the mechanism that is developed to check, verify or correct the postulates of the hypotheses that have been created.

Factorial Design: Experimental design that serves to study the individual effect and interaction of several factors on one or several responses.

Flatness: The surface is considered flat when it is totally comprised between two parallel planes, separated from one another by tolerances.

Frequency: It is the magnitude that measures the number of repetitions per unit of time of any phenomenon or periodic event.

Interaction Effect: Two factors interact significantly on the response variable when the effect of one depends on the level in which the other is.

Parameters: Characteristics that, through their numerical value, describe a set of elements or individuals.

Pressure: It is a scalar physical quantity that measures the force in a perpendicular direction per unit of surface and serves to characterize how a determined force is applied on a surface.

Process: It is a unique combination of machines, tools, methods, materials and human being.

Radiofrequency: Also called radio frequency spectrum or RF, it is applied to the least energetic portion of the electromagnetic spectrum, located between about 3 Hz and about 300 Ghz. The Hertz is the unit of measurement of the frequency of the waves and corresponds to one cycle per second.

Randomization: It is the process, in experimental studies, by which the subjects are randomly assigned to the treatment and control groups.

Representative Sample: It is a part of a population, properly selected, that preserves the key aspects of the population.

Variation: Action and effect of varying. Modification, change, or transformation.

Voltage: Electrical voltage or potential differential is a physical quantity that quantifies the electrical potential difference between two points.

APPENDIX

Figure 9. Appendix

DOE RUNS

1	2	3	4	5	6	7	8
303.9481	323.1564	570.7111	409.597	392.9725	419.3773	344.5938	389.8701
381.232	467.0651	292.2478	402.402	559.6674	481.2402	327.6341	476.0842
201.7099	492.4224	651.8754	400.7516	301.3535	415.5923	294.4491	473.6624
482.8273	444.3718	300.7099	406.3881	487.0939	376.7959	331.174	371.1576
249.5279	495.3682	334.3874	418.8082	330.6449	452.9653	375.6789	397.5635

Variable	Total Count	Mean	SE Mean	StDev	Variance	Sum	Minimum	Maximum	Variable	Range	Skewness	Kurtosis
1	5	325.0	49.6	110.9	12298.6	1625.2	201.7	482.8	1	281.1	0.55	-0.61
2	5	444.5	31.7	70.9	5029.4	2222.4	323.2	495.4	2	172.2	-1.79	3.31
3	5	430.0	75.5	168.7	28465.4	2149.9	292.2	651.9	3	359.6	0.70	-2.54
4	5	407.59	3.20	7.16	51.25	2037.95	400.75	418.81	4	18.06	1.09	0.93
5	5	414.3	48.3	108.0	11660.1	2071.7	301.4	559.7	5	258.3	0.47	-1.69
6	5	429.2	17.7	39.7	1575.2	2146.0	376.8	481.2	6	104.4	0.05	-0.32
7	5	334.7	13.2	29.4	864.9	1673.5	294.4	375.7	7	81.2	0.07	1.12
8	5	421.7	22.1	49.5	2452.0	2108.3	371.2	476.1	8	104.9	0.43	-3.02

Chapter 2
Design of Experiments to Optimize Soxhlet-HTP Method to Establish Environmental Diagnostics of Polluted Soil:
Optimization of the Soxhlet-HTP Method by DOE

Carlos Mario Morales-Bautista
Universidad Juárez Autónoma de Tabasco, Mexico

Candelario Méndez-Olán
División Académica de Ciencias Básicas, Universidad Juárez Autónoma de Tabasco, Mexico

Sugey López-Martínez
https://orcid.org/0000-0002-3681-4001
División Académica de Ciencias Biológicas, Universidad Juárez Autónoma de Tabasco, Mexico

Marcia E. Ojeda-Morales
https://orcid.org/0000-0003-0360-4654
División Académica de Ingeniería y Arquitectura, Universidad Juárez Autónoma de Tabasco, Mexico

ABSTRACT

One of the great challenges of the environmental diagnoses of soils contaminated with hydrocarbons is the optimization of analytical determinations. For this reason, this chapter evaluates the extraction of hydrocarbons by the Soxhlet method through the design of experiments (DOE), variating three different solvents, three soils, and three extraction times. Soil was experimentally contaminated at different concentrations, and hydrocarbons totals relying on conditions organic matter, electrical conductivity,

DOI: 10.4018/978-1-7998-1518-1.ch002

pH and textures, amount of sample, solvents, and NaSO4, were studied. The variables were evaluated by means of an analysis of the Taguchi design and a factorial design, with the results the significant and optimal parameters of the process were determined, which were solvent type and time (10 hours and dielectric constant of 9). Also, the model discards the soil properties. These results will save time and resources, and they reduce errors.

INTRODUCTION

During the last decade, Mexico has been involved in some major changes, mainly focused in the public policies, for example, recently the Mexican government enacted a law to reform the constitution, this law consisted on the allowance of private capital investment in the energy sector in order to maximize the use of our natural resources, also to improve the economy of our country. After that, institutions such as the National Hydrocarbons Commission (CNH), and the Safety Energy and Environment Agency (ASEA) were formed. Currently, these organizations regulate the activities of the petroleum industry and determine the mechanisms for evaluating the impacts that they have and will have on our ecological systems (Hernández, Santillan, Sabelkin, and Parra, 2003; Barrientos & Añorve, 2014, Merchand, 2015; Martínez-Lara & Páez Melo, 2017).

Within the regulations that were established by these newly created public agencies it is specified that analyzes and studies performed should be based on scientific studies, an example is the environmental-base-lines made for reopening or development in the hydrocarbons sector, which specify that the proposed analytical methods must be supported by rigorous scientific studies, mainly, if these are not regulated at least they should be validated, this is the reason why some tools such as design experiments (DOE) and experimental mathematical models have a high application and potential, including some that have already been successfully tested (Morales-Bautista, Méndez-Olán, Hernández-Jiménez, and Adams, 2018).

Currently, one of the great challenges of environmental analytical chemistry is the complexity of the behavior of analytes in the different matrices, in the case of soils, there are still few basic studies in tropical areas that have been subject to constant oil spills, many of them, have been restored but some of the areas with agricultural use vocation continue with problems to establish crops and have resulted in negative impacts in the primary sector, leading to legal and social conflicts for many decades, especially in the southeast states (Castro, Acevedo-Berruecos, Urbieta, Iturbe and, Delgado-Rodríguez, 2012; Adams, Álvarez-Ovando, and Castañón, 2015). In these sites, residual hydrocarbons associated with those effects have been identified, some studies mention that there is a different effect between each type of soil, since these can vary from one site to another and, therefore, have different capacity to respond to a similar pollutant (Zamora, Ramos y Arias, 2012; Palma-Cruz, Pérez-Vargas, Casado, Guzmán, and Calva-Calva, 2016).

Different studies have shown that omitting this parameter causes high uncertainties in declared environmental diagnoses, besides recently was found that the type of hydrocarbon can also influence the level of affectation, in general, it was observed that depending on the fraction of hydrocarbon present in the environment is the type of remediation to be used (García-López, Zavala-Cruz, and Palma-López,

2006; Ojeda-Morales, Hernández-Rivera, Martínez-Vazquez, Diaz-Flores, and Rivera-Cruz, 2012; Guzmán-Osorio & Adams, 2015).

Some research performed in restored sites has established the hypothesis that the presence of residual hydrocarbons causes adverse effects, such as, water repellency, which has a direct relationship with the polar fraction of crude oil derivatives whose effects are reflected in the low field capacity and, therefore, in the decrease of primary productivity. Many experts are concerned about this phenomenon, especially since at some point it will gain access to soils near farmland and therefore our food quality is not guaranteed in the future (Pérez-Hernández, Ochoa-Gaona, Schroeder, Rivera-Cruz, and Geissen, 2013; Marín-García, Adams, and Hernández-Barajas, 2016).

In this context, there are laboratory-scale studies in which environmental variables (rainfalls and temperature variations) are controlled and hydrocarbon spills are simulated. Have been defined that the level of affectation will usually be delimited by three factors: the type of soil, the analytical used methods in the determination or evaluations and the abundance of SARA fractions (Saturated, Aromatics, Resins and Asphaltenes) of crude oil and its derivatives which are determined from the extract of the total hydrocarbons or HTP present on the soil by various methods (Pons-Jiménez, Guerrero-Peña, Zavala-Cruz, and Alarcón, 2011; Díaz-Ramírez, Escalante-Espinosa, Schroeder, Fócil-Monterrubio, and Ramírez-Saad, 2013; Soto, Botello, Licea-Durán, Lizarraga-Partida, and Yañez-Arancibia, 2014; García, Bautista, Olán, and Schroeder, 2016).

The perspective of this type of evaluations applied to the normative analytical methods is discussed in this work. Including the detected errors reported in the literature, such as the type of soil and the solvents compatibility with the components of crude petroleum the main goal of this work is the optimization of the HTP determination method by implementing the analysis of the design of experiments (DOE), specifically, orthogonal arrangements of Taguchi.

BACKGROUND

In edaphological terms, the Mexican republic have different types of reliefs with specific geological characteristics which including other factors within a hydrological basin define the classification of the different soil types of each region. In particular, each state have different sources of pollution, for example, the effects of the mining sector are located mostly in the northern and some states in the center of the country, however, these places are also immersed in different activities, mainly, maquiladoras that still represent an anthropogenic source. Instead, although the entire country is crossed for hydrocarbon piping, most of this sector is distributed especially in the demarcations bordering the Gulf of Mexico (Ruiz-Álvarez, Arteaga-Ramírez, Vazquez-Peña, Ontiveros, y López-López, 2012; Gardi et al., 2014; Soto et al., 2014).

In this geographical area one of the basins with the highest production of crude oil and its derivatives is the southeast basin, within tabasco state is located, this territory has a tropical climate with the predominance of alluvial soils and some hillocks very fertile for planting and grazing- however, there are a lot of oil facilities so it is common that the activities of the primary sector are combined with industrial, so the pollution can bio-accumulate or bio-magnified (Palma-López, Cisneros, Moreno, and Rincón-Ramírez, 2007; Ortiz-Salinas, Cram, and Sommer, 2012; Vázquez-Luna, 2014).

In terms of protecting ecosystems, there are several methods that evaluate the contamination of soils by hydrocarbons, several of them specify that some may or may not have negative impacts depending on the amount of organic matter (MO), the amount of clay (% r), ph and cationic exchange capacity (CIC), likewise the spill extension will depend on the viscosity of the hydrocarbons and the environmental conditions of the site, it has been observed that in cold climates heavier hydrocarbons are usually less mobile than in tropical ecosystems where warn environments and torrential rains predominate (Liu, Lou, Teng, Li, and Ma, 2010; García-Torres et al., 2011; Adams, Kanga-Leyva, Guzmán-Osorio, and Escalante-Espinoza, 2011; Pindado, Pérez, y García, 2014).

Several researches have emphasized that the most important component of the soil in relation to the persistence and migration of toxic substances is the clay structure, which vary in their properties such as surface area, acidity and porosity among themselves, therefore, the absorption or desorption of the contaminant is intrinsically related to the soil geomorphology. However, there are no specific studies base on to the clay type that relate the disposal or mobility of the contaminant towards the environment (Adams, Zavala-Cruz, and Morales-García, 2008; Maldonado-Chávez, Rivera-Cruz, Izquierdo-Reyes, and Palma-López, 2010). For example, research conducted in tropical areas have reported that some soils of hillside or terraced areas have fertility problems even with concentrations below the regulatory limit, instead, it has been found that in different alluvial zones with high levels of hydrocarbons in which, forage grass and other plant species have been established (Gutiérrez & Zavala, 2002; Zavala et al., 2005; Morales-Bautista, Adams, Hernández-Barajas, Lobato-García, and Torres-Torres, 2016).

However, scientific information has been generated most of the Mexican environmental legislation is not updated or does not considered these studies, so in the matter of hydrocarbons it is not entirely specific. The ASEA lacks regulations instead list conditions to perform environmental diagnosis, and indicates the guidelines according to field application, for these reasons, most of environmental managers still use the methods of nom-138-semarnat-ssa1-2012, especially due its infrastructure and analytical equipment which this regulation is based and although this has been modified through the years, its applicability goes back decades, so still has some uncertainties (sample (Vale, Silva, Damin, Sánchez, and Welz, 2008; Dof, 2012).

In this context, many investigations agree that considering only the normed limits to declare a contaminated site, restored or as an environmental passives, it is not entirely adequate due the level of impact depends on more parameters and, mainly, the methods of determination have variations according to the different constituents in the sample, among are the characteristics of the pollutant, soil type, as well as the chemical and physical interaction between both and the treatment conditions (Meléndez et al., 2012, Vázquez-Luna, 2014).).

There are different methods to determine total hydrocarbons of petroleum or htp based on various analytical fundaments, for example, some are spectrophotometric, such as infrared, with ranges between 1600-2800 nm and, uv-vis, with ranges between 200-320 nm, both are widely used in analysis of contaminated sediments, waters and soils. However, some of its limitations are the amount of sample analyzed, interferences for the low purity of the analyte (because it is filtered on silica or cellulose) and samples where there is saturation of the solvent (dilutions generation much residues). In addition to the ranges of readings are only associated with aromatic and saturated fractions, leaving out polar groups related to water repellency (Morales-Bautista et al., 2018; Arguijo-Portillo, Guerrero-Peña, Domínguez-Rodríguez, Carrillo-Ávila, and Zavala-Cruz, 2019).

Other used techniques are continuous extractions such as ultrasonic baths and soxhlet, although the first is one have methodologies that shown excellent purifications, it is still discussed whether the amount of sample used is representative and if the type of solvent used is compatible with all the analyte, since there is a selective method with problems in mixtures, such as petroleum hydrocarbons, plus the non-standardized methods by some Mexican regulations for this type of pollutant (Pino, Ayala, Afonso, and González, 2001, Navarro, Etxebarria, and Arana, 2009, Rotola, Lacorte, Barceló, and Alves, 2009).

Due to the response variables involved, one of the most commonly used techniques standardized is continuous extraction by soxhlet method. However, in some studies in which different soils have been experimentally contaminated with this method, lower yields have been observed with respect to the expected, based on these results, several reports have established two hypotheses, the first is that the solvent does not extract all hydrocarbon families, as resins and asphaltenes (these vary according to the type of hydrocarbon and weathering conditions) and the second is that the extraction time is not sufficient to remove the hydrocarbons from the soil matrix (interactions between soil properties and hydrocarbon fractions), consequently, the restorations are not entirely effective (Riveroll-Larios, Escalante-Espinoza, Fócil-Monterrubio, and Díaz-Ramírez, 2015; García et al., 2016; Lee, Liong, and Jemain, 2017).

Also, it has been observed that the chemical interaction between the solvent and hydrocarbons is intrinsically related to the type of soil and the °API, depending on the properties each one possesses the processes in the absorption and desorption of the contaminant will also be different, specifically, for the compatibility of each fraction towards characteristics of soil, such as cation contents in clays, the components of organic matter, the ph and the polarity of the solvent (Islas-Flores, Buenrostro-Gonzales, and Lira-Galeana, 2005; Urpí, 2012; Wang et al., 2013; Careghini, Mastorgio, Saponaro, and Sezenna, 2015).

MAIN FOCUS OF THE CHAPTER

Although the design of experiments (DOE) is a very effective tool in the analysis of methods and errors associated with the response variables, there are very few studies that consider include them with environmental regulations, as mentioned above, in Mexico there are very scarce studies that consider the classification of the soil type and the properties of crude oil. Pons-Jiménez et al. (2011), evaluated different methods of HTP determination, they found that in the Soxhlet method presents greater advantages compared to other techniques, mainly due that better results were observed when solvent mixtures are considered according to the soil type, they express that the chemical compatibility between hydrocarbon-soil-solvent must be investigated if it is desired to reduce errors by diffusion and desorption processes.

In this sense, one of the most used methods to evaluate and optimize soil treatments that establish these relationships is Taguchi, authors as Castorena-Cortéz et al. (2009), Gómez & Sartaj (2014) and Khayati & Barati (2017) specify that orthogonal design and considerations such as signal-noise and the evaluation of an extract of the total experiments contributes to cost reduction. This is why they must be implemented environmental diagnostics or evaluations of soil with lower uncertainties, if the method can be optimized, not only these errors are reduced, also the operating costs and the response time to emergencies in the case of spills.

In this context, in this paper we follow by De Freitas et al. (2013) and Silva et al. (2014), who specified the following steps for implementing Taguchi experimental design:

1. Selection of response variable to be optimized
2. Identify input factors affecting the response variable
3. Choose the numbers of levels to be tested
4. Decide the orthogonal array
5. Assign factors and interactions to the columns of the array

6. Perform experiments, carry out a statistical analysis and the signal-to-noise ratio and determine the optimal conditions to adjust factor levels
7. Perform confirmatory experiment.

Under this approach, three possible factors affecting Soxhlet analysis were identified, which were the soil classification, the polarity of the extraction solvents and their compatibility with the heavy fraction of the oil as well as the quantities of reagents used and the extraction time; according with the later the concentration of hydrocarbons in HTP was considered as response variable.

Control Soils

Soils adjacent to oil facilities the Tabasco state were collected, each site was selected according to geographical position and textural classification reported by Zavala-Cruz et al. (2005) and Palma-López et al. (2007), geological strata information were obtained from three municipalities, Cárdenas (near structures of the Samaria-Luna Asset) considered an alluvial zone and with clay-loam texture, another near the coast of Paraíso (near the maritime oil terminal wo mouths of the Bellota-Jujo Asset), which presented a clay-sandy texture and, a third collected in Macuspana, located in the area of hills with slopes no greater than 32° (near the Shishito battery of Muspac-Macuspana Asset), this it was identified with a franc-clay-sandy texture. The location points of the soils sampling are expressed in table 1.

Contaminated Soils

Soil pre-treatment consisted of removing roots and rocks, then drying them in an oven at 60 °C, after this step samples were ground and sieved, later they were characterized according to NOM-021-SEMARNAT-2000 (DOF, 2002). Each soil was experimentally contaminated with crude oil at 45,000 $mg{\cdot}kg^{-1}$ (the hydrocarbon was constituted by a mixture of different samples obtained from several wells of the San Magallanes Oil Field located within the Cinco Presidentes Asset). °API was determined to the hydrocarbon by the method proposed by Morales-Bautista et al. (2013), also it was determined SARA analysis using mixtures of solvents for each fraction as specified Marín-García et al. (2016).

Table 1. Benchmarks of soil samples.

Municipality Soil	UTM 15 Q ± 2	
Cárdenas	430597 m E	1992822 m N
Paraíso	475593 m E	2030615 m N
Macuspana	551975 m E	1974505 m N

Where: UTM is Universal Transverse Mercator, E is East and, N is North. *Source: (Own Edition)*

SOXHLET-HTP DETERMINATION METHOD AND ASSOCIATED VARIABLES TO CONSIDERER

Subsequently, it was treated by means of a Soxhlet extraction to the each contaminated soil varying the amount of sample (5, 7.5 and 10 g), the amount of $NaSO_4$ (0.5, 1 and 2 g), the volume of extraction solvents (50, 75 and 100 ml) classified by their dielectric constant (2, 9 and 33) and at three different times (6, 8 and 10 h). Finally, of each test HTP extract was obtained and concentration was calculated by the gravimetric method (DOF, 2012; Pons-Jiménez et al., 2011; Serrano, Torrado, and Pérez, 2013; Castellanos, Isaza, and Torres, 2015; Martínez-Chávez, Morales-Bautista, and Alor-Chávez et al., 2017)

OPTIMIZATION OF THE HTP EXTRACTION METHOD BY TAGUCHI METHODS

Taking into account that there are different parameters that possibly cause errors in the extraction process, it was searched for a method that would gather the best results with respect to the best performance, so it was decided to use two designs: Taguchi (orthogonal arrangement), to evaluate roughly the effects of the various factors taken as relevant within the process and, a factorial design analysis, which considered the parameters that showed to have an important effect in the extractions of HTP.

The variables evaluated were the solvent and sample quantity, organic matter (MO), sodium sulfate added amount, electrical conductivity (CE), pH, extraction time (t), solvent dielectric constant, sand (% A) and clay (% R) content. The evaluation was based on establishing an orthogonal array L27 with 10 factors each at three levels and an analysis of means was carried out in Minitab 18 and the signal-to-noise ratio larger-the-better (equation 1), later factor analysis was performed to evaluate the best yields (Polewczyk, Marchut-Mikolajczyk, and Smigielski, 2015; Khayati & Barati, 2017; Wahla, Iqbal, Anwar, Firdous, and Mueller, 2019).

$$\frac{S}{N} = -10\log\left[\frac{1}{n}\sum_{i=1}^{n}\frac{1}{y_1^2}\right] \tag{1}$$

Most of the reports that use this tool agree that it has the virtue of determining the effect of different factors by analyzing an orthogonal matrix and an approach based on statistical techniques, in such a way that it facilitates the study of the system by a set of independent variables, on a specific region of interest (Mandal, Mohan, and Hemalatha, 2008: Garapati & Mishra, 2012; Sohrabi, Jamshidi, and Esmaeilifar, 2012; Zapata & Sarache, 2014).

SOLUTIONS AND RECOMMENDATIONS

In Table 2, are listed the °API of crude petroleum results and the SARA fractions percentage. Secondly, in table 3 the properties of the reference soils are listed.

According to the results shown in Table 2, the heaviest fraction (Resins + Asphaltenes) of crude oil is equivalent to 40%, these results coincide with those from some heavy oils from the Gulf of Mexico region and the Caribbean Sea reported in the literature. However, the chemical composition of each is

Table 2. Properties of crude petroleum.

Classification	°API	%Saturated/ Aliphatic	%Aromatics	%Resins/Polar	%Asphaltenes
Heavy	13	37 ± 0.7	23 ± 0.5	21 ± 0.8	19 ± 0.2

Source: (Edited by authors)

not known since in the refineries it is usually analyzed and will determine the type of pretreatment to be carried out (Fan & Buckley, 2002; Wang & Buckley, 2003; Merkl, Schultze-Kraft, and Infante, 2005; Gaona, Manrique, JPB, and Majé. 2010; Infante & Morales, 2012; Meléndez et al., 2012; Marín-García et al., 2016).

Also, when comparing the results of Table 3 with the reports of Palma-López et al. (2007), Trujillo-Narcía et al. (2012), Barba-Marcía et al. (2014) and Zavala et al. (2014), it was found that the characteristics such as the proportionality of the textures (% A and % R), as well as the acidic pH and CIC means, the Cárdenas soil correspond to a Fluvisols, on the other hand, the soil of Paradise area has properties similar to those of an Arenosols-Gleysols (due to the amount of MO, high CE and the proportions of %

Table 3. Characteristics of reference soils

Municipality Soil	pH	CE	Textures			CIC	CC	DA	DR	MO
			%A	%R	%L					
Cárdenas	5.08	1.08	45	35	20	23.1	32	0.98	2.29	1.0
Paraíso	7.30	3.88	58	42	0	32.6	20	0.91	2.17	3.0
Macuspana	6.47	2.01	52	38	10	11.3	13	1.04	2.48	2.0

Where: CE is Electrical Conductivity in $dS{\cdot}m^{-1}$, % A, % L and% R, are the weight percentage of sands, clays and silts in the textures, CIC is Cationic Exchange Capacity in $Cmol^{+}{\cdot}kg^{-1}$, CC is field capacity in %, DA is bulk density and DR is real density, both in $g{\cdot}cm^{-3}$, MO is organic matter in % weight.

Source: (Edited by authors)

Table 4. Variables considered in the orthogonal array of Taguchi Method design

Factor	Levels (1, 2, 3)		
Amount of sample (g)	5	7.5	10
Amount of solvent (ml)	50	75	100
Organic matter (%)	1	2	3
Anhydrous $NaSO_4$ (g)	0.5	1	2
Electrical Conductivity ($dS{\cdot}m^{-1}$)	1.08	2.01	3.88
pH	5.08	6.47	7.30
Extraction time (h)	6	8	10
Dielectric constant	2	9	33
Sandy (%A)	45	52	58
Clay (%R)	35	38	42

Source: (Edited by authors)

A and % R observed) and, as for the Macuspana samples, by the moderately acid pH,% R > 50% and high densities correspond to a Luvisols.

On the other hand, it was observed that the soils properties that varied among themselves were mainly MO, CE, PH and textures, so these parameters were added to the applied variables during the extraction process, and finally, were all added to the proposed model, which is established in Table 4.

With the data presented in table 4, an analysis was carried, Taguchi Methods in Minitab 18, the results show an S/N of 92% for the model, considering as response variable the TPH in mg•kg^{-1}, it is important to mention that the sequence of the runs was randomized to reduce the bias, the results of the analysis of variance and response for signal-to-noise ratios for Taguchi model proposed, is listed in table 5 and 6, respectively.

In the results of Table 5, it was observed that the time variables and the dielectric constant have a level of significance since the values of P are less than 0.05, that is why the model's responses establish

Table 5. Analysis of variance of means (ANOVA) of the Taguchi design.

Source	GL	SC	MC Ajust.	F	P
Amount of sample (g)	2	79514473	39757236	2.21	0.191
Amount of solvent (ml)	2	34609900	17304950	0.96	0.434
Organic matter (%)	2	34613204	17306602	0.96	0.434
Anhydrous $NaSO_4$ (g)	2	92741263	46370631	2.58	0.156
Electrical conductivity (dS·m^{-1})	2	16546602	8273301	0.46	0.652
pH	2	12686615	6343307	0.35	0.717
Extraction time (h)	2	149877050	74938525	4.16	0.043
Dielectric constant	2	785764775	392882387	21.82	0.002
Sandy (%A)	2	40974343	20487171	1.14	0.381
Clay (%R)	2	9516578	4758289	0.26	0.776
Residual Error	6	108011127	18001855		
Total	26	1364855929			

Where: GL are the degrees of freedom, SC is the sum of sequential squares, MC the average of squares adjusted, F calculated for the hypothesis test and, P the percentile.

Source: (Edited by authors)

Table 6. Response for signal-to-noise ratios (bigger is better).

Level	Quanty		MO	$NaSO_4$	CE	pH	Time	Dielectric Constant	%A	%R
	Sample	Solvent								
1	86.58	86.82	87.25	86.23	86.97	87.65	86.33	88.07	86.82	87.14
2	87.86	87.66	87.08	87.94	87.74	87.56	87.66	89.49	87.44	87.41
3	87.83	87.8	87.95	88.11	87.57	87.06	88.3	84.72	88.02	87.74
Delta	1.28	0.98	0.87	1.88	0.77	0.59	1.97	4.77	1.2	0.6
Classify	4	6	7	3	8	10	2	1	5	9

Source: (Edited by authors)

them as 1 and 2 as variables of priority, shown in Table 6, discarding properties such as pH and % R (in order 9 and 10 respectively). Some works mention that the soils have compatibility with polar compounds especially if these are halogenated, for this study, the solvent with a constant of ~ 9 is the most feasible for the type of crude oil analyzed, it can also be said that the extractions improve over time, despite this halogenated solvents are considered as dangerous for the environment and public health. Also, there is the possibility that field capacity or bulk density can be better indicators to evaluate extractions, because it is so related with % A and MO, that the first one are in the order of priority 5 and 6, in comparison with the second one that are in the order of 9 and 10 (Pavel & Gavrilescu, 2008; Rauckyte, Zák, Pawlak, and Oloyede, 2010; Pons-Jiménez et al., 2011, Morales-Bautista et al., 2018).

In addition, a run of the Taguchi prediction model was performed considering the best conditions shown, these were 10 h of extraction, 5 g of sample, 100 ml of solvents with constant ~ 9, varying the pH of the soil and the textures, resulting with better yields observed for the more acidic soils with sandy textures and lower for neutral soils with clay texture, both with S / R ratios greater than 90 (the bigger the better) and with yields between 60% and 80%, which coincides with other methods reported in the literature (Wan et al., 2013; Adams et al., 2015, Khayati & Barati, 2017).

On the other hand, because it is known that the extraction time and the type of solvent can influence, a factorial experiment was carried out to determine the effect that they have only by varying these parameters on the TPH values of the three types of soil and leaving the rest of the conditions constant according to the Taguchi prediction model. In general, there are 3 factors with 3 levels each, so a factorial design 33 was analyzed, 3 replications were also made for each combination of treatment levels, resulting in a total of 81 observations. The null hypotheses to prove are that the average extraction is the same for the three types of soils; the three types of solvents; and the three times, against the alternative hypothesis that at least one of the three types of soils, solvents and time has a different average extraction, then the variance analysis method (ANOVA) was performed, which is shown in table 7 (Tang et al., 2007, Pandiarajan, Kumaran, Kumaraswamidhas, and Saravanan, 2016).

The table shows that all the main factors, double and then triple their interactions at the level of 5%; from which it is then concluded that there is evidence of a significant effect on the soil (A), solvent (B), time (C) and double and triple interactions on the response variable average concentration. In addition to F, it is observed that B has a more important effect, followed by the effect C and at the end A, this

Table 7. ANOVA for the linear model of the three factors with interaction.

Source	SC	GL	MC	F	P
A	74096629	2	37048314	27832	0.0000
B	2383068867	2	1191534433	895110	0.0000
C	786218236	2	393109118	295313	0.0000
AB	251093941	4	62773485	47157	0.0000
AC	199846894	4	49961723	37532	0.0000
BC	310028298	4	77507075	58225	0.0000
ACB	196804290	8	24600536	18481	0.0000
ERROR	71883	54	1331		
TOTAL	4201229036	80			

Source: (Edited by authors)

coincides with the order of priority shown in table 6. It is also observed that the double interaction BC has a more important effect followed by AB, at the end AC. In this sense, a contrast was made between the solvent-time and soil-time interactions (for the HTP results of solvent with dielectric constant ~ 9), which are shown in figure 1 and 2, respectively.

As shown in Figure 1, the treatment with better results is the dielectric constant of 9 and a time of 10 hours, which differ from those established by the norm, however, although in Figure 2 it is observed that for these conditions all the soils have had good yields, it is also observed that none reach 100%, in this case, the one from Paraíso is the one who presents the highest extractions followed by the soil sample from Cárdenas with lower extractions. Although both models allow the simulation of the results, it must be evaluated at longer times and evaluate significant differences among them to be able to establish the optimal extraction time.

Figure 1. Solvent-time interaction
Source: (Edited by authors)

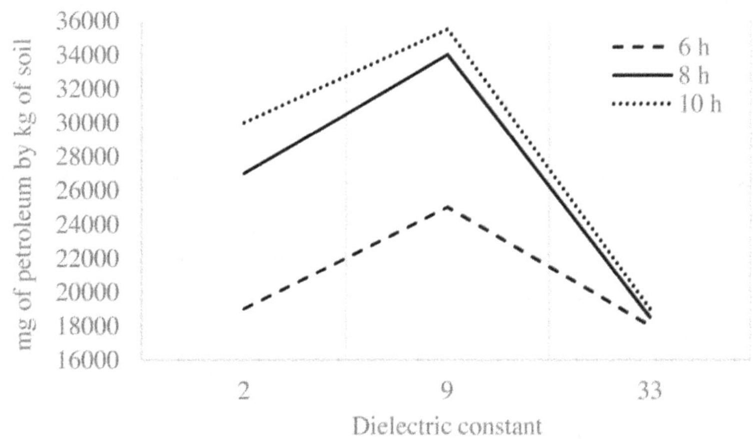

Figure 2. Soil-time interaction
Source: (Edited by authors)

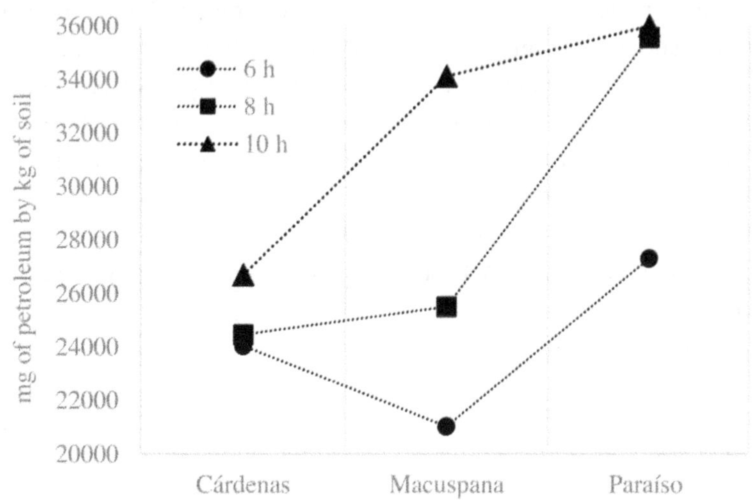

When these results were contrasted, was observed that they coincide with the HTP analyzes reported by Martínez-Chavez et al. (2017) performed on sandy soils and with García et al. (2016) in both cases, they used solvents with dielectric constant of 9, but could not optimize extraction time, by limiting your studies with eight hours of treatment and reaching a maximum of 75% yield. Also, it was observed that the extraction percentages obtained for soils with high clay content require a longer extraction time coincide with the established by Pons-Jiménez et al. (2011), the extraction times do not coincide because in the results of these authors they present treatments soil, contaminated or/and has been outdoors for long time, by which, pollutant-soil interaction is different and could influence extraction times (Pérez-Hernández et al., 2017).

In comparison with other methods, some results found in this study show similarity with the analyzes performed by spectrophotometric methods such as those from Gómez, Pandiyan, Iris, Figueroa, and Bazúa, (2004) and Schwartz, Eshel, and Ben-Dor, (2011), which may be related to the fact that these authors used solvents Halogenated to prepare their calibration curves and in their extractions, in addition that the latter were performed in columns with natural sorbets, which allow a better purification of the analyte and, consequently, reduction of uncertainties (Mažeikiene et al., 2005).

FUTURE RESEARCH DIRECTIONS

Design perspectives of experiments (DOE) through the taguchi design are high, due to the fact that it allows the evaluation based on an orthogonal design discarding the variables that have no relationship with the evaluated data, although this tool is currently untiled, an area of opportunity are the impact diagnoses in sites with environmental liabilities of the oil industry, although the effects on the soils are known, not we have been evaluated if there is any relationship with sara fractions and water repellency, in both cases, not considered in Mexican environmental legislation.

CONCLUSION

The implementation of the Taguchi and factorial designs allowed the optimization of the hydrocarbon extraction process by the Soxhlet method, in which it was determined that the best solvent is that with a dielectric constant of 9 and that the highest extractions are at 10 hours, contrary to what is marked by the NOM-138-SEMANARNAT-SSA1-2012 in soils contaminated by hydrocarbons, to the above it is added that the model optimized the treatment with 5 g of samples, 100 ml of solvents, 1 g of $NaSO_4$. Finally, it was verified that the soil parameters such as pH, CE and MO, are not determinants in the method, however changes in treatments were observed according to the textures, which gave variations in recovery percentages between 60 and 80%. Therefore, although the results show the variables that allow the optimization of the method, validation in the future by is necessary, mainly, adding more soils and other hydrocarbons.

ACKNOWLEDGMENT

To the Universidad Juárez Autónoma de Tabasco for the financing of this project through the funds allocated for the infrastructure of the soil and hydrocarbon analysis laboratory.

REFERENCES

Adams, R. H., Álvarez-Ovando, A. L., & Castañón, N. (2015). Efecto de la concentración de hidrocarburos sobre la producción del pasto (*Brachiaria humidicola*) en Texistepec, Veracruz. *Phyton (Buenos Aires)*, *84*(1), 222–232.

Adams, R. H., Kanga-Leyva, K., Guzmán-Osorio, F. J., & Escalante-Espinosa, E. (2011). Comparison of moisture management methods for the bioremediation of hydrocarbon contaminated soil. *African Journal of Biotechnology*, *10*(3), 394–404.

Adams, R. H., Zavala-Cruz, J., & Morales-García, F. (2008). Concentración residual de hidrocarburos en suelo del trópico. II: Afectación a la fertilidad y su recuperación. *Interciencia*, *33*(7), 483–489.

Arguijo-Portillo, N. V., Guerrero-Peña, A., Domínguez-Rodríguez, V. I., Carrillo-Ávila, E., & Zavala-Cruz, J. (2019). Modelos de calibración para la cuantificación espectrofotométrica de hidrocarburos totales del petróleo en suelo. *Revista Internacional de Contaminación Ambiental*, *35*(2), 469–479. doi:10.20937/RICA.2019.35.02.17

Barba-Macías, E., Rangel-Mendoza, J., & Ramos-Reyes, R. (2014). Clasificación de los humedales de Tabasco mediante sistemas de información geográfica. *Ecosistemas Recursos Agropecuarios*, *22*(2).

Barrientos del Monte, F., & Añorve Añorve, D. (2014). México 2013: Acuerdos, reformas y descontento. *Revista de ciencia política (Santiago)*, *34*(1), 221–247. doi:10.4067/S0718-090X2014000100011

Careghini, A., Mastorgio, A. F., Saponaro, S., & Sezenna, E. (2015). Bisphenol A, nonylphenols, benzophenones, and benzotriazoles in soils, groundwater, surface water, sediments, and food: A review. *Environmental Science and Pollution Research International*, *22*(8), 5711–5741. doi:10.100711356-014-3974-5 PMID:25548011

Castellanos, M. L., Isaza, R. J., & Torres, J. M. (2015). Evaluación de los hidrocarburos totales de petróleo (TPH) sobre suelos urbanos en Maicao, Colombia. *Revista Colombiana de Quimica*, *44*(3), 11–17. doi:10.15446/rev.colomb.quim.v44n3.55605

Castorena-Cortés, G., Roldán-Carrillo, T., Zapata-Peñasco, I., Reyes-Avila, J., Quej-Ake, L., Marin-Cruz, J., & Olguín-Lora, P. (2009). Microcosm assays and Taguchi experimental design for treatment of oil sludge containing high concentration of hydrocarbons. *Bioresource Technology*, *100*(23), 5671–5677. doi:10.1016/j.biortech.2009.06.050 PMID:19635663

Castro, A., Acevedo-Berruecos, D., Urbieta, J., Iturbe, R., & Delgado-Rodríguez, O. (2013). Caracterización geoquímica y geoeléctrica de un sitio contaminado por hidrocarburos en el estado de Puebla, México. *Boletín de la Sociedad Geológica Mexicana*, *65*(2), 405–418. doi:10.18268/BSGM2013v65n2a20

De Freitas, A. P. B. R., de Freitas, L. V., Loures, C. C. A., dos Santos, M. A. R., Ricardo, G. D., Marins, F. A. S., & Silva, M. B. (2013). Taguchi method applied to environmental engineering. In *Design of experiments-applications*. IntechOpen. doi:10.5772/56531

Díaz-Ramírez, I., Escalante-Espinosa, E., Schroeder, R. A., Fócil-Monterrubio, R., & Ramírez-Saad, H. (2013). Hydrocarbon biodegradation potential of native and exogenous microbial inocula in Mexican tropical soils. In R. Chamy, & F. Rosenkranz (Eds.), In biodegradation of hazardous and special products (pp. 155-178). InTech. doi:10.5772/56233

DOF. Diario oficial de la Federación. (2002). *Norma oficial Mexicana NOM-021-SEMARNAT-2000. Que establece las especificaciones de fertilidad, salinidad y clasificación de suelos, estudio, muestreo y análisis*. Retrieved from http://biblioteca.semarnat.gob.mx/janium/Documentos/Ciga/libros2009/DO2280n.pdf

DOF. Diario oficial de la Federación. (2012). *NOM-138-SEMARNAT/SSA1-2012. Límites máximos permisibles de hidrocarburos en suelos y lineamientos para el muestreo en la caracterización y especificaciones para la remediación* Retrieved from http://www.dof.gob.mx/nota_detalle.php?codigo=5313544&fecha=10/09/2013

Fan, T., & Buckley, J. S. (2002). Rapid and accurate SARA analysis of medium gravity crude oils. *Energy & Fuels*, *16*(6), 1571–1575. doi:10.1021/ef0201228

Gaona, J. A. S., Manrique, J. P. B., & Majé, Y. M. (2010). Predicción de la estabilidad de los asfaltenos mediante la utilización del análisis SARA para petróleos puros. *Ingeniería y Región*, *7*, 103–110. doi:10.25054/22161325.805

Garapati, V. K., & Mishra, S. (2012). Hydrocarbon degradation using fungal isolate: Nutrients optimized by combined grey relational analysis. *Int. J. Eng. Res. Appl*, *2*(2), 390–399.

García, M. A. H., Bautista, C. M. M., Olán, C. M., & Schoreder, R. H. A. (2016). Extracción de Hidrocarburo Pesado en Suelo Acrisol. *Journal of Basic Sciences*, *2*(5).

García-López, E., Zavala-Cruz, J., & Palma-López, D. J. (2006). Caracterización de las comunidades vegetales en un área afectada por derrames de hidrocarburos. *Terra Latinoamericana*, *24*(1), 17–26.

García-Torres, R., Ríos-Leal, E., Martínez-Toledo, Á, Ramos-Morales, F. R., Cruz-Sánchez, J. S., & Cuevas-Díaz, M. D. C. (2011). Uso de cachaza y bagazo de caña de azúcar en la remoción de hidrocarburos en suelo contaminado. *Revista Internacional de Contaminación Ambiental*, *27*(1), 31–39.

Gardi, C., Angelini, M., Barceló, S., Comerma, J., Cruz Gaistardo, C., Encina Rojas, A., . . . Vargas, R. (Eds.). (2014). Atlas de suelos de América Latina y el Caribe, Comisión Europea Oficina de Publicaciones de la Unión Europea, L-2995 Luxembourg.

Gomez, F., & Sartaj, M. (2014). Optimization of field scale biopiles for bioremediation of petroleum hydrocarbon contaminated soil at low temperature conditions by response surface methodology (RSM). *International Biodeterioration & Biodegradation*, *89*, 103–109. doi:10.1016/j.ibiod.2014.01.010

Gómez, R. S. G., Pandiyan, T., Iris, V. A., Figueroa, M. R., & de Bazúa, C. D. (2004). Identificación y cuantificación de hidrocarburos aromáticos policíclicos en el suelo usando métodos espectroscópicos. *Tecnología, Ciencia. La Educación, 19*(2), 69–82.

Gutiérrez, M. D. C., & Zavala, J. (2002). Rasgos hidromórficos de suelos tropicales contaminados con hidrocarburos. *Terra Latinoamericana, 20*(2).

Guzmán-Osorio, F. J., & Adams, R. H. (2015). Mitigation of water repellency in the treatment of contaminated muds using the chemical–biological stabilization process. *International Journal of Environmental Science and Technology, 12*(6), 2071–2078. doi:10.100713762-014-0606-z

Hernández, M. A., Santillán, S., Sabelkin, V., & Parra, M. L. (2003). Caso de aplicación del método Taguchi en el diseño de herramientas de muestreo de Hidrocarburos. *Ingeniería Mecánica, Tecnología y Desarrollo, 1*(3), 91–101.

Infante, C., & Morales García, F. A. (2012). Evaluación de la toxicidad en desechos y suelos petrolizados empleando semillas de Lactuca sativa L. *Interciencia, 37*(10).

Islas-Flores, C. A., Buenrostro-Gonzalez, E., & Lira-Galeana, C. (2005). Comparisons between open column chromatography and HPLC SARA fractionations in petroleum. *Energy & Fuels, 19*(5), 2080–2088. doi:10.1021/ef050148+

Khan, A. B., & Kathi, S. (2014). Evaluation of heavy metal and total petroleum hydrocarbon contamination of roadside surface soil. *International Journal of Environmental Science and Technology, 11*(8), 2259–2270. doi:10.100713762-014-0626-8

Khayati, G., & Barati, M. (2017). Bioremediation of petroleum hydrocarbon contaminated soil: optimization strategy using taguchi design of experimental (DOE) methodology. *Environmental Processes, 4*(2), 451–461. doi:10.100740710-017-0244-9

Lee, L. C., Liong, C. Y., & Jemain, A. A. (2017). A contemporary review on Data Preprocessing (DP) practice strategy in ATR-FTIR spectrum. *Chemometrics and Intelligent Laboratory Systems, 163*, 64–75. doi:10.1016/j.chemolab.2017.02.008

Liu, W., Luo, Y., Teng, Y., Li, Z., & Ma, L. Q. (2010). Bioremediation of oily sludge-contaminated soil by stimulating indigenous microbes. *Environmental Geochemistry and Health, 32*(1), 23–29. doi:10.100710653-009-9262-5 PMID:19363671

Maldonado-Chávez, E., Rivera-Cruz, M. C., Izquierdo-Reyes, F., & Palma-López, D. J. (2010). Efectos de rizósfera, microorganismos y fertilización en la biorremediación y fitorremediación de suelos con petróleos crudo nuevo e intemperizado. *Universidad y ciencia, 26*(2), 121-136.

Mandal, V., Mohan, Y., & Hemalatha, S. (2008). Microwave assisted extraction of curcumin by sample–solvent dual heating mechanism using Taguchi L9 orthogonal design. *Journal of Pharmaceutical and Biomedical Analysis, 46*(2), 322–327. doi:10.1016/j.jpba.2007.10.020 PMID:18309573

Marín-García, D. C., Adams, R. H., & Hernández-Barajas, R. (2016). Effect of crude petroleum on water repellency in a clayey alluvial soil. *International Journal of Environmental Science and Technology, 13*(1), 55–64. doi:10.100713762-015-0838-6

Martínez, A. M. L., Marcos, J. M. F., Manzanares, M. V., Urbina, E. M. C., Iliana, M., & Hernández, J. A. (2018). Cuantificación de etanol en gasolina por microextracción líquido-líquido y valoración potenciométrica. *Revista de Ciencia y Tecnología, 34*(1).

Martínez-Chávez, C. G., Morales-Bautista, C. M., & Alor-Chávez, M. D. J. (2017). Extracción de hidrocarburo pesado en suelo arenoso. *Avances en Ciencias e Ingeniería, 8*(1).

Martínez-Lara, J. M., & Páez Melo, M. I. (2017). Diseño de experimentos aplicado en la optimización del método de extracción QuEChERS para la determinación de plaguicidas organoclorados y organofosforados en suelos. *Revista Internacional de Contaminación Ambiental, 33*(4), 559–573. doi:10.20937/RICA.2017.33.04.02

Mažeikiene, A., Rimeika, M., Valentukevičiene, M., Oškinis, V., Paškauskaite, N., & Brannvall, E. (2005). Removal of petroleum products from water using natural sorbent zeolite. *Journal of Environmental Engineering and Landscape Management, 13*(4), 187–191. doi:10.3846/16486897.2005.9636870

Melendez, L. V., Lache, A., Orrego-Ruiz, J. A., Pachon, Z., & Mejía-Ospino, E. (2012). Prediction of the SARA analysis of Colombian crude oils using ATR–FTIR spectroscopy and chemometric methods. *Journal of Petroleum Science Engineering, 90*, 56–60. doi:10.1016/j.petrol.2012.04.016

Merchand, M. A. (2015). Estado y reforma energética en México. *Problemas del Desarrollo, 46*(183), 117–139. doi:10.1016/j.rpd.2015.10.006

Merkl, N., Schultze-Kraft, R., & Infante, C. (2005). Assessment of tropical grasses and legumes for phytoremediation of petroleum-contaminated soils. *Water, Air, and Soil Pollution, 165*(1-4), 195–209. doi:10.100711270-005-4979-y

Morales-Bautista, C., Méndez-Olán, C., Hernández-Jiménez, G., & Adams-Schroeder, R. (2018). Normalization of calibration curves and their application in the determination of total oil hydrocarbons (HTP). *Revista Técnica de la Facultad de Ingeniería. Universidad del Zulia, 41*(3), 116–124.

Morales-Bautista, C. M., Adams, R. H., & Guzmán-Osorio, F. D. Marín-García. (2013). Dilution-extrapolation hydrometer method for easy determination of API gravity of heavily weathered hydrocarbons in petroleum contaminated soil. Energy and Environmental Research, 3(1), 115–124. doi:10.5539/eer.v3n1p115 doi:10.5539/eer.v3n1p115

Morales-Bautista, C. M., Adams, R. H., Hernández-Barajas, J. R., Lobato-García, C. E., & Torres-Torres, J. G. (2016). Characterization of fresh and weathered petroleum for potential impacts to soil fertility. *International Journal of Environmental Science and Technology, 13*(11), 2689–2696. doi:10.100713762-016-1097-x

Navarro, P., Etxebarria, N., & Arana, G. (2009). Development of a focused ultrasonic-assisted extraction of polycyclic aromatic hydrocarbons in marine sediment and mussel samples. *Analytica Chimica Acta, 648*(2), 178–182. doi:10.1016/j.aca.2009.06.062 PMID:19646582

Ojeda-Morales, M. E., Hernández-Rivera, M. Á., Martínez-Vázquez, J. G., Díaz-Flores, L. L., & Rivera-Cruz, M. D. C. (2012). Tres dosis inoculantes con Proteus sp en la biodegradación de petróleo crudo. *Revista Internacional de Contaminación Ambiental, 28*(4), 323–332.

Ortiz-Salinas, R., Cram, S., & Sommer, I. (2012). Hidrocarburos aromáticos policíclicos (HAPs) en suelos de la llanura aluvial baja del estado de Tabasco, México. *Universidad y ciencia, 28*(2), 131-144.

Palma-Cruz, F. D. J., Pérez-Vargas, J., Casado, N. A. R., Guzmán, O. G., & Calva-Calva, G. (2016). Phytoremediation potential and ecological and phenological changes of native pioneer plants from weathered oil spill-impacted sites at tropical wetlands. *Environmental Science and Pollution Research International, 23*(16), 16359–16371. doi:10.100711356-016-6675-4 PMID:27164872

Palma-López, D. J., Cisneros, D. J., Moreno, C. E., & Rincón-Ramírez, J. A. (2007). *Suelos de Tabasco: su uso y manejo sustentable* (p. 195). Villahermosa, México: Colegio de Postgraduados-ISPROTAB-FUPROTAB.

Pandiarajan, S., Kumaran, S. S., Kumaraswamidhas, L. A., & Saravanan, R. (2016). Interfacial microstructure and optimization of friction welding by Taguchi and ANOVA method on SA 213 tube to SA 387 tube plate without backing block using an external tool. *Journal of Alloys and Compounds, 654*, 534–545. doi:10.1016/j.jallcom.2015.09.152

Pavel, L. V., & Gavrilescu, M. (2008). Overview of ex situ decontamination techniques for soil clean-up. [EEMJ]. *Environmental Engineering and Management Journal*, *7*(6), 815–834. doi:10.30638/eemj.2008.109

Pérez-Hernández, I., Ochoa-Gaona, S., Adams, R. H., Rivera-Cruz, M. C., Pérez-Hernández, V., Jarquín-Sánchez, A., & Martínez-Zurimendi, P. (2017). Growth of four tropical tree species in petroleum-contaminated soil and effects of crude oil contamination. *Environmental Science and Pollution Research International, 24*(2), 1769–1783. doi:10.100711356-016-7877-5 PMID:27796985

Pérez-Hernández, I., Ochoa-Gaona, S., Schroeder, R. A., Rivera-Cruz, M. C., & Geissen, V. (2013). Tolerance of four tropical tree species to heavy petroleum contamination. *Water, Air, and Soil Pollution, 224*(8), 1637. doi:10.100711270-013-1637-7

Pindado, O., Pérez, R. M., & García, S. (2014). *Análisis del TPH y las fracciones de hidrocarburos alifáticos y aromáticos en matrices de interés medioambiental.* Editorial CIEMAT.

Pino, V., Ayala, J. H., Afonso, A. M., & González, V. (2001). Ultrasonic micellar extraction of polycyclic aromatic hydrocarbons from marine sediments. *Talanta, 54*(1), 15–23. doi:10.1016/S0039-9140(00)00622-6 PMID:18968221

Polewczyk, A., Marchut-Mikolajczyk, O., & Smigielski, K. B. (2015). Diesel fuel extraction from soil. *Acta Scientiarum Polonorum. Formatio Circumiectus, 14*(2), 165–173. doi:10.15576/ASP.FC/2015.14.2.165

Pons-Jiménez, M., Guerrero-Peña, A., Zavala-Cruz, J., & Alarcón, A. (2011). Extracción de hidrocarburos y compuestos derivados del petróleo en suelos con características físicas y químicas diferentes. *Universidad y ciencia, 27*(1), 1-15.

Ratola, N., Lacorte, S., Barceló, D., & Alves, A. (2009). Microwave-assisted extraction and ultrasonic extraction to determine polycyclic aromatic hydrocarbons in needles and bark of Pinus pinaster Ait. and Pinus pinea L. by GC–MS. *Talanta, 77*(3), 1120–1128. doi:10.1016/j.talanta.2008.08.010 PMID:19064100

Rauckyte, T., Żak, S., Pawlak, Z., & Oloyede, A. (2010). Determination of oil and grease, total petroleum hydrocarbons and volatile aromatic compounds in soil and sediment samples. *Journal of Environmental Engineering and Landscape Management*, *18*(3), 163–169. doi:10.3846/jeelm.2010.19

Riveroll-Larios, J., Escalante-Espinosa, E., Fócil-Monterrubio, R. L., & Díaz-Ramírez, I. J. (2015). Biological activity assessment in Mexican tropical soils with different hydrocarbon contamination histories. *Water, Air, and Soil Pollution*, *226*(10), 353. doi:10.100711270-015-2621-1 PMID:26478633

Ruiz-Álvarez, O., Arteaga-Ramírez, R., Vázquez-Peña, M. A., Ontiveros Capurata, R. E., & López-López, R. (2012). Balance hídrico y clasificación climática del estado de Tabasco, México. *Universidad y ciencia*, *28*(1), 1-14.

Schwartz, G., Eshel, G., & Ben-Dor, E. (2011). Reflectance spectroscopy as a tool for monitoring contaminated soils. In Soil contamination. InTechOpen.

Serrano Guzmán, M. F., Torrado Gómez, L. M., & Pérez Ruiz, D. D. (2013). Impact of oil spills on the mechanical properties of sandy soils. *Revista Científica General José María Córdova*, *11*(12), 233–244.

Silva, M. B., Carneiro, L. M., Silva, J. P. A., dos Santos Oliveira, I., Izário Filho, H. J., & de Oliveira Almeida, C. R. (2014). An application of the Taguchi method (robust design) to environmental engineering: Evaluating advanced oxidative processes in polyester-resin wastewater treatment. *American Journal of Analytical Chemistry*, *5*(13), 828–837. doi:10.4236/ajac.2014.513092

Sohrabi, M. R., Jamshidi, S., & Esmaeilifar, A. (2012). Cloud point extraction for determination of Diazinon: Optimization of the effective parameters using Taguchi method. *Chemometrics and Intelligent Laboratory Systems*, *110*(1), 49–54. doi:10.1016/j.chemolab.2011.09.009

Soto, L. A., Botello, A. V., Licea-Durán, S., Lizárraga-Partida, M. L., & Yáñez-Arancibia, A. (2014). The environmental legacy of the Ixtoc-I oil spill in Campeche Sound, southwestern Gulf of Mexico. *Frontiers of Materials Science*, *1*, 57.

Tang, S. H., Tan, Y. J., Sapuan, S. M., Sulaiman, S., Ismail, N., & Samin, R. (2007). The use of Taguchi method in the design of plastic injection mould for reducing warpage. *Journal of Materials Processing Technology*, *182*(1-3), 418–426. doi:10.1016/j.jmatprotec.2006.08.025

Trujillo-Narcía, A., Rivera-Cruz, M. D. C., Lagunes-Espinoza, L. D. C., Palma-López, D. J., Soto-Sánchez, S., & Ramírez-Valverde, G. (2012). Efecto de la restauración de un Fluvisol contaminado con petróleo crudo. *Revista Internacional de Contaminación Ambiental*, *28*(4), 360–374.

Urpí, J. L. (2012). *Tecnología y margen de refino del petróleo*. Ediciones Díaz de Santos.

Vale, M. G. R., Silva, M. M., Damin, I. C., Sanches Filho, P. J., & Welz, B. (2008). Determination of volatile and non-volatile nickel and vanadium compounds in crude oil using electrothermal atomic absorption spectrometry after oil fractionation into saturates, aromatics, resins and asphaltenes. *Talanta*, *74*(5), 1385–1391. doi:10.1016/j.talanta.2007.09.009 PMID:18371794

Vázquez-Luna, D. (2014). Chronic toxicity of weathered oil-contaminated soil. In M. C. Hernandez-Soriano (Ed)., Environmental risk assessment of soil contamination (pp. 87-104). InTech. doi:10.5772/57253

Wahla, A. Q., Iqbal, S., Anwar, S., Firdous, S., & Mueller, J. A. (2019). Optimizing the metribuzin degrading potential of a novel bacterial consortium based on Taguchi design of experiment. *Journal of Hazardous Materials*, *366*, 1–9. doi:10.1016/j.jhazmat.2018.11.054 PMID:30500692

Wang, J., & Buckley, J. S. (2003). Asphaltene stability in crude oil and aromatic solvents the influence of oil composition. *Energy & Fuels*, *17*(6), 1445–1451. doi:10.1021/ef030030y

Wang, Y., Feng, J., Lin, Q., Lyu, X., Wang, X., & Wang, G. (2013). Effects of crude oil contamination on soil physical and chemical properties in Momoge wetland of China. *Chinese Geographical Science*, *23*(6), 708–715. doi:10.100711769-013-0641-6

Wang, Y., Feng, J., Lin, Q., Lyu, X., Wang, X., & Wang, G. (2013). Effects of crude oil contamination on soil physical and chemical properties in Momoge wetland of China. *Chinese Geographical Science*, *23*(6), 708–715. doi:10.100711769-013-0641-6

Zamora, A., Ramos, J., & Arias, M. (2012). Efecto de la contaminación por hidrocarburos sobre algunas propiedades químicas y microbiológicas de un suelo de sabana. *Bioagro-*, *24*(1), 5–12.

Zapata Gómez, A., & Sarache Castro, W. A. (2014). Mejoramiento de la calidad del café soluble utilizando el método Taguchi. *Ingeniare. Revista Chilena de Ingeniería*, *22*(1), 116–124. doi:10.4067/S0718-33052014000100011

Zavala-Cruz, J., Gavi-Reyes, F., Adams-Schroeder, R. H., Ferrera-Cerrato, R., Palma-López, D. J., Vaquera-Huerta, H., & Domínguez-Ezquivel, J. M. (2005). Derrames de petróleo en suelos y adaptación de pastos tropicales en el Activo Cinco Presidentes, Tabasco, México. *Terra Latino Americana*, *23*(3), 293–302.

Zavala Cruz, J., Salgado García, S., Marín Aguilar, Á., Palma López, D. J., Castelán Estrada, M., & Ramos Reyes, R. (2014). Transecto de suelos en terrazas con plantaciones de cítricos en Tabasco. *Ecosistemas Recursos Agropecuarios*, *1*(2), 123–137.

ADDITIONAL READING

Adams, R. H., Guzmán-Osorio, F. J., & Domínguez-Rodríguez, V. I. (2014). Field-scale evaluation of the chemical–biological stabilization process for the remediation of hydrocarbon-contaminated soil. *International Journal of Environmental Science and Technology*, *11*(5), 1343–1352. doi:10.100713762-013-0321-1

Adams, R. H., Olán-Castro, D., Guzmán-Osorio, F. J., & Díaz-Ramírez, I. J. (2009). Relationship between geomorphology and contamination with weathered hydrocarbons in an old river levee/marsh association. *International Journal of Environmental Science and Technology*, *6*(4), 527–538. doi:10.1007/BF03326092

Arias-Trinidad, A., Rivera-Cruz, M. D. C., & Trujillo Narcía, A. (2017). Fitotoxicidad de un suelo contaminado con petróleo fresco sobre Phaseolus vulgaris L.(LEGUMINOSAE). *Revista Internacional de Contaminación Ambiental*, *33*(3), 411–419. doi:10.20937/RICA.2017.33.03.05

Doerr, S. H., Llewellyn, C. T., Douglas, P., Morley, C. P., Mainwaring, K. A., Haskins, C., ... Ferreira, A. (2005). Extraction of compounds associated with water repellency in sandy soils of different origin. *Soil Research (Collingwood, Vic.)*, *43*(3), 225–237. doi:10.1071/SR04091

Domínguez-Rodríguez, V. I., Adams-Schroeder, R. H., Goñi-Arévalo, J. A., & López-Vidal, I. (2018). Water lily (Nymphaea sp.): An alternative organic amendment for treatment of hydrocarbon-contaminated soil by chemical–biological stabilization. *International Journal of Environmental Science and Technology*, 1–8.

Jahanbakhshi, S. (2014). Optimization of phytoremediation in Cd-contaminated soil by using Taguchi method in Spinacia oleracea. *Proceedings of the International Academy of Ecology and Environmental Sciences*, *4*(4), 185.

Mohan, S. V., Sirisha, K., Rao, R. S., & Sarma, P. N. (2007). Bioslurry phase remediation of chlorpyrifos contaminated soil: Process evaluation and optimization by Taguchi design of experimental (DOE) methodology. *Ecotoxicology and Environmental Safety*, *68*(2), 252–262. doi:10.1016/j.ecoenv.2007.06.002 PMID:17640730

Morales-Bautista, C. M., Lobato García, C. E., Méndez-Olán, C., & Alor Chávez, M. D. J. (2016). Evaluación del tratamiento del intercambio catiónico en dos suelos aluviales contaminados con aguas congénitas. *Interciencia*, *41*(10).

Travis, M. J., Weisbrod, N., & Gross, A. (2008). Accumulation of oil and grease in soils irrigated with greywater and their potential role in soil water repellency. *The Science of the Total Environment*, *394*(1), 68–74. doi:10.1016/j.scitotenv.2008.01.004 PMID:18280539

Wang, W., Simonich, S. L. M., Xue, M., Zhao, J., Zhang, N., Wang, R., ... Tao, S. (2010). Concentrations, sources and spatial distribution of polycyclic aromatic hydrocarbons in soils from Beijing, Tianjin and surrounding areas, North China. *Environmental Pollution*, *158*(5), 1245–1251. doi:10.1016/j.envpol.2010.01.021 PMID:20199833

KEY TERMS AND DEFINITIONS

Active Petroleum: Considered as a number of wells and oil facilities of a region, it is generally part of a paleontological sub-basin of reserves of this natural resource.

ASEA: Agency responsible for conducting, monitoring and enforcing the environmental regulations of the energy sector in Mexico.

Edaphology: Science that studies the classification of soil according to the study of the properties of the horizons of a profile.

Chapter 3
Design of Experiments for Evaluation of Variables Involved in the Removal of Heavy Metal Ions From Water Using Agro-Industrial Waste-Based Adsorbents

Gabriela Islas Guerrero
Universidad Politécnica de Francisco I. Madero, Mexico

Evelin Gutiérrez Moreno
https://orcid.org/0000-0001-7610-9318
Universidad Politécnica Metropolitana de Hidalgo, Mexico

ABSTRACT

The design of experiments (DOE) is a useful tool to define the most significative variables of a process, and the optimal operation conditions, or reduce the noise caused by uncontrollable variables. An advantage of the employ of DOE is the reduction of time required for analysis and costs associated. The Factorial, surface response methodology, and Taguchi have been employed to analyze the variables involved in the removal of heavy metal ions coming from wastewater by adsorption employing low-cost adsorbents, which include agro-industrial waste and biomass. The most important variables associated to increase of adsorption capacity and evaluated by researchers include temperature, pH, adsorbent dose, initial concentration of metal, and particle size.

DOI: 10.4018/978-1-7998-1518-1.ch003

INTRODUCTION

The industrial activity historically has caused an increase in the concentration of heavy metal ions in water associated with the incorporation of wastewater into the environment (Barakat, 2011). Heavy metal ions are toxic and half of these, when are emitted into the environment, are a risk factor to human health, animals and ecosystems (Wu et al., 2019). Most studied heavy metal ions in water are cobalt (Co), iron (Fe) lead (Pb), cadmium (Cd), copper (Cu), nickel (Ni), hexavalent chromium Cr(VI) and arsenic (As) (Fu & Wang, 2011). Different techniques have been employed for the removal of heavy metals, such as precipitation, adsorption, ion exchange, and membrane filtration as reverse osmosis.

Low-cost adsorbents are obtained from agro-industrial wastes as raw materials, such as rice husk, nutshell, wood, coconut shell, bean hulls, artichoke, seed, banana pith, eggshell, crimp husk and tree leaves, the employ of these can help to reduce environmental pollution. An advantage associated with the use of low-cost adsorbents for heavy metal removal is that at the end of metal recovery, the used biomass can be filtered of an aqueous solution, dried and burned for energy generation. A disadvantage associated with the use of low-cost adsorbents can be related to low removal capacity, however, the use of designs of experiments (DOE) is a useful tool that has been employed to define the experimental parameters, their levels, and interactions that represent a high contribution in adsorption processes for heavy metals removal. The DOE allows reducing the analysis time and costs with a lower number of better-designed trials.

Some types of DOE employed for the improvement of the heavy metal ions removal process include factorial design (full factorial and fractional factorial), Taguchi design, response surface methodology; which also considers Box-Behnken, and central composite design. The type of DOE is generally selected by researchers in order to take advantage of the sorbents derived from abundant renewable resources, agro-industrial by-products or the waste plant material which are an economic and environmentally friendly source of raw material.

The present chapter is focused on the application of experiments design for the determination of optimal conditions of controllable factors and their influence on the removal of heavy metal ions employing low-cost adsorbents.

BACKGROUND

Heavy Metal Pollution

Currently, there is a growing interest in environmental care. The industrial activity is responsible for the increase in the concentration of heavy metal into the environment. There is not a clear definition of heavy metals, however, the term heavy metals is often used to name metals and metalloids characterized by a density that exceeds 5 $g \cdot cm^{-3}$, most of them are highly water soluble, present high toxicity when accumulating in living organisms and ecosystems and some of these have been identified as carcinogenic agents. The heavy metal ions definition includes those that represent a threat associated with exposure, however, some of these are essential elements for the growth, reproduction and/or survival of living organisms (Wu et al, 2019). The heavy metal most studied are: Hg, Pb, Cd, Cu, Ni, Cr and As, being the last a metalloid usually classified as a heavy metal (Fu & Wang, 2011).

Water pollution is a worldwide problem which has been an object of study in the last years. Among the main contributions to water pollution is the presence of metals coming from natural and mainly anthropogenic sources such as mining operations, tanneries, electronics, radiator manufacturing, alloy industries, electroplating, dying industry, fertilizers, batteries manufacture and petrochemical industries, as well as in textile mill products (Fu & Wang, 2011). Table 1 includes some anthropogenic sources of heavy metals and their effects on health. The toxicity of heavy metals depends on their mobility in the medium, chemical species, persistence, and tendency to bioaccumulation in the environment. The heavy metals under certain environmental conditions can accumulate at a toxic concentration and cause ecological damage (Barakat, 2011). In consideration of the threat that represents the presence of heavy metals in discharged wastewater, the United States Environmental Protection Agency (USEPA) established regulations to minimize human and environmental exposure to hazardous chemicals, this includes the maximum contaminant level (MCL) in heavy metals that may be present in discharged waters.

Agro-Industrial Wastes

Agroindustry has been defined as an economic activity relating agricultural and industrial production processes, which objective is to obtain food or semi-finished raw materials intended for market purposes (Saval, 2012). The agro-industrial sector as a result of its activities generates an important quantity of wastes during the industrial processing of agricultural or animal products, these are considered an environmental problem when they do not have appropriate dispositions or treatment. Some agro-industrial wastes include straw, roots, stalk, leaves, seeds, fruit/vegetable peel, cereals such as rice and wheat, as well as bagasse coming from sugarcane. The composition of agro-industrial wastes consists of sugars,

Table 1. Sources of pollution and effects on the health of heavy metal ions

Heavy Metal	Anthropogenic Sources	Toxicities	MCL ($mg{\cdot}L^{-1}$)
Arsenic (As)	Pyrometallurgy industry, foundry, wood preservation, pesticides	Skin manifestations, visceral cancers, vascular disease	0.050
Cadmium (Cd)	Mining operations, metal manufacturing, batteries manufacturing, plastics, and microelectronics industry	Kidney damage, renal disorder, a human carcinogen, bone fracture	0.01
Copper (Cu)	Microelectronics industry, wood treatment, pyrometallurgy industry, pesticides, and waste deposits	Damage in liver and kidney, insomnia	0.25
Chromium (Cr)	Industrial refrigerants, manufacture of chromium salts, leather tanning, and refineries, pyrometallurgy industry	Headache, diarrhea, nausea, vomiting, carcinogenic, reproductive damage	0.05
Lead (Pb)	Paint industry, exhaust gases of automobiles, lead plumbing, mining, refineries	Damage the brain and kidney, circulatory system, and nervous system, decrease fertility	0.006
Mercury (Hg)	Electrolytic and plastic industries, mining operations, paper industry, fungicides	Rheumatoid arthritis, and diseases of the kidneys, circulatory system, hypertension, and nervous system	0.00003
Nickel (Ni)	Iron and steel industry, mining, metal manufacturing, microelectronics industry, batteries manufacturing	Dermatitis, chronic asthma, coughing, a human carcinogen	0.20

(Barakat, 2011)

fibers, minerals and proteins which make these an interesting option to be considered as raw material for new industrial processes.

Agro-industrial wastes have been employed for animal food production, as well as substrates for solid-state fermentation. The employ of agro-industrial wastes as adsorbents for recovery of dyes and heavy metals from wastewaters has also been reported (Yu, Wang & Wu, 2018). Low-cost adsorbents can be generated from agro-industrial waste or by modification of commercial carbons employing agro-industrial waste (Table 2). These are an interesting option for removal of heavy metal ions present in wastewater as a result of industrial activity (Tounsadi et al., 2016c). Some agro-industrial wastes that have been employed for the preparation of activated carbon include: banana peels, palm kernel, grape stalk, coffee residue, pomelo skin, olive kernel, corn cobs, rapeseed stalks, soya stalks, olive seed residue, almond shells, nutshells, apricot stones, cherry stones, grape seeds, coconut husk and peanut husk. Tovar-Gómez et al. (2015) has worked on the modification of commercial carbons with agro-industrial waste. Specifically, Tovar-Gómez et al. (2015) employed calcium extracted from eggshell wastes as a low-cost activating agent in combination with H_3PO_4; in the study was observed that the increase of adsorption capacities of commercial carbons was the result of the formation of calcium phosphate on the adsorbent surface.

Experiment Design for Chemical Applications

The design of experiments (DOE) is a useful tool to increase the quality of processes and products. Different methodologies of DOE have been applied to perform proper planning and employment of resources available in the removal of pollutants from wastewater through the adsorption process. Specifically, DOE has been used to define the experimental parameters, their levels, and interactions that represent a high contribution to adsorption processes.

Different designs of experiments, such as factorial designs, fractional factorial design, response surface methods, and Taguchi designs have been selected by researchers to take advantage of the sorbents derived from the abundant renewable resources, agro-industrial by-products or the waste plant material which are an economic and environmentally friendly source of raw material. The factorial and fractional design are useful for screening the variables in a process to determine those that are the most important, while the response surface methods are designed to move the operational conditions of a process in an optimal direction and to define the parameters and levels that optimize a quality characteristic, within the range of operability of a process. Taguchi design allows reducing the variability in the process or products, increasing the robustness by minimizing the effects on the output caused by the variables that are difficult to control during the process operation (Montgomery, 2009). The use of such designs has allowed obtaining information to optimize the operation conditions for removal of metals heavy metal ions from wastewater employing adsorption processes; for this purpose, activated carbon obtained from different natural sources has been employed as an adsorbent; commercial activated carbons have also been modified with agro-industrial waste and thermally treated biomass.

DOE is carried out through a test or series of tests, changing the input variables of a process to identify the effect caused in the quality characteristic. Figure 1 shows that the quality characteristic can be altered as a result of the effect of controllable and uncontrollable factors (noise factors) associated with the process. The main objectives of the DOE include: to define the variables that have highest influence on the quality characteristic, to select the set of controllable factors to reach a nominal requirement of

Table 2. Agro-industrial wastes and biomass employed for the generation of the low-cost adsorbent

Low-cost Adsorbent	Pollutant (Metallic ions/dye)	Reference
Aspergillus niger	Cd, Ni, Pb	Amini & Younesi, 2009
Aspergillus sp	Cu, Ni	Pundir et al., 2018
Rhizopus arrhizus	Cu	Preetha & Viruthagiri, 2007
Rice husk ash	Cd, Ni, Zn	Srivastava, Mall, & Mishra, 2008
Cynara scolymus L. (artichoke)	Pb, Cu, Cd	Fernández-López et al., 2019
Sargassum muticum brown alga	Cr(VI)	Bermúdez et al., 2012
Sargassum vulgaris and *Padina pavonia* (brown marine algae)	Cd, Ni	Ofer et al., 2003
Bagasse fly ash	Cd, Ni, Zn	Srivastava et al., 2007
Glebionis coronaria L.	Cd, Co	Tounsadi et al., 2016c
Neem leaves	Cu	Kumar et al., 2013
Coffee grounds and pomelo skin	Volatile organic compounds	Ma, & Ouyang, 2013
Sugarcane bagasse	Cr(VI)	Cronje, Chetty, Carsky, Sahu, & Meikap, 2011
Sugarcanne bagasse	Ni	Garg, Kaur, Garg, & Sud, 2008
Sugarcanne bagasse chemically modified	Cu, Cd, Pb	Karnitz et al., 2007
Saccharomyces cerevisiae	Pb, Cd	Hadiani, Darani, Rahimifard, & Younesi, 2018
Saccharomyces cerevisiae	As(III), As(V)	Hadiani, Khosravi-Darani, & Rahimifard, 2019

Figure 1. Variables of a process

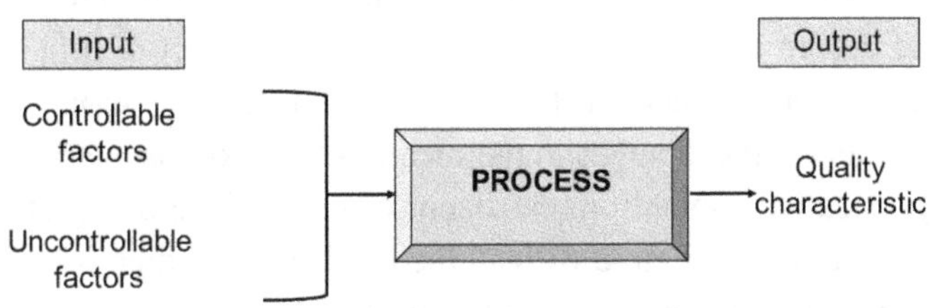

the quality characteristic, to determine the levels were the controllable factors reduces the variability of the quality characteristic and minimize the effect of the uncontrollable factors (Montgomery, 2009).

Experiment designs has allowed obtaining information to optimize the operation conditions for removal of metals such as Cd, Cu, Ni, Zn, and Pb from wastewater employing adsorption processes; for these purpose has been employed as adsorbents activated carbon obtained from different natural sources, commercial activated carbon modified with agro-industrial waste and biomass thermally treated.

The removal of heavy metal ions through the adsorption process is carried out employing agro-industrial waste and biomass, which requires an exhaustive analysis to define the conditions of the process that produces the maximum sorption capacity (q), that is the most referenced quality characteristic in sorption process. The sorption capacity, q (expressed in units of mg $_{metal}$/g $_{adsorbent}$), is calculated from a mass balance (Equation 1) with data obtained employing different analysis techniques (Hibbert, 2012).

$$q = \frac{\left(C_0 - C_f\right) \cdot V}{w}$$

where C_0 and C_f are the initial and final metal concentration (mg·L^{-1}), V; volume of metal solution used for adsorption experiments (L) and w; mass of sorbent used (g).

The controllable factors employed in the DOE design to improve the sorption capacity are generally: temperature, pH, adsorbent dose, concentration initial of metal, agitation rate, conditions for activation of sorbent, contact time and particle size of the sorbent.

The employ of DOE is more efficient than a traditional strategy of experimentation, which consists of varying one factor at a time and held the remaining factors at a baseline level. The most important disadvantage associated with this process is that the possible interaction between the factors is not considered.

Factorial Design

Factorial designs were proposed in the 1920s by R.A Fisher and were developed in the 1940s for their use in industrial experiments. A factorial design is a type of experiment design that allows studying the effects that several factors may have on an output variable or quality characteristic of a process. When experimenting, the levels of all the factors are varied at the same time rather than one at a time, allowing the study of interactions between factors. Factorial designs can be classified into two types: full factorial design and fractional factorial design (Hibbert, 2012).

A factorial design aims to study the effect of several factors on one or more responses when the same interest is had on all factors. The factors may be qualitative or quantitative. To study how each factor influences the response variable is necessary to choose at least two test levels for each of them. The effects of a factorial design are classified in three ways (1) effect of a factor, (2) main effect and (3) interaction effect. The effect of a factor is the change generated in response to the level change of a factor, while the main effects are the changes in the mean of the response variable due to the individual action of each factor. Significative interactions can occur when the difference in the response between the levels of one factor changes at the same time with all the levels of the other factors (Montgomery, 2009).

The main effects and interactions are represented in Figure 2, where the x-axis corresponds to the level of the factor and the y-axis shows the value of quality characteristic at the conditions evaluated. In Figure 2, lines with similar slopes, which are approximately parallel are the result of the absence of interaction, while the presence of interaction corresponds to lines with different slopes. This analysis is done to estimate the effect of major contribution on the response variable, considering the presence of main effects and interaction effect, to find the optimal conditions of the system. To explain the behavior of the response, an ANOVA analysis of variance is performed to know if the effects are statistically significant (Davim, 2015).

Figure 2. Representation of the interaction effect

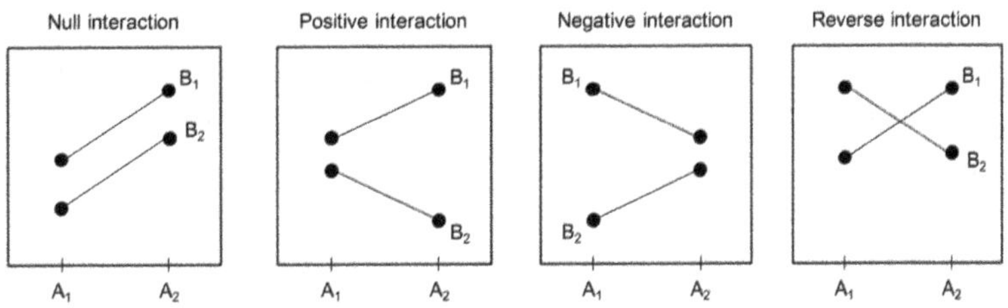

Figure 3. Full factorial designs. (a) Two-factor, two-level design and (b) two-factor three-level design

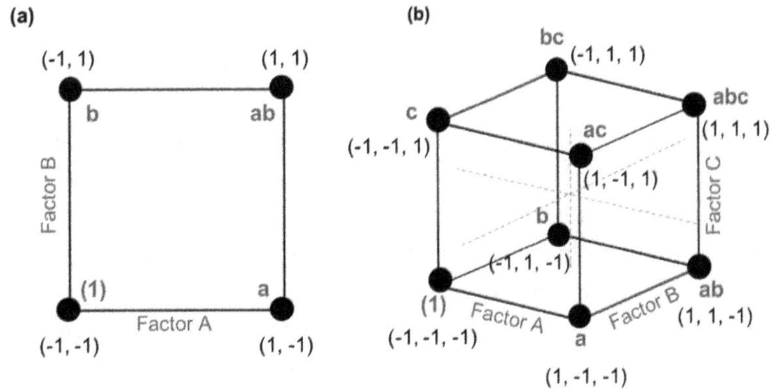

Full Factorial Design

A full factorial design is a DOE where is measured the response of the selected quality characteristic applying all possible combinations of factors and their levels. The design matrix or factorial arrangement is the set of experimental points or treatments that can be formed, considering all possible combinations of levels of the factors. Full factorial 2^k is used when the number of selected factors is between two and five, at a greater number of factors is recommended a fractional factorial 2^{k-p}, which is a fraction $1/2^p$ of the complete factorial design 2^k.

A full factorial design at two levels is represented as 2^k, where k is the number of factors. The number of experiments required is determined by $2 \times 2 = 2^2$, which consists of four combinations or experimental points. The notational system used in this type of design is +1 and -1 for the high and low level of the 2^k (Davim, 2015).

The factorial design 3^k consists of three levels of each factor, these can be referred to as lower, intermediate and higher levels, employing the notation -1, 0, +1, respectively. Figure 3 (a) and (b) show the geometric representation of the 2^k and 3^k factorial designs at k = 2, respectively. Each vertex on geometric representation corresponds to an experimental treatment.

The notation of signs + and - combined with Yates notation, allows calculating the effects of interest. The full factorial design according to the number of levels and factors requires a different number of experiments, for example, a design 2^k with k=6 requires 64 experiments, to k=9 are required 512 experiments. In a 3^k design with three levels with k=4 are required 81 experiments. The fractional factorial design allows reducing the number of trials in the experiment (Davim, 2015).

Fractional Factorial Design

Fractional factorial design is a class of factorial design, where the number of experiments is performed using a selected subset or "fraction" of the experimental runs included in the factorial design. The notation 2^{k-1} corresponds to a fraction half of the full factorial design 2^k. Fractional factorial designs are a suitable option when resources are limited, or the number of factors included in the design is great. In a full factorial design 2^3 seven effects can be estimated: A, B, C, AB, AC, BC, and ABC, generally the lower effect corresponds to ABC interaction, this effect is used to generate fraction half, losing a minimum of information (Ou, Zhang, & Qin, 2019).

The generation of the fraction is made based on the signs of ABC contrast: the "+" signs of the ABC contrast point to treatments that make the call "main fraction", and the signs "-" indicate the "complementary fraction" generating two fractional factorial designs 2^{3-1}, both fractions provide the same quality of information about the effects. The not estimable effect ABC is called "generator of the fraction". Figure 4 is the geometric representation of both fractions (principal and complementary) of full factorial design (2^3), corresponding to a 2^{3-1} design, where the response is measured only in four of the eight vertex points of the full factorial design (Davim, 2015; Mongomery, 2009).

Taguchi Design

This robust parameter design has its origin with quality engineering and product improvement, this is a result of the ideas of Dr. Genichi Taguchi. Taguchi design consists of adjusting a combination parameter and their level to avoid the variation in an objective characteristic. Taguchi design reaches its objective considering the loss due to imperfect production and environmental factors. The typical reduction of variations respect to a target during the manufacturing phase of the process includes the inspection, screening, and salvaging, while Taguchi method is focused on product design phase (*off-line quality*

Figure 4. Representation of fractional factorial designs 2^{3-1}

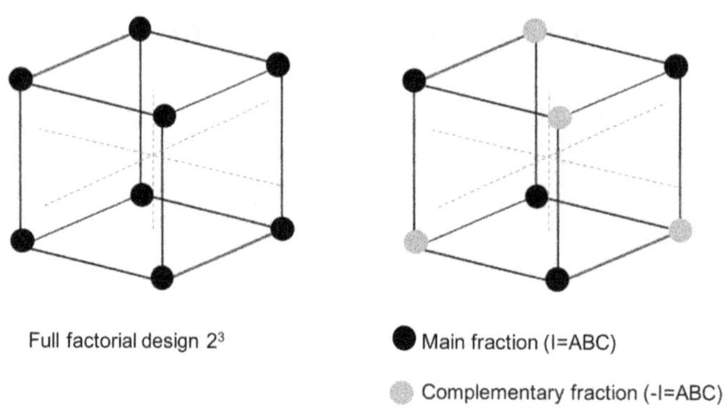

control), to make products and process insensitive to hard control factors (uncontrollable factors) such as environmental conditions, wear and different external factors (Roy,2010).

Taguchi design classify the factors that have influence on a target characteristic of a product or process, as controllable and uncontrollable/noise factors, being the first those that can be fixed easily to desired levels, while the levels of the uncontrollable factors can be unknown through the life of a product or their control represents a high cost; however, sometimes these can be controlled for experimental purposes (Taguchi, 2005). Controllable factors can be classified according to their effect on the performance characteristic: those that affect the media value, the variability, both or neither (Antony & Kaye, 2012).

To apply the Taguchi design it is necessary to select the output variable to be optimized, identify the key factors (controllable and uncontrollable), as well as to define a combination of levels for the controllable factors; that allows to reduce the variation on the performance characteristic despite the presence of noise factors, select the adequate orthogonal array, to perform the experiments, to realize the statistical analysis to obtain the optimum levels of controllable factors and finally to carry on a confirmatory experiment (Barrado, Vega, Pardo, Grande, & Del Valle, 1996). Analysis of variability of a product or process by the Taguchi design takes into consideration a tool named *quality loss function* (Figure 5), associated with the process capability index. The *quality loss function L(y)*, estimates the monetary loss associated to bad quality, as a result of deviations of a target value (t) from the quality characteristic (y) and a constant value (k) associated to lower (LL) and upper (UL) limits specification (tolerance) (Pulido, De la Vara Salazar, González, Martínez, & Pérez, 2008).

The developing of a robust design requires the presence of at least a noise factor. Taguchi proposed the design with an external and inner array, that consists of two orthogonal arrays, each one formed by controllable (internal array) and uncontrollable (external array) factors and their levels. The inner and outer orthogonal arrays employed for the construction of Taguchi experimental designs can be complete, fractional factorial designs or mixed. Typical orthogonal arrays (OA) employed in Taguchi design includes L_4, L_8, L_9, L_{12}, L_{16} and L_{18}, where the subscript indicates the number of experimental runs of the design, each of these can be employed for different number of factors, as example the orthogonal array L_8 is useful for 4, 5, 6, or 7 factors at two levels (Roy, 2010). To construct the matrix of experiments,

Figure 5. Taguchi's loss function

$$L(y) = k(y - t)^2$$

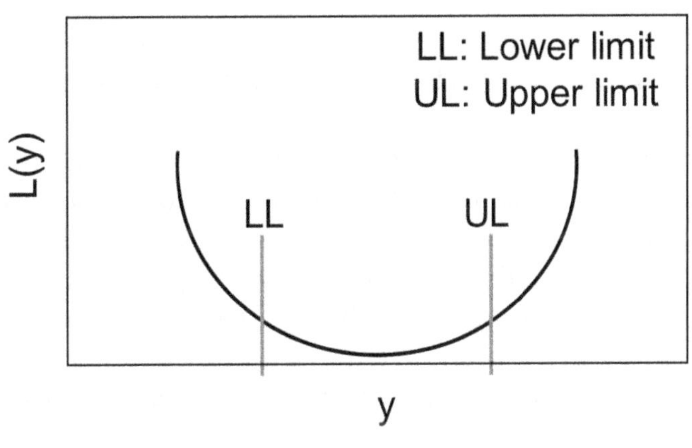

Table 3. Common orthogonal arrays

Run	Factors			Factors							Factors			
	A	B	C	A	B	C	D	E	F	G	A	B	C	D
1	1	1	1	1	1	1	1	1	1	1	1	1	1	1
2	1	2	2	1	1	1	2	2	2	2	1	2	2	2
3	2	1	2	1	2	2	1	1	2	2	1	3	3	3
4	2	2	1	1	2	2	2	2	1	1	2	1	2	3
5	$L_4(2^3)$ array			2	1	2	1	2	1	2	2	2	3	1
6				2	1	2	2	1	2	1	2	3	1	2
7				2	2	1	1	2	2	1	3	1	3	2
8				2	2	1	2	1	1	2	3	2	1	3
9				$L_8(2^7)$ array							3	3	2	1
											$L_9(3^4)$ array			

Figure 6. Signal to noise

Description	S/N
The lower the better (LTB)	$-10\log\left[\frac{1}{n}\sum_{i=1}^{n} y_i^2\right]$
The nominal is the better (NTB)	$-10\log\left[S^2\right]$
The higher the better (HTB)	$-10\log\left[\frac{1}{n}\sum_{i=1}^{n}\frac{1}{y_i^2}\right]$

the levels corresponding to noise factors are placed on the row as opposed to the levels of the controllable factors that are placed on the column. Some examples of the orthogonal array (L_4, L_8, and L_9) are included in Table 3.

Taguchi design employs a logarithmic function to optimize the variance (minimize or maximize), named signal to noise (S/N) ratio, which depends on the criterion for the quality characteristic to be optimized. The main categories of the S/N ratio that are useful in different applications are shown in Figure 6 (Gupta & Lataye, 2018). The S/N is then the ratio of the mean signal to the standard deviation (noise).

where n is the number of repetitions in each of the trials, the y_i are the experimental values obtained for each trial, and S^2 is the sample variance. An example of Taguchi design is shown in Table 4, it was proposed by Barrado et al. (1996), to optimize a precipitation method to purify wastewater polluted with

Table 4. Orthogonal array $L_9(3^4)$ with three level of the noise factor, and remaining metal concentration (R) as quality characteristic (Barrado et at, 1996)

Trial	Controllable Factors and Levels				R ($mg{\cdot}L^{-1}$)			S/N Ratio
	T	P	F	H	N_1	N_2	N_3	
1	1	1	1	1	2.24	5.29	155.04	-39.35
					0.59	1.75	166.27	
2	1	2	2	2	1.75	1.05	0.38	-7.05
					5.07	0.41	0.48	
3	1	3	3	3	5.32	0.40	0.51	-7.05
					0.65	1.07	0.36	
4	2	1	2	3	0.37	0.34	4.31	-5.19
					0.32	0.68	0.65	
5	2	2	3	1	7.20	0.48	0.80	-9.54
					0.49	0.44	0.88	
6	2	3	1	2	39.17	46.54	138.08	-39.34
					27.05	25.77	165.61	
7	3	1	3	2	0.57	0.61	0.91	0.28
					1.26	0.70	1.42	
8	3	2	1	3	3.88	22.74	92.80	-36.20
					7.85	36.33	120.33	
9	3	3	2		15.42	35.27	67.56	-33.79
					25.52	48.61	72.73	

Controllable factors: Fe(II)/metal ratio (F), temperature, °C (T), ageing time, h (H) and pH (P), uncontrollable factor: $KMnO_4$, mol L^{-1} (N).

metal ions. The method consists of an orthogonal array with four controllable factors at three levels, which was analyzed through nine experimental runs $L_9(3^4)$, employing as a noise factor, the variability of solution composition simulated by potassium permanganate ($mol{\cdot}L^{-1}$) at three levels (N_1, N_2, N_3); the performance characteristic determined was the remaining metal concentration R ($mg{\cdot}L^{-1}$) obtained in duplicate.

To obtain information about the effect of the analyzed factors on the performance characteristic can be applied different approaches employing both, the raw data, as well as the S/N signals, can be applied. Variance analysis (ANOVA) is useful to identify the significant parameters and the most important interactions, while the graphics of controllable factors and their levels *vs* performance characteristic or S/N ratio are an excellent option to identify the factors which are responsible for changes on media value or high variability for performance characteristic, respectively. An analysis realized by the example of main effects graphs (Figure 7) shows the effect of controllable factors on recovery of metals (Cd, Ni, and Zn) and S/N ratio, employing as adsorbent rice husk ash (RHA) (Srivastava et al. 2008). In Figure 7 the combination of optimal levels of some controllable factors considering reaching a high average of removal of metal ions ($mg{\cdot}g^{-1}$) and a maximal S/N ratio, includes D3 (Temperature 40°C), E2 (pH 6), F1 (adsorbent dose 5 g_{RHA}/L), G2 (contact time 60 min) (Srivastava et al., 2008).

Figure 7. Effect of controllable factors vs performance characteristics and S/N ratio

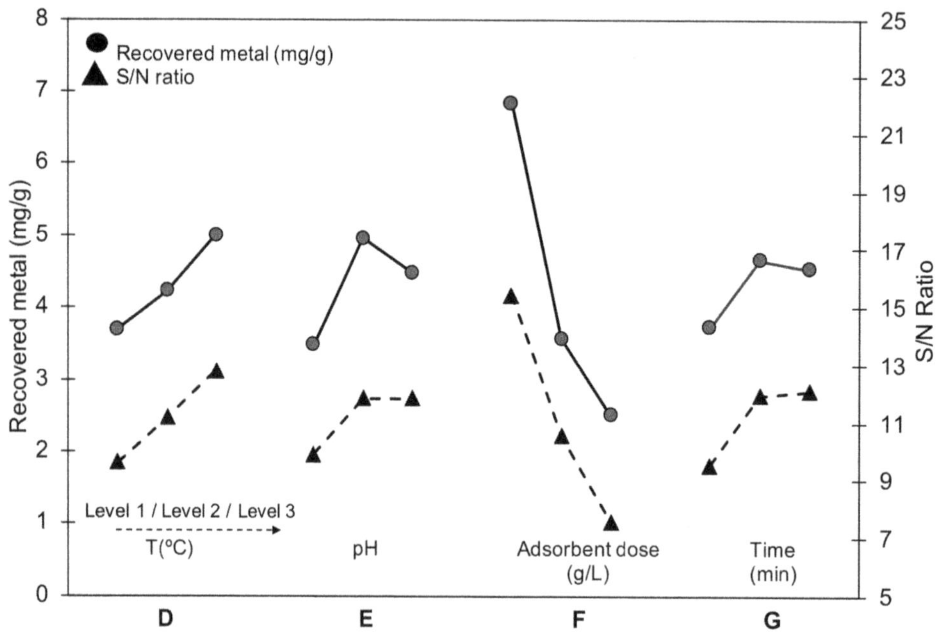

The S/N ratio combines the average level of the performance characteristic and the variance associated. Higher values of the S/N ratio are related to a signal much higher than the random effects caused by noise factors. The employment of the levels of controllable factors with high values of S/N ratio produces an increase in quality and a reduction of variation (robust process) (Srivastava et al., 2007).

Response Surface Methodology

The response surface methodology (RSM) is a collection of mathematical and statistical techniques useful in the analysis of the effect of several independent variables on the quality characteristic, it was introduced by George E. P. Box in the 1950s (Amini et al., 2008; Montgomery, 2001). RSM is useful to search a maximum, minimum or goal value of the quality characteristic, analyzing the variation of the response as a function of the different treatments. The relation between desired response and independent variables can be written as:

$$y = f\left(x_1, x_2, x_3, \ldots, x_n\right) + \varepsilon$$

where y represents the response, x are the independent variables, n is the number of studied factors and ε is the experimental error. This methodology has a sequential character and allows move the experimental region in an appropriate direction to reach the target response value once the initial analysis was realized, as well as to analyze in detail the proposed experimental region (Figure 8).

In RSM, the full or fractional design is employed to explore the experimental region selected initially. At the initial stage of the analysis employing RSM, it is common that the form of the relationship between the response and the independent variables are unknown, thus it is possible to realize an approximation

Figure 8. The sequential character of the RSM
(Montgomery, 2004)

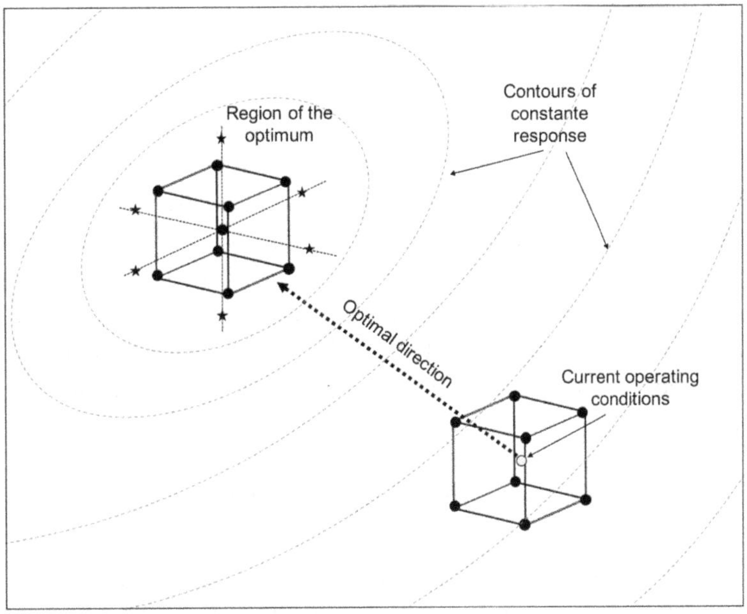

Figure 9. Mathematical models

$$y = \beta_0 + \sum_{i=1}^{k} \beta_i x_i + \varepsilon$$

First order model

$$y = \beta_0 + \sum_{i=1}^{k} \beta_i x_i + \sum_{i=1}^{k} \beta_{ii} x_i^2 + \sum_{i=1<}^{k} \sum_{j=1}^{k} \beta_{ij} x_i x_j + \varepsilon$$

Second-order model

employing a first or second-order mathematical model for the *k* factors (Figure 9), although this strategy employs generally a reduced number of factors (Montgomery, 2009). An alternative to the presence of a large number of factors is to carry out a fractional factorial design and to determine those factors that represent an important effect on the response variable through experimentation and data analysis. In this methodology, it is necessary to consider that the surfaces cannot be completely graphed from 3 factors. For k=3 factors, for example, it is possible to make 3 graphs with two factors at the same time.

The aspects related to the methodology of the response surface include the design, modeled, and optimization. The design and model are performed in the initial stage and depend on the type of system response. The data can be fixed to a first-order (flat) or second-order (curvature) model. This methodology requires the use of multiple linear regressions and the selection of a DOE that can be used to adjust to that model. To model the information, it is required to have knowledge about the parameters of the model, model adjustment, significance, lack of adjustment tests, residuals, predicted, confidence intervals and determination coefficient. Considering that the RSM is a sequential procedure it is possible to employ initially a first-order model to reach the vicinity of the optimum rapidly, and so continue the analysis with a second-order model. The ANOVA can be employed to define the most significative factors and the adjust of data and the selected DOE for the analysis.

The optimization employing the RSM once a first-order model has been applied can be carried out by the method of steepest ascent where it is employed a combination of variables for the sequential analysis, the objective of the method of steepest ascent is to reach a new experimental region closest to the optimum (Kuehl, 2001). The combination of variables for sequential analysis are defined according to the coefficients of the linear equation estimated, when a change on the tendency of response occurs, there is possible to design a new experiment employing as center the combination of variables where the variation was observed, at this point it is possible to adjust the experimental data to a second-order model, which can be verified with the ANOVA. To explore the optimal region and to have a best adjust, the set data can be extended by the employ of central composite design (DCC), which is the most popular class of response surface methodology and has been highly recommended (Figure 10). The optimization of RSM once that the region of the optimum has been found can be carried out by canonical analysis or ridge systems (Montgomery, 2001).

DOE Applied for Heavy Metal Ions Removal Using Low-Cost Adsorbents

In the next sections are briefly described some DOE, useful for the determination of the best operating conditions for the sorption process of heavy metal ions from wastewater by material obtained from agro-industry and natural sources. In Table 5 are included some examples of the employ of DOE in the selection of the best combination of factors and their levels to increase the sorption capacity of heavy metal ions using agro-industrial waste.

Authors as Tounsadi et al. (2016 a, 2016c), have employed a full factorial design (2^4) to define the effect of factors involved in the preparation of activated carbon from *Glebionis coronaria L* and *Diplotaxis Harra,* which were activated by phosphoric acid (H_3PO_4) and applied for the removal of heavy metals (Cd and Co) from aqueous solution. The researchers proposed four controllable factors, that

Figure 10. Representation and construction of the DCC (k=2 and k=3)
(Pulido et al., 2008)

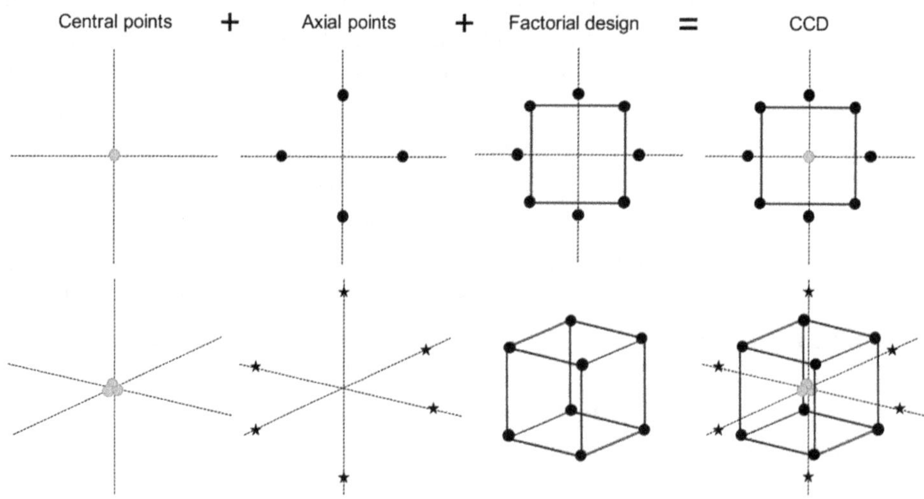

involve the activation process, such as carbonization temperature (500-600°C), activation temperature (400-500°C), activation time (1-2h) and impregnation ratio (g H_3PO_4·g^{-1} carbon) (1.5-2). The results revealed that *Glebionis coronaria L.* is a good precursor to produce activated carbon; they observed that a key factor in the preparation of this type of sorbents is the carbonization temperature, which was the most significant factor in the research. To corroborate the effectiveness of the adsorption capacity of carbon the iodine number and methylene blue index was measured. The removal of Cd and Co onto activated carbons is influenced by the interaction between carbonization temperature and impregnation ratio. The sorption capacities observed were 57.87 mg·g^{-1} for Cd and 45.75 mg·g^{-1} for Co. Employing the optimal conditions obtained in the design, the performance of a commercial activated carbon was evaluated, the sorption capacities of Cd and Co ions from water, were 0.85 mg·g^{-1} and 21.24 mg·g^{-1}, respectively. According to this analysis, the employ of DOE allows increasing the sorption capacity of biomass, respect to commercial carbons.

A 2^3 full factorial design was used by Mitic et al. (2019) for the determination of the effect of factors involved in the heavy metal ions removal (Cu, Cr(VI), Fe, and Ni) from water samples, using biomass of garden sage (*Salvia officinalis L.*). The variables involved in the design were: liquid/solid ratio (15 and 30 mL·g^{-1}), extraction temperature (40 and 100 °C) and extraction time (20 and 80 min). The interaction of factors was evaluated by the analysis of variance and linear regression model. At 95% confidence level, all the factors are associated with Fe, Cu, and Cr(VI) removal. The researchers observed that the extraction time was the most important factor and represents a high contribution to the heavy metal ions removal.

The Taguchi design has the objective to define the optimal values of each factor considered and to develop a robust process, that is, a process affected minimally by external sources of variability (Montgomery, 2001). Guijarro-Aldaco et al. (2011) have employed the Taguchi design (L_9 OA) to improve the adsorption capacity of commercial carbons (coconut shell, bituminous or lignite carbon) for multiple ions removal (Zn, Cd and Ni), realizing a pretreatment with acid solution (HCl, H_3PO_4 or none) and a surface modification with low-cost activating agent (calcium impregnation, obtained from hen eggshell wastes) and finally a thermal treatment. The evaluated factors were carbon type, acid treatment, calcium concentration and temperature. They observed that a key factor in the preparation of this type of sorbents is the acid treatment, because of the formation of calcium phosphate on the carbon surface, which is associated with the use of H_3PO_4.

Pundir, et al. (2018) investigated the removal of heavy metals (Cu and Ni) employing fungal biomass (*Aspergillus sp.*). The controllable factors evaluated were the inoculum concentration, pH, temperature and initial concentration of metal. They observed that pH is an important factor in the adsorption process which should be evaluated at different levels. For the pH selection should be considered that low pH values cause the presence of positive charges in the functional groups on the outer layer of biomass, which inhibit the interactions of the metallic ions, by electrostatic repulsion, resulting in low removal capacity. The high pH value (alkaline region) also affects the heavy metal removal as a result of the formation of metal hydroxide and precipitation, then in the alkaline region the removal of cationic metal ions is not possible due to precipitation; a correct pH selection implies increases on percentage removal of heavy metal ions as a result of the increase in the number of surface negative charges.

Among the most important factors to consider in the adsorption process are temperature, adsorbent dose, contact time and pH. In accordance with the exothermic character of the sorption process, it is expected that the increase in temperature results in a reduction of sorption capacity, however, in processes controlled by diffusion, the adsorption capacity results increases at higher temperatures. When biomass is employed as adsorbent, the factor temperature affects microorganism growth. High temperatures in an adsorption process can promote the damage to proteins and affect the metabolic activity of fungal biomass (Pundir et al., 2018; Srivastava et al., 2007).

Fernandez-Lopez et al. (2019) carried out the analysis for the removal of Pb, Cu, and Cd employing agro-industrial waste biomass from artichoke. They employed an L_9 OA, considering the factors pH, initial concentration of metal, adsorbent dose and temperature. They realized the analysis for metallic ions in a separate way, pH 5 and 30 °C were defined as optimal values for three types of ions, while the initial concentration of metal and adsorbent dose was not the same. They observed important differences in adsorption capacity values for the individual removal of heavy metal ions after optimization with Taguchi design. The removal capacities were 86.2 $mg \cdot g^{-1}$ for Pb, 35.8 $mg \cdot g^{-1}$ for Cd and 24.4 $mg \cdot g^{-1}$ for Cu. They mentioned that the removal efficiency is influenced by both, the nature of the metal ion and the parametric conditions of the experiment.

The response surface methodology is a technique that involves experimental designs such as Box-Behnken and composite central design to increase data point of analysis, this technique also allows move the operating conditions in the best direction to reach an optimal region of operating values.

There is an important number of authors that have reported the optimization of heavy metal removal by RSM, employing second-order mathematical models, where are considered the effect of factors and their interactions. Khalifa et al. (2019) employed the RSM methodology to define the optimal conditions for Cr(VI) removal from wastewater employing orange peel. In the adsorption processes the effect of adsorbent amount, pH and temperature were analyzed. The analysis of pH of zero charges suggested them, that pH values lower than 2.5 are associated with a positive charge of the surface. Then, low pH values are associated with the presence of $HCrO_4^-$ as the predominant species of chromium, which are attracted by the positively charged biosorbent. At optimum conditions, adsorbent dose 1.12 g, pH 2 and 34.17 °C was reached 97% of Cr(VI) removal. According to researchers, the temperature effect in sorption processes is related to the increase of active sites available for adsorption, as a result of the changes in pore size and increase in the diffusion of ions due to a decrease of solution viscosity. This makes the temperature a key factor to be considered in DOE associated with the removal of heavy metal ions by adsorption processes.

Hadiani et al. (2018) worked in the biosorption of low concentration levels of Pb and Cd ions by *Saccharomyces cerevisiae* fungae. They clarify that there are different factors that affect biosorption yield and that the approach in experimental planning requires an initial stage of screening where different factors can be taken into account and only some factors are chosen to continue with an experimental optimization design. They selected initially seven factors associated with Pb and Cd biosorption: biomass dose, pH, temperature, contact time, heavy metal initial concentration, shaking rate and biomass viability (live or non-living) each one evaluated at two levels. A Taguchi design L_8 was proposed for the determination of the optimum experimental conditions associated with the maximal removal of Pb and Cd. They employed Minitab (version 18.1) statistical software, which allows identifying four significative variables: pH, adsorbent dose, heavy metal initial concentration, and biomass viability. To optimize the four chosen variables a central composite design (CCD) was employed under response surface methodology, using only live yeast and analyzing the three remaining factors. *S. Cerevisiae* has high-affinity for Pb and Cd ions at low concentration, the maximal biosorption reached was 91.6 and 95.3%, respectively. To reduce the number of trials there is useful to employ other DOE to select the variables with a major contribution. Throughout this section, it has been possible to analyze that the use of DOE is a useful tool that allows to reduce the number of trials as well as to define the critical variables and the optimal combination of these to reach a high removal capacity of heavy metal ions employing low-cost adsorbents.

Table 5. Examples of research conducted to determine critical factors and optimize the removal process of heavy metal ions in water

Metallic Ion	Adsorbent	Treatment	DOE	Factors (Optimal)	Removal	Reference
Cr(VI)	*Sargassum muticum*	Biomass was washed with deionized water to remove Ca^{2+} or Na^{+}. It was dried at 60ºC and stored in a desiccator.	Full factorial 3^3	C_0: 20 mg·L^{-1} m: 2.0 g·L^{-1} T: 50ºC	84%	Bermúdez et al., 2012
Cd, Co	*Glebionis coronaria L.*(ac)	Stems of the plant were crushed and sieved, carbonized and impregnated with H_3PO_4. The carbons were thermally activated and washed with HCl, NaOH and deionized water until neutral pH. The powder obtained was dried, powdered and sieved.	Full factorial 2^4	$T_{carbonization}$: 600ºC $T_{activation}$: 400ºC $t_{activation}$: 1 h $r_{impregnation:}$ $1.5g_{acid}·g_{carbon}^{-1}$	Cd: 57.87 mg·g^{-1} Co: 45.75 mg·g^{-1}	Tounsadi et al., 2016c
Pb	*Aspergillus niger*	Biomass was deactivated by heating, grinding and screening, it was mixed in 0.5N NaOH, 20min at T_b, washed with deionized water, centrifuged and dried at T=50ºC.	CCD RSM	pH: 3.44 C_0: 19.28 mg·L^{-1} m: 3.74 g·L^{-1}	96.21%	Amini et al., 2008
Cd, Ni, Pb	*Aspergillus niger*	Commercial *Aspergilus niger* was cultured, propagated in nutrient agar (sucrose, NH_4NO_3, KH_2PO_4, $MgSO_4$), 30ºC, pH=5 for 5 days, filtered and stored at 4ºC.	CCD RSM	pH: 6.01 C_0: 89.93 mg·L^{-1} m: 5.22 g·L^{-1}	Cd: 98% Ni: 80% Pb: 99%	Amini & Younesi, 2009
Cu	*Rhizopus arrhizus*	The fungus was grown at 25ºC in potato extract and dextrose solution. pH=5.3 before sterilization. The cell suspension was separated, dried, homogenized and stored.	Factorial 2^4 (CCD) RSM	pH: 4.14 C_0: 53.84 ppm m: 8.17 g·L^{-1} T: 37.75ºC	98%	Preetha & Viruthagiri, 2007
Cd, Ni, Zn	Rice husk ash (RHA)	RHA was obtained from a paper mill, it was only sieving.	Taguchi (L_{27})	pH: 6, C_0(Cd, Ni, Zn): 100 mg·L^{-1} m: 5 g·L^{-1}, t: 60 min, T: 40ºC	9.46 mg·g^{-1}	Srivastava et al., 2008
Pb, Cu, Cd	*Cynara Scolymus L.*	Biomass was dried at 70ºC for 24 h, milled and sieved.	Taguchi (L_9)	**Pb** (pH:5, C_0:50 mg·L^{-1}, m: 0.5 g·L^{-1}, T:30ºC) **Cu** (pH:5, C_0:30 mg·L^{-1}, m: 1.0 g·L^{-1}, T:30ºC) **Cd** (pH:5, C_0:50 mg·L^{-1}, m: 1.0 g·L^{-1}, T:30ºC)	Pb: 86.2 mg·g^{-1} Cd: 35.8 mg·g^{-1} Cu: 24.4 mg·g^{-1}	Fernández-López et al., 2019
Cu, Ni	*Aspergillus sp*	*Aspergillus sp* was cultured over potato agar.	Taguchi (L_9)	pH: 4, C_0: 50.0 mg·L^{-1} T: 30ºC, Inoculum v/v (%): 15	Cu: 98.8% Ni. 97.9%	Pundir et al., 2018
Cd, Ni, Zn	Baggase fly ash	The adsorbent was obtained from sugar mill and it was only sieved.	Taguchi (L_{27})	pH: 6, C_0 (Cd, Ni, Zn):100 mg·L^{-1} m: 5 g·L^{-1}, t: 60 min· T: 40ºC	16.01 mg·g^{-1}	Srivastava et al., 2007
Cu	Neem leaves	Mature leaves were collected, washed, dried under sunlight at 80ºC until it became crisp. The dried leaves were crushed and blended. The material was sieved and treated with HNO_3, 24h, washed with deionized water until pH 7.2, stored in a desiccator.	Taguchi (L_{16})	pH: 7 C_0: 5 mg·L^{-1} t: 20 min s: 75 µm rate: 150 rpm	Not analyzed at optimal conditions.	Kumar et al., 2013

continued on following page

Table 5. Continued

Metallic Ion	Adsorbent	Treatment	DOE	Factors (Optimal)	Removal	Reference
Cd	*Carpinus betulus* leaves	Leaves were washed with distilled water, dried, pulverized and sieved.	Taguchi (L_{16})	pH: 4.8, C_0: 140 mg·L^{-1} m: 2.0 g·L^{-1}, T: 25°C, s: 0.12-0.17 mm	35.87 mg·g^{-1}	Zolgharnein et al., 2013
Cr(VI)	Orange peel	Orange peels were collected from a waste of an orange juice shop, these were washed, dried at 80°C, ground and sieved.	Doehlert design RSM	pH: 2.07 m: 1.12 g T: 34.17°C	97.0%	Khalifa et al., 2019
Cd	*Cystoseria myricaas*	Biomass was collected, washed, dried and powdered. The dry biomass was chopped and sieved.	RSM	pH: 3, C_0: 150 mg·L^{-1}, m: 7 g·L^{-1} t: 75 min, T: 40°C	95.37%	Niad, Zaree, & Tahanzadeh, 2016
Pb, Cd	*Saccharomyces cerevisiae*	Commercial *S. Cerevisiae* was cultured in sterilized media.	Taguchi (L_8) CCD-RSM	pH: 5 C_0: 52.5 µg·L^{-1} m: 32.5 (×10^7 CFU)	Pb: 91.6% Cd: 95.3%	Hadiani et al., 2018
Cu	Pomegranate peel	Pomegranate peel waste was obtained from the juice factory. The waste was washed, dried, milled and sieved.	CCD (2^4) RSM	pH: 5, C_0: 125 mg·L m: 0.2 g·L^{-1}, t: 97.4 min	Cu: 103 mg·g^{-1}	Ali et al., 2018
Cr(VI)	*Helianthus annuus*	Sunflowers were collected from agricultural fields, treatment was previously reported.	Box-Behnken RSM	pH: 2 C_0: 40 mg·L^{-1} m: 0.5 g·(100mL^{-1})	90.8%	Jain et al., 2011
Cu, Ni, Pb	Banana peel (ac)	Banana peels were collected from local markets, washed, dried and ground. The material was carbonized, soaked with KOH solution, N_2 at 500°C for 0.5 h. Finally, the material was washed and dried.	CCD RSM	pH: 6.5, 6.4, 6.1 C_0: 85, 90.3, 74.4 ppm m: 2.4, 1.8, 0.9 g·L^{-1} for Cu, Ni and Pb, respectively.	Cu: 14.3 mg·g^{-1} Ni: 27.4 mg·g^{-1} Pb: 34.5 mg·g^{-1}	Thuan, Quynh, Nguyen, & Bach, 2017
Ni	Sugarcane bagasse	The raw material was collected from a sugarcane crushing plant. Bagasse was boiled, dried, ground and sieved.	CCD RSM	pH: 7.52 m: 1.5 g·L^{-1} rate: 150 rpm	79% (confirmatory experiment)	Garg et al., 2008
Cu, Pb	Papaya seed	Papaya seed was sun-dried, crushed. The material was defatted by soxhlet extraction with hexane.	CCD RSM	C_0: 150 mg·L^{-1} m: 0.30 g rate: 180 rpm	Cu: 97.54% Pb: 99.96%	Garba et al., 2016
Pb	*Aeromonas hydrophila*	*Hydrophila* was cultivated in nutrient broth containing beef extract agar.	CCD (2^2) RSM	Flow rate: 2.0 mL·min^{-1} Bed height: 19.0cm	88.27%	Hasan, Srivastava, & Talat, 2010
Fe(III), Zn	*Typha domingensis* phytomass	*Typha domingensis* leaves were cut into small pieces then dried by in an oven at 70°C, 24 h, subsequently, the plant materials were grounded till use.	Full factorial 2^3	pH: 6, 2.5 T: 25°C (Fe/Zn) m: 1, 0.5 g for Fe and Zn, respectively	Fe: 97.26%. Zn:93.49%	Abdel-Ghani, Hegazy, El-Chaghaby, & Lima, 2009
Cd, Cu, Fe	Pectin and cellulose microfibers (orange waste)	Pectin and cellulose microfibers were extracted from orange waste, the pectin aqueous solution was stirred with adequate amounts of cellulose microfibers. The spherical beads were formed instantaneously due to ionic crosslinking of Pectin with Ca ions.	CCD (2^3) RSM	pH: 4 Content of cellulose microfibers into the beads: 5.5% m: 150 mg	Cd: 58% Cu: 77% Fe: 94%	Lessa, Medina, Ribeiro, & Fajardo, 2017

continued on following page

Table 5. Continued

Metallic Ion	Adsorbent	Treatment	DOE	Factors (Optimal)	Removal	Reference
Cd, Co	*Diplotaxis harra*	Steams were crushed, afterward, the biomass dried, carbonized and the samples the carbon were impregnated with phosphoric acid at phosphoric acid to carbon ratio (g H_3PO_4 / g carbon) of 1.5-2, 6 h. Later, heated at 110°C for 24 h to remove excess moisture.	Full factorial 2^4	$T_{carbonization}$: 500°C $T_{activation}$: 500°C $t_{activation}$: 2 h $r_{impregnation}$: 2g H_3PO_4/g carbon	Cd: 31.6 mg·g^{-1} Co: 25.9 mg·g^{-1}	Tounsadi et al., 2016a
Cd, Co	*Glebionis coronaria L.*	The steams of the plant were crushed and sieved. The biomass was carbonized at 500°C after the samples were impregnated with potassium hydroxide at KOH to carbon ratio 1-2 (g KOH / g carbon). Activated carbons were washed with HCl and deionized water. The obtained samples were dried and sieved.	Full factorial 2^4	$T_{carbonization}$: 600°C $T_{activation}$: 500°C $t_{activation}$: 2 h $r_{impregnation}$: 2gKOH/g carbon	Cd: 70.67 mg·g^{-1} Co: 36.62 mg·g^{-1}	Tounsadi et al., 2016b
Cu, Cr, Fe, Ni	*Salvia officinalis* L.	The maceration procedure was employed for the extraction of minerals from *S. officinalis*. The samples were extracted by different volumes of water, at a different temperature and extraction time. The extraction process was carried out using a bath thermostat.	Full factorial 2^3	$r_{impregnation}$: 15 mL·g^{-1} $T_{extraction}$: 100°C $t_{extraction}$: 80 min	Not determined	Mitić et al. 2019
Co, Cu	Apple pulp	Apple pulp was air-dried, crushed and sieved. Carbonization process was carried out at 550°C. Apple pulp and apple pulp carbon were used as sorbents.	Full factorial 2^5	pH: 5.0 (Co/Cu), m: 0.4 g·50 mL^{-1} (Co/Cu) C_0: 20, 10 mg·L^{-1}, T: 20, 40°C, t: 120, 60 min for Co and Cu respectively	Co: 65.11% Cu: 90.49%	Ozbay & Yargic 2018
Cd, Pb	Acacia Nilotica (natural Gum)	The adsorbent of acacia gum was cleaned for removing the sticky peel and other impurities, later the adsorbent was dry and stored for use.	CCD (2^3) RSM	pH: 5.0, 4.0 t: 25, 20 min T: 70, 50°C for Cd and Pb respectively	Cd: 100% Pb: 60%	Bouazizi, Jamoussi, & Bousta 2016
Cu, Cd, Pb	*Tricholoma lobayense*	The biomass was obtained from a mushroom growing, it was washed with deionized water, dried in an oven at 50°C and ground with a pulverizing mill.	CCD RSM	T: 30°C, pH: 5.0, C_0: 60 mg/L, M: loading 2 g·L^{-1}, t: 9h, rate: 200rpm	Cu: 6.0-91.9% Cd: 10.1-84.1% Pb: 6.4-97.9%	Cao, Liu, Cheng, Jing & Xu, 2010

CCD: composite central design, RSM: response surface methodology, C_0: concentration initial of metal, T: temperature, T_b: boiling temperature, t: time,

ac: activated carbon, t: time, r: ratio, rate: agitation rate, s: particle size, m: sorbent dose.

CONCLUSION

The employment of low-cost adsorbents, mainly coming from agro-industrial waste is an excellent option to take advantage of natural resources that can be problematic when they are not adequately disposed. These types of adsorbents have attracted the attention of researchers because they are an alternative to carry out the removal of heavy metal ions coming from wastewater.

The adequate employment of DOE to increase the removal capacity of heavy metal ions using low-cost adsorbents has allowed high percentages of removal for different types of ions. Some authors have realized the comparative study of commercial activated carbons and low-cost adsorbents, their studies allowed to observe that the correct selection of parameters of the processes, is responsible for the increase of removal percentages comparing low-cost adsorbents versus commercial adsorbents. The use of DOE is useful to analyze the main factors involved in adsorption processes with a minimal number of trials. It is important to modify the variables involved in adsorption processes before using chemical activators that may represent an environmental risk.

REFERENCES

Abdel-Ghani, N. T., Hegazy, A. K., El-Chaghaby, G. A., & Lima, E. C. (2009). Factorial experimental design for biosorption of iron and zinc using *Typha domingensis*. Phytomass. *Desalination*, *249*(1), 343–347. doi:10.1016/j.desal.2009.02.065

Ali, M. E. M., Abdelsalam, H., Ammar, N. S., & Ibrahim, H. S. (2018). Response surface methodology for optimization of the adsorption capability of ball-milled pomegranate peel for different pollutants. *Journal of Molecular Liquids*, *250*, 433–445. doi:10.1016/j.molliq.2017.12.025

Amini, M., & Younesi, H. (2009). Biosorption of Cd (II), Ni (II) and Pb (II) from aqueous solution by dried biomass of *Aspergillus niger*: Application of response surface methodology to the optimization of process parameters. *Clean-Soil, Air, Water (Basel)*, *37*(10), 776–786.

Amini, M., Younesi, H., Bahramifar, N., Lorestani, A. A. Z., Ghorbani, F., Daneshi, A., & Sharifzadeh, M. (2008). Application of response surface methodology for optimization of lead biosorption in an aqueous solution by Aspergillus niger. *Journal of Hazardous Materials*, *154*(1-3), 694–702. doi:10.1016/j.jhazmat.2007.10.114 PMID:18068898

Antony, J., & Kaye, M. (2012). *Experimental quality: a strategic approach to achieve and improve quality. New York*: Springer Science Business Media.

Barakat, M. A. (2011). New trends in removing heavy metals fromindustrial wastewater. *Arabian Journal of Chemistry*, *4*(4), 361–377. doi:10.1016/j.arabjc.2010.07.019

Barrado, E., Vega, M., Pardo, R., Grande, P., & Del Valle, J. L. (1996). Optimisation of a purification method for metal-containing wastewater by use of a Taguchi experimental design. *Water Research*, *30*(10), 2309–2314. doi:10.1016/0043-1354(96)00119-4

Bermúdez, Y. G., Rico, I. L. R., Guibal, E., de Hoces, M. C., & Martín-Lara, M. A. (2012). Biosorption of hexavalent chromium from aqueous solution by *Sargassum muticum* brown alga. Application of statistical design for process optimization. *Chemical Engineering Journal*, *183*, 68–76. doi:10.1016/j.cej.2011.12.022

Bouazizi, S., Jamoussi, B., & Bousta, D. (2016). Application of response surface methodology for optimization of heavy metals biosorption on natural gum of Acacia Nilotica. *International Journal of Engineering Research & Technology*, *5*, 2278–0181.

Cao, Y., Liu, Z., Cheng, G., Jing, X., & Xu, H. (2010). Exploring single and multi-metal biosorption by immobilized spent Tricholoma lobayense using multi-step response surface methodology. *Chemical Engineering Journal*, *164*(1), 183–195. doi:10.1016/j.cej.2010.08.055

Cronje, K. J., Chetty, K., Carsky, M., Sahu, J. N., & Meikap, B. C. (2011). Optimization of chromium (VI) sorption potential using developed activated carbon from sugarcane bagasse with chemical activation by zinc chloride. *Desalination*, *275*(1-3), 276–284. doi:10.1016/j.desal.2011.03.019

Davim, J. P. (2015). *Design of experimental in production engineering*. New York.

Fernández-López, J. A., Angosto, J. M., Roca, M. J., & Miñarro, M. D. (2019). Taguchi design-based enhancement of heavy metals bioremoval by agroindustrial waste biomass from artichoke. *The Science of the Total Environment*, *653*, 55–63. doi:10.1016/j.scitotenv.2018.10.343 PMID:30404069

Fu, F., & Wang, Q. (2011). Removal of heavy metal ions from wastewaters: A review. *Journal of Environmental Management*, *92*(3), 407–418. doi:10.1016/j.jenvman.2010.11.011 PMID:21138785

Garba, Z. N., Bello, I., Galadima, A., & Lawal, A. Y. (2016). Optimization of adsorption conditions using central composite design for the removal of copper (II) and lead (II) by defatted papaya seed. *Karbala International Journal of Modern Science*, *2*(1), 20–28. doi:10.1016/j.kijoms.2015.12.002

Garg, U. K., Kaur, M. P., Garg, V. K., & Sud, D. (2008). Removal of nickel (II) from aqueous solution by adsorption on agricultural waste biomass using a response surface methodological approach. *Bioresource Technology*, *99*(5), 1325–1331. doi:10.1016/j.biortech.2007.02.011 PMID:17383868

Guijarro-Aldaco, A., Hernandez-Montoya, V., Bonilla-Petriciolet, A., Montes-Morán, M. A., & Mendoza-Castillo, D. I. (2011). Improving the adsorption of heavy metals from water using commercial carbons modified with egg shell wastes. *Industrial & Engineering Chemistry Research*, *50*(15), 9354–9362. doi:10.1021/ie2006627

Gupta, T. B., & Lataye, D. H. (2018). Adsorption of indigo carmine and methylene blue dye: Taguchi's design of experiment to optimize removal efficiency. *Sadhana*, *43*(10), 170. doi:10.100712046-018-0931-x

Hadiani, M. R., Darani, K. K., Rahimifard, N., & Younesi, H. (2018). Biosorption of low concentration levels of Lead (II) and Cadmium (II) from aqueous solution by Saccharomyces cerevisiae: Response surface methodology. *Biocatalysis and Agricultural Biotechnology*, *15*, 25–34. doi:10.1016/j.bcab.2018.05.001

Hadiani, M. R., Khosravi-Darani, K., & Rahimifard, N. (2019). Optimization of As (III) and As (V) removal by *Saccharomyces cerevisiae* biomass for biosorption of critical levels in the food and water resources. *Journal of Environmental Chemical Engineering*, *7*(2). doi:10.1016/j.jece.2019.102949

Hasan, S. H., Srivastava, P., & Talat, M. (2010). Biosorption of lead using immobilized Aeromonas hydrophila biomass in up flow column system: Factorial design for process optimization. *Journal of Hazardous Materials*, *177*(1-3), 312–322. doi:10.1016/j.jhazmat.2009.12.034 PMID:20036460

Hibbert, D. B. (2012). Experimental design in chromatography: A tutorial review. *Journal of Chromatography. B, Analytical Technologies in the Biomedical and Life Sciences*, *910*, 2–13. doi:10.1016/j.jchromb.2012.01.020 PMID:22333438

Jain, M., Garg, V. K., & Kadirvelu, K. (2011). Investigation of Cr (VI) adsorption onto chemically treated *Helianthus annuus*: Optimization using response surface methodology. *Bioresource Technology*, *102*(2), 600–605. doi:10.1016/j.biortech.2010.08.001 PMID:20739179

Karnitz Jr., O., Gurgel, L. V. A., De Melo, J. C. P., Botaro, V. R., Melo, T. M. S., de Freitas Gil, R. P., & Gil, L. F. (2007). Adsorption of heavy metal ion from aqueous single metal solution by chemically modified sugarcane bagasse. *Bioresource Technology*, *98*(6), 1291–1297. doi:10.1016/j.biortech.2006.05.013 PMID:16843656

Khalifa, E. B., Rzig, B., Chakroun, R., Nouagui, H., & Hamrouni, B. (2019). Application of response surface methodology for chromium removal by adsorption on low-cost biosorbent. *Chemometrics and Intelligent Laboratory Systems*, *189*, 18–26. doi:10.1016/j.chemolab.2019.03.014

Kuehl, R. O. (2001). *Diseño de experimentos: principios estadísticos para el diseño y análisis de investigaciones*. Thomson Learning.

Kumar, S. V., Pai, K. V., Narayanaswamy, R., & Sripathy, M. (2013). Experimental optimization for cu removal from aqueous solution using neem leaves based on Taguchi method. *International Journal of Science, Environmental Technology*, *2*(1), 103–114.

Lessa, E. F., Medina, A. L., Ribeiro, A. S., & Fajardo, A. R. (2017). Removal of multi-metals from water using reusable pectin/cellulose microfibers composite beads. *Arabian Journal of Chemistry*. doi:10.1016/j.arabjc.2017.07.011

Ma, X., & Ouyang, F. (2013). Adsorption properties of biomass-based activated carbon prepared with spent coffee grounds and pomelo skin by phosphoric acid activation. *Applied Surface Science*, *268*, 566–570. doi:10.1016/j.apsusc.2013.01.009

Mitić, M., Tošić, S., Pavlović, A., Mašković, P., Kostić, D., Mitićc, J., & Stevanović, V. (2019). Optimization of the extraction process of minerals from *Salvia officinalis* L. using factorial design methodology. *Microchemical Journal*, *145*, 1224–1230. doi:10.1016/j.microc.2018.12.047

Montgomery, D. C. (2001). *Design and analysis of experiments*. New York: John Wiley & Sons.

Montgomery, D. C. (2004). *Design and analysis of experiments*. Arizona: Limusa Wiley.

Montgomery, D. C. (2009). *Statistical quality control*. New York: Wiley.

Niad, M., Zaree, S., & Tahanzadeh, N. (2016). Response surface methodology for optimization of Cd (II) biosorption by *Cystoseria myricaas*. *Journal of Biomedicine and Health*, *1*(2), 6264.

Ofer, R., Yerachmiel, A., & Shmuel, Y. (2003). Marine macroalgae as biosorbents for cadmium and nickel in water. *Water Environment Research, 75*(3), 246–253. doi:10.2175/106143003X141033 PMID:12837031

Ou, Z., Zhang, M., & Qin, H. (2019). Tripling of fractional factorial designs. *Journal of Statistical Planning and Inference, 199*, 151–159. doi:10.1016/j.jspi.2018.06.002

Ozbay, N., & Yargic, A. S. (2018). Statistical analysis of Cu(II) and Co(II) sorption by apple pulp carbon using factorial design approach. *Journal of Industrial and Engineering Chemistry, 57*, 275–283. doi:10.1016/j.jiec.2017.08.033

Preetha, B., & Viruthagiri, T. (2007). Application of response surface methodology for the biosorption of copper using *Rhizopus arrhizus*. *Journal of Hazardous Materials, 143*(1-2), 506–510. doi:10.1016/j.jhazmat.2006.09.077 PMID:17084526

Pulido, H. G., De la Vara Salazar, R., González, P. G., Martínez, C. T., & Pérez, M. D. C. T. (2008). *Análisis y diseño de experimentos*. McGraw-Hill.

Pundir, R., Chary, G. H. V. C., & Dastidar, M. G. (2018). Application of Taguchi method for optimizing the process parameters for the removal of copper and nickel by growing *Aspergillus sp. Water Resources and Industry, 20*, 83–92. doi:10.1016/j.wri.2016.05.001

Roy, R. K. (2010). *A primer on the Taguchi method*. USA: Society of Manufacturing Engineers.

Sadh, P. K., Duhan, S., & Duhan, J. S. (2018). Agro-industrial wastes and their utilization using solid state fermentation: A review. *Bioresources and Bioprocessing, 5*(1), 1–15. doi:10.118640643-017-0187-z

Saval, S. (2012). Aprovechamiento de residuos agroindustriales: Pasado, presente y futuro. *BioTecnología, 16*, 14–46.

Srivastava, V. C., Mall, I. D., & Mishra, I. M. (2007). Multicomponent adsorption study of metal ions onto bagasse fly ash using Taguchi's design of experimental methodology. *Industrial & Engineering Chemistry Research, 46*(17), 5697–5706. doi:10.1021/ie0609822

Srivastava, V. C., Mall, I. D., & Mishra, I. M. (2008). Optimization of parameters for adsorption of metal ions onto rice husk ash using Taguchi's experimental design methodology. *Chemical Engineering Journal, 140*(1-3), 136–144. doi:10.1016/j.cej.2007.09.030

Taguchi, G., Chowdhury, S., & Wu, Y. (2005). *Taguchi's quality engineering handbook*. Hoboken, NJ: Wiley.

Thuan, V. T., Quynh, B. T. P., Nguyen, T. D., & Bach, L. G. (2017). Response surface methodology approach for optimization of Cu^{2+}, Ni^{2+} and Pb^{2+} adsorption using KOH-activated carbon from banana peel. *Surfaces and Interfaces, 6*, 209–217. doi:10.1016/j.surfin.2016.10.007

Tounsadi, H., Khalidi, A., Abdennouri, M., & Barka, N. (2016a). Activated carbon from *Diplotaxis Harra* biomass: Optimization of preparation conditions and heavy metal removal. *Journal of the Taiwan Institute of Chemical Engineers, 59*, 348–358. doi:10.1016/j.jtice.2015.08.014

Tounsadi, H., Khalidi, A., Farnane, M., Abdennouria, M., & Barka, N. (2016b). Experimental design for the optimization of preparation conditions of highly efficient activated carbon from *Glebionis coronaria* L. and heavy metals removal ability. *Process Safety and Environmental Protection*, *102*, 710–723. doi:10.1016/j.psep.2016.05.017

Tounsadi, H., Khalidi, A., Machrouhia, A., Farnane, M., & Elmoubarkia, R. (2016c). Highly efficient activated carbon from *Glebionis coronaria L.* biomass: Optimization of preparation conditions and heavy metals removal using experimental design approach. *Journal of Environmental Chemical Engineering*, *4*(4), 4549–4564. doi:10.1016/j.jece.2016.10.020

Tovar-Gómez, R., Moreno-Virgen, M., Moreno-Pérez, J., Bonilla-Petriciolet, A., Hernández-Montoya, V., & Durán-Valle, C. J. (2015). Analysis of synergistic and antagonistic adsorption of heavy metals and acid blue 25 on activated carbon from ternary systems. *Chemical Engineering Research & Design*, *93*, 755–772. doi:10.1016/j.cherd.2014.07.012

Wu, Y., Pang, H., Liu, Y., Wang, X., Yu, S., Fu, D., ... Wang, X. (2019). Environmental remediation of heavy metal ions by novel-nanomaterials: A review. *Environmental Pollution*, *246*, 608–620. doi:10.1016/j.envpol.2018.12.076 PMID:30605816

Yu, D., Wang, L., & Wu, M. (2018). Simultaneous removal of dye and heavy metal by banana peels derived hierarchically porous carbons. *Journal of the Taiwan Institute of Chemical Engineers*, *93*, 543–553. doi:10.1016/j.jtice.2018.08.038

Zolgharnein, J., Asanjarani, N., & Shariatmanesh, T. (2013). Taguchi L16 orthogonal array optimization for Cd(II) removal using *Carpinus betulus* tree leaves: Adsorption characterization. *International Biodeterioration & Biodegradation*, *85*, 66–77. doi:10.1016/j.ibiod.2013.06.010

Chapter 4
Application of Design of Experiments in Biofuel Production:
A Review

Jesús Andrés Tavizón-Pozos
https://orcid.org/0000-0002-8788-4034
Cátedras CONACYT, Universidad Autónoma del Estado de Hidalgo, Mexico

Israel S. Ibarra
https://orcid.org/0000-0001-8449-9774
Universidad Autónoma del Estado de Hidalgo, Mexico

Alfredo Guevara-Lara
Universidad Autónoma del Estado de Hidalgo, Mexico

Carlos Andrés Galán-Vidal
Universidad Autónoma del Estado de Hidalgo, Mexico

ABSTRACT

Biofuels emerge as an alternative to mitigate climate change. In this sense, four biofuels generations have been proposed to produce clean and renewable fuels. To achieve this, the development of these fuels requires an extensive and rigorous experimental work that will bring optimal results in short time periods. Hence, to accelerate the development of clean fuels, the Design of Experiments (DoE) methodologies are a useful tool to improve the operational conditions such as temperature, time, pressure, and molar ratios. Several authors have studied and optimized the different biofuel production systems using Factorial Designs and Response Surface Design methods and statistical analysis with reliable results. This chapter reviews and classifies the results obtained by these investigations and demonstrates the scopes and limitations of the application of DoE.

DOI: 10.4018/978-1-7998-1518-1.ch004

Figure 1. World TFC by fuel (♦) Oil, (∗) Electricity, (▲) Natural Gas, (●) Biofuels and Waste, (■) Coal, (▼) Other, from 2000 to 2016. Adapted from (IEA, 2019b)

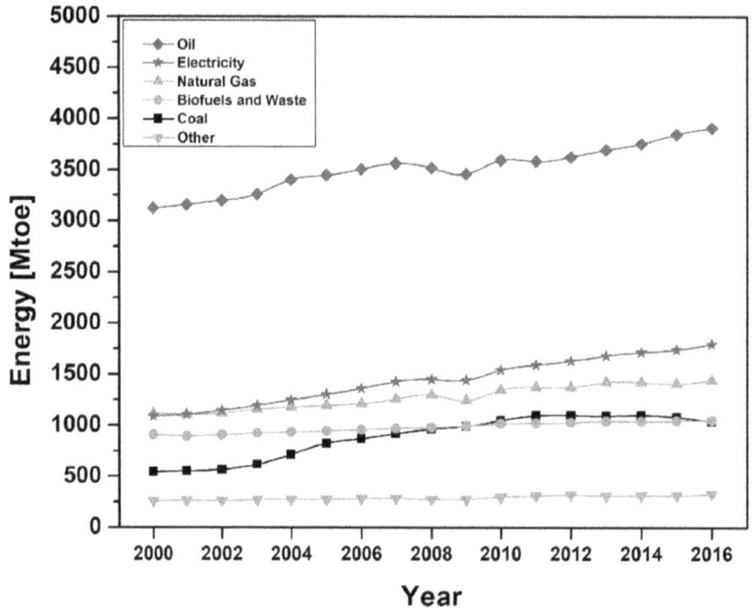

INTRODUCTION

Biofuels

The constant growth of the world population has caused the increase on demand for energy. Industry and transportation sectors are the most end-use energy consumption in the world which require a large amount of liquid fuels. In this way, the combustion of fossil fuels has led to the emission and accumulation of greenhouse gases (GHG) causing negative effects to environment, such as global warming. Therefore, to mitigate the GHG emissions and eventual depletion of petroleum, new technologies and infrastructure have been developed to produce sustainable, efficient, economically viable and renewable sources of liquid fuels (H. Chen et al., 2017).On this basis, the use of renewable energy sources has presented a growth of 2.8% annually since 2008 and it is expected to increase according to the objectives of The Paris Agreement in 2016 (Paris Agreement, 2016). However, the global energy and economics are still based in fossil fuel utilization. Petro-oil represents the highest consumption of energy available for final use in industry transport and home (Total Final Consumption, TFC), compared with other energy sources, followed by electricity and natural gas as Figure 1 exhibits.

Within the alternative energy sources, biomass transformation into energy is an important option to consider. Biomass from crops and agroindustry residues can be converted into biofuels which are a suitable and ecological friendly option since they are inexpensive, clean, sustainable and they could improve the rural economy (Sunde, Brekke, & Solberg, 2011). Also, the emitted CO_2 can be reintegrated into the carbon cycle by photosynthesis and it can be mixed with fossil fuels as additives which would reduce SOx and NOx emissions. According to the International Energy Agency (IEA), it is expected that biofuels may provide the 27% of the world transportation fuel by 2050, which means that nearly 3 billion

Figure 2. Global historical (2010-2016), forecast (2017-2023) and SDS targets (2025-2030) biofuels production (IEA, 2019a)

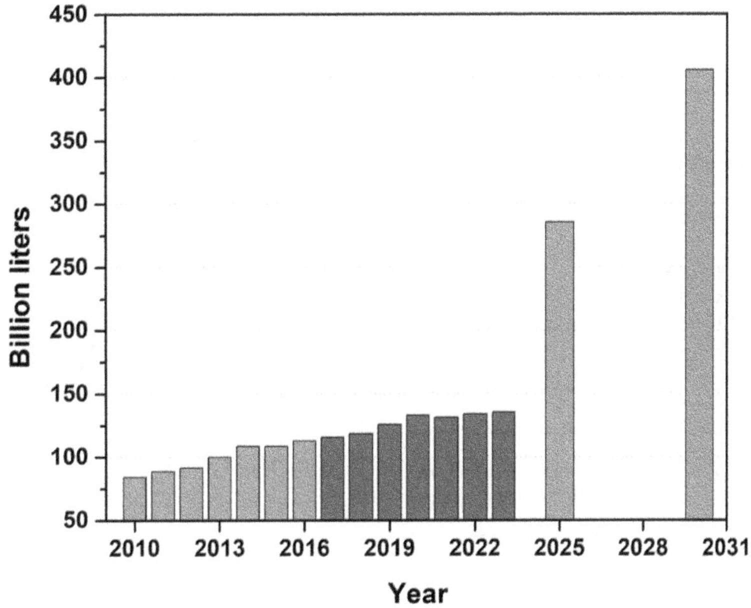

tons of biomass will be needed annually (IEA, 2011; IEA, 2019). Even so, the production of biofuels requires to triple -from 130 to 400 billion liters per year; Figure 2- its growth to achieve the 2030 under the Suitable Development Scenario (SDS) (IEA, 2019a). It is important to point out that biofuels may not replace the total transportation fossil fuel demand (Furimsky, 2013). Nonetheless, the potential of these fuels resides in that they can be blended with typical petroleum-based fuels, in such way that they can be distributed and used in the current infrastructure and combustion systems

In this sense, four generations of biofuels have been developed based on the raw material and method of production to overcome the agricultural issues and to supply fuel depending in the resources of each ecosystem (Correa, Beyer, Possingham, Thomas-Hall, & Schenk, 2017). Figure 3 summarizes the biofuel generations and main products.

First generation biofuels (1G) are produced from sugar and vegetable oils to produce bioethanol and biodiesel. In contrast, second generation biofuels (2G) are generated form the organic waste from agroindustry into fuel. Third and fourth biofuel generations (3G and 4G respectively) produce biomass (sugars, waste and triglycerides) from micro and macro algae, which is transformed into fuel by similar processes used in 1G and 2G. The 4G biofuels have been developed by improvement of the photosynthetic microorganisms to produce sustainable and renewable fuels by biological engineering. Table 1 displays the pros and cons of each biofuel generation.

Biorefineries appear as conceptual models for biofuel production which integrate different upgrading conversion units of biomass. It is necessary to adapt them into biomass feedstocks according to the local needs to improve the scientific and technological advancements to build an economically and sustainable transformation systems (S. K. Maity, 2015). Hence, new challenges for optimization in the production capacity of the biorefinery and technical advances are ahead. Several authors have agreed that integration of biorefineries with fossil fuel industries and renewable energy sources can be made

Figure 3. Biofuels and generation classification and main products

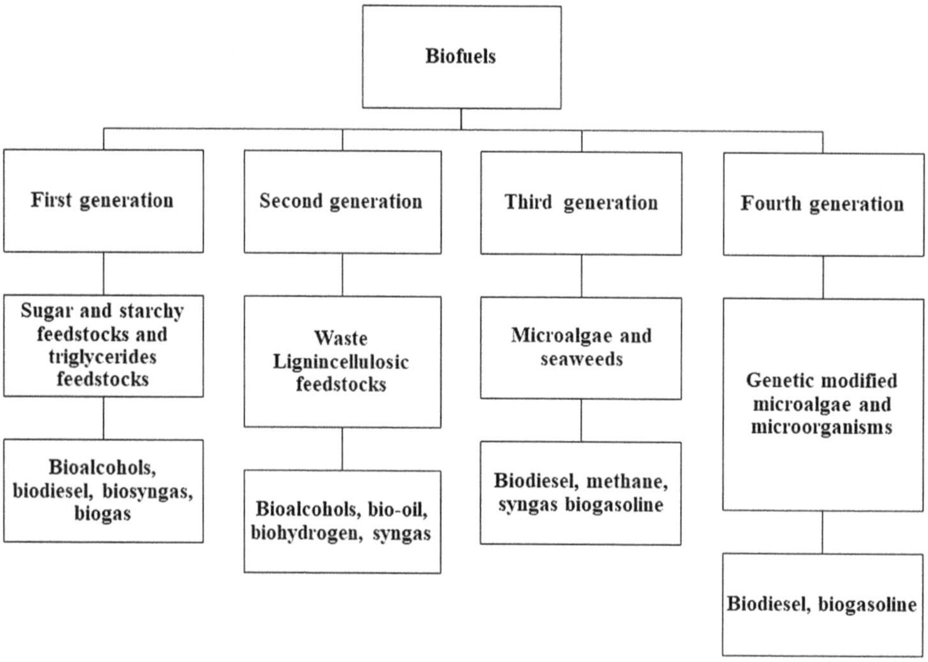

Table 1. Comparison between biofuel generations (Adapted from (Abdullah et al., 2019))

Topic	Biofuels			
	1G	2G	3G	4G
Competition with food	Yes	No	No	No
Land footprint	Large area	Large area	Small area	Small area
Conversion to biofuels	Easy	Difficult	Easy	Easy
Environment impact	Use of fertilizers	Deforestation	Marine eutrophication	Release GM organisms
Commercialization	Produced	Produced	Non produced	Non produced
Financial input	Low	Low	Large initial and cultivation costs	Large initial and cultivation costs

to simplify the scalation and reduce investment costs (Palmeros Parada, Osseweijer, & Posada Duque, 2017; Yang & Yu, 2013). These advanced plants are still in demonstration or pilot-plant stage an requires experiments to settle down the optimal conditions and operation variables. Thus, the essential optimization experiments and predicting models are needed to accelerate the understanding of the multifaceted biofuels production systems

Figure 4. First generation biofuels problems, benefits and processes

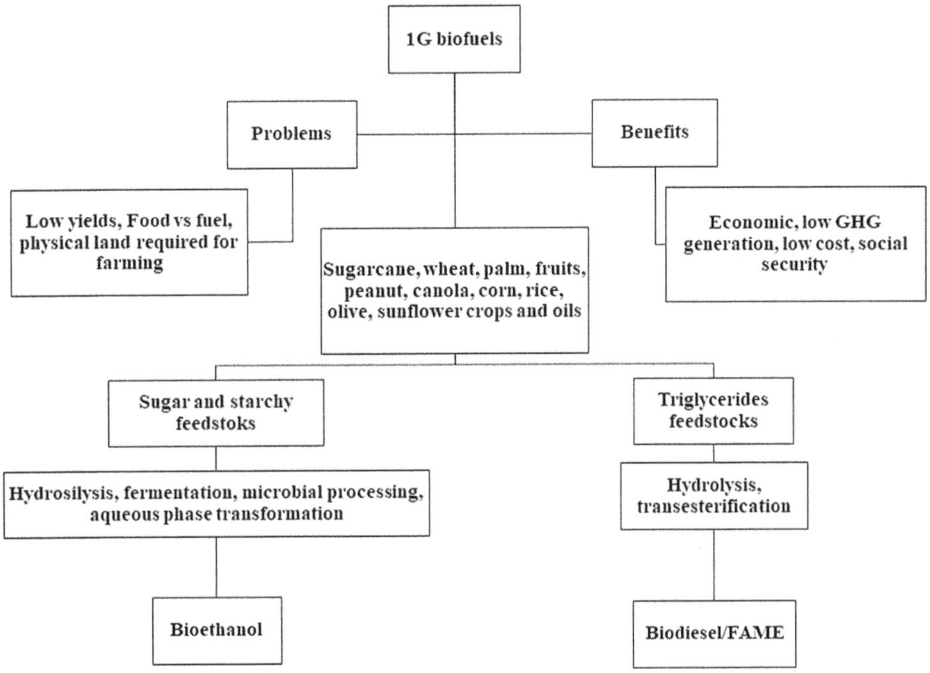

The use of Design of Experiments (DoE) in the biofuel industry has been applied to obtain critical information from the system in order to improve the production including the catalytic synthesis, the profits and capability, the performance and the manufacturing costs and reduce design and development time (Cox & Reid, 2000). The attention on application of DoE in biofuels development is directed to the screening and optimization of combining factors such as the harvesting biomass and optimal operational conditions. The present chapter is focused on the review, comparation and discussion of the application of different DoE methodologies in the biofuel production optimization to analyze and compile information of the factors and optimal results published by different authors.

First, Second and Third-Generation Biofuels

First-generation (1G) bioethanol is primarily produced from fermentation of sugar containing crops like sugarcane, wheat, rice and corn (Erlandsson, & Taherzadeh, 2014). Whereas, 1G biodiesel is obtained from food and non-edible vegetable oils, e.g., palm, canola, rapeseed and olive oils (Joshi, Pandey, Rana, & Rawat, 2017). The triglycerides present in these oils react with alcohol in presence of an alkali catalyst to produce fatty acid methyl esters (FAME) by transesterification reactions (Joshi et al., 2017). On this regard, temperature, pressure, catalysts concentration and composition, alcohol/oil molar ratio and time of reaction are the main operational variables to consider in the optimization by DoE (Dhawane, Karmakar, Ghosh, & Halder, 2018). However, since their production must consider the so-called "food vs fuel" controversy, there are uncertainties in their environmental and social impacts. The agriculture crops priority is to supply the population alimentation instead of fuel and developing countries may not be suitable to produce these fuels. This is summarized in Figure 4.

Figure 5. Second generation biofuel production processes

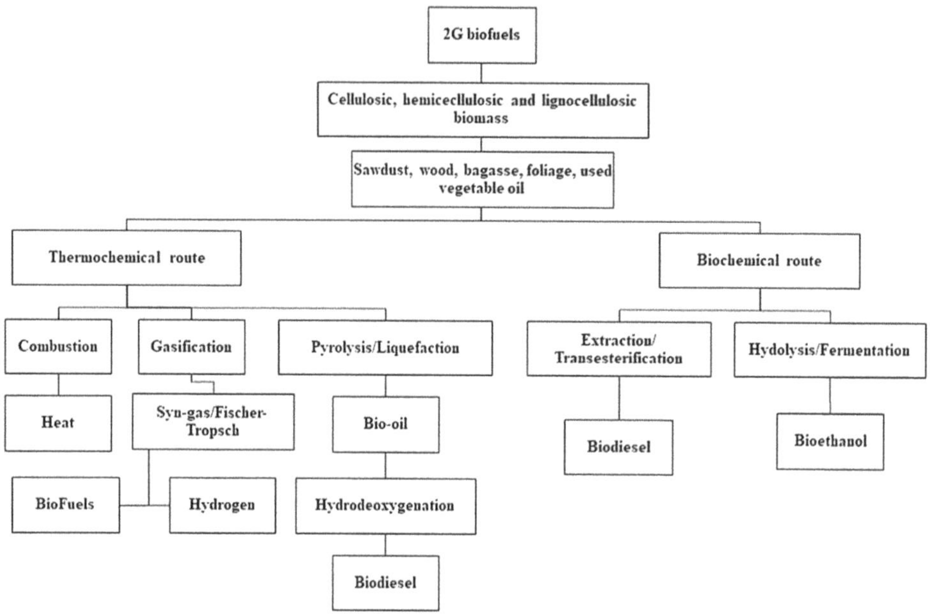

The second-generation biofuels (2G) are produced from inexpensive and abundant cellulosic biomass waste such as cellulose, hemicellulose and lignin materials (Furimsky, 2013). 2G biofuels include a wide variety of processes to obtain different products such as bioethanol, biodiesel, bio-oil, butanol, etc. There are two classifications in the conversion pathway to transform biomass into fuel: biochemical and thermochemical pathway (Figure 5) (Damartzis & Zabaniotou, 2011). Biochemical pathway consists in the use of catalytic hydrolysis to transform polysaccharides into sugar or oils to produce biodiesel (Hamze, Akia, & Yazdani, 2015). The thermochemical pathway involves pyrolysis, liquefaction and gasification of lignin rich biomass to produce bio-oil that can be transformed into biodiesel by conventional refinery units, e.g., hydrotreatment (Furimsky, 2013). Gasification of the biomass produces syngas (H_2 and CO mix) which subsequently is treated by Fischer-Tropsch (FT) (Ail & Dasappa, 2016).

Catalytic systems present different and complex correlations between the activity/selectivity and synthesis. Consequently, the development of catalysts is slow process and it is necessary the application of DoE improve their production.

Third generation biofuels (3G) have been produced by using microalgae and seaweeds (macroalgae) to produce biodiesel without competing with food industry. Due to their aquatic nature, macroalgae produce higher amounts of cellulosic biomass than terrestrial lignocellulosic biomass in less physical space and can grow fast if nutrients (nitrogen and phosphorus containing compounds), light, water and CO_2 are available (van Hal, Huijgen, & López-Contreras, 2014). Algae biomass is composed by different compounds which are synthetized by metabolic routes and their weight is principally constituted by proteins and triacylglycerol-based lipids (TAGs, 20-50%) (Mathimani & Pugazhendhi, 2019). In the harvesting of algae, biomass is obtained by flocculation, filtration, centrifugation and sedimentation (Dutta, Daverey, & Lin, 2014). In this sense, 3G biomass can be transformed into fuel by different methods: thermochemical and biochemical as Figure 6 exhibits.

Figure 6. Third generation biofuel production processes

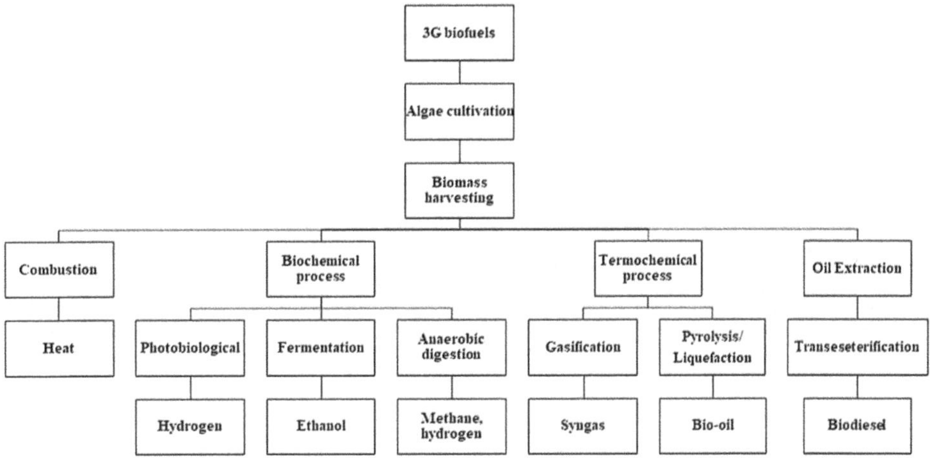

Thermochemical methods involve gasification, liquefaction and pyrolysis; whereas biochemical methods include anaerobic digestion, fermentation and transesterification of algal lipids (Tsukahara & Sawayama, 2005). In recent years, several methods of microalgae cultivation had been developed to maximize their biomass production (Slade & Bauen, 2013). Hence, different configurations, conditions, costs, biodiesel quality requirements, energy saving schemes, scale-up and supply of resources are issues to consider in the implementation of this technology (Brennan & Owende, 2010; Slade & Bauen, 2013).

DESIGN OF EXPERIMENTS

The DoE objective is to simplify the development of mathematical models. It contributes to reduce the variation and supports the planning of experiments to avoid the unnecessary ones which will reduce the required number of them to perform the optimization (Mandenius, 2016). It elucidates the influence of the interactions between factors involved and their responses without a complex interpretation providing reliable results by consideration of randomization, replication and blocking to orientate the experimentation faster to the true optimum (J. Antony, 2014b). The structuration of a DoE in most of the cases are in terms of coded factors (-1, +1; -1, 0, +1 or 1, 2, 3) indicating in each coded value the relative magnitude of the levels from the factors. The use of coded or uncoded factors is defined by mathematical reasons and the orthogonality from the system. However, in some cases, DoE does not employ the coded factors so, the factors are established independently (uncoded) for the respective evaluation (Brereton, 2007). The selection of the factors and levels and defining the objectives and targets that the study is aimed to, depends on the analytical matrix nature, biofuel process or the type synthesis. Another important aspect to consider is to choose an adequate DoE has been the experiment randomization in order to evaluate the system repeatability (Islas et al., 2018). In this regard, different DoE types are available to apply into the analysis; they are listened in Table 2.

Table 2. Comparison of objectives between types of DoE (Adapted from (NIST/SEMATECH, 2013))

Type of DoE	Objectives
Comparative designs	To choose between alternative factors with a narrow or broad scope. Make conclusions about if the changes in one important factor change the response for different levels of that factor.
Screening designs	To identify and extract which factor is the most important to measure. Select or screen out the few important main effects from the many less important ones. It is useful when there are 2-5 factors that can be performed.
Response surface methods	To estimate the interaction and quadratic effects of factors and levels. They are used to find improved or optimal process settings, weak points, reduce variation and make a process more robust against external influences.
Regression modeling	To model a response as a mathematical function of a few continuous factors with unbiased and minimum variance. It estimates a precise model, quantifying the dependence of response variables on process inputs.

Figure 7. Experimental design stages

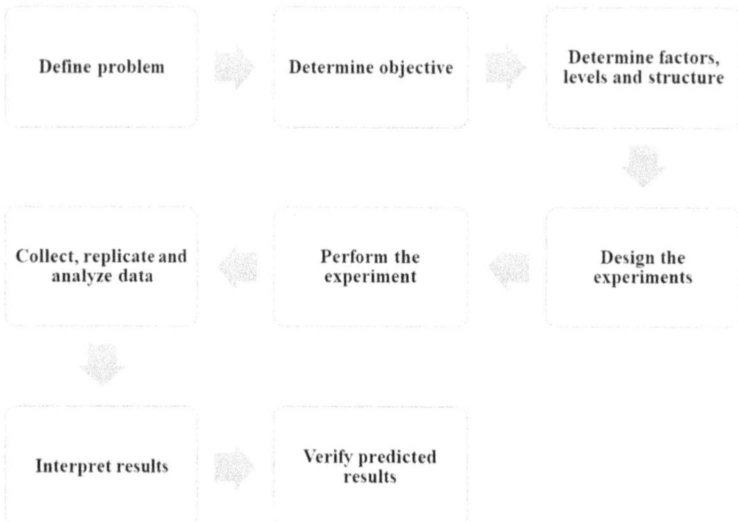

The selection of DoE allows the interpretation of the system based on a mathematical model, explaining the contribution of each factor in the system. However, depending on the DoE nature the interaction of the factors is not evaluated (Brereton, 2007; Islas et al., 2018). In addition, the statistical data is used to improve the system allowing to determine the best values in function of the output factor. In summary DoE methodology benefits mainly in:

- Organize the approaches that relate experiments to each other with the interpretation.
- Elucidate the interactions between qualitative and quantitative factors.
- Acquisition of information with less experiments.
- Results include the multiplicity of the systems.
- Support decision making of the systems.

Figure 8. Graphical representation of the respond (output factor) in OVAT design. White points: factors to evaluate in several levels (input factor), dark grey point: best experimental condition and black point: true optimal experimental conditions

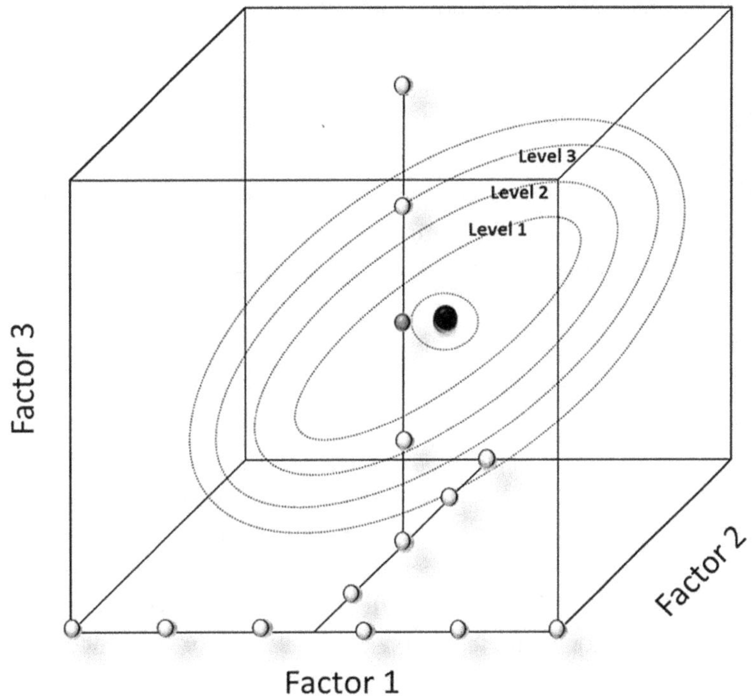

One Variable at a Time

The multiple process variables (input factors) and quality characteristics (output performance characteristics) in the biorefinery systems are related by highly complex effects. Qualitative inputs can be the raw materials, catalyst, conditions, etc. Whereas, outputs are the qualitative characteristics such as a composition, octane number, viscosity, etc. (Anderson & Whitcomb, 2015). Therefore, experiments must be performed to enhance the knowledge about the process comportment and variability by the estimation of the input factor effects. Finally, the data acquisition is determined by the effects of the process variables on the output performance and are verified by a comparison with the predicted result to bring robustness to the study. This is summarized in the Figure 7.

In process development, the initial step to follow is to plan an exploration by designed experiments. For example, if a new catalyst is under development (input factor), it is needed to test it under different conditions to observe the variation sensibility (response). It is necessary to keep in mind that outputs can be sensitive to one or several input variables. Therefore, the first suggestion is to use the One-Variable-at-a-Time (OVAT) which consist in modify one factor values at a time and keeping all other fixed. The data interpretation is analyzed by graphical methods where the factors show the possible effects in the system (Figure 8) (Weissman & Anderson, 2014).

The use of the OVAT design in 1-3G biofuel production has been a recurrent practice on the determination of signification level of each factor (Hu et al., 2008; Meneghetti, Meneghetti, Serra, Barbosa, & Wolf, 2007). The critical factors in biofuel production has been focused principally in reaction temperature, catalyst concentration, alcohol/oil molar ratio, reaction time and agitation (Hu et al., 2008; Li et al., 2008). The data analysis of the output factor is on terms of fatty acid ethyl ester (FAEE) and FAME yield. However, the OVAT provides disadvantages, for example, it does not lead to a non-optimized final process and the studies do not determine the degree of interactions between the factors, requires an exhaustive number of experiments, time-consuming and in some cases expensive (Weissman & Anderson, 2014). Consequently, this type of design could provide experimental factors closed to the optimal but do not guarantee to obtain the optimal experimental conditions (Rozet, Lebrun, Hubert, Debrus, & Boulanger, 2013; Toms et al., 2017). In this sense, statistical planning may link the inputs and outputs variables and report the correct method to understand multiplicity to collect data to make decisions about the bioprocess.

Factorial Design

Factorial design has been considered an important tool of design, employed for the development of several biofuel production systems or in preliminary studies for subsequent optimization. In factorial designs, each evaluated factor is considered as an independent factor, allowing the simultaneous evaluation in two or more factors levels (Hibbert, 2012). The DoE configuration allows to obtain information about the interactions between the factors and the response. The effects of each factor and their interactions are determined by graphical method. The Y-axis represents the factor levels while the X-axis output factor to the corresponding levels; the principal factor is represented by the maximal response. The interaction effects: X-axis represents the output magnitude, Y-axis the evaluated levels, each plotted point represent the average of the output factor obtained in each level, the interaction is defined by slopes different (Hibbert, 2012).

Factorial design is classified as full factorial and fractional factorial design. The factorial analysis is based on the use of coded or uncoded factors, however, in most of the cases it is recommended the use of coded factors where the model coefficients are dimensionless and can be adequate to compare and determine the significance levels (Ridzuan, Adam, & Yaacob, 2016; Veličković, Stamenković, Todorović, & Veljković, 2013).

Full Factorial Design

In a full factorial design, the methods are described as l^k, where l is defined as the number of factor levels and k is the number of factors to evaluate, where each combination is analyzed (J. Antony, 2014a). According to Figure 9, the full factorial design shows two typical configurations for two-, and three-factors, defined as square (Figure 9A), or a cube in the factor space (Figure 9B) respectively, the number of experiments are defined in function of the k value, where each factor is tested at two levels: (-1) for low and (+1) for high levels (Hanrahan, Montes, & Gomez, 2008).

Figure 9. General configurations for a full factorial: A) Two factor two-level factorial design (2^2 design) and B) Three-factor two-level factorial design (2^3 design)

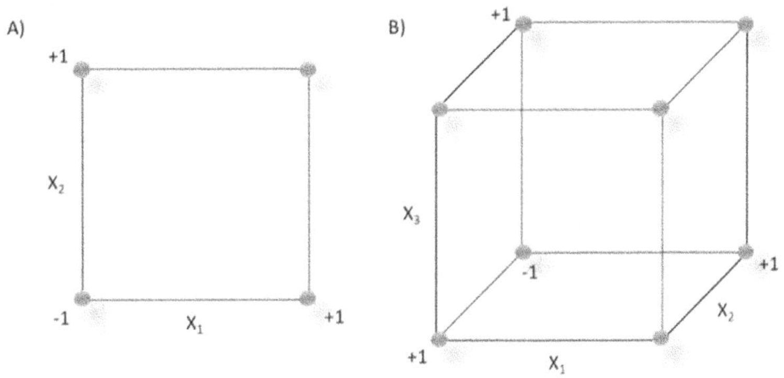

Figure 10. Graphical representation of: A) 2^3 (full factorial) and B) 2^{3-1} (factorial fractioned); X1, X2, and X3 as variables representation

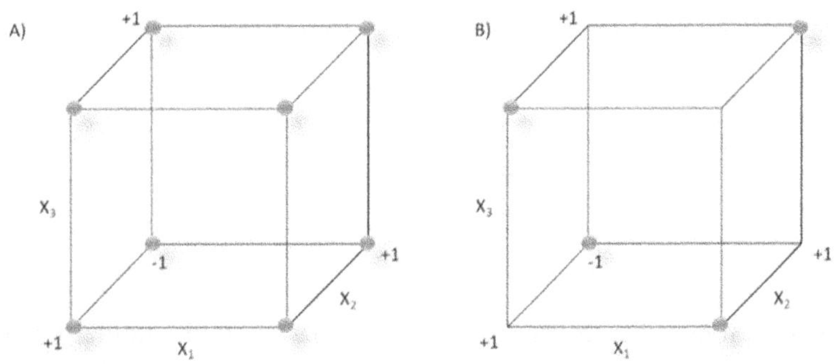

Fractional Factorial Design

The fractional factorial is a type of orthogonal array, determined by coded factors as -1 or +1 to define the low and high levels employed in the evaluation of specific factors, interactions and their main effects in biofuel production and catalytic synthesis. A fractional factorial design is an alternative to the full factorial during expensive performance, risky or in preliminary studies to determine the first approximation of the behavior of the factors involved in each system. The output factor is a representative function to the full factorial design (Hibbert, 2012). Fractional factorial design has been accepted by the advantages offered in function of the number of experiments in comparison with OVAT design. The number of experiments in a fractional factorial design is determined using the notation l^{k-p}, where l is defined as the number of factor levels, k is the number of factors to evaluate, and p is defined as the size of the faction of the full factorial, reducing the number of experiments (Sakkas et al., 2010). Figure 10 shows a comparison between the full factorial l^{k} (Figure 10A) and the fractional factorial design l^{k-p} (Figure 10B), the number of experiments for the fractional factorial design is represented as $1/2p$ of the full factorial. However, the use of fractional factorial design origin the losses of information due to several effects that cannot be estimated such the variable interactions (Hanrahan et al., 2008; Sakkas et

Table 3. Full factorial design methodology for optimization of biofuel production

Optimization System (matrix/ via)	Factorial Design (Experimental run)	Input Factor	Output Factor	Regression Model	Optimal Conditions	Reference
Biodiesel production (sunflower oil)	3^3; 27	X1: Temp. (°C) X2: Ethanol/oil X3: Catalyst loading (wt.%)	-Fatty acid ethyl ester (%FAEE)	Quadratic model	X1: 74.6 °C X2: 12:1 X3: 1.25 wt.%	(Veličković et al., 2013)
Biodiesel production (lard)	3^2: 18	X1: Agitation speed (rpm) X2: Catalyst concentration (wt.%)	-Fatty acid methyl esters (%FAME)	Linear model	X1: 600 rpm X2: 0.9 wt.%	(Berrios et al., 2009)
Biodiesel production (sunflower oil)	2^2: 8	X1: Temp. (°C) X2: Catalyst concentration (wt.%)	-Transesterification Triglyceride to Methyl Ester	Linear model	X1: > 60°C X2: > 1.5 wt.%	(Vicente et al., 1998)
Biofuel production (microalgae Chlorella)	2^3: 24	X1: Wastewater sludge X2: Amount of technical glycerol X3: Amount of salt in the BG11 medium	-Concentration of Chlorella sp. Biomass (g L^{-1})	Quadratic model	X1: 0.12 X2: 2.00 X3: 0.00	(Skorupskaite et al., 2015)
Biodiesel production (Thermomyces lanuginosus)	3^2: 18	X1: Temp. (°C) X2: pH X3: Ethanol/oil	-Conversion (%) -Kinematic viscosity (mm^2 s^{-1})	Quadratic model	X1: 20°C X2: 12.0 X3: 3.4:1	(Verdugo et al., 2011)

al., 2010). A Pareto chart represents the quotient effects between their standard error, considered as a critical tool to provide fundamental information on the factors with high significance in biofuel production, using simple graphics or modified version of histogram (Omar & Amin, 2011).

Factorial design has been applied to investigate the effect of critical factors (input factors) in biofuel employing sunflowers oil, and lard to obtain biodiesel of first generation the catalyst concentration, alcohol/oil molar ratio and reaction temperature, agitation and pH.

The optimal conditions were 74.6°C with a molar ratio ethanol/oil molar ratio of 12:1 and catalyst concentration of 1.25 wt % of NaOH in a 3^3 system, with an adjustment at a quadratic model in the biodiesel production using sunflowers oil (Veličković et al., 2013). This study presented a 95% of accuracy between the experimental and the theoretical results. In 3^2 systems Berrios et al., evaluated the agitation speed in biodiesel production from lard obtaining %FAME satisfactory at 600 rpm using a 0.9 wt.% of KOH as optimal concentration using lard as principal source in the biodiesel production (Berrios et al., 2009). On the other hand, Verdugo at al., evaluated the behavior of the conversion and the kinematic viscosity by the evaluation of the temperature, pH value and the ethanol/oil molar ratio. The pH value played a role important in the system as catalyst. However, in this case the highest conversion was obtained at 20°C with an ethanol/oil molar ratio of 3.4:1, with an adjustment at a quadratic model (Verdugo et al., 2011). Vicente et al. through a 2^2 system, determined a temperature >60°C with a catalyst concentration >1.5% as optimal conditions (95% confidence) for the transesterification process with 8 experimental runs compared with 27 carried out by Veličković et al. (Vicente et al., 1998). In general, the optimal conditions obtained were catalyst concentration (0.9 to > 1.5 wt%), alcohol/oil molar ratio (ethanol/oil 3.4:1 to 12:1), reaction temperature (20 to 74.6°C), agitation (600 rpm), pH (12.0), wastewater sludge (0.12), amount of technical glycerol (2.0) and the amount of salt in the BG11 medium (0.0). According to the DoE analysis, the models have been adjusted at two regression models to describe the effect of each factor or their interactions in each process, and the prediction of the value of the output factor in different factor levels studied. In 1G biofuel production and in catalyst synthesis the use of factorial has demonstrated be more efficient to obtain crucial systems information in comparison with the OVAT design which usually employs a large number of experiments without guaranteeing the system optimization.

Taguchi Design

Taguchi design has been applied in quality assessments of the industrial process due at their robustness and resistant to variation from the noise factors, with minimal experiments. It employs an orthogonal array, which provides a simple experimental design in comparison with traditional DoE (Robinson, Borror, & Myers, 2004). Taguchi is focused on the identification of the controlled factors or first factors (*parameters*) to minimize the effect of uncontrolled factors (*noise*) by the significant effects of each factor on the output through the analysis of the corresponding and standard deviations (Kharia & Singhai, 2013). The number of experiments is defined by the basic Taguchi array as L4, L8, L9, L12, L16 and L32 for two-, and three-level factors, where the number indicates the total number of experiments by design. L_n array is defined as a l^{n-1} in a fractional factorial design, where l is the number of factor levels represented as 1, 2 and 3 (high) factors levels. However, Taguchi array cannot estimate the interaction between the factors on the response, each factor is evaluated independently. Also, there is the absence of regression model adjustment to determine the interaction mode between the factors (Kharia & Singhai, 2013; Vinodh, Karthik Bharathi, & Gopi, 2016).

According to Table 4, Taguchi design has been focused on the evaluation of three and four factors: reaction time, catalyst concentration, alcohol/oil molar ratio and temperature on transesterification process (output factors) determined by the %FAEE, %FFA, %RSME, %AKOME and %HBME with 9, 16 and 30 experimental runs (Adewale et al., 2017; Dhawane et al., 2017; Kumar et al., 2015; Tan et al., 2017). In general, these studies found that optimal values of catalyst concentration (0.7 to 10.0 wt%), alcohol/oil molar ratio (6.0:1.0 to 20.0-1.0), reaction time (0.45 to 16 h) with temperature reaction (40° to 65°C) and they highly depend in the used raw materials. Tan et al. compared Taguchi method with Response Surface and showed that the optimum conditions of catalyst concentration, methanol/oil molar ratio, temperature and the reaction time for the transesterification of waste cooking oil presented only about the 2% of error. They concluded that Taguchi is efficient only at specific level and requires less experimental data to analyse (Tan et al., 2017). Karabas concluded that the signal-to-noise ratio can be used to measure accurately the deviation of the desired value (Karabas, 2013). Kumar et al., have evaluated the agitation speed and the enzyme dosage in the biodiesel synthesis from rubber seed oil and crude tall oil with 1250 rpm and 1.0 wt% respectively. Their results had a 94.83% of accuracy with the predicted optimum. It was assumed that these estimations provide an optimum in the enzyme dosage to produce biodiesel (Kumar et al., 2015; Adewale et al., 2017). The use of Taguchi different in biofuel processes has demonstrated be an excellent and efficient tool to improve any process through the adequate data treatment to ensure the identification and selection of the optimal conditions for the process and catalysts synthesis (Buasri et al., 2014; Dhawane et al., 2017).

Table 4. Taguchi design methodology for optimization of biofuel production and catalyst synthesis

Optimization System (matrix/via)	Orthogonal Arrangement (Experimental Run)	Input Factor	Output Factor	Regression Model	Optimal Conditions	Reference
Biodiesel production (Waste cooking oil catalyzed by solid ostrich and chicken-eggshell)	L30: 30	X1: Catalyst concentration (wt.%) X2: Methanol/oil X3: Temp. (°C) X4: Time (min)	Biodiesel yield (%FFA)	-	X1: 1.50 wt.% X2: 10:01 X3: 65°C X4: 2 h	(Tan, Abdullah, Nolasco-Hipolito, & Zauzi, 2017)
Biodiesel production (Manilkara zapota (L.) seed oil)	L9: 9	X1: Methanol/oil X2: Catalyst concentration (wt.%) X3: Time (min) X4: Temp.(°C)	Biodiesel yield (%FFA)	-	X1: 6:1 X2: 1.0 wt.% X3: 90 min X4: 50°C	(Kumar, Sureshkumar, & Velraj, 2015)
Biodiesel production (Rubber seed oil)	L9: 9	X1: Catalyst concentration (wt.%) X2: Temp. (°C) X3: Methanol/oil X4: Agitation speed (rpm)	Rubber seed oil methyl ester yield (%RSME)	-	X1: 4.5 wt.% X2: 9:1 X3: 60°C X4: 1250 rpm	(Dhawane et al., 2017)
Biodiesel production (Enzyme-catalyzed-crude tall oil)	L9: 9	X1 Temp. (°C) X2 Time (min) X3 Methanol/oil X4 Enzyme dosage (wt.%)	Fatty acid methyl esters Yield (%FAME)	-	X1: 40°C X2: 16 h X3: 1:1.5 X4: 1.0 wt.%	(Adewale, Vithanage, & Christopher, 2017)
Biodiesel production (Crude acorn Quercus frainetto L. kernel oil)	L9: 9	X1: Catalyst concentration (wt.%) X2: Alcohol/oil X3: Time (min) X4: Temp. (°C)	The acorn kernel oil methyl ester yield (%AKOME)	-	X1: 0.7 wt.% X2: 8:1 X3: 40 min X4: 50°C	(Karabas, 2013)
Biodiesel production (Hevea brasiliensis oil)	L9: 9	X1: Time (min) X2: Temp. (°C) X3: Catalyst concentration (wt.%) X4: Alcohol/oil	Hevea brasiliensis methyl ester Yield (%HBME)	-	X1: 60 min X2: 55°C X3: 3.5 wt.% X4: 15:1	(Dhawane et al., 2016)
Biodiesel Production (Scallop Waste Shell from Palm Oil)	L9: 9	X1: Time (h) X2: Temp. (°C) X3: Catalyst concentration (wt.%) X4: Methanol/oil	Fatty acid methyl esters Yield (%FAME)	-	X1: 3 h X2: 65°C X3: 10.0 wt% X4: 9:1	(Buasri, Worawanitchaphong, Trongyong, & Loryuenyong, 2014)
Biodiesel production (Castor oil)	L16: 16	X1: Temp. (°C) X2: Time (h) X3: Catalyst concentration (wt.%) X4: Methanol/oil X5: Agitation Speed (rpm)	Biodiesel yield (%FFA)	-	X1: 50 °C X2: 1 h X3: 1.0 wt.% X4: 20:1 X5: 700 rpm	(Karmakar, Dhawane, & Halder, 2018)
Biodiesel production (Waste cooking oil)	L9: 9	X1: Temp. (°C) X2: Methanol/Oil X3: Time (h) X4: Catalyst concentration (wt.%)	Fatty acid methyl esters yield (%FAME)	Linear model	X1: 3 h X2: 65°C X3: 10.0 wt.% X4: 9:1	(Dhawane et al., 2018)
Catalyst synthesis (Carbon nanotubes)	L16: 16	X1: Annealing time (min) X2: Hydrogen flow rate (sccm) X3: Annealing temperature (°C) X4: Argon flow rate (sccm)	- Roughness **(A)** - Catalyst size **(B)**	-	**A** X1: 2.5 min X2: 65 sccm X3: 730°C X4: 0 sccm **B** X1: 1.0 min X2: 65 sccm X3: 730°C X4: 40 sccm	(Pander, Hatta, & Furuta, 2016)

Figure 11. Graphical representation of the trial points used A) two-, and B) three-factor central composite design

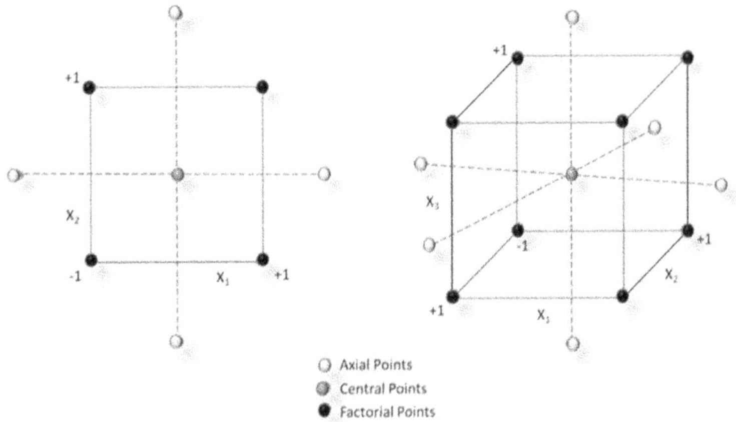

Response Surface Design (RSD)

The response surface design (RSD) is a DoE focused to understand and optimize any process by the analysis of the output factor. The optimization process is a mathematical and statistical method based on the fit of a polynomial equation to the experimental data. In the biofuel and catalyst process, the analysis of the surface equation allows to predict the behavior of the system, identify the effect on the response when a critical factor is modified, finding the levels of factors that optimize response and the most important aspect is the selection of the optimal conditions (Baş & Boyacı, 2007). The model is represented as:

$$Y = \beta_0 + \sum \beta_{1\text{-}n} X_{i\text{-}n} + \sum \beta_{ii} X_i^2 + \sum \beta_{1j} X_i X_j + \varepsilon$$

where y: define the output factor, β_0: a constant, $X_{i\text{-}n}$: the factor numbers employed in each system, $\beta_{1\text{-}n}$: the unknown function of Y and ε: the statistical errors not considered.

Central Composite Design

A central composite design (CCD) has been most used RSD in industrial and analytical processes since it is a modification of factorial or fractional factorial designs. In general, CCD uses a system of coded factors (-1, 0, +1) to represent the low, center and high levels to evaluate the design which includes center and axial points (star points) in the design (Figure 11) (Yadav, Singh, Balan, Pareek, & Vivekanand, 2019).

CCD allows to estimate the effects of each factor and the curvature of the model ($l^k + lk + nc$) where l is the number of factor levels, k is the number of factors and nc represents the number of replicate central points designed in function of the factorial design, while the axial points are employed to improve the precision. The axial and factorial points are situated equidistant from the central points (circumscribed central composite design), or within the factorial design (inscribed central composite design) (Hibbert, 2012; Ridge & Kudenko, 2010).

Table 5. Central composite design methodology for optimization of biofuel production

Optimization System (matrix/via)	CCD/Experimental Run	Input Factor	Output Factor	Regression Model	Optimal Conditions	Reference
Biodiesel production (soybean oil)	(30) 2^4 factorial experiments, 8 axial points and 6 center points	X1: Alcohol/oil X2: Catalyst concentration (wt.%) X3: Temp. (°C) X4: Time (min)	Ethyl esters (%) Glycerol (%)	Quadratic model	X1: 9:1 X2: 1.3 wt.% X3: 40.0°C X4: 80 min	(Silva, Camargo, & Ferreira, 2011)
Biodiesel production (Castor oil)	(12) 2^3 factorial experiments and 4 center points	X1: Temp. (°C) X2: Catalyst concentration (wt.%) X3 Methanol/oil	Product Yield (%)	Quadratic model	X1: 65°C X2: 1.5 wt.% X3: 7:1.	(Kılıç, Uzun, Pütün, & Pütün, 2013)
Heterogeneous catalyst (Hevea brasiliensis oil)	(13) 2^2 factorial experiments and, 4 axial points, 1 center point, and 4 replicated at the center point.	X1: Activation temperature (°C) X2: Activation time (h)	- Surface area (m^2/g) -Total pore volume (cm^3/g)	Quadratic model	X1: 350°C X2: 1.5 h	(Dhawane et al., 2015)
Biodiesel production (Edible and nonedible vegetable oils)	(30) 2^4 factorial experiments, 8 axial points and 6 center points	X1: Temp.(°C) X2: Alcohol/oil X3: Catalyst concentration (wt%) X4: Time (min)	Fatty acid ethyl ester (%FAEE)	Quadratic model	X1: 43.50°C X2: 8.8:1, X3: 1.9 wt.% X4: 58.4 min.	(Gupta, Agarwal, & Dalai, 2016)
Biodiesel production (Shea butter)	(30) 2^4 factorial experiments, 8 axial points and 6 center points	X1: Temp. (°C) X2: Agitation (rpm) X3: Alcohol/oil X4: Catalyst concentration (wt.%)	%FFA and %yield of shea biodiesel (SBD)	Quadratic model	X1: 40.19°C X2: 807.98 rpm X3: 7.25 X4: 1.00 wt.%	(Ajala, Aberuagba, Olaniyan, Ajala, & Sunmonu, 2017)
Biodiesel production (Waste cooking oil catalyzed by solid ostrich and chicken-eggshell)	(30) 2^4 factorial experiments, 8 axial points and 6 center points	X1: Catalyst concentration (wt.%) X2: Methanol/oil X3: Temp. (°C) X4: Time (h)	Biodiesel yield (%FFA)	Quadratic model	X1: 1.57 wt.% X2: 11:01 X3: 65°C X4: 1.9 h	(Tan, Abdullah, Nolasco-Hipolito, & Zauzi, 2017)
Biodiesel production (Waste cooking palm oil)	(26) 2^4 factorial experiments, 6 axial points and 2 center points	X1: Methanol/oil X2: Catalyst concentration (wt.%) X3: Time (min) X4: Temp. (°C)	Biodiesel yield (%FFA)	Quadratic model	X1: 29:1 X2: 2.7 wt.% X3: 169 min X4: 115.5°C	(Omar & Amin, 2011)
Biodiesel production (Coconut shell)	(26) four-factor-three-level central composite design (CCD)	X1: Carbonization Temp. (°C) X2: Carbonization Time (h) X3: Sulfonation Temp. (°C) X4: Sulfonation Time (h)	Fatty acid methyl esters (%FAME)	Quadratic model	X1: 422°C X2: 4 h X3: 100°C X4: 15 h	(Endut et al., 2017)
Biodiesel production (Cottonseed oil)	(30) 2^4 factorial experiments, 8 axial points and 6 center points	X1: Methanol/oil X2: Temp. (°C) X3: Time (min) X4: Catalyst concentration (wt.%)	Biodiesel yield (%FFA)	Quadratic model	X1: 6:1 X2: 55°C X3: 60 min X4: 0.6 wt.%	(Onukwuli, Emembolu, Ude, Aliozo, & Menkiti, 2017)
Biodiesel production (lard oil)	30 2^4 factorial experiments, 8 axial points and 6 center points	X1: Reaction temp (°C) X2: Catalyst concentration (wt.%) X3: Time (min) X4: Methanol/oil	Fatty acid methyl esters (%FAME)	Quadratic model	X1: 65°C X2: 1.25 wt.% X3:40 min X4: 6:1	(Ezekannagha, Ude, & Onukwuli, 2017)

CCD is considered a rotatable design according to the analysis of variance (ANOVA), where the information obtained is of main importance to determine the effect or effects of each factor through the mathematical model (linear, square and two-way interaction) and the corresponding response surface and contour plot in order to understand the behavior of the system (Ridge & Kudenko, 2010). According to Table 5, the CCD has been employed in biofuel process with at least three or more factors such as: alcohol/oil molar ratio, catalyst concentration, reaction temperature, agitation and reaction time (Ajala et al., 2017; Silva et al., 2011). The main outputs are the activation temperature on transesterification

Figure 12. Graphical representation of the trial points in a three-factor Box–Behnken design

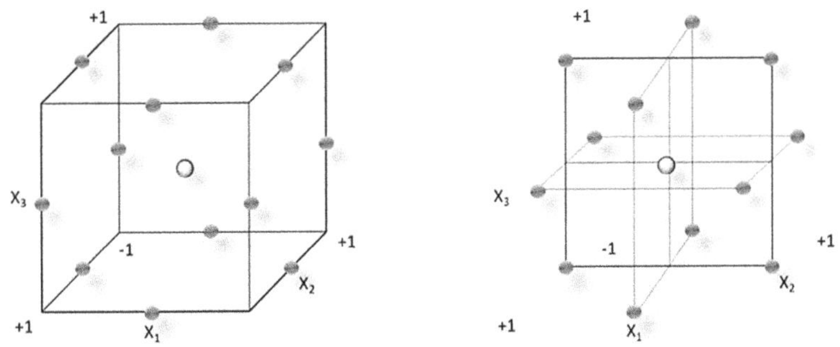

process determined as the ethyl esters, glycerol, %FFA, %SBD, %FAME, surface area and total pore volume (Dhawane et al., 2015; Kılıç et al., 2013; Silva et al., 2011).

In biofuel production, the studies have employed castor oil, a mixture vegetable oils, shea butter, waste cooking oils, refined cottonseed oil and lard oil. The corresponding evaluation has allowed to obtain better conditions with 13, 26 and 30 experimental runs (Gupta et al., 2016; Omar & Amin, 2011; Tan et al., 2017). Based on these results, the optimal conditions have presented a wide range of values: alcohol/oil molar ratio (6:1 to 29:1), reaction temperature (40 to 115°C), reaction time (40-120 min) and catalyst concentration (0.6 to 7.5 wt%). Despite this, these authors have mentioned that CCD is a good statistical tool since its predictions presented an of 98%. In comparison with Taguchi method, CCD have shown better results (Tan et al., 2017). Ajala et al., concluded that the agitation might not to be considered as fundamental factor to ensure the complete transesterification (Ajala et al., 2017). According to the biofuel process and catalyst synthesis, the quadratic model explains the significance level of each factor and their interactions in the systems through their mathematical equation by the corresponding contour plot, surface responds to obtain the optimal values (Ezekannagha et al., 2017; Gupta et al., 2016; Onukwuli et al., 2017). However, despite the advantages of CCD, two important aspects must be considered: the experimental array and experimental run. The incorrect levels assignation in the axial points affects negatively the data analysis and the costs.

Box-Behnken Design

Box-Behnken Design (BBD) is based in midpoints in the experimental structure. In contrast, it reduces the experiment numbers in comparison with CCD, which implements vertices point with factorial or fractional factorial design (Figure 12) (Hibbert, 2012). Box-Behnken design has been employed in the development of several analytical and industrial systems. However, the design is conditioned for the process with at least three factors. BBD is a class of rotatable or nearly rotatable design based in the use of coded factors system (-1, 0, +1) to identify the low, middle and high levels of the factors (Aslan, 2008).

The number of experiments in BBD is defined as $N=l[k(k-1)]+C_0$, where the number of central points is C_0. In comparison with CCD, BBD does not evaluate combinations where the factors are at their higher or lower levels according to the experimental array (Hibbert, 2012). A BBD in biofuel and catalyst process has been considered an important statistic tool to establish the adequate conditions in their process and limits of each factor. In addition, it allows to determine the variance of the process by the ANOVA, where the information obtained defines the influence of each effect by a mathematical

Table 6. Box-Behnken Design methodology for optimization of biofuel production

Optimization System (matrix/via)	Experimental Run (N) N=l k(k−1) +C_0	Input Factor	Output Factor	Regression Model	Optimal Conditions	Reference
Biodiesel production (waste cooking oil)	15	X1: Temp. (oC) X2: Catalyst concentration (wt.%) X3: Methanol/oil	Biodiesel yield (%)	Quadratic	X1: 65°C X2: 1.4 wt.% X3: 7.5:1	(Hamze et al., 2015)
Biodiesel and fuel production (continuous supercritical process)	27 with replica	X1: Temp. (°C) X2: Residence time (min) X3: Ethanol/oil X4: Iso-propanol/ oil	Fatty acid ethyl ester (%FAEE)	Quadratic	X1: 375°C X2: 7 min X3: 23:1 X4: 0.1:1	(Akkarawatkhoosith et al., 2019)
Biodiesel production (waste cooking oil catalyzed by Zn)	29 with replica	X1: Zn doping X2: Catalyst loading (wt.%) X3: Methanol/oil X4: Time (h)	Fatty acid methyl esters (%FAME)	Quadratic	X1: 1-Zn/CaO X2: 5.0 wt.% X3: 20:1 X4: 4 h	(Borah et al., 2019)
Biodiesel production (soybean oil)	15	X1: Temp. (°C) X2: Catalyst concentration (wt.%) X3: Methanol/oil	Fatty acid methyl esters (%FAME)	Quadratic	X1: 60°C X2: 1.2 wt.% X3: 9:1	(Rahimi et al., 2014)
Biodiesel production (Palmitic acid from Schizochytrium)	29 with replica	X1: Cassava (g L^{-1}) X2: Yeast extract (g L^{-1}) X3: Soybean meal (g L^{-1}) X4: Na_2SO_4 (g L^{-1})	Palmitic acid (PA) Yield %	Quadratic	X1: 122.2 gL^{-1} X2: 10.0 g L^{-1} X3: 7.0 g L^{-1} X4: 3.3 g L^{-1}	(W. Chen et al., 2015)
Biodiesel production (ultrasound power and reactor dimension)	17 with replica	X1: Height (mm) X2: Reactor diameter (mm) X3: Ultrasonic amplitude (%)	Fatty acid methyl esters (%FAME)	Quadratic	X1: 110 mm X2: 75 mm X3: 62%	(Mostafaei et al., 2016)
Bioethanol production (Indian bamboo Dendrocalamus)	27 with replica	X1: Biomass loading (wt.%) X2: Enzyme loading (FPU/g) X3: Surfactant (wt.%) X4: Incubation time (h)	Reducing sugar (g/g)	Quadratic	X1: 15.0 wt.% X2: 50.0 X3: 0.2 wt.% X4: 42	(Sindhu, Kuttiraja, Binod, Sukumaran, & Pandey, 2014)
Biofuel production (Cultivation of Arthrobacter AK19)	No reported	X1 KNO_3 (g) X2: NaCl (g) X3: $FeSO_4.7H_2O$ (mg) X4: $ZnSO_4.7H_2O$ + $CoSO_4.7H_2O$ (1:1) (g)	Biomass concentration (g L^{-1})	Quadratic	X1: 2.09 g X2: 1.18 g X3: 5.7 mg X4: 0.41 g	(Srinophakun, Thanapimmetha, Rattanaphanyapan, Sahaya, & Saisriyoot, 2017)
Bioethanol production (chili post-harvest residue)	31 with replica	X1: Biomass loading (wt.%) X2: Sonication time (min) X3: H_2SO_4 conc. (wt.%) X4: Incubation time (min)	Reducing sugar (g g^{-1})	Quadratic	X1: 20 wt.% X2: 4 wt.% X3: 4 wt.% X4: 60 min	(Sindhu, Binod, & Pandey, 2016)

model in RSD, trough the evaluation of linear, square, two-way interaction and the lack of fit (Akkarawatkhoosith, Kaewchada, & Jaree, 2019; Borah, Das, Das, Bhuyan, & Deka, 2019). The effectiveness on transesterification process has been the evaluation of FAEE, FAME and palmitic acid (PA) yields, sugar reduction and biomass concentration using a system of 15 and 31 experiments with three or four factors such as alcohol/oil molar ratio using, catalyst concentration, reaction temperature and reaction time. These studies had shown optimal values of methanol or ethanol/oil molar ratio of 7.5:1 to 23.1 and 0.1:1 using isopropanol/oil, with 1.2 to 5.0 wt% and 60 to 375°C for 7 to 240 min respectively (Akkarawatkhoosith et al., 2019; W. Chen et al., 2015; Mostafaei, Javadikia, & Naderloo, 2016; Rahimi, Aghel, Alitabar, Sepahvand, & Ghasempour, 2014).

Borah et al. has evaluated Zn doping on the transesterification of waste cooking oil for biodiesel using Zn substituted waste eggshell derived CaO nanocatalyst (Borah et al., 2019). Srinophakun et al. evaluated the mass of KNO_3, NaCl, and $FeSO_4$ in feedstock production to produce 3G biofuels through a cultivation system (Srinophakun et al., 2017). Sindhu et al, used as factors, the biomass loading, sonication time, H_2SO_4 concentration and incubation time (min) as a novel sonoassited acid pretreatment of chili post-harvest residue for bioethanol production (Sindhu et al., 2016). Chen et al. evaluated cassava, yeast, soybean meal and Na_2SO_4 in feedstock for biodiesel production (palmitic acid from Schizochytrium). They concluded that BBD overcome the limitations of conventional methods and successfully optimized the components of the fermentation (W. Chen et al., 2015). Mostafaei et al, studied the height, reactor diameter ultrasonic amplitude to model the effects of ultrasound power and dimension on the biodiesel production by comparison between BBD and adaptive neuro-fuzzy inference system (ANFIS) model (Mostafaei et al., 2016). This group confirmed that even with the BBD benefits, the ANFIS model was more robust and accurate in its predictions. In this sense, this latter methodology is expected to be used in future and more complex biofuel developments.

CONCLUSION

The application of DoE is an important tool for the recent investigations and development of biofuels. The reviewed works showed the potential use of these methodologies. Nonetheless, despite the wide variety of biofuel production processes, 1G bioethanol and biodiesel are the most studied by application of DoE. This is due to the facility to obtain the raw materials and the simplicity of the process. Furthermore, basically all the investigation groups have opted to use the same factors to optimize the systems which are temperature, alcohol/oil ratio, catalyst and reaction time. In contrasts, the use of DoE in 2G and 3G biofuels is not that frequent since these systems seem to be more complex. 2G biofuels depend in more factors to consider. For example, first is necessary to develop an efficient catalyst and then factors to develop the unit and the process. On the other hand, 3G biofuels require to control de metabolism and the growth rate conditions of the algae. It is for this reason that the studies focus in the cultivation conditions since they are easier to control than biological factors. The application of DoE in biofuel production, has demonstrated that RSD are more efficient than other Factorial designs since their predictions 99% accurate compared to 88-97%. This was expected since factorial designs are considered screening methods and surface response are used to estimate in a deeper way the optimum values. In this sense, Factorial designs are recommended to be use as a first step to identify the most important factor to be measure. After this, the application of an RSD is needed to adjust and find the optimal values. CCD and Taguchi are useful tools to contemplate depending the systems. In this sense, these latter methods have shown 90 to 97% of accuracy. However, BBD has exhibited more precision in their predictions. Even so, Factorial designs and RSD have been widely employed in the optimization process focused on analytical techniques. Finally, it is possible to affirm that DoE is capable to accelerate the optimization studies to the implementation of biofuels in a short time. The correct application enhances the integration of factors in such way that it is expected that more complex systems begin to use it. More investigation and investment are needed to commercialize biofuels globally.

REFERENCES

Abdullah, B., Syed Muhammad, S. A. F., Shokravi, Z., Ismail, S., Kassim, K. A., Mahmood, A. N., & Aziz, M. M. A. (2019). Fourth generation biofuel: A review on risks and mitigation strategies. *Renewable & Sustainable Energy Reviews*, *107*, 37–50. doi:10.1016/j.rser.2019.02.018

Adewale, P., Vithanage, L. N., & Christopher, L. (2017). Optimization of enzyme-catalyzed biodiesel production from crude tall oil using Taguchi method. *Energy Conversion and Management*, *154*, 81–91. doi:10.1016/j.enconman.2017.10.045

Ail, S. S., & Dasappa, S. (2016). Biomass to liquid transportation fuel via Fischer Tropsch synthesis – Technology review and current scenario. *Renewable & Sustainable Energy Reviews*, *58*, 267–286. doi:10.1016/j.rser.2015.12.143

Ajala, E. O., Aberuagba, F., Olaniyan, A. M., Ajala, M. A., & Sunmonu, M. O. (2017). Optimization of a two stage process for biodiesel production from shea butter using response surface methodology. *Egyptian Journal of Petroleum*, *26*(4), 943–955. doi:10.1016/j.ejpe.2016.11.005

Akkarawatkhoosith, N., Kaewchada, A., & Jaree, A. (2019). Simultaneous development of biodiesel synthesis and fuel quality via continuous supercritical process with reactive co-solvent. *Fuel*, *237*, 117–125. doi:10.1016/j.fuel.2018.09.077

Anderson, M. J., & Whitcomb, P. J. (2015). *DOE simplified: Practical tools for effective experimentation (3rd Ed.)*. Boca Raton, FL: CRC Press Taylor & Francis Group.

Antony, J. (2014a). 6 - Full factorial designs. In S. E. Antony (Ed.), *J. B. T.-D. of E. for E. and S* (pp. 63–85). Oxford, UK: Elsevier;

Antony, J. (2014b). *Design of experiments for engineers and scientists* (2nd ed.). London, UK: Elsevier.

Aslan, N. (2008). Application of response surface methodology and central composite rotatable design for modeling and optimization of a multi-gravity separator for chromite concentration. *Powder Technology*, *185*(1), 80–86. doi:10.1016/j.powtec.2007.10.002

Baş, D., & Boyacı, I. H. (2007). Modeling and optimization I: Usability of response surface methodology. *Journal of Food Engineering*, *78*(3), 836–845. doi:10.1016/j.jfoodeng.2005.11.024

Berrios, M., Gutiérrez, M. C., Martín, M. A., & Martín, A. (2009). Application of the factorial design of experiments to biodiesel production from lard. *Fuel Processing Technology*, *90*(12), 1447–1451. doi:10.1016/j.fuproc.2009.06.026

Borah, M. J., Das, A., Das, V., Bhuyan, N., & Deka, D. (2019). Transesterification of waste cooking oil for biodiesel production catalyzed by Zn substituted waste egg shell derived CaO nanocatalyst. *Fuel, 242*, 345–354. doi:10.1016/j.fuel.2019.01.060

Brennan, L., & Owende, P. (2010). Biofuels from microalgae—A review of technologies for production, processing, and extractions of biofuels and co-products. *Renewable & Sustainable Energy Reviews, 14*(2), 557–577. doi:10.1016/j.rser.2009.10.009

Brereton, R. G. (2007). *Applied chemometrics for scientists*. John Wiley & Sons. doi:10.1002/9780470057780

Buasri, A., Worawanitchaphong, P., Trongyong, S., & Loryuenyong, V. (2014). Utilization of scallop waste shell for biodiesel production from palm oil–optimization using Taguchi method. *APCBEE Procedia, 8*, 216–221. doi:10.1016/j.apcbee.2014.03.030

Chen, H., Liu, J., Chang, X., Chen, D., Xue, Y., & Liu, P., … Han, S. (2017). A review on the pretreatment of lignocellulose for high-value chemicals. *Fuel Processing Technology, 160*, 196–206.

Chen, W., Ma, L., Zhou, P., Zhu, Y., Wang, X., Luo, X., ... Yu, L. (2015). A novel feedstock for biodiesel production: The application of palmitic acid from Schizochytrium. *Energy, 86*, 128–138. doi:10.1016/j.energy.2015.03.110

Correa, D. F., Beyer, H. L., Possingham, H. P., Thomas-Hall, S. R., & Schenk, P. M. (2017). Biodiversity impacts of bioenergy production: Microalgae vs. first generation biofuels. *Renewable & Sustainable Energy Reviews, 74*, 1131–1146. doi:10.1016/j.rser.2017.02.068

Cox, D. R., & Reid, N. (2000). *The theory of the design of experiments* (1st ed.). New York: Chapman and Hall/CRC. doi:10.1201/9781420035834

Damartzis, T., & Zabaniotou, A. (2011). Thermochemical conversion of biomass to second generation biofuels through integrated process design—A review. *Renewable & Sustainable Energy Reviews, 15*(1), 366–378. doi:10.1016/j.rser.2010.08.003

Dhawane, S. H., Bora, A. P., Kumar, T., & Halder, G. (2017). Parametric optimization of biodiesel synthesis from rubber seed oil using iron doped carbon catalyst by Taguchi approach. *Renewable Energy, 105*, 616–624. doi:10.1016/j.renene.2016.12.096

Dhawane, S. H., Karmakar, B., Ghosh, S., & Halder, G. (2018). Parametric optimisation of biodiesel synthesis from waste cooking oil via Taguchi approach. *Journal of Environmental Chemical Engineering, 6*(4), 3971–3980. doi:10.1016/j.jece.2018.05.053

Dhawane, S. H., Kumar, T., & Halder, G. (2015). Central composite design approach towards optimization of flamboyant pods derived steam activated carbon for its use as heterogeneous catalyst in transesterification of Hevea brasiliensis oil. *Energy Conversion and Management, 100*, 277–287. doi:10.1016/j.enconman.2015.04.083

Dhawane, S. H., Kumar, T., & Halder, G. (2016). Biodiesel synthesis from Hevea brasiliensis oil employing carbon supported heterogeneous catalyst: Optimization by Taguchi method. *Renewable Energy, 89*, 506–514. doi:10.1016/j.renene.2015.12.027

Dutta, K., Daverey, A., & Lin, J.-G. (2014). Evolution retrospective for alternative fuels: First to fourth generation. *Renewable Energy*, *69*, 114–122. doi:10.1016/j.renene.2014.02.044

Endut, A., Abdullah, S. H. Y. S., Hanapi, N. H. M., Hamid, S. H. A., Lananan, F., & Kamarudin, M. K. A., … Khatoon, H. (2017). Optimization of biodiesel production by solid acid catalyst derived from coconut shell via response surface methodology. *International Biodeterioration & Biodegradation, 124*, 250–257.

Ezekannagha, C. B., Ude, C. N., & Onukwuli, O. D. (2017). Optimization of the methanolysis of lard oil in the production of biodiesel with response surface methodology. *Egyptian Journal of Petroleum*, *26*(4), 1001–1011. doi:10.1016/j.ejpe.2016.12.004

Furimsky, E. (2013). Hydroprocessing challenges in biofuels production. *Catalysis Today*, *217*(0), 13–56. doi:10.1016/j.cattod.2012.11.008

Ghosh, D., Dasgupta, D., Agrawal, D., Kaul, S., Adhikari, D. K., Kurmi, A. K., ... Negi, M. S. (2015). Fuels and chemicals from lignocellulosic biomass: an integrated biorefinery approach. *Energy & Fuels*, *29*(5), 3149–3157. doi:10.1021/acs.energyfuels.5b00144

Gupta, J., Agarwal, M., & Dalai, A. K. (2016). Optimization of biodiesel production from mixture of edible and nonedible vegetable oils. *Biocatalysis and Agricultural Biotechnology*, *8*, 112–120. doi:10.1016/j.bcab.2016.08.014

Hamze, H., Akia, M., & Yazdani, F. (2015). Optimization of biodiesel production from the waste cooking oil using response surface methodology. *Process Safety and Environmental Protection*, *94*, 1–10. doi:10.1016/j.psep.2014.12.005

Hanrahan, G., Montes, R., & Gomez, F. A. (2008). Chemometric experimental design based optimization techniques in capillary electrophoresis: A critical review of modern applications. *Analytical and Bioanalytical Chemistry*, *390*(1), 169–179. doi:10.100700216-007-1619-y PMID:17909758

Hibbert, D. B. (2012). Experimental design in chromatography: A tutorial review. *Journal of Chromatography. B, Analytical Technologies in the Biomedical and Life Sciences*, *910*, 2–13. doi:10.1016/j.jchromb.2012.01.020 PMID:22333438

Hu, Q., Sommerfeld, M., Jarvis, E., Ghirardi, M., Posewitz, M., Seibert, M., & Darzins, A. (2008). Microalgal triacylglycerols as feedstocks for biofuel production: Perspectives and advances. *The Plant Journal*, *54*(4), 621–639. doi:10.1111/j.1365-313X.2008.03492.x PMID:18476868

IEA. (2011). Biofuels can provide up to 27% of world transportation fuel by 2050, IEA report says - IEA 'roadmap' shows how biofuel production can be expanded in a sustainable way, and identifies needed technologies and policy actions. Retrieved from https://www.iea.org/newsroom/news/2011/april/biofuels-can-provide-up-to-27-of-world-transportation-fuel-by-2050-iea-report-.html

IEA. (2019a). Biofuels for transport: Tracking clean energy progress. Retrieved from https://www.iea.org/tcep/transport/biofuels/#

IEA. (2019b). Key world energy statistics. Retrieved from https://www.iea.org/statistics/kwes/consumption/

Islas, G., Rodriguez, J. A., Perez-Silva, I., Miranda, J. M., & Ibarra, I. S. (2018). Solid-phase extraction and large-volume sample stacking-capillary electrophoresis for determination of tetracycline residues in milk. *Journal of Analytical Methods in Chemistry*. PMID:29675286

Joshi, G., Pandey, J. K., Rana, S., & Rawat, D. S. (2017). Challenges and opportunities for the application of biofuel. *Renewable & Sustainable Energy Reviews*, *79*, 850–866. doi:10.1016/j.rser.2017.05.185

Karabas, H. (2013). Biodiesel production from crude acorn (Quercus frainetto L.) kernel oil: An optimisation process using the Taguchi method. *Renewable Energy*, *53*, 384–388. doi:10.1016/j.renene.2012.12.002

Karmakar, B., Dhawane, S. H., & Halder, G. (2018). Optimization of biodiesel production from castor oil by Taguchi design. *Journal of Environmental Chemical Engineering*, *6*(2), 2684–2695. doi:10.1016/j.jece.2018.04.019

Kharia, A. A., & Singhai, A. K. (2013). Screening of most effective variables for development of gastroretentive mucoadhesive nanoparticles by Taguchi design. *ISRN Nanomaterials*, 2013.

Kılıç, M., Uzun, B. B., Pütün, E., & Pütün, A. E. (2013). Optimization of biodiesel production from castor oil using factorial design. *Fuel Processing Technology*, *111*, 105–110. doi:10.1016/j.fuproc.2012.05.032

Kumar, R. S., Sureshkumar, K., & Velraj, R. (2015). Optimization of biodiesel production from Manilkara zapota (L.) seed oil using Taguchi method. *Fuel*, *140*, 90–96. doi:10.1016/j.fuel.2014.09.103

Lennartsson, P. R., Erlandsson, P., & Taherzadeh, M. J. (2014). Integration of the first and second generation bioethanol processes and the importance of by-products. *Bioresource Technology*, *165*, 3–8. doi:10.1016/j.biortech.2014.01.127 PMID:24582951

Maity, S. K. (2015). Opportunities, recent trends and challenges of integrated biorefinery: Part I. *Renewable & Sustainable Energy Reviews*, *43*, 1427–1445. doi:10.1016/j.rser.2014.11.092

Mandenius, C.-F. (2016, March 29). Challenges for bioreactor design and operation. *Bioreactors*.

Mathimani, T., & Pugazhendhi, A. (2019). Utilization of algae for biofuel, bio-products and bio-remediation. *Biocatalysis and Agricultural Biotechnology*, *17*, 326–330. doi:10.1016/j.bcab.2018.12.007

Meneghetti, S. M. P., Meneghetti, M. R., Serra, T. M., Barbosa, D. C., & Wolf, C. R. (2007). Biodiesel production from vegetable oil mixtures: Cottonseed, soybean, and castor oils. *Energy & Fuels*, *21*(6), 3746–3747. doi:10.1021/ef070039q

Mostafaei, M., Javadikia, H., & Naderloo, L. (2016). Modeling the effects of ultrasound power and reactor dimension on the biodiesel production yield: Comparison of prediction abilities between response surface methodology (RSM) and adaptive neuro-fuzzy inference system (ANFIS). *Energy*, *115*, 626–636. doi:10.1016/j.energy.2016.09.028

NIST/SEMATECH. (2013). *How do you select an experimental design?* Retrieved from https://www.itl.nist.gov/div898/handbook/pri/section3/pri33.htm#DesignSelection

Omar, W. N. N. W., & Amin, N. A. S. (2011). Optimization of heterogeneous biodiesel production from waste cooking palm oil via response surface methodology. *Biomass and Bioenergy*, *35*(3), 1329–1338. doi:10.1016/j.biombioe.2010.12.049

Onukwuli, D. O., Emembolu, L. N., Ude, C. N., Aliozo, S. O., & Menkiti, M. C. (2017). Optimization of biodiesel production from refined cotton seed oil and its characterization. *Egyptian Journal of Petroleum, 26*(1), 103–110. doi:10.1016/j.ejpe.2016.02.001

Palmeros Parada, M., Osseweijer, P., & Posada Duque, J. A. (2017). Sustainable biorefineries, an analysis of practices for incorporating sustainability in biorefinery design. *Industrial Crops and Products, 106*, 105–123. doi:10.1016/j.indcrop.2016.08.052

Pander, A., Hatta, A., & Furuta, H. (2016). Optimization of catalyst formation conditions for synthesis of carbon nanotubes using Taguchi method. *Applied Surface Science, 371*, 425–435. doi:10.1016/j.apsusc.2016.02.216

Pandey, M. P., & Kim, C. S. (2011). Lignin depolymerization and conversion: a review of thermochemical methods. *Chemical Engineering & Technology, 34*(1), 29–41. doi:10.1002/ceat.201000270

Paris Agreement, U. N. (2016). *The Paris Agreement.* Retrieved from https://unfccc.int/process-and-meetings/the-paris-agreement/the-paris-agreement

Rahimi, M., Aghel, B., Alitabar, M., Sepahvand, A., & Ghasempour, H. R. (2014). Optimization of biodiesel production from soybean oil in a microreactor. *Energy Conversion and Management, 79*, 599–605. doi:10.1016/j.enconman.2013.12.065

Ridge, E., & Kudenko, D. (2010). Tuning an algorithm using design of experiments. In *Experimental methods for the analysis of optimization algorithms* (pp. 265–286). Springer. doi:10.1007/978-3-642-02538-9_11

Ridzuan, N., Adam, F., & Yaacob, Z. (2016). Screening of factor influencing wax deposition using full factorial experimental design. *Petroleum Science and Technology, 34*(1), 84–90. doi:10.1080/10916466.2015.1122625

Robinson, T. J., Borror, C. M., & Myers, R. H. (2004). Robust parameter design: A review. *Quality and Reliability Engineering International, 20*(1), 81–101. doi:10.1002/qre.602

Rozet, E., Lebrun, P., Hubert, P., Debrus, B., & Boulanger, B. (2013). Design spaces for analytical methods. *TrAC Trends in Analytical Chemistry, 42*, 157–167. doi:10.1016/j.trac.2012.09.007

Sakkas, V. A., Islam, M. A., Stalikas, C., & Albanis, T. A. (2010). Photocatalytic degradation using design of experiments: A review and example of the Congo red degradation. *Journal of Hazardous Materials, 175*(1–3), 33–44. doi:10.1016/j.jhazmat.2009.10.050 PMID:19931983

Silva, G. F., Camargo, F. L., & Ferreira, A. L. O. (2011). Application of response surface methodology for optimization of biodiesel production by transesterification of soybean oil with ethanol. *Fuel Processing Technology, 92*(3), 407–413. doi:10.1016/j.fuproc.2010.10.002

Sindhu, R., Binod, P., & Pandey, A. (2016). A novel sono-assisted acid pretreatment of chili post harvest residue for bioethanol production. *Bioresource Technology, 213*, 58–63. doi:10.1016/j.biortech.2016.02.079 PMID:26949055

Sindhu, R., Kuttiraja, M., Binod, P., Sukumaran, R. K., & Pandey, A. (2014). Bioethanol production from dilute acid pretreated Indian bamboo variety (Dendrocalamus sp.) by separate hydrolysis and fermentation. *Industrial Crops and Products*, *52*, 169–176. doi:10.1016/j.indcrop.2013.10.021

Skorupskaite, V., Makareviciene, V., & Levisauskas, D. (2015). Optimization of mixotrophic cultivation of microalgae Chlorella sp. for biofuel production using response surface methodology. *Algal Research*, *7*, 45–50. doi:10.1016/j.algal.2014.12.001

Slade, R., & Bauen, A. (2013). Micro-algae cultivation for biofuels: Cost, energy balance, environmental impacts and future prospects. *Biomass and Bioenergy*, *53*, 29–38. doi:10.1016/j.biombioe.2012.12.019

Srinophakun, P., Thanapimmetha, A., Rattanaphanyapan, K., Sahaya, T., & Saisriyoot, M. (2017). Feedstock production for third generation biofuels through cultivation of Arthrobacter AK19 under stress conditions. *Journal of Cleaner Production*, *142*, 1259–1266. doi:10.1016/j.jclepro.2016.08.068

Sunde, K., Brekke, A., & Solberg, B. (2011). Environmental impacts and costs of woody Biomass-to-Liquid (BTL) production and use — A review. *Forest Policy and Economics*, *13*(8), 591–602. doi:10.1016/j.forpol.2011.05.008

Tan, Y. H., Abdullah, M. O., Nolasco-Hipolito, C., & Zauzi, N. S. A. (2017). Application of RSM and Taguchi methods for optimizing the transesterification of waste cooking oil catalyzed by solid ostrich and chicken-eggshell derived CaO. *Renewable Energy*, *114*, 437–447. doi:10.1016/j.renene.2017.07.024

Toms, D., Deardon, R., & Ungrin, M. (2017). Climbing the mountain: Experimental design for the efficient optimization of stem cell bioprocessing. *Journal of Biological Engineering*, *11*(1), 35. doi:10.118613036-017-0078-z PMID:29213303

Tsukahara, K., & Sawayama, S. (2005). Liquid fuel production using microalgae. *Journal of the Japan Petroleum Institute*, *48*(5), 251–259. doi:10.1627/jpi.48.251

van Hal, J. W., Huijgen, W. J. J., & López-Contreras, A. M. (2014). Opportunities and challenges for seaweed in the biobased economy. *Trends in Biotechnology*, *32*(5), 231–233. doi:10.1016/j.tibtech.2014.02.007 PMID:24767734

Veličković, A. V., Stamenković, O. S., Todorović, Z. B., & Veljković, V. B. (2013). Application of the full factorial design to optimization of base-catalyzed sunflower oil ethanolysis. *Fuel*, *104*, 433–442. doi:10.1016/j.fuel.2012.08.015

Verdugo, C., Luna, D., Posadillo, A., Sancho, E. D., Rodríguez, S., Bautista, F., ... Romero, A. A. (2011). Production of a new second generation biodiesel with a low-cost lipase derived from Thermomyces lanuginosus: Optimization by response surface methodology. *Catalysis Today*, *167*(1), 107–112. doi:10.1016/j.cattod.2010.12.028

Vicente, G., Coteron, A., Martinez, M., & Aracil, J. (1998). Application of the factorial design of experiments and response surface methodology to optimize biodiesel production. *Industrial Crops and Products*, *8*(1), 29–35. doi:10.1016/S0926-6690(97)10003-6

Vinodh, S., Bharathi, S. K., & Gopi, N. (2016). Parametric optimization of submerged arc welding using Taguchi Method BT - Design of experiments in production engineering. In J. P. Davim (Ed.), Design of experiments in production engineering (pp. 183–194). Cham, Switzerland: Springer; doi:10.1007/978-3-319-23838-8_8

Weissman, S. A., & Anderson, N. G. (2014). Design of experiments (DoE) and process optimization. A review of recent publications. *Organic Process Research & Development*, *19*(11), 1605–1633. doi:10.1021/op500169m

Yadav, M., Singh, A., Balan, V., Pareek, N., & Vivekanand, V. (2019). Biological treatment of lignocellulosic biomass by Chaetomium globosporum: Process derivation and improved biogas production. *International Journal of Biological Macromolecules*, *128*, 176–183. doi:10.1016/j.ijbiomac.2019.01.118 PMID:30684578

Yang, S.-T., & Yu, M. (2013, July 12). Integrated biorefinery for sustainable production of fuels, chemicals, and polymers. *Bioprocessing Technologies in Biorefinery for Sustainable Production of Fuels, Chemicals, and Polymers.*

ADDITIONAL READING

Mathews, P. G. (2005). *Design of Experiments with MINITAB*. ASQ Quality Press.

Kennard, R. W., & Stone, L. A. (1969). Computer aided design of experiments. *Technometrics*, *11*(1), 137–148. doi:10.1080/00401706.1969.10490666

Pukelsheim, F. (2006). *Optimal design of experiments*. Society for Industrial and Applied Mathematics. doi:10.1137/1.9780898719109

Sims, R. E., Mabee, W., Saddler, J. N., & Taylor, M. (2010). An overview of second-generation biofuel technologies. *Bioresource Technology*, *101*(6), 1570–1580. doi:10.1016/j.biortech.2009.11.046 PMID:19963372

Bullen, R. A., Arnot, T. C., Lakeman, J. B., & Walsh, F. C. (2006). Biofuel cells and their development. *Biosensors & Bioelectronics*, *21*(11), 2015–2045. doi:10.1016/j.bios.2006.01.030 PMID:16569499

Wojcik, E. Z., Singleton, C., Chapman, L. N., Parker, D. A., & Love, J. (2001). Plant Biomass as Biofuels. *eLS*, 1-11.

Lackner, M. (2014). 3 rd-Generation Biofuels: Bacteria and Algae as Sustainable Producers and Converters. Handbook of Climate Change Mitigation and Adaptation, 1-32.

Pittock, A. B. (2017). Climate change: turning up the heat. Routledge. Anderson, V. L., & McLean, R. A. (2018). Design of experiments: a realistic approach. Routledge. doi:10.4324/9781315065847

Myers, R. H., Montgomery, D. C., & Anderson-Cook, C. M. (2016). *Response surface methodology: process and product optimization using designed experiments*. John Wiley & Sons.

KEY TERMS AND DEFINITIONS

Biodiesel: Diesel obtained from triglycerides.

Bioethanol: Term used for ethanol obtained from fermentation of biological sugars to be used as a fuel.

Biofuel: Fuel obtained from biological sources.

Biorefinery: Factory which transforms the biomass into fuel.

Input Factors: Independent variables of the studied system; they are introduced as the first approach in the optimization process.

Orthogonality: Forms of comparison that ensures that all the parameters can be estimated independently.

Output Factors: Dependent variables or response of the studied system.

Regression Model: Statistical method to determine the correlations between the factors.

Chapter 5
The Role of Multivariant Analysis on the Interpretation of FTIR and Raman Spectra

Sefa Celik
Istanbul University, Turkey

Ali Tugrul Albayrak
Istanbul University-Cerrahpasa, Turkey

Sevim Akyuz
Istanbul Kultur University, Turkey

Aysen E. Ozel
Istanbul University, Turkey

ABSTRACT

FTIR and Raman spectroscopy are complementary spectroscopic techniques that play an important role in the analysis of molecular structure and the determination of characteristic vibrational bands. Vibrational spectroscopy has a wide range of applications including mainly in physics and biology. Its applications have gained tremendous speed in the field of biological macromolecules and biological systems, such as tissue, blood, and cells. However, the vibrational spectra obtained from the biological systems contain a large number of data and information that make the interpretation difficult. To facilitate the analysis, multivariant analysis comprising the reduction of the dimension of spectrum data and classification of them by eliminating redundancy data, which are obtained from the spectra and does not have any role, becomes critical. In this chapter, the applications of Principal Component Analysis (PCA), Linear Discriminant Analysis (LDA), and their combination PCA-LDA, which are widely used among multivariant techniques on biological systems will be disclosed.

DOI: 10.4018/978-1-7998-1518-1.ch005

INTRODUCTION

Infrared Spectroscopy

Infrared spectroscopy, which characterizes the chemical structure of the molecules and is based on molecular vibrations, is one of the most important techniques commonly used (Christy et al., 2001; Smith, 2011; Ciurczak and Igne, 2015; Alvarez-Ordonez and Prieto, 2012; Stuart, 2004). When a molecular system is irradiated by infrared (IR) radiation, that has all IR frequencies, the molecules absorb characteristic frequencies, that correspond to energies equal to the difference between their vibrational energy levels. There are selection rules for absorbing infrared radiation: The obvious selection rule for vibrational transitions is that the electric dipole moment of the molecule must change during the vibrational motion. Identification of the structure of molecules in the sample is performed by the determination of functional groups and chemical bonds, using IR spectroscopy. In addition, the surroundings of the functional group can also be examined. In IR transmission spectroscopy, the sample is placed on the path through which the beam passes, and the molecules in the sample are excited by incident beam at different wavenumbers in the range of 4000 to 400 cm^{-1}. The wavenumber at which the radiation is absorbed is measured and a spectrum in which x-axis is wavenumber and y-axis is percent transmittance (or percentage absorbance) is obtained.

One of the most commonly used sampling techniques for FTIR is ATR based on the presence of evanescent wave in the solid or liquid sample and ATR unit contains a crystal with high refractive index. In ATR technique, information on the surface structure of the material to be analyzed is obtained by the high-quality spectra in a short time without sample preparation for materials which have generally strong absorption (ie low transmittance in the IR range) or are too thick (Crompton, 2006; Theophanides, 2012; Larkin, 2011; Bart, 2006; Settle, 1997).

One of the sampling methods used for FTIR analysis of solid materials is the KBr disk method; the sample is ground into powder form and then ca. 1 mg sample is mixed well with powdered 100 mg KBr which is fully transparent in the IR range down to 400 cm^{-1}, and is finally pressed into a pellet under to 10 tones to prepare the sample for analysis. Because of high hygroscopic nature of KBr, the presence of water in KBr can mask peaks in the IR spectrum, therefore KBr should be dried in a vacuum oven before the pellet preparation (Griffiths and Haseth, 2007; Yadav, 2005; Nyquist, 2001; Kenkel, 2014, Hansen et al., 2012). Solid samples can also be analyzed by mull technique; For this process, a small amount of nujol (mineral oil) is dropped into the crushed sample in a agate mortar, and a paste is prepared. After then a thin film of nujol paste is applied onto the salt plates (mostly KBr plates are used), and the plates are mounted in a path of IR beam for recording the IR spectrum. (Anderson et al., 2004; Hayes, 2008). Liquid and solution samples can also be analyzed by FTIR by placing them between NaCl or KBr plates which are transparent to IR above 700 cm^{-1} or 400 cm^{-1}, respectively (Hollas, 2004; Kelsall et al., 2005; Ghatak, 2010; Pavia et al., 2018; Dean, 2003).

Raman Spectroscopy

In Raman spectroscopy, when the sample is illuminated in the UV-visible region, the incident photons are scattered from the sample molecules (Ferraro and Nakamoto, 1994; Lewis and Edwards, 2001; Tu, 1982; Vandenabeele, 2013; Smith and Dent, 2005; Gardiner and Graves, 1989). Depending on the elastic or inelastic collisions between the photons and the molecules, the scattered light is called Rayleigh or

Raman scattering, respectively. In Rayleigh scattering, the scattered light has the same frequency as the incident light, in other words, the incident light has the same energy as the scattered light. However, the energy of the incident light is different from that of the scattered light in Raman scattering. Raman scattering is also divided into two types: 1) Stokes Raman scattering in which the scattered light is scattered with a lower frequency (υ_0 - υ_m) than the frequency of the incident light (υ_0) and 2) anti-stokes Raman scattering in which the scattered light is scattered with a greater frequency (υ_0 + υ_m) than the frequency of the incident light (υ_0). The υ_m is a vibration frequency of the molecule. There is a significant difference in the level of theory between IR spectroscopy and Raman spectroscopy. In IR spectroscopy, resonance absorption takes place when the incident beam is equal to the difference between the ground vibrational energy of the molecule and the vibrational energy of the molecule in the excited state, whereas there is no such requirement in Raman spectroscopy. In Raman spectroscopy, Raman scattering occurs as a result of a molecular vibration causing a change in polarizability of the molecule, while IR spectroscopy occurs as a result of molecular vibrations that alter the electrical dipole moment of the molecule during vibration. Rayleigh scattering is the elastic scattering of photons and is not the desired data in the analysis process because they do not provide information about matter molecules, whereas Stokes Raman and anti-Stokes Raman scattering is non-elastic scattering of photons and gives information about material molecules.

Advantages and Disadvantages of IR Spectroscopy

One of the important advantages of IR spectroscopy is that the sample is analyzed nondestructively since the infrared light has low energy.

IR spectra of many substances such as solids, liquids, gases, organics, inorganic compounds, polymer composites, polymers can be measured. In the IR spectrum, peak position and peak intensity gives information about the structure and the concentration of a molecule, respectively. The larger the peak intensity, the higher the concentration of that molecule in the sample. In addition, if the molecules in the sample to be analyzed interact with the matrix, ie other components, the resulting spectrum will change.

Nowadays, time is of considerable importance especially in continuous processes in industry. It is desirable that the product quality be controlled quickly and thus there not be time loss that will affect production. By IR spectroscopy, analysis can be done in short times such as few minutes. However, with the use of the Attenuated Total Reflectance (ATR) apparatus that allows examination without the preparation of liquid and solid sample, the measurement time can be reduced to a few seconds and useful spectra can be obtained. This is a great advantage over the LC(Liquid Chromatography) and GC(Gas Chromatography) methods by which analyses take long. While the analysis time of a desired compound for gas chromatography and liquid chromatography varies depending on the analysis method chosen, this is not the case in FTIR. For example, the analysis of thiophene in benzene by GC-PFPD takes about 10 min. (OI Analytical, a xylem brand) and dibenzothiophene analysis in n-decane by HPLC-UV-Vis detector takes about 9 min (Bal and Bhasarkar, 2019), whereas the analysis time of organic compounds is in seconds by FTIR spectroscopy.

The IR spectrometer is relatively cheap compared to LC-MS, GC-MS, ICP-MS, GC-SCD and NMR. By FT-IR spectrometer, the spectra of the samples at the level of few miligrams can be measured.

Since the homonuclear diatomic molecules such as N_2 and O_2 have a symmetrical stretching, they do not have variable dipole moment during the vibration, therefore they do not yield IR peaks.

The more unique molecules a sample contains, the more complicated the spectra measured will be. In a word, it will be difficult to determine which peak belongs to which molecule. Some separation operations may be required to solve these problems. If the sample is a liquid mixture, the components in the sample may be separated by distillation, depending on the difference in their boiling points. If there are solids in the liquid sample, the sample can be isolated from the solids by a simple filtration process. If it is desired to analyze a desired component in a solid mixture containing different components, crystallization can be carried out by dissolving the mixture in a heated solvent, followed by cooling. Consequently, other components are separated from the mixture.

After these separation processes, the number of compounds in the sample will decrease and the measured spectra will be simplified.

By scanning the FT-IR library database or by searching similar spectra on the internet, possible molecules in the sample can be estimated. If it is known previously what some molecules in sample are, a simpler spectrum can be obtained by subtracting the pure spectra of the known compounds from the spectra of the mixture using a computer software.

In IR spectroscopy, if the sample contains water, large water bands arise a major problem because they will shield the bands of other molecules. In this case, the sample must be dried in a vacuum oven prior to analysis and stored in a desiccator in order not to be affected by the humidity of the air (Christy et al., 2001; Smith, 2011; Alvarez-Ordóñez and Prieto, 2012), If these are not sufficient, the subtraction process is carried out and the water bands are removed with the help of a computer program.

Advantages and Disadvantages of Raman Spectroscopy

Since IR spectroscopy depends on direct absorption of photons by molecules, but Raman spectroscopy depends on inelastic scattering of photons by molecules, the IR and Raman activity conditions, i.e. observation of the vibrational frequencies of the molecules by IR and Raman spectroscopy, are different. For this reason, some vibrational modes of the molecules can be only IR or Raman active whereas some modes can be both IR and Raman active, depending on the symmetry.

Water molecules are weak Raman scatterers, and therefore aqueous solutions does not arise a problem in Raman spectroscopy, unlike IR spectroscopy. Small sample quantities, as in IR spectroscopy, are sufficient for analysis. In the case of Raman spectroscopy, since the sample is irradiated by an intensive laser beam, it can be degraded. Another disadvantage of Raman spectroscopy appears in fluorescence samples. In such samples, huge fluorescence background shields the Raman bands. However, in biological samples Raman spectroscopy may be more advantageous than IR spectroscopy as the bands are sharper. (Ferraro and Nakamoto, 1994; Tu, 1982). Raman spectrometers are much more expensive than a FT-IR spectrometer.

Usage Areas of IR Spectroscopy

Infrared spectroscopy has been commonly used in the determination of organic compounds (aliphatic hydrocarbons and aromatic compounds), oxygen-containing compounds (alcohols, phenols, ethers, aldehydes, ketones, esters, carboxylic acids and anhydrides), nitrogen-containing compounds (amines and amides), halogenated compounds, heterocyclic compounds, boron compounds, silicon compounds, phosphorus compounds and sulfur compounds in the wavenumber range of 4000 to 400 cm^{-1}. It is also used in the identification of common inorganic ions (CO_3^{2-}, SO_4^{2-}, NO_3^-, PO_4^{3-}, SiO_4^{2-}, NH_4^+, MnO_4^-), diatomic, triatomic, four-atom, five-atom and octahedral inorganic molecules, coordination compounds, ligands forming linkage isomers, metal carbonyls, organometallic compounds and minerals. It is also used to identify lipids, proteins and peptides, nucleic acids, microbial cells (Barth 2007, Goormaghtigh et al. 2009, Gasper et al. 2009), and the structures of plants (Kavkler et al. 2011, Akyuz et al. 2013, Abidi et al. 2014, Akyuz et al. 2018, Zeng et al 2011) Thus it is used in pharmaceutical applications, in food science, agricultural applications, pulp and paper industries, paint industry and environmental applications (Stuart, 2004).

Infrared spectroscopy is also used in cancer research (Moss, 2011, Bel'skaya 2019). In a study examining cancerous tissues in the esophagus, spectra of normal and cancerous tissues were measured, and in the range of 1800-1400 cm^{-1}, bands in normal tissues were shown to be more intense and sharper than bands in cancerous tissues (Wang et al., 2003).

It has been reported that by FTIR spectroscopy using fiber optic probes compatible with endoscopies, it is possible to examine the structures of pre-malignant and malignant tissues, in live (Mackanos et al., 2010).

In a study comparing 894 FT-IR spectra in cancer diagnosis analysis, it was reported that cancer tissues and cells can be successfully characterized by using Support Vector Algorithms and the method must also be applied to larger data set (Lim and Cohenford, 2014).

In a review, it has been mentioned that the comparison and the distinction by identifying biomarkers in white blood cells and blood cancers can be accomplished by FTIR spectroscopy (Mostaço-Guidolin and Bachmann, 2011).

It has been suggested that FTIR spectroscopy can be very useful for the rapid and safe identification of cancerous tissues (colon, breast and uterine cancer) for the diagnosis of cancer at very early stages and consequently a more effective treatment (Khanmohammadi and Garmarudi, 2011).

In another review, it was emphasized that using Mid-IR biospectroscopy with chemometric based on mathematical modeling can be effective in identifying different types of cancers and in solving a large number of complex data groups (Siqueira and Lima, 2016). It has been shown that prostate cancer and normal prostate tissues can be observed and detected quickly in high accuracy by FTIR spectroscopy (Siqueira and Lima, 2016).

In a review emphasizing the importance of IR spectroscopy for fast and reliable results in early diagnosis to clarify the molecular structure of healthy and cancerous tissue samples, it is stated that it is necessary to examine the spectra carefully by using statistical analyses not to misinterpret (Kumar et al., 2018).

It has been reported in clinical laboratories that the results obtained by examining body fluids, cells and tissues for cancer diagnosis by traditional and expensive medical laboratory devices can be obtained in a short time using low cost mid-infrared spectroscopy (De Bruyne et al., 2018).

By FT-IR spectroscopy, cancerous and healthy tissues can be distinguished with high accuracy by comparing their measured spectra and also changes in cancerous tissues can be monitored over time (Kumari et al., 2018).

Infrared spectroscopy also has applications in the field of microbiology. It has been shown that the identification of the structures of bacterial strains and changes in their structures according to environmental conditions are analyzed by FTIR spectroscopy (Ojeda and Dittrich, 2012).

It is stated that it is appropriate to examine the bacteria in pure water and in biofilms formed on surfaces by FTIR technique (Schmitt and Flemming, 1998).

It has been reported it may be possible to discriminate each type of microorganism quickly by classifying by Raman spectroscopy as complementary to FTIR spetroscopy, thus obtaining more precise information about the structure of microorganisms (Naumann, 2000).

It has been stated that FTIR techniques could be faster than molecular biology techniques to characterize new bacterial types (Quintelas et al., 2018).

It is emphasized that bacterial discrimination can be performed more easily by using FTIR spectroscopy which has low operating cost and uses databases (Wenning and Scherer, 2013).

It has been explained that changes in the species antibiotic-resistant Staphylococcus aureus can be defined quickly and reproducibly by FTIR spectroscopy (Becker et al., 2006).

It was stated that microbial degradation which are of importance in terms of product quality control in agricultural and food products can be determined by FTIR spectroscopy (Kandpal and Cho, 2014).

It has been explained that biofilm bacteria can be identified by classifying by micro-FTIR. It has been revealed that the similarities and differences between the bacterial species and the database comprising 139 bacterial reference spectra can be determined (Helm et al., 1991).

It has been indicated that both Raman and IR imaging could be used to obtain information on the characterization and chemical structures of biological tissue samples (Harrison and Berry, 2017).

It is reported that chemical changes in medium induced by Lactobacillus plantarum and Lactobasillus brevis causing fermentation and by yarrowia lipolytica fungus acting as enzyme can be observed by FTIR spectroscopy (Vannini et al., 1996).

It has been demonstrated that the formation of pathogenic bacteria in fluids can be determined rapidly using multivariate techniques with FTIR spectroscopy (Rodriguez-Saona et al., 2001). Furthermore, in food microbiology, harmful bacteria in food can be classified according to their types and quantitative measurements of bacteria can be realized (Davis and Mauer, 2010).

Using FT-IR hyperspectral imaging with neural network-based image segmentation analysis, pathogenic bacteria can be classified and distinguished with high precision (Lasch et al., 2018). Esherichia Coli strains can be distinguished using low-cost and easy-to-use FTIR spectrometer with microscope and multivariate statistics (Carlos et al., 2011).

In a study in which a total of 1570 FTIR spectra were obtained from 164 gram positive and gram negative bacteria isolated from patients for bacterial identification, a classification of 100% for gram positive bacteria and specificity of 80% for gram negative bacteria were obtained using the multivariate technique (Sandt et al., 2006).

In a study in which dynamic changes in bacterial cell walls were examined, gram positive bacteria were discriminated from gram negative bacteria as a result of the measured AFM-IR spectra (Kochan et al., 2018).

In a study on the identification of propionibacteria strains, it was shown that 93% of the bacteria could be correctly identified using artificial neural networks developed from FTIR spectra (Dziuba, 2013).

Usage Areas of Raman Spectroscopy

It is used in the analysis of polymer films, synthetic and natural polymer fibers and composites, emulsion polymerization systems, liquid crystals, foams and food (Nguyen, 2015; Young et al., 2009; Wang et al., 1992; Amer, 2009). In addition to medical applications (bone imaging and cancer studies) (Amer, 2009), it is also used in pharmaceutical applications as in IR (Ekins, 2008).

There are many studies related to cancer diagnosis by Raman spectroscopy.

Using discrimination algorithms and spectral maps, it has been shown that skin, breast, gastrointestinal tract and uterine cancers have the potential to be quickly and effectively diagnosed by Raman spectroscopy (Keller et al., 2006). It has been revealed that statistically successful results were obtained in the diagnosis of breast cancer by Raman spectroscopy (Abramczyk et al., 2008).

In addition to early detection of cancer, it has been reported that the status of cancerous tissues can be monitored by Raman spectroscopy after radiotherapy (Devpura et al., 2014). It has been reported that surface enriched Raman spectroscopy can be adapted to clinical applications in order to diagnose in vitro and in vivo using nanoparticles to increase the sensitivity and specificity of cancer diagnosis (Ravanshad et al., 2018).

It is stated that mathematical models of detailed data groups in the near future for non-invasive cancer diagnosis (breast cancer, skin cancer, lung cancer, esophageal cancer, head and neck cancer, brain cancer, colon cancer, uterine cancer, vulval cancer, bladder cancer, prostate cancer) and Raman spectroscopy can be used as a complement to clinical applications (Santos et al., 2017).

Since there are many different organs in the human body, modern instrumental devices such as Raman spectroscopy are needed to identify each type of cancer and to observe changes in the molecular structure of tissues. It has the potential to obtain reliable results in a short time by differentiating the types of cancers and by modeling data by use of various algorithms (Auner et al., 2018). It has been reported that promising results can be obtained with high sensitivity and specificity using SRS (stimulated Raman scattering) microscopy for cancer diagnosis (Cui et al., 2018).

Because Raman spectrometers are expensive, it is stated that low cost Raman devices are needed to be widely used in clinical applications, and it is also necessary to preprocess the data obtained from Raman spectra and to perform data analysis (Ramírez-Elías and González, 2018). The comparison of the SERS spectra of urine specimens from breast cancer patients and healthy people resulted in the differentiation of the spectra of cancer patients using multivariate analysis techniques with a sensitivity of 81% and a specificity of 95% (Moisoiu et al., 2019).

Raman spectroscopy is also used in the field of microbiology. It has been shown that qualitative and quantitative information about microbes can be obtained rapidly by tip-enhanced Raman spectroscopy and that this type of spectroscopy can be used in clinical applications in the future (Ashton et al., 2011).

Micro Raman spectroscopy technique was used since classical methods are very time consuming to determine the Brucella bacterium in the milk with a mixed matrix containing protein, carbohydrate and fat and may lead to misidentifications which do not conform to the bacterial classification. Brucella bacteria in milk using chemometrics can be defined quickly with a specificity of 94% (Meisel et al., 2012).

In a review on the identification of bacteria, it was revealed that the study of bacteria by Raman spectroscopy did not require the cultivation of bacteria (Lorenz et al., 2017).

Since bacteria in food can lead to food poisoning, highly accurate and fast identification of bacteria has been possible by using surface enhanced Raman spectroscopy (Chauvet et al., 2017).

Due to the increasing food poisoning caused by pathogenic bacteria and the threats of bioterrorism on the earth, it has become important to determine the microorganisms on-site quickly. Some bacteria are also used to reduce environmental pollution. Therefore, it is emphasized that Raman spectroscopy can be very effective for accurate analysis in very low amounts integratedly with other techniques such as robust MNPs and RACE (Chisanga et al., 2018).

In a study in which six different microbial species were analyzed with SERS at the single cell level, it was stated that SERS can be used although the distinctive properties of some microorganisms could not be identified and that even the metabolite activities of the cells complicated the spectra (Weiss et al., 2019).

In a study in which 88 peritoneal samples were taken from 45 patients, It has been reported that the fungi Candida spp. which lead to peritonitis is identified with a specifity of 90% in 12 to 24 hours, whereas the identification takes 3 to 4 days by conventional method (Ibelings et al., 2005). A series of Escherichia Coli bacteria were identified by using SERS, DFA and HCA and it was stated that this method has the potential to be used for all organism fingerprints (Jarvis and Goodacre, 2004).

It is reported the database size should be sufficient depending on the technique used to evaluate in detail the statistical results and close to the truth and to obtain high-precision definitions for the identification of bacteria by Raman spectroscopy (Stöckel et al., 2016). Complex microorganisms could be determined easily and rapidly in varying environmental conditions using PCA and PLS-DA by Raman spectroscopy (Mobili et al., 2010).

PLS-DA is a dimension reduction technique used to classify high dimensional data. PCA-LDA is a multivariate data analysis method used for dimension reduction and feature extraction (Yang et al., 2017; Fordellone et al., 2018; Lee et al., 2018).

Although it is stated that bacteria in complex samples to be identified by Raman spectroscopy should be isolated, one of the biggest advantages of Raman spectroscopy is that no cultivation is needed for days, analyzes can be performed within 2-3 hours, single cells can be examined, sensitivity and high specificity which is obtained when the spectroscopy is used in combination with chemometrics (Pahlow et al., 2015).

The interactions between drug and pathogenic bacteria, and hence the efficacy of the antibiotics used can be evaluated in a short time (<3.5 h) by Raman spectroscopy (Neugebauer et al., 2015).

It has been reported that bacteria can be determined and identified with a higher precision than SERS as a result of obtaining improved spectra using EC-SERS (Lynk et al., 2018).

It has been demonstrated that very low amounts of pathogenic bacteria in blood are determined by using micro-Raman spectrometer (De Plano et al., 2019).

By using Raman spectroscopy in combination with deep learning, it was shown that a bacterial identification of about 99% specificity was performed taking advantage of only 10 bacterial spectra from 25 isolates (Ho et al., 2019).

As a result of changes in the intensity of the vibration bands and the emergence of new bands, the effect of UV irradiation time on bacteria was assessed by determining the dead and live bacteria by Raman spectroscopy (Li et al., 2019).

In a study examining pathogenic bacteria in human serum, lysis filtration did not affect the Raman spectra of hydrophilic bacteria, while lysis filtration greatly hindered the identification of hydrophobic bacteria by Raman spectra (Kotanen et al., 2016).

A Brief History of Multivariant Analysis

Multivariate analysis is a statistical tool used to determine the single contribution of various factors. The principal component analysis (PCA) of the multivariate observation is based on an article published by Pearson in 1901; study related to factor analysis study by Spearman in 1904, about multivariate analysis of variance in 1932 by Wilks, the study of canonical correlation by Hotelling in 1933 were published. In 1936, Hotelling found a new version of the power method to find Principle Components (PCs). In 1939, Girshick investigated the coefficients of PCs and the asymptotic sampling distribution of its variances. (Macnamara, 1969; Hotelling, 1933; Pearson, 1901; Spearman, 1904, Wilks, 1932; Hotelling, 1936; Girshick, 1939).

The data set analyzed in the principal component analysis should be multivariate. If this data set has a very large size, by virtue of this analysis, the number of dimensions in the data set without loss of information is reduced. The main components commonly used in areas such as Biology, Medicine, Chemistry, Physics and Economics are a multivariate statistical method especially used to eliminate the dependence structure between dimension reduction and variables. PCA is also used in data compression.

In statistics, Linear Discriminant Analysis (LDA), also known as Canonical Variate Analysis (CVA), is used to determine the differences between objects. This method uses the linear combination of properties to characterize the classes. This method, developed by Fisher in 1936, is simple, but the models it produces are as good as those produced by complex algorithms (Fisher, 1936). The LDA method has two important purposes, such as finding a rule that allows objects to be divided into predefined classes, and creating a model that can help the user find traces and sequence in the data (Elston et al., 2002; Smilauer et al., 2014).

Multivariate Analysis in IR and Raman Spectroscopy

Nowadays, there is a need for computational and appropriate statistical programs that can use the data generated from the IR or Raman spectra and give information about the structure of the samples. One is a multivariate analysis, which provides detailed information on the physical and chemical properties of the samples. It also provides the identification and quantification of a substance in a sample. Using multivariate analysis, what the additive in an olive oil sample is and its amount can be easily determined from the IR spectra (Ozaki et al., 2006). Multivariate analysis (MVA) is used in medical fields in IR spectroscopy, whereas MVA is generally used in biological and biomedical applications in Raman spectroscopy. In addition, MVA can be used to characterize broad overlapping peaks that are difficult to distinguish, which can provide information about the structure of the sample in the IR region (Gemperline, 2006).

While univariate analyzes were used 30-40 years ago, high reproducibility results have been obtained in short periods of time by using computer and statistical programs along with developing technology nowadays. For example, aspirin, acetaminophen, caffeine in solution and also protein conformation can be determined using MVA (Gremlich and Yan, 2000).

By using MVA from the IR spectra obtained in a short time, the concentrations of the samples can be determined, batch processes can be observed, the samples can be classified by identifying them and product quality can be checked online (O'Donnell et al., 2014).

MVA from the NIR data is also used in the agricultural area (determination of nitrogen content in cereals) (Varmuza and Filzmoser, 2009).

In a review, it has been reported that water concentrations and sugar concentration, which obstructs the quantitative determination of small amounts of other compounds in the fruit, can be reliably determined in a short time by the use of NIR reflectance spectroscopy and MVA combination (Cozzolino et al., 2011).

It has been shown that co-use of non-invasive Raman spectroscopy and multivariate analysis to identify lichen sclerosus reduces the number of conventional invasive tissue diagnostics and replaces tissue biopsies (Frost et al., 2017).

The concentration of dipyrone used as a drug using Raman spectroscopy and MVA for the purpose of quality control was found to be close to the concentration of the label indicated in the drug formulations (a difference of less than 5%) (Guimarães et al., 2018).

In a study on cancer diagnosis, Raman spectrum of pure paraffin which is used to prevent the deterioration of tissues was first measured for the characterization of paraffin-embedded cancerous tissues, and it has been demonstrated the structure of the cancerous tissue can be identified clearly by subtracting the pure paraffin spectrum from the paraffinic-containing tissue spectra measured using the ICA, PLS and IC-PLS methods (Meksiarun et al., 2017).

Compounds in drug using Raman imaging and SMCR analysis can be qualitatively estimated, including concentration profiles (Shinzawa et al., 2009).

It has been reported that cancer development in tissue samples from the liver can be defined at the molecular level using Raman spectroscopy and PCA-LDA method, and that this technique can be used with a specificity of 100% and a sensitivity of 100% instead of other conventional methods for early diagnosis of liver cancer (Zhang et al., 2018).

It has been shown that seven different bacterial species, which lead to food spoilage, can be identified with an error margin of 3.5% using PCA and CDA in combination with Raman spectroscopy (Klein et al., 2019).

In 2018, in a study conducted by Kaznowska and his colleagues to determine the classification of lung cancer and the degree of malignancy, FTIR spectra of the samples were taken and PCA-LDA analysis was performed. As a result of this analysis, the authors found that the sensitivity and the specificity of the model are between 78 and 99% and between 65 and 99%, respectively. According to the results obtained from the authors, the models could be helpful not only in the classification of tumor types but also in the evaluation of malignancy. (Kaznowska et al., 2018).

In the study conducted by Liu et al. in 2016, Raman spectroscopy and PCA-LDA and PLS-DA models were used for Colorectal Cancer Diagnosis. In this study, it is reported the accuracy of PLS-DA modeling (84.3%) was significantly higher than that of PCA-LDA (79.2%). Therefore, it has been found that the PLS-DA model is preferred to distinguish cancer from normal tissues. (Liu et al., 2016).

In 2014, Li et al. combined Raman spectroscopy with multivariate statistical analysis for rapid detection and differentiation of nasopharyngeal cancer from normal nasopharynx tissue. The posterior probability of the nasopharyngeal cancer groups was calculated from the LDA model, and the diagnostic sensitivity and specificity were found to be 81 and 87%, respectively, for the detection of nasopharyngeal cancer. With PLS-DA multivariate statistical technique, these rates increased to 85% and 88%, respectively. (Li et al., 2014).

In the study aimed at the characterization and classification of the serum SERS spectra between bladder cancer patients and normal volunteers using genetic algorithms combined with linear discriminate analysis, an improved diagnostic sensitivity of 90.9% and improved diagnostic specificity of 100% were obtained. . Six diagnostic Raman bands related to proteins, nucleic acids and lipids are picked out with the genetic algorithms combined with linear discriminate analysis (Li et al., 2015).

In the study conducted by Li et al. in 2012, saliva SERS spectra of 21 lung cancer patients and 20 normal individuals were measured. Most of Raman peak intensities decreased in lung cancer patients compared to normal people. Principal component analysis (PCA) and linear discriminant analysis (LDA) were used to reduce and differentiate the two data groups, and it resulted in an accuracy of 80%, a sensitivity of 78% and a specificity of 83% (Li et al., 2012).

In 2017, Elmi et al. analyzed the characteristics of serums of breast cancer patients by combining PCA-LDA analysis with FTIR spectroscopy. The diagnostic accuracy, sensitivity and specificity of PCA-LDA analysis in 3000-3600 cm^{-1} (N-H stretching) were found to be 83, 84, 74% for control and 80, 76, 72% for breast cancer patients, respectively (Elmi et al., 2017).

Santos et al. in 2017 also used ATR-FTIR technique in combination with multivariate analysis, to identify and differentiate the infections of dengue serotype 3 (DENV-3) virüs in diluted serum and blood samples. It was found to be a rapid and non-destructive approach to quantitative analysis of dengue virus. For classification of the ATR-FTIR spectra, techniques such as principal component analysis-linear separation analysis (PCA-LDA), sequential projection algorithm-linear separation analysis (linear discriminant analysis SPA-LDA) and genetic algorithm - linear separation analysis and genetic algorithm-linear discriminant (GA-LDA) were applied. They obtained better results with 100% sensitivity and specificity values in blood samples for PCN-LDA, SPA-LDA and GA-LDA classification models for comparison of DENV-3 samples in serum and blood (Santos et al., 2017).

In 2018, Depciuch and his colleagues compared normal and neoplastic (FTC) thyroid tissues via Fourier Transform Infrared (FTIR) spectroscopy. As a result of FTIR spectra and Basic Component Analysis (PCA), they concluded that there were differences in the FTIR spectrum of these tissues. In addition, their results showed a decrease in the number of functional groups that form cellular and tissue structures in tumor tissues (Depciuch et al., 2018).

In the study performed by Fujioka and his colleagues in 2004, malignant and normal human stomach tissue were examined by Fourier transform infrared (FTIR) spectroscopy. Significant increases were observed in the 10 bands of malignant tissues compared to normal tissue. To distinguish malignant tissues from normal stomach tissues, malignant gastric tissue and normal tissue spectra were subjected to discriminant analysis. The 10IR absorption bands for which the P values were less than 0.05 were used for discriminant analysis using the linear combination (Fujioka et al., 2004).

In 2017, Ketty and his colleague tried to estimate the geographical origin of raw milk using the ATR-FTIR together with the multivariate PCA method. They found that by forming the PCA model, the possible source of the raw milk sample could be determined. The combination of these two methods is non-destructive and the analysis time is fast (Ketty et al., 2017).

In a thesis made by Tas in 2008, the FTIR spectra of pure olive oil samples obtained from oil production plants in Aydın, cotton oil samples from other similar production sites, pure corn oil samples and sunflower oil samples obtained from markets were examined. Mixtures of olive oil and other vegetable oils were prepared in specific proportions and classified by principal component analysis which is chemometric method using the SİMCA-P statistical program. It has been shown that the infrared spectrum and PCA can easily separate the pure oils from each other (Tas, 2008).

In 2018, Nesakumar et al. tried to determine the moisture content and the age of beet by using principal component analysis (PCA) combined with infrared spectroscopy (FTIR). The validation test results confirmed that the accuracy of the beet quality estimation model reached 97.5%. This study shows that FTIR spectroscopy combined with basic component analysis and beet quality estimation models can be an effective tool to distinguish the moisture content in fresh, semi and completely unspoiled stages of beet samples (Nesakumar et al., 2018).

In a study carried out by Gajjar and his colleagues in 2013, infrared (IR) and/or Raman spectroscopy combined with multivariate analysis were used to differentiate between normal brain tissue and different tumor types (meningioma, glioma, and brain metastasis). They observed a clear distinction between normal and different tumor subtypes by ATR-FTIR spectroscopy. They also observed the differentiation of tumor classes by Raman spectroscopy (Gajjar et al., 2013).

In a study performed by Li et al. in 2014, it is revealed that the diagnostic sensitivities for differentiation between normal, nodular and malignant thyroid tissue samples by the SERS spectrum combined with PCA and LDA were 92, 75, 87.5%, and diagnostic specificities 82.6, 89.4, 84.4%, respectively (Li et al., 2014).

In a study fulfilled by Sil et al. in 2017 to classify four E. coli species, Raman microscopy in combination with the chemometric method was used. In this study, hierarchical cluster analysis, principal component analysis and linear discrimination analysis were performed to acquire the classification of Raman spectra obtained from four species of E. coli. Linear discrimination analysis using PC scores (PC-LDA) was found to be very good with 100% accuracy (Sil et al., 2017).

It is emphasized that glycerol and methyl alcohol in a yeast culture can be observed online and offline using the ATR probe and FTIR spectrometer, hence enabling rapid detection of additives that can be harmful to humans used in the food industry compared to conventional methods (You, 2017). In a review, it was stated that the quality of olive oil can be determined by optimizing the process variables in olive oil production by artificial neural networks (Gonzalez-Fernandez et al., 2019). It has been reported that the quality of liquid foods can be quickly determined using chemometrics by IR spectroscopy (Su and Sun, 2019). Fatty acids present in Copaiba oils, Andiroba and Brazilian nuts in the Brazilian region have been shown to be determined rapidly using the Raman and IR spectroscopy (Martins et al., 2019). In this study the results were discussed with DFT calculations complementing those obtained by FTIR and Raman. It is stated that Raman, NIR and FTIR spectroscopy can be widely used in the near future in determining the quality of foods quickly with high accuracy (Fakayode et al., 2019). In another review, it is mentioned that the concentration of chemicals in fruits and vegetables could be determined faster at lower cost by FTIR spectroscopy in combination with chemometrics, than conventional methods (Bureau et al., 2019). It has been shown that the components in milk, which is consumed very frequently in daily life and is a basic liquid food for human beings, can be determined by Raman spectroscopy without the challenge of sample preparation (He et al., 2019). It has been reported that by the use of Raman spectroscopy and the PLS model, determination of the main components in extra virgin olive oil and cheap adulterants that lower the purity of the olive oil may be a solution against food fraud (Duraipandian et al., 2019). The spectra of 150 freshwater fish were obtained by NIR reflectance spectroscopy in the wavelength range of 1000-1799 nm and their freshness was determined by using the CARS algorithm combined with the statistical method PLSR (Zhou et al., 2019). It has been reported that Raman and FTIR spectroscopy can be used in the investigation of petroleum inclusions (Volk and George, 2019). It is predicted that information about the structure of microalgae lipids can be obtained from FTIR and NIR spectra (Challagulla et al., 2017). It has been reported that microplastics causing environmental pollution in the world in the recent times can be identified by Raman and FTIR spectroscopy (Hanvey et al., 2017). It has been reported that chemicals in liquid and gaseous fuels can be quantitatively and qualitatively determined by FTIR, NIR and Raman spectroscopy using Chemometrics, thus making the characterization of these fuels possible (Kiefer, 2015). It is stated that FTIR and Raman spectroscopy can be used in chemical analysis of artifacts (Perez-Alonso et al., 2006).

In a review, the potential for future use of handheld FTIR and Raman spectrometers for quality control in the laboratory and on-line inspection (eg characterization of plastic sheet samples, determination of drug formulations, determination of concentration of components in alcohol mixtures, determination of organic substances harmful to health in the soil) were discussed (Sorak et al., 2012). Point-of-care diagnostic by vibrational spectroscopy using artificial neural networks and chemometrics has been reported to be much faster and cheaper than conventional diagnostic methods (Pahlow et al., 2018). It has been shown that iodine values of edible oils and thus their oil qualities can be determined in situ using MATLAB software for combined data analysis with portable IR and Raman spectrometer (Yan et al., 2018). The potential of using portable and hand-held NIR and Raman spectrometers for rapid and qualitative and quantitative determination of biopharmaceutical products that have a growing market in the pharmaceutical sector has been mentioned (Deidda et al., 2019). Since increasing spectral data as a result of the introduction of handheld and portable spectrometers can make it difficult for chemometric approaches to use, it has been shown that, along with these approaches, deep learning can be very useful for feature selection and robustness of models from many spectral data (Yang et al., 2019). Artificial neural network assisted FTIR spectroscopy has been successfully applied in the identification and classification of bacillus cereus group, which is difficult to identify by conventional diagnosis and causes food poisoning (Bağcıoğlu et al., 2019). Using the FTIR spectra of 64 hydrocarbon fuels drawn in the IR range of 3300-3550 nm in the vapor phase, the physicochemical properties of those fuels were determined with high precision (Wang et al., 2019). Palmitic acid quantities in palm oil samples were predicted with high precision in a short time using low-cost portable NIR spectrometer and multivariate statistical analysis(Kaufmann et al., 2019). It has been mentioned that small and portable NIR spectroscopy can be used to quickly determine whether meat and meat products are spoiled for quality control purposes (Kademi et al., 2019).

It has been mentioned that DOE can be applied to optimize system variables to obtain high quality spectra prior to the application of chemometrics (Santos et al., 2019). In pharmaceutical applications, calibration models have been established with the data obtained by NIR spectrometer and the effect of DOE on the performance of the relevant models has been tested (Bondi et al., 2012). The composition and density of biodiesel and biodiesel blends were determined using NIR spectroscopy and some chemometric methods and which chemometric methods gave more accurate predictions was evaluated by DOE (Oliveira et al., 2012).

With the aim of preventing meat fraud, DOE was used for classification of a series of different meats such as beef, chicken, horse meat etc. by NIR spectrometer as well as the mixtures of these meats in certain proportions and for the prediction accuracy of the established models (Wiedemair et al., 2018).

CONCLUSION

Online information about the purity of the products produced in the industry or about the pollution in the products can be obtained quickly by FTIR and Raman spectroscopy and they are both cheaper and less time consuming as compared to more detailed analysis techniques such as GC, GC-MS, HPLC, NMR. This is particularly useful for controlling the production process and, if there is a problem in production, to take precaution.

In this chapter, FTIR and Raman spectroscopy has been shown to be effective tool in the investigation of molecular systems, especially biomolecular systems and disease diagnosis when used in conjunction with multivariate analyzes. It is within this context that we expect to gain information the most from molecular spectra and increase our understanding on a wide variety of molecular processes. That means that Multivariate techniques attempt to model on the IR and Raman spectra where each situation, structure and decision involves more than a single factor. It is expected that the applications of multivariate analysis on the interpretation of FTIR and Raman Spectra will greatly expand in the future.

REFERENCES

Abidi, N., Cabrales, L., & Haigler, C. H. (2014). Changes in the cell wall and cellulose content of developing cotton fibers investigated by FTIR spectroscopy. *Carbohydrate Polymers*, *100*, 9–16. doi:10.1016/j.carbpol.2013.01.074 PMID:24188832

Abramczyk, H., Placek, I., Brożek-Płuska, B., Kurczewski, K., Morawiec, Z., & Tazbir, M. (2008). Human breast tissue cancer diagnosis by Raman spectroscopy. *Spectroscopy (Springfield, Or.)*, *22*(2-3), 113–121. PMID:20678135

Akyuz, S., Akyuz, T., Celik, O., & Atak, Ç. (2013). FTIR and EDXRF investigations of salt tolerant soybean mutants. *Journal of Molecular Structure*, *1044*, 67–71. doi:10.1016/j.molstruc.2013.01.025

Akyuz, S., Akyuz, T., Celik, O., & Atak, C. (2018). FTIR spectroscopy of protein isolates of salt-tolerant soybean mutants. *Journal of Applied Spectroscopy*, *84*(6), 1019–1023. doi:10.100710812-018-0580-1

Alvarez-Ordóñez, A., & Prieto, M. (2012). Technical and methodological aspects of fourier transform infrared spectroscopy in food microbiology research. In *Fourier Transform Infrared Spectroscopy in Food Microbiology* (pp. 1–18). Boston, MA: Springer. doi:10.1007/978-1-4614-3813-7_1

Amer, M. S. (Ed.). (2009). *Raman spectroscopy for soft matter applications*. Hoboken, NJ: Wiley. doi:10.1002/9780470475997

OI Analytical, a xylem brand, Analysis of Thiophene in Benzene using ASTM Method D7011 and a Pulsed Flame Photometric Detector (PFPD), Publication 43600916.

Anderson, R. J., Bendell, D. J., & Groundwater, P. W. (2004). *Organic spectroscopic analysis* (Vol. 22). Royal Society of Chemistry.

Ashton, L., Lau, K., Winder, C. L., & Goodacre, R. (2011). Raman spectroscopy: Lighting up the future of microbial identification. *Future Microbiology*, *6*(9), 991–997. doi:10.2217/fmb.11.89 PMID:21958140

Auner, G. W., Koya, S. K., Huang, C., Broadbent, B., Trexler, M., Auner, Z., ... Brusatori, M. A. (2018). Applications of Raman spectroscopy in cancer diagnosis. *Cancer and Metastasis Reviews*, *37*(4), 691–717. doi:10.100710555-018-9770-9 PMID:30569241

Bağcıoğlu, M., Fricker, M., Johler, S., & Ehling-Schulz, M. (2019). Detection and identification of Bacillus cereus, Bacillus cytotoxicus, Bacillus thuringiensis, Bacillus mycoides and Bacillus weihenstephanensis via machine learning based FTIR Spectroscopy. *Frontiers in Microbiology*, *10*, 902. doi:10.3389/fmicb.2019.00902 PMID:31105681

Bal, D. K., & Bhasarkar, J. B. (2019). Mechanistic investigation of Sono–Phosphotungstic acid/phase transfer agent assisted oxidative desulfurization of liquid fuel. *Asia-Pacific Journal of Chemical Engineering*, *14*(1). doi:10.1002/apj.2271

Bart, J. C. J. (2006). *Plastics additives: advanced industrial analysis*. IOS Press.

Barth, A. (2000). The infrared absorption of amino acid side chains. *Progress in Biophysics and Molecular Biology*, *74*(3-5), 141–173. doi:10.1016/S0079-6107(00)00021-3 PMID:11226511

Barth, A. (2007). Infrared spectroscopy of proteins. *Biochimica et Biophysica Acta (BBA)-. Bioenergetics*, *1767*(9), 1073–1101. doi:10.1016/j.bbabio.2007.06.004

Becker, K., Al Laham, N., Fegeler, W., Proctor, R. A., Peters, G., & von Eiff, C. (2006). Fourier-transform infrared spectroscopic analysis is a powerful tool for studying the dynamic changes in Staphylococcus aureus small-colony variants. *Journal of Clinical Microbiology*, *44*(9), 3274–3278. doi:10.1128/JCM.00847-06 PMID:16954260

Bel'skaya, L. V. (2019). Use of IR spectroscopy in cancer diagnosis. A review. *Journal of Applied Spectroscopy*, *86*(2), 187–205. doi:10.100710812-019-00800-w

Bondi, R. W. Jr, Igne, B., Drennen, J. K. III, & Anderson, C. A. (2012). Effect of experimental design on the prediction performance of calibration models based on near-infrared spectroscopy for pharmaceutical applications. *Applied Spectroscopy*, *66*(12), 1442–1453. doi:10.1366/12-06689 PMID:23231907

Bureau, S., Cozzolino, D., & Clark, C. J. (2019). Contributions of Fourier-transform mid infrared (FT-MIR) spectroscopy to the study of fruit and vegetables: A review. *Postharvest Biology and Technology*, *148*, 1–14. doi:10.1016/j.postharvbio.2018.10.003

Carlos, C., Maretto, D. A., Poppi, R. J., Sato, M. I. Z., & Ottoboni, L. M. (2011). Fourier transform infrared microspectroscopy as a bacterial source tracking tool to discriminate fecal E. coli strains. *Microchemical Journal*, *99*(1), 15–19. doi:10.1016/j.microc.2011.03.002

Challagulla, V., Nayar, S., Walsh, K., & Fabbro, L. (2017). Advances in techniques for assessment of microalgal lipids. *Critical Reviews in Biotechnology*, *37*(5), 566–578. doi:10.1080/07388551.2016.1206058 PMID:27417693

Chauvet, R., Lagarde, F., Charrier, T., Assaf, A., Thouand, G., & Daniel, P. (2017). Microbiological identification by surface-enhanced raman spectroscopy. *Applied Spectroscopy Reviews*, *52*(2), 123–144. doi:10.1080/05704928.2016.1209760

Chisanga, M., Muhamadali, H., Ellis, D. I., & Goodacre, R. (2018). Surface-enhanced raman scattering (SERS) in microbiology: Illumination and enhancement of the microbial world. *Applied Spectroscopy*, *72*(7), 987–1000. doi:10.1177/0003702818764672 PMID:29569946

Christy, A. A., Ozaki, Y., & Gregoriou, V. G. (2001). *Modern Fourier transform infrared spectroscopy* (Vol. 2001). Amsterdam, The Netherlands: Elsevier.

Ciurczak, E. W., & Igne, B. (2015). Pharmaceutical and medical applications of near-infrared spectroscopy. Boca Raton, FL: CRC Press.

Cozzolino, D., Cynkar, W. U., Shah, N., & Smith, P. (2011). Multivariate data analysis applied to spectroscopy: Potential application to juice and fruit quality. *Food Research International*, *44*(7), 1888–1896. doi:10.1016/j.foodres.2011.01.041

Crompton, T. R. (2006). *Polymer reference book*. iSmithers Rapra Publishing.

Cui, S., Zhang, S., & Yue, S. (2018). Raman spectroscopy and imaging for cancer diagnosis. *Journal of Healthcare Engineering*, 2018. PMID:29977484

Daoud, S., Bou-maroun, E., Dujourdy, L., Waschatko, G., Billecke, N., & Cayot, P. (2019). Fast and direct analysis of oxidation levels of oil-in-water emulsions using ATR-FTIR. *Food Chemistry*, *293*, 307–314. doi:10.1016/j.foodchem.2019.05.005 PMID:31151616

Davis, R., & Mauer, L. J. (2010). Fourier transform infrared (FT-IR) spectroscopy: a rapid tool for detection and analysis of foodborne pathogenic bacteria. *Current research, technology, and education topics in applied microbiology and microbial biotechnology, 2*, 1582-1594.

De Bruyne, S., Speeckaert, M. M., & Delanghe, J. R. (2018). Applications of mid-infrared spectroscopy in the clinical laboratory setting. *Critical Reviews in Clinical Laboratory Sciences*, *55*(1), 1–20. doi:10.1080/10408363.2017.1414142 PMID:29239240

De Oliveira, R. R., das Neves, L. S., & de Lima, K. M. G. (2012). Experimental design, near-infrared spectroscopy, and multivariate calibration: an advanced project in a chemometrics course. *Journal of Chemical Education*, *89*(12), 1566–1571. doi:10.1021/ed200765j

De Plano, L. M., Fazio, E., Rizzo, M. G., Franco, D., Carnazza, S., Trusso, S., ... Guglielmino, S. P. (2019). Phage-based assay for rapid detection of bacterial pathogens in blood by Raman spectroscopy. *Journal of Immunological Methods*, *465*, 45–52. doi:10.1016/j.jim.2018.12.004 PMID:30552870

Dean, J. R. (2003). *Methods for environmental trace analysis* (Vol. 12). John Wiley & Sons. doi:10.1002/0470863250

Deidda, R., Sacre, P. Y., Clavaud, M., Coïc, L., Avohou, H., Hubert, P., & Ziemons, E. (2019). Vibrational spectroscopy in analysis of pharmaceuticals: Critical review of innovative portable and handheld NIR and Raman spectrophotometers. *TrAC Trends in Analytical Chemistry*.

Depciuch, J., Stanek-Widera, A., Lange, D., Biskup-Fruzynska, M., Stanek-Tarkowska, J., Czarny, W., & Cebulski, J. (2018). Spectroscopic analysis of normal and neoplastic (WI-FTC) thyroid tissue. *Spectrochimica Acta. Part A: Molecular and Biomolecular Spectroscopy*, *204*, 18–24. doi:10.1016/j.saa.2018.06.010 PMID:29902767

Devpura, S., Barton, K. N., Brown, S. L., Palyvoda, O., Kalkanis, S., Naik, V. M., ... Chetty, I. J. (2014). Vision 20/20: The role of Raman spectroscopy in early stage cancer detection and feasibility for application in radiation therapy response assessment. *Medical Physics*, *41*(5). doi:10.1118/1.4870981 PMID:24784365

Duraipandian, S., Petersen, J. C., & Lassen, M. (2019). Authenticity and concentration analysis of extra virgin olive oil using spontaneous Raman spectroscopy and multivariate data analysis. *Applied Sciences*, *9*(12), 2433. doi:10.3390/app9122433

Dziuba, B. (2013). Identification of propionibacteria to the species level using Fourier transform infrared spectroscopy and artificial neural networks. *Polish Journal of Veterinary Sciences*, *16*(2), 351–357. doi:10.2478/pjvs-2013-0047 PMID:23971204

Ekins, S. (2008). *Pharmaceutical applications of Raman spectroscopy*. John Wiley & Sons.

Elmi, F., Movaghar, A. F., Elmi, M. M., Alinezhad, H., & Nikbakhsh, N. (2017). Application of FT-IR spectroscopy on breast cancer serum analysis. *Spectrochimica Acta. Part A: Molecular and Biomolecular Spectroscopy*, *187*, 87–91. doi:10.1016/j.saa.2017.06.021 PMID:28666157

Elston, R. C., Olson, J. M., & Palmer, L. (Eds.). (2002). *Biostatistical genetics and genetic epidemiology*. John Wiley & Sons.

Fakayode, S. O., Baker, G. A., Bwambok, D. K., Bhawawet, N., Elzey, B., Siraj, N., ... Warner, I. M. (2019). Molecular (Raman, NIR, and FTIR) spectroscopy and multivariate analysis in consumable products analysis. *Applied Spectroscopy Reviews*, 1–77. doi:10.1080/05704928.2019.1631176

Ferraro, J. R., & Nakamoto, K. (1994). *Introductory Raman spectroscopy*. Academic Press, Inc.

Fisher, R. A. (1936). The use of multiple measurements in taxonomic problems. *Annals of Eugenics*, *7*(2), 179–188. doi:10.1111/j.1469-1809.1936.tb02137.x

Fordellone, M., Bellincontro, A., & Mencarelli, F. (2018). Partial least squares discriminant analysis: A dimensionality reduction method to classify hyperspectral data. *arXiv preprint arXiv:1806.09347*.

Frost, J., Ludeman, L., Hillaby, K., Gornall, R., Lloyd, G., Kendall, C., ... Stone, N. (2017). Raman spectroscopy and multivariate analysis for the noninvasive diagnosis of clinically inconclusive vulval lichen sclerosus. *Analyst (London)*, *142*(8), 1200–1206. doi:10.1039/C6AN02009G PMID:27827479

Fujioka, N., Morimoto, Y., Arai, T., & Kikuchi, M. (2004). Discrimination between normal and malignant human gastric tissues by Fourier transform infrared spectroscopy. *Cancer Detection and Prevention*, *28*(1), 32–36. doi:10.1016/j.cdp.2003.11.004 PMID:15041075

Gajjar, K., Heppenstall, L. D., Pang, W., Ashton, K. M., Trevisan, J., Patel, I. I., ... Martin, F. L. (2013). Diagnostic segregation of human brain tumours using Fourier-transform infrared and/or Raman spectroscopy coupled with discriminant analysis. *Analytical Methods*, *5*(1), 89–102. doi:10.1039/C2AY25544H PMID:24098310

Gardiner, D. J., & Graves, P. R. (1989). *Practical Raman spectroscopy*. Springer. doi:10.1007/978-3-642-74040-4

Gasper, R., Dewelle, J., Kiss, R., Mijatovic, T., & Goormaghtigh, E. (2009). IR spectroscopy as a new tool for evidencing antitumor drug signatures. *Biochimica et Biophysica Acta (BBA)-. Biomembranes, 1788*(6), 1263–1270. doi:10.1016/j.bbamem.2009.02.016 PMID:19250921

Gemperline, P. (2006). Practical guide to chemometrics. Boca Raton, FL: CRC Press. doi:10.1201/9781420018301

Ghatak, K. L. (2010). *Techniques and methods in biology*. PHI Learning Pvt. Ltd.

Girshick, M. A. (1939). On the sampling theory of roots of determinantal equations. *Annals of Mathematical Statistics, 10*(3), 203–224. doi:10.1214/aoms/1177732180

Gonzalez-Fernandez, I., Iglesias-Otero, M. A., Esteki, M., Moldes, O. A., Mejuto, J. C., & Simal-Gandara, J. (2019). A critical review on the use of artificial neural networks in olive oil production, characterization and authentication. *Critical Reviews in Food Science and Nutrition, 59*(12), 1913–1926. doi:10.1080/10408398.2018.1433628 PMID:29381389

Goormaghtigh, E. (2016). Infrared imaging in histopathology: Is a unified approach possible? *Biomedical Spectroscopy and Imaging, 5*(4), 325–346. doi:10.3233/BSI-160151

Goormaghtigh, E., Gasper, R., Bénard, A., Goldsztein, A., & Raussens, V. (2009). Protein secondary structure content in solution, films and tissues: Redundancy and complementarity of the information content in circular dichroism, transmission and ATR FTIR spectra. *Biochimica et Biophysica Acta (BBA)-. Proteins and Proteomics, 1794*(9), 1332–1343. doi:10.1016/j.bbapap.2009.06.007 PMID:19540367

Gremlich, H. U., & Yan, B. (2000). Infrared and Raman spectroscopy of biological materials. Boca Raton, FL: CRC Press. doi:10.1201/b16936

Griffiths, P. R., & De Haseth, J. A. (2007). *Fourier transform infrared spectrometry* (Vol. 171). John Wiley & Sons. doi:10.1002/047010631X

Guimaraes, L. L., Moreira, L. P., Lourenço, B. F., Toma, W., Zangaro, R. A., Pacheco, M. T. T., & Silveira, L. (2018). Multivariate method based on Raman spectroscopy for quantification of dipyrone in oral solutions. *Journal of Spectroscopy*, 2018.

Hansen, S. H., Pedersen-Bjergaard, S., & Rasmussen, K. (2012). *Introduction to pharmaceutical chemical analysis*. John Wiley & Sons.

Hanvey, J. S., Lewis, P. J., Lavers, J. L., Crosbie, N. D., Pozo, K., & Clarke, B. O. (2017). A review of analytical techniques for quantifying microplastics in sediments. *Analytical Methods, 9*(9), 1369–1383. doi:10.1039/C6AY02707E

Harrison, J. P., & Berry, D. (2017). Vibrational spectroscopy for imaging single microbial cells in complex biological samples. *Frontiers in Microbiology, 8*, 675. doi:10.3389/fmicb.2017.00675 PMID:28450860

Hayes, A. W. (2008). Principles and methods of toxicology. Boca Raton, FL: CRC Press.

He, H., Sun, D. W., Pu, H., Chen, L., & Lin, L. (2019). Applications of Raman spectroscopic techniques for quality and safety evaluation of milk: A review of recent developments. *Critical Reviews in Food Science and Nutrition, 59*(5), 770–793. doi:10.1080/10408398.2018.1528436 PMID:30614242

Helm, D., Labischinski, H., Schallehn, G., & Naumann, D. (1991). Classification and identification of bacteria by Fourier-transform infrared spectroscopy. *Microbiology, 137*(1), 69–79. PMID:1710644

Ho, C. S., Jean, N., Hogan, C. A., Blackmon, L., Jeffrey, S. S., Holodniy, M., . . . Dionne, J. (2019). Rapid identification of pathogenic bacteria using Raman spectroscopy and deep learning. *arXiv preprint arXiv:1901.07666*.

Hollas, J. M. (2004). *Modern spectroscopy*. John Wiley & Sons.

Hotelling, H. (1933). Analysis of a complex of statistical variables into principal components. *Journal of Educational Psychology, 24*(6), 417–441. doi:10.1037/h0071325

Hotelling, H. (1936). Simplified calculation of principal components. *Psychometrika, 1*(1), 27–35. doi:10.1007/BF02287921

Ibelings, M. S., Maquelin, K., Endtz, H. P., Bruining, H. A., & Puppels, G. J. (2005). Rapid identification of Candida spp. in peritonitis patients by Raman spectroscopy. *Clinical Microbiology and Infection, 11*(5), 353–358. doi:10.1111/j.1469-0691.2005.01103.x PMID:15819860

Jarvis, R. M., & Goodacre, R. (2004). Discrimination of bacteria using surface-enhanced Raman spectroscopy. *Analytical Chemistry, 76*(1), 40–47. doi:10.1021/ac034689c PMID:14697030

Kademi, H. I., Ulusoy, B. H., & Hecer, C. (2019). Applications of miniaturized and portable near infrared spectroscopy (NIRS) for inspection and control of meat and meat products. *Food Reviews International, 35*(3), 201–220. doi:10.1080/87559129.2018.1514624

Kandpal, L. M., & Cho, B. K. (2014). A review of the applications of spectroscopy for the detection of microbial contaminations and defects in agro foods. *Journal of Biosystems Engineering, 39*(3), 215–226. doi:10.5307/JBE.2014.39.3.215

Kaufmann, K. C., Favero, F. D. F., de Vasconcelos, M. A. M., Godoy, H. T., Sampaio, K. A., & Barbin, D. F. (2019). Portable NIR spectrometer for prediction of palm oil acidity. *Journal of Food Science, 84*(3), 406–411. doi:10.1111/1750-3841.14467 PMID:30758058

Kavkler, K., Gunde-Cimerman, N., Zalar, P., & Demšar, A. (2011). FTIR spectroscopy of biodegraded historical textiles. *Polymer Degradation & Stability, 96*(4), 574–580. doi:10.1016/j.polymdegradstab.2010.12.016

Kaznowska, E., Depciuch, J., Łach, K., Kołodziej, M., Koziorowska, A., Vongsvivut, J., ... Cebulski, J. (2018). The classification of lung cancers and their degree of malignancy by FTIR, PCA-LDA analysis, and a physics-based computational model. *Talanta, 186*, 337–345. doi:10.1016/j.talanta.2018.04.083 PMID:29784370

Keller, M. D., Kanter, E. M., & Mahadevan-Jansen, A. (2006). Raman spectroscopy for cancer diagnosis. *Spectroscopy-Springfield then eugene then duluth-, 21*(11), 33.

Kelsall, R. W., Hamley, I. W., & Geoghegan, M. (Eds.). (2005). *Nanoscale science and technology* (p. 108). Chichester, UK: Wiley. doi:10.1002/0470020873

Kenkel, J. (2014). Analytical chemistry for technicians. Boca Raton, FL: CRC Press.

Ketty, G. S. L., Khairunnisak, M., Faridah, I., Roosnoor, F. H., Izwan, I., Norakmar, I., ... Saipul, B. A. R. (2017). Preliminary study on classification of raw bovine's milk using atr-ftir coupled with PCA from peninsular Malaysia. *Malaysian Journal of Veterinary Research*, *8*(1), 95–101.

Khanmohammadi, M., & Garmarudi, A. B. (2011). Infrared spectroscopy provides a green analytical chemistry tool for direct diagnosis of cancer. *TrAC Trends in Analytical Chemistry*, *30*(6), 864–874. doi:10.1016/j.trac.2011.02.009

Kiefer, J. (2015). Recent advances in the characterization of gaseous and liquid fuels by vibrational spectroscopy. *Energies*, *8*(4), 3165–3197. doi:10.3390/en8043165

Klein, D., Breuch, R., von der Mark, S., Wickleder, C., & Kaul, P. (2019). Detection of spoilage associated bacteria using Raman-microspectroscopy combined with multivariate statistical analysis. *Talanta*, *196*, 325–328. doi:10.1016/j.talanta.2018.12.094 PMID:30683371

Kochan, K., Perez-Guaita, D., Pissang, J., Jiang, J. H., Peleg, A. Y., McNaughton, D., ... Wood, B. R. (2018). In vivo atomic force microscopy–infrared spectroscopy of bacteria. *Journal of the Royal Society, Interface*, *15*(140). doi:10.1098/rsif.2018.0115 PMID:29593091

Kotanen, C. N., Martinez, L., Alvarez, R., & Simecek, J. W. (2016). Surface enhanced Raman scattering spectroscopy for detection and identification of microbial pathogens isolated from human serum. *Sensing and bio-sensing research, 8*, 20-26.

Kumar, S., Srinivasan, A., & Nikolajeff, F. (2018). Role of infrared spectroscopy and imaging in cancer diagnosis. *Current Medicinal Chemistry*, *25*(9), 1055–1072. doi:10.2174/0929867324666170523121314 PMID:28545365

Kumari, A., Kaur, J., & Bhattacharyya, S. (2018). Application of fourier transform-infrared spectroscopy as a tool for early cancer detection. *American Journal of Biomedical Sciences*, *10*(3).

Larkin, P. (2011). *Infrared and Raman spectroscopy: principles and spectral interpretation*. Elsevier.

Lasch, P., Stämmler, M., Zhang, M., Baranska, M., Bosch, A., & Majzner, K. (2018). FT-IR hyperspectral imaging and artificial neural network analysis for identification of pathogenic bacteria. *Analytical Chemistry*, *90*(15), 8896–8904. doi:10.1021/acs.analchem.8b01024 PMID:29944341

Lee, L. C., Liong, C. Y., & Jemain, A. A. (2018). Partial least squares-discriminant analysis (PLS-DA) for classification of high-dimensional (HD) data: A review of contemporary practice strategies and knowledge gaps. *Analyst (London)*, *143*(15), 3526–3539. doi:10.1039/C8AN00599K PMID:29947623

Lewis, I. R., & Edwards, H. (2001). Handbook of Raman spectroscopy: from the research laboratory to the process line. Boca Raton, FL: CRC Press. doi:10.1201/9781420029253

Li, R., Dhankhar, D., Chen, J., Krishnamoorthi, A., Cesario, T. C., & Rentzepis, P. M. (2019). Identification of live and dead bacteria: A Raman spectroscopic study. *IEEE Access: Practical Innovations, Open Solutions*, *7*, 23549–23559. doi:10.1109/ACCESS.2019.2899006

Li, S., Li, L., Zeng, Q., Zhang, Y., Guo, Z., Liu, Z., ... Liu, S. (2015). Characterization and noninvasive diagnosis of bladder cancer with serum surface enhanced Raman spectroscopy and genetic algorithms. *Scientific Reports*, *5*(1), 9582. doi:10.1038rep09582 PMID:25947114

Li, X., Yang, T., & Lin, J. (2012). Spectral analysis of human saliva for detection of lung cancer using surface-enhanced Raman spectroscopy. *Journal of Biomedical Optics, 17*(3). doi:10.1117/1.JBO.17.3.037003 PMID:22502575

Li, Y., Huang, W., Pan, J., Ye, Q., Lin, S., Feng, S., ... Chen, R. (2015). Rapid detection of nasopharyngeal cancer using Raman spectroscopy and multivariate statistical analysis. *Molecular and Clinical Oncology, 3*(2), 375–380. doi:10.3892/mco.2014.473 PMID:25798270

Li, Z., Li, C., Lin, D., Huang, Z., Pan, J., Chen, G., ... Chen, R. (2014). Surface-enhanced Raman spectroscopy for differentiation between benign and malignant thyroid tissues. *Laser Physics Letters, 11*(4). doi:10.1088/1612-2011/11/4/045602

Lim, S., & Cohenford, M. (2014, September). FT-IR spectra analysis toward cancer detection. In *Proceedings 2014 25th International Workshop on Database and Expert Systems Applications* (pp. 77-81). IEEE. 10.1109/DEXA.2014.31

Liu, W., Sun, Z., Chen, J., & Jing, C. (2016). Raman spectroscopy in colorectal cancer diagnostics: Comparison of PCA-LDA and PLS-DA models. *Journal of Spectroscopy*, 2016.

Lorenz, B., Wichmann, C., Stöckel, S., Rösch, P., & Popp, J. (2017). Cultivation-free Raman spectroscopic investigations of bacteria. *Trends in Microbiology, 25*(5), 413–424. doi:10.1016/j.tim.2017.01.002 PMID:28188076

Lynk, T. P., Sit, C. S., & Brosseau, C. L. (2018). Electrochemical surface-enhanced Raman spectroscopy as a platform for bacterial detection and identification. *Analytical Chemistry, 90*(21), 12639–12646. doi:10.1021/acs.analchem.8b02806 PMID:30350616

Mackanos, M. A., & Contag, C. H. (2010). Fiber-optic probes enable cancer detection with FTIR spectroscopy. *Trends in Biotechnology, 28*(6), 317–323. doi:10.1016/j.tibtech.2010.04.001 PMID:20452071

Macnamara, J. (1969). *The Irish Journal of Education / Iris Eireannach an Oideachais, 3*(1), 71-73. Retrieved from http://www.jstor.org/stable/30076687

Martins, Q. S., Aguirre, C. A., & Farias, J. L. B. (2019). Approach by Raman and infrared spectroscopy in three vegetable oils from the Brazilian Amazon. *Revista Mexicana de Física, 65*(4 Jul-Aug), 328-332.

Meisel, S., Stöckel, S., Elschner, M., Melzer, F., Rösch, P., & Popp, J. (2012). Raman spectroscopy as a potential tool for detection of Brucella spp. in milk. *Applied and Environmental Microbiology, 78*(16), 5575–5583. doi:10.1128/AEM.00637-12 PMID:22660699

Meksiarun, P., Ishigaki, M., Huck-Pezzei, V. A., Huck, C. W., Wongravee, K., Sato, H., & Ozaki, Y. (2017). Comparison of multivariate analysis methods for extracting the paraffin component from the paraffin-embedded cancer tissue spectra for Raman imaging. *Scientific Reports, 7*(1), 44890. doi:10.1038rep44890 PMID:28327648

Mobili, P., Londero, A., De Antoni, G., Gomez-Zavaglia, A., Araujo-Andrade, C., Avila-Donoso, H., ... Frausto-Reyes, C. (2010). Multivariate analysis of Raman spectra applied to microbiology: Discrimination of microorganisms at the species level. *Revista Mexicana de Física, 56*(5), 378–385.

Moisoiu, V., Socaciu, A., Stefancu, A., Iancu, S. D., Boros, I., Alecsa, C. D., ... Socaciu, C. (2019). Breast cancer diagnosis by Surface-Enhanced Raman Scattering (SERS) of urine. *Applied Sciences*, *9*(4), 806. doi:10.3390/app9040806

Moss, D. (2011). *Biomedical applications of Synchrotron Infrared Microspectroscopy: A practical approach*. Royal Society of Chemistry.

Mostaço-Guidolin, L. B., & Bachmann, L. (2011). Application of FTIR spectroscopy for identification of blood and leukemia biomarkers: A review over the past 15 years. *Applied Spectroscopy Reviews*, *46*(5), 388–404. doi:10.1080/05704928.2011.565534

Naumann, D. (2000). Infrared spectroscopy in microbiology. Encyclopedia of analytical chemistry.

Nesakumar, N., Baskar, C., Kesavan, S., Rayappan, J. B. B., & Alwarappan, S. (2018). Analysis of moisture content in beetroot using fourier transform infrared spectroscopy and by principal component analysis. *Scientific Reports*, *8*(1), 7996. doi:10.103841598-018-26243-5 PMID:29789563

Neugebauer, U., Roesch, P., & Popp, J. (2015). Raman spectroscopy towards clinical application: Drug monitoring and pathogen identification. *International Journal of Antimicrobial Agents*, *46*, S35–S39. doi:10.1016/j.ijantimicag.2015.10.014 PMID:26612228

Nguyen, V. H. (2015). *Scanning angle Raman spectroscopy in polymer thin film characterization (No. IS-T 3192). Ames Laboratory*. Ames, IA: AMES. doi:10.2172/1409182

Nyquist, R. A. (2001). *Interpreting infrared, Raman, and nuclear magnetic resonance spectra*. Academic Press.

O'Donnell, C. P., Fagan, C., & Cullen, P. J. (Eds.). (2014). *Process analytical technology for the food industry*. Springer. doi:10.1007/978-1-4939-0311-5

Ojeda, J. J., & Dittrich, M. (2012). Fourier transform infrared spectroscopy for molecular analysis of microbial cells. In *Microbial Systems Biology* (pp. 187–211). Totowa, NJ: Humana Press. doi:10.1007/978-1-61779-827-6_8

Ozaki, Y., McClure, W. F., & Christy, A. A. (Eds.). (2006). *Near-infrared spectroscopy in food science and technology*. John Wiley & Sons. doi:10.1002/0470047704

Pahlow, S., Meisel, S., Cialla-May, D., Weber, K., Rösch, P., & Popp, J. (2015). Isolation and identification of bacteria by means of Raman spectroscopy. *Advanced Drug Delivery Reviews*, *89*, 105–120. doi:10.1016/j.addr.2015.04.006 PMID:25895619

Pahlow, S., Weber, K., Popp, J., Bayden, R. W., Kochan, K., Rüther, A., ... Gardner, B. (2018). Application of vibrational spectroscopy and imaging to point-of-care medicine: A review. *Applied Spectroscopy*, *72*(101), 52–84. PMID:30265133

Pavia, D. L., Lampman, G. M., Kriz, G. S., & Engel, R. G. (2018). *A microscale approach to organic laboratory techniques*. Cengage Learning.

Pearson, K. (1901). LIII. On lines and planes of closest fit to systems of points in space. *The London, Edinburgh, and Dublin Philosophical Magazine and Journal of Science, 2*(11), 559–572. doi:10.1080/14786440109462720

Pérez-Alonso, M., Castro, K., & Madariaga, J. M. (2006). Vibrational spectroscopic techniques for the analysis of artefacts with historical, artistic and archaeological value. *Current Analytical Chemistry, 2*(1), 89–100. doi:10.2174/157341106775197385

Quintelas, C., Ferreira, E. C., Lopes, J. A., & Sousa, C. (2018). An overview of the evolution of infrared spectroscopy applied to bacterial typing. *Biotechnology Journal, 13*(1). doi:10.1002/biot.201700449 PMID:29090857

Ramirez-Elias, M. G., & González, F. J. (2018). Raman Spectroscopy for in vivo medical diagnosis. In *Raman Spectroscopy*. IntechOpen. doi:10.5772/intechopen.72933

Ravanshad, R., Karimi Zadeh, A., Amani, A. M., Mousavi, S. M., Hashemi, S. A., Savar Dashtaki, A., ... & Zare, B. (2018). Application of nanoparticles in cancer detection by Raman scattering based techniques. *Nano Reviews & Experiments, 9*(1).

Rodriguez-Saona, L. E., Khambaty, F. M., Fry, F. S., & Calvey, E. M. (2001). Rapid detection and identification of bacterial strains by Fourier transform near-infrared spectroscopy. *Journal of Agricultural and Food Chemistry, 49*(2), 574–579. doi:10.1021/jf000776j PMID:11261995

Sandt, C., Madoulet, C., Kohler, A., Allouch, P., De Champs, C., Manfait, M., & Sockalingum, G. D. (2006). FT-IR microspectroscopy for early identification of some clinically relevant pathogens. *Journal of Applied Microbiology, 101*(4), 785–797. doi:10.1111/j.1365-2672.2006.02969.x PMID:16968290

Santos, I. P., Barroso, E. M., Schut, T. C. B., Caspers, P. J., van Lanschot, C. G., Choi, D. H., ... Hegt, V. N. (2017). Raman spectroscopy for cancer detection and cancer surgery guidance: Translation to the clinics. *Analyst (London), 142*(17), 3025–3047. doi:10.1039/C7AN00957G PMID:28726868

Santos, M. C., Nascimento, P. A. M., Guedes, W. N., Pereira Filho, E. R., Filletti, É. R., & Pereira, F. M. V. (2019). Chemometrics in analytical chemistry–an overview of applications from 2014 to 2018. *Eclética Química, 44*(2), 11–25. doi:10.26850/1678-4618eqj.v44.2.2019.p11-25

Santos, M. C., Nascimento, Y. M., Araújo, J. M., & Lima, K. M. (2017). ATR-FTIR spectroscopy coupled with multivariate analysis techniques for the identification of DENV-3 in different concentrations in blood and serum: A new approach. *RSC Advances, 7*(41), 25640–25649. doi:10.1039/C7RA03361C

Schmitt, J., & Flemming, H. C. (1998). FTIR-spectroscopy in microbial and material analysis. *International Biodeterioration & Biodegradation, 41*(1), 1–11. doi:10.1016/S0964-8305(98)80002-4

Settle, F. A. (1997). *Handbook of instrumental techniques for analytical chemistry*. Prentice Hall PTR.

Shinzawa, H., Awa, K., Kanematsu, W., & Ozaki, Y. (2009). Multivariate data analysis for Raman spectroscopic imaging. *Journal of Raman Spectroscopy: An International Journal for Original Work in all Aspects of Raman Spectroscopy, Including Higher Order Processes, and also Brillouin and Rayleigh Scattering, 40*(12), 1720-1725.

Sil, S., Mukherjee, R., Kumar, N. S., Aravind, S., Kingston, J., & Singh, U. K. (2017). Detection and classification of bacteria using Raman spectroscopy combined with multivariate analysis. *Defence Life Science Journal, 2*(4), 435–441. doi:10.14429/dlsj.2.12275

Siqueira, L. F., & Lima, K. M. (2016). MIR-biospectroscopy coupled with chemometrics in cancer studies. *Analyst (London), 141*(16), 4833–4847. doi:10.1039/C6AN01247G PMID:27433557

Siqueira, L. F., & Lima, K. M. (2016). A decade (2004–2014) of FTIR prostate cancer spectroscopy studies: An overview of recent advancements. *TrAC Trends in Analytical Chemistry, 82*, 208–221. doi:10.1016/j.trac.2016.05.028

Smilauer, P., & Leps, J. (2014). *Multivariate analysis of ecological data using CANOCO 5*. Cambridge University Press. doi:10.1017/CBO9781139627061

Smith, B. C. (2011). Fundamentals of Fourier transform infrared spectroscopy. Boca Raton, FL: CRC Press. doi:10.1201/b10777

Smith, E., & Dent, G. (2005). *Modern Raman spectroscopy: a practical approach*. Hoboken, NJ: John Wiley & Sons.

Sorak, D., Herberholz, L., Iwascek, S., Altinpinar, S., Pfeifer, F., & Siesler, H. W. (2012). New developments and applications of handheld Raman, mid-infrared, and near-infrared spectrometers. *Applied Spectroscopy Reviews, 47*(2), 83–115. doi:10.1080/05704928.2011.625748

Spearman, C. (1904). General Intelligence, objectively determined and measured. *The American Journal of Psychology, 15*(2), 201–292. doi:10.2307/1412107

Stöckel, S., Kirchhoff, J., Neugebauer, U., Rösch, P., & Popp, J. (2016). The application of Raman spectroscopy for the detection and identification of microorganisms. *Journal of Raman Spectroscopy: JRS, 47*(1), 89–109. doi:10.1002/jrs.4844

Stuart, B. H. (2004). *Infrared spectroscopy: Fundamentals and applications*. John Wiley & Sons. doi:10.1002/0470011149

Su, W. H., & Sun, D. W. (2019). Mid-infrared (MIR) spectroscopy for quality analysis of liquid foods. *Food Engineering Reviews*, 1–17.

Tas, T. (2008). *Katışıklı zeytinyağlarında katıştırılan yağ cinsi ve oranının ftır spektroskopisi ve kemometrik yöntemlerle belirlenmesi*. Yüksek Lisans Tezi.

Theophanides, T. (Ed.). (2012). *Infrared spectroscopy: Materials science, engineering, and technology*. Published by InTech. doi:10.5772/2055

Tu, A. T. (1982). *Raman spectroscopy in biology. Principles and Applications*. John Wiley & Sons.

Vandenabeele, P. (2013). *Practical Raman spectroscopy: an introduction*. Chichester, UK: Wiley. doi:10.1002/9781119961284

Vannini, L., Lanciotti, R., Gardini, F., & Guerzoni, M. E. (1996). Evaluation of Fourier-transform infrared spectroscopy for data capture in predictive microbiology. *World Journal of Microbiology & Biotechnology, 12*(1), 85–90. doi:10.1007/BF00327808 PMID:24415096

Varmuza, K., & Filzmoser, P. (2009). *Introduction to multivariate statistical analysis in chemometrics*. Taylor & Francis Group.

Volk, H., & George, S. C. (2019). Using petroleum inclusions to trace petroleum systems–a review. *Organic Geochemistry*, *129*, 99–123. doi:10.1016/j.orggeochem.2019.01.012

Wang, C., Vickers, T. J., Schlenoff, J. B., & Mann, C. K. (1992). In situ monitoring of emulsion polymerization using fiber-optic Raman spectroscopy. *Applied Spectroscopy*, *46*(11), 1729–1731. doi:10.1366/0003702924926961

Wang, J. S., Shi, J. S., Xu, Y. Z., Duan, X. Y., Zhang, L., Wang, J., ... Wu, J. G. (2003). FT-IR spectroscopic analysis of normal and cancerous tissues of esophagus. *World Journal of Gastroenterology*, *9*(9), 1897. doi:10.3748/wjg.v9.i9.1897 PMID:12970871

Wang, Y., Ding, Y., Wei, W., Cao, Y., Davidson, D. F., & Hanson, R. K. (2019). On estimating physical and chemical properties of hydrocarbon fuels using mid-infrared FTIR spectra and regularized linear models. *Fuel*, *255*, 115715. doi:10.1016/j.fuel.2019.115715

Weiss, R., Palatinszky, M., Wagner, M., Niessner, R., Elsner, M., Seidel, M., & Ivleva, N. P. (2019). Surface-enhanced Raman spectroscopy of microorganisms: Limitations and applicability on the single-cell level. *Analyst (London)*, *144*(3), 943–953. doi:10.1039/C8AN02177E PMID:30574650

Wenning, M., & Scherer, S. (2013). Identification of microorganisms by FTIR spectroscopy: Perspectives and limitations of the method. *Applied Microbiology and Biotechnology*, *97*(16), 7111–7120. doi:10.100700253-013-5087-3 PMID:23860713

Wiedemair, V., De Biasio, M., Leitner, R., Balthasar, D., & Huck, C. W. (2018). Application of design of experiment for detection of meat fraud with a portable near-infrared spectrometer. *Current Analytical Chemistry*, *14*(1), 58–67. doi:10.2174/1573411013666170207121113

Wilks, S. S. (1932). Certain generalizations in the analysis of variance. *Biometrika*, *24*(3-4), 471–494. doi:10.1093/biomet/24.3-4.471

Yadav, L. D. S. (2005). *Organic spectroscopy*. Springer Science & Business Media. doi:10.1007/978-1-4020-2575-4

Yan, H., Zhang, J., Gao, J., Huang, Y., Xiong, Y., & Min, S. (2018). Towards improvement in prediction of iodine value in edible oil system based on chemometric analysis of portable vibrational spectroscopic data. *Scientific Reports*, *8*(1), 14729. doi:10.103841598-018-33022-9 PMID:30283065

Yang, B., Yao, L., & Pan, T. (2017). Near-infrared spectroscopy combined with partial least squares discriminant analysis applied to identification of liquor brands. *Engineering*, *9*(2), 181–189. doi:10.4236/eng.2017.92009

Yang, J., Xu, J., Zhang, X., Wu, C., Lin, T., & Ying, Y. (2019). Deep learning for vibrational spectral analysis: Recent progress and a practical guide. *Analytica Chimica Acta*, *1081*, 6–17. doi:10.1016/j.aca.2019.06.012 PMID:31446965

You, Z. (2017). Application of infrared Raman spectroscopy in analysis of food agricultural products. *Chemical Engineering Transactions*, *59*, 763–768.

Young, R. J., & Eichhorn, S. J. (2009). Raman applications in synthetic and natural polymer fibers and their composites. Raman spectroscopy for soft matter applications, 63-94.

Zeng, H. Y., Cai, L. H., Cai, X. L., Wang, Y. J., & Li, Y. Q. (2011). Structure characterization of protein fractions from lotus (Nelumbo nucifera) seed. *Journal of Molecular Structure*, *1001*(1-3), 139–144. doi:10.1016/j.molstruc.2011.06.031

Zhang, K., Hao, C., Man, B., Zhang, C., Yang, C., Liu, M., ... Chen, C. (2018). Diagnosis of liver cancer based on tissue slice surface enhanced Raman spectroscopy and multivariate analysis. *Vibrational Spectroscopy*, *98*, 82–87. doi:10.1016/j.vibspec.2018.07.010

Zhou, J., Wu, X., Chen, Z., You, J., & Xiong, S. (2019). Evaluation of freshness in freshwater fish based on near infrared reflectance spectroscopy and chemometrics. *LWT*, *106*, 145–150. doi:10.1016/j.lwt.2019.01.056

Section 2
Pharmaceutical

Chapter 6
Optimizing the Size of Drug-Loaded Nanoparticles Using Design of Experiments:
Solid Lipid Nanoparticles

Paola Cervantes-Covarrubias
Universidad Autónoma de Baja California, Mexico

Ayla Vea-Barragan
https://orcid.org/0000-0002-1125-1652
Universidad Autónoma de Baja California, Mexico

Aracely Serrano-Medina
https://orcid.org/0000-0002-6598-7016
Universidad Autónoma de Baja California, Mexico

Eugenia Gabriela Carrillo-Cedillo
https://orcid.org/0000-0003-4239-5904
Universidad Autónoma de Baja California, Mexico

José Manuel Cornejo-Bravo
https://orcid.org/0000-0002-0013-8937
Universidad Autónoma de Baja California, Mexico

ABSTRACT

Nanoparticles formed from lipids are currently applied successfully to deliver drugs. The particle size of the nanoparticle system is an essential characteristic to enhance the entrance of the drugs inside tissues and cells. Using design of experiment is appealing to find the specific conditions to optimize particle size of drug-loaded nanoparticles. Authors of this chapter applied a fractional factorial design of half fraction 2^{4-1} with levels between continue factors, finding statistically significant differences for two factors such as concentrations of drugs and type of solvent where the organic phase is dissolved. This design shows the optimization of a formulation of capsaicin in solid lipid nanoparticles. The chapter also includes information on methods to prepare solid lipid nanoparticles (SLN), the variables involved, and a selection of studies about optimization of SLN formulations.

DOI: 10.4018/978-1-7998-1518-1.ch006

INTRODUCTION

Nanoparticles made with biocompatible materials have become very popular as drug delivery systems due to the ability to control drug release and to the possibility of their administration by parenteral routes. Modification of nanoparticle surface with ligands can provide site-specific drug delivery for the treatment of tumors or inflamed tissue (Siafaka, Üstündağ Okur, Karavas, & Bikiaris, 2016; Wissing, Kayser, & Müller, 2004).

Passive vectorization is possible taking advantage of the enhanced permeation retention effect consisting in an increased permeability into tumors and inflamed tissue driven by bigger fenestration than healthy tissue, and the lack of lymphatic drainage, which retains the nanoparticles in the tumors where loaded antineoplastic drugs can be delivered (Nakamura, Mochida, Choyke, & Kobayashi, 2016).

Solid lipid nanoparticles (SLN) are a type of nanomaterial that has shown to increase the bioavailability of orally administered drugs with poor water solubility and enhancer of drug concentration on dermal delivery, since the nanometric scale increases dissolution velocity and saturation solubility, resulting in improved absorption of loaded drugs. Routes of administration for SLN have been extended to dermal, transdermal, ocular, pulmonary and rectal (Din et al., 2015; Hu, Jia, & Ding, 2010; Sharif Makhmal Zadeh, Niro, Rahim, & Esfahani, 2018; Sharma, Jindal, Aggarwal, & Jain, 2010).

The chapter presents general information about the preparation and application of SLN, and a summary of researches that applied Design of Experiments (DoE) to optimize the size and another parameters of drug-loaded SLN. In most studies, lipid and surfactant used for the preparation of SLN stand out as independent factors, evaluating the influence over particle size and entrapment efficient (EE), principally (Table 1). The chapter provides information about the principal methods to produce SLN. Subsequently, a formulation of capsaicin-loaded SLN is describe as a case of study applying DoE. The purpose is to optimize the particle size of the drug-loaded SLN using a fractional factorial design $2^{(k-p)}$. The design has two levels and IV resolution to estimate main effects, plus two central points per block and two blocked replicates (n=3); selecting as experimental factors: drug contained, lipid content, stabilizer content, and type of solvent. This design was selected to study the effect of the experimental factors on the response on particle size, combinate with the optimization of the process. In order to determine the best concentrations of capsaicin (CAP), Gelucire® 44/14 (lipid) and polyvinyl alcohol (PVAL) (surfactant) to generate the smaller particle size to be used in topical formulations, since particle's size less than 500 nm is crucial to transdermal delivery (Al-Kassas et al., 2016). The SLN were synthesized by an adaptation of the solvent emulsion-diffusion method, for the suitability to make small particles at room temperature (RT), in a few steps, using just one solid lipid, and stabilized with PVAL.

BACKGROUND

Lipid-based carriers are particularly suited for topical application (Mahant, Rao, & Nanda, 2018). Liposomes are spherical vesicles composed of one or more phospholipid bilayers, representing the first generation of the novel lipid colloidal carriers after submicron emulsion-bases products were developed in 1960s (Joshi & Müller, 2009; Mehnert & Mäder, 2012). Liposomes offered encapsulation of hydrophobic and hydrophilic drugs but have many disadvantages, including short shelf life, poor stability, low encapsulation efficacy, and cell interactions (Czajkowska-Kośnik, Szekalska, & Winnicka, 2019).

In 1990s Müeller started exploring the potential of nanoparticles-based on solid lipids. This drug delivery system was developed as an alternative system for poorly water-soluble drugs. The lipid phase of an SLN is solid at body and room RT. These lipid carriers minimize problems connected with traditional drug formulations and have some advantages, such as ease of preparation, good biocompatibility, lower cytotoxicity, avoidance of organic solvents, and wide possibilities of applications (Czajkowska-Kośnik et al., 2019; Müller, Mder, & Gohla, 2000).

The second generation of lipid nanoformulations is nanostructured lipid carriers, developed in 1999, composed of binary mixture of solid lipid and a spatially different liquid lipid as the carrier, dispersed in the aqueous phase with surfactants; the advantage of this system is characterized by high drug load due to unordered structure (Czajkowska-Kośnik et al., 2019; Joshi & Müller, 2009).

Lipid drug conjugates (LDC) were developed especially for the loading of hydrophilic drug molecules. LDC are lipidic prodrugs which contain the drug covalently or non- covalently linked to a lipid moiety, such as a fatty acid, a diglyceride or a phosphoglyceride. LDC helps to improve the penetration of the active ingredients through the physiological membranes (Adhikari et al., 2017; Joshi & Müller, 2009).

The carrier system of SLN was chosen in the case of study because the use of solid lipids prevents the drug from immediate release and allows the physical stability of the preparation, also small particles are suitable to obtain, these are relevant features in topical release (Lasa-Saracibar, Estella-Hermoso De Mendoza, Guada, Dios-Vieitez, & Blanco-Prieto, 2012). This system of nanoparticles use solid lipids as a matrix material for drug delivery. It can be formulated using a wide variety of lipids which are well tolerated by body (e.g. fatty acids, mono-, di-and triglycerides, glycerides mix, cetyl palmitate, waxes, etc.), being stabilized with biocompatible surfactants (ionic or non-ionic) agents, such as lecithin, biliary salts, poloxamers, polysorbates, PVAL, and another compounds of ester from fatty acids (Albano et al., 2019; Ekambaram & Abdul, 2011; Müller et al., 2000).

SLN have a core-shell structure (Figure 1); the outer shell is stabilized with surfactants, the core has a lipid composition where the inclusion of drugs is available (Ding, Pyo, & Muller, 2017). Some models assume this inclusion of substrates either in the core, the outer shell, as a colloidal dispersion in the lipid matrix or in a homogeneous distribution (Saeidpour et al., 2017); sustained drug release from SLN is possible due to this solid matrix. The lipid matrix employed in lipid nanoparticles is usually a physiological lipid (Das, Ng, Kanaujia, Kim, & Tan, 2011). The inclusion of drugs inside the particle matrix being more protective for chemically labile molecules. Nanoparticles with a positive surface charge are capable of being attached more rapidly than negative charged surface by the cells and can be endocytosed easily by them (Siafaka et al., 2016).

Drug adsorption enhancement using carriers which inhibit P-glycoprotein-mediated drugs efflux and pre-absorptive metabolism involving gut membrane-bound cytochrome enzymes and promotions of lymphatic transport. These mechanisms deliver the drug to the blood circulation system and avoid the hepatic first-pass metabolism (Siafaka et al., 2016).

The SLN has been introduced successfully to the skin by percutaneous absorption in occlusion conditions; therefore, the nanoparticles could promote the penetrations of the drug through the epidermis as shown on Figure 2. The skin permeation abilities of the system evaluated by *ex-vivo* permeation studies show increases in permeation of SLN observed over time; the nanoparticles were much more efficient than the free drug solutions. The SLN showed better skin permeation capacity (You, Yuan, & Chen, 2017). The catchment of SLN less than 180 nm inside keratinocytes was studied, and it is suggested that these particles cross the cellular membrane easily, without toxic effects (Prow et al., 2011).

Figure 1. Structure of solid lipid nanoparticle

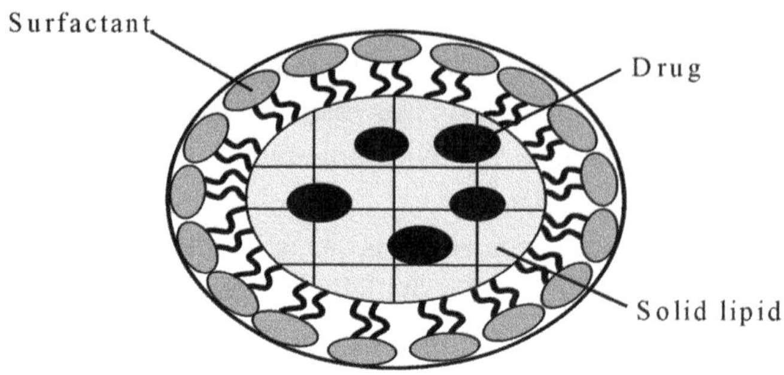

Figure 2. Schematic representation of penetration of drugs throughout the skin

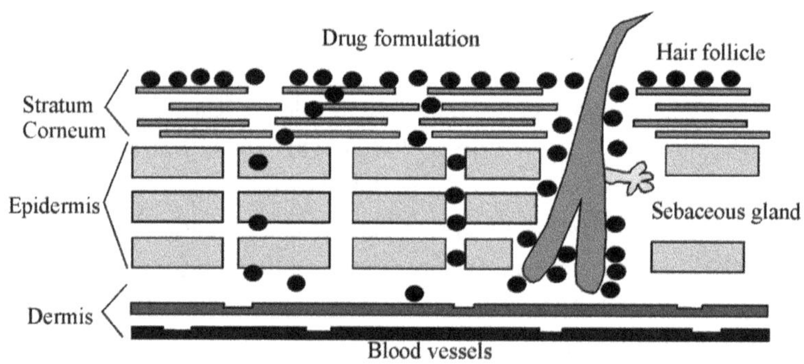

Different hydrophobic drugs have already been formulated successfully in SLN, for instance ramipril (Ekambaram & Abdul, 2011), repaglinide (Ebrahimi, Javadzadeh, Hamidi, & Barzegar Jalali, 2016; Rawat, Jain, & Singh, 2011), methotrexate (Ferreira et al., 2016), docetaxel (Singh, Swami, Jeengar, Khan, & Sistla, 2015), curcumin (Jiang et al., 2017; S. Kim, Diab, Joubert, Canilho, & Pasc, 2016), rifampicin (Singh et al., 2015), calendula officinalis (Arana et al., 2015), doxorubicin, verapamil and propranolol clorhidrate (Wong, Bendayan, Rauth, & Wu, 2004), paclitaxel (Cavalli, Caputo, & Gasco, 2000), cyclosporine (Urbán-Morlán et al., 2010), spironolactone (Kelidari et al., 2015), guggul (Gaur, Mishra, & Purohit, 2013), montelukast (Priyanka & Sathali, 2012) and capsaicin. The later drug has a especial interest due to powerful analgesic effect and the requirement of encapsulation to avoid local adverse effects (Agrawal, Gupta, & Vyas, 2015; J. H. Kim et al., 2014).

Capsaicin ((6E)-N-(4-Hydroxy-3-methoxybenzyl)-8-methyl-6-nonenamide) is a natural alkaloid, an active component responsible for the pungent of chili peppers, plants that belong to Capsicum genre. It has part of the vanilla (4-hydroxy-3-methoxybenzyl), and this confers its biological activity. Structurally it has a benzene ring and long hydrophobic chains with an amide group (Hayman & Kam, 2008). It is irritating to animals and produces a burning sensation in the tissues in which it comes into contact (Levy, Abraham, & Tomlinson, 1991). It is known by its analgesic properties, mainly used in neuropathic an arthritic pain administered topically (Hayman & Kam, 2008). Its effect by initial neuronal excitation following loss sensibilization of nerve fibers and finally at high concentrations of drug or repeating doses can have a local effect describe as defunctionalization, constituted by reduction of spontaneous activity and a loss of responsiveness to a wide range of sensory stimuli (Anand & Bley, 2011). This desensitization is the basis of its analgesic effect (Levy et al., 1991). The limiting condition of capsaicin is on the step initial neuronal excitation where appear sensations of heat, burning, stinging or itching attributed to the release of substance P, a neurotransmitter released by stimulated nerve fibers. An immune-histochemical study using antibodies to protein gene product 9.5 or other nerve fibre proteins provide evidence that capsaicin can produce a highly localized loss of the nociceptive nerve fibre terminals in the epidermis and dermis (Anand & Bley, 2011).

To use capsaicin as an analgesic is very important to carry the drug through the superficial layer of the skin into deep dermal region in order to reduce irritation (Wang et al., 2017). For this purpose, SLN are the adequate vehicle due to their capacity to penetrate in the hair follicles and deliver drug after fusion with the lipids in the skin (Prow et al., 2011).

Some factors such as type of solvent, lipid concentration, lipid type, surfactant concentration, surfactant type and drug concentration can influence on the particle size, polydispersity index (PDI, a measure of size distribution) and drug loading (DL) of SLN. Even experimental conditions such as temperature, homogenization time, sonication time, stirring rate are evaluated (Malam, Loizidou, & Seifalian, 2009; Quintanar-Guerrero, Tamayo-Esquivel, Ganem-Quintanar, Allémann, & Doelker, 2005).

An optimal formulation is essential to obtain a nanocarrier with enhanced permeation of drugs to target site. The parameters to optimize usually are solvent employed, concentrations of surfactant, lipid and drug (Das et al., 2011; Taveira et al., 2012; Urbán-Morlán et al., 2010). This information is better described on section 2. Applications of DoE on the optimization of nanoparticles.

DoE has been used to investigate the role of different factors affects the size of nanoparticles. Two-level factorial and fractional factorial designs are widely used to identify significant effects in industrial processes, the 2^p treatment combinations can be applied to the experimental units in 2^{p-1} blocks of size two so that 2^{p-1} factorial effects are orthogonal to blocks, and are estimable, and the remaining effects are completely confounded with blocks, and are inestimable. Such an arrangement will be described as a blocked replicate. It is assumed that blocks do not interact with factors (Godolphin, 2019).

PREPARATION AND OPTIMIZATION METHODS FOR DRUG-LOADED SLN

Methods to Produce SLN

Several methods have been used for the preparation of SLN. Among the more used are high shear homogenization, ultrasonication/high-speed homogenization, emulsion-evaporation technique, and emulsion-diffusion technique (Mukherjee, Ray, & Thakur, 2009).

High Shear Homogenization

In high shear homogenization a pre-emulsion of the drug, lipids (trimyristin, tripalmitin, a mixture of mono, di and triglycerides, etc.), and steric stabilizers (e.g. glycerol behenate, poloxamer 188, etc), are passed through a high shear homogenizer to produce nanoparticles. This method allows scaling up the production of nanoparticles. One disadvantage is that dispersion quality is often compromised by the presence of microparticles. Emulsification time, stirring rate and cooling condition are among the parameters that affect particle size, PDI, and zeta potential (a dispersion stability parameter).

The technique can be performed as hot homogenization or cold homogenization. Hot homogenization is carried out at temperatures above the melting point of the lipid. A pre-emulsion of the drug dispersed in lipid melt and the aqueous emulsifier phase (same temperature) is obtained by high-shear mixing device.

High-Pressure Homogenization

High-pressure homogenization of the pre-emulsion is performed above the lipid melting point. Usually, smaller particle sizes are obtained at higher processing temperatures because of the lowered viscosity of the lipid phase (Lander et al., 2000). Typically, 3-5 passes through the high-pressure homogenizer are required to obtain good quality nanoparticles.

On the other hand, cold homogenization process is carried out with the solid lipid. The lipid must remain in the solid state during homogenization, through effective temperature regulation. In the first step, the drug is solubilized or dispersed of in the lipid melt, as in hot homogenization procedure. Then the drug containing melt is rapidly cooled to obtain homogenous drug distribution in the lipid matrix. The dispersion is ground by ball/mortar milling to obtain microparticles. The SLN are dispersed in a cold emulsifier solution. The dispersion is subjected to high-pressure homogenization at or below room temperature. Cold homogenization was developed to avoid the temperature-mediated degradation of the drug, and else to overcome partitioning of drug into the aqueous phase during homogenization. However, larger particle sizes with high PDI are typical of cold homogenization process.

Ultrasonication or High-Speed Homogenization

SLN can also be produced by high speed stirring or sonication (Eldem, Speiser, & Hincal, 1991). The main advantage of this method is that instrumentation is less expensive and more common in the labs that a high-pressure homogenizer. The disadvantage is this method produces broader particle size distribution, which leads to physical instability during storage.

Solvent Emulsification/Evaporation Technique

For this technique the lipid and the drug are dissolved in water-immiscible organic solvent (e.g. chloroform, ethyl acetate, cyclohexane, etc.), then the solution is emulsified in an aqueous phase containing and steric stabilizer (e.g. PVAL, polyethylenglycol, poloxamers, etc.) or an emulsifier (e.g. lecithin, sodium glycocholate, etc.) (Sjöström & Bergenståhl, 1992). The emulsion is subject to high-speed stirring or ultrasound. Upon evaporation of the solvent by stirring at atmospheric pressure or under vacuum nanoparticle dispersion is formed by precipitation of the drug-loaded lipid. The method has the advantage of simple setup and it has also been applied to the production of nanoparticles of biodegradable polymers (Lemos-senna, Wouessidjewe, Lesieur, & Duche, 1998).

Solvent Emulsification-Diffusion Technique

The method solvent emulsification-diffusion previously development (Quintanar-Guerrero et al., 2005) consists of dissolving the lipid in a partially water-miscible solvent, such as ethyl acetate or methyl acetate (previously saturated with water) at RT, or at controlled temperature depending on lipid solubility. This organic phase is emulsified in an aqueous solution of a stabilizing agent such as PVAL (saturated with solvent) by conventional stirring at the same temperature used to dissolve the lipid. This oil/water emulsion is then diluted with an excess of water at controlled temperature in order to provoke the diffusion from the internal phase into the external phase, thereby causing lipid aggregation in the form of SLN. This technique has the advantages of being efficient and versatile, easy implementation and scaling up, high reproducibility and narrow size distribution; avoids long exposure to high temperatures and to mechanical dispersion. Particles with a diameter under 100 nm can be produced by increasing the process temperature, the stirring rate, the amount of stabilizer, and by lowering the amount of lipid. Control of the preparative variables allowed to obtain SLN with diameters under 100 nm. In terms of stability, only PVAL was able to preserve the physical stability of the dispersion for long periods after preparation. This effect was attributed to the ability of PVAL chains to form a strongly attached layer on the nanoparticle surface with an excellent repulsion effect.

Applications of DoE on the Optimization of SLN

As presented, during the manufacturing process, many parameters appear to have a marked influence on the physicochemical properties of SLN. It is, therefore, essential to have a clear understanding of how preparation conditions determine particle characteristics and, in particular, how these characteristics are influenced by potential interactions between variables in the preparation process.

Physicochemical characterizations such as particle size, zeta potential, transmission electron microscopy (TEM), scanning electron microscopy (SEM), differential screening calorimetric (DSC), X-ray powder diffraction (XRD), Fourier infrared spectroscopy (FTIR), near-infrared spectroscopy (NIR) are usually performed.

To evaluate the effect of a large number of formulation variables usually requires many experiments, which are often costly and time consuming. It is therefore prudent to minimize the total number of experiments in the optimization process, without sacrificing final product quality.

For instance, DoE is a very useful tool to elucidate the effect of the variables and allows the selection of the optimal formulation to attain the optimized physicochemical properties. Such properties will translate in optimized *in-vivo* performance of the nanoparticles. One of the main advantages of nanoparticles is the improved permeation through physiological barriers. As a rule of thumb smaller particles gave better diffusion characteristics, for example for intestinal, transdermal, ocular, nasal, and also to cross de blood brain barrier introducing drugs into the brain. Beside the requirement of small size, have a monodisperse distribution is important to assure batch-to-batch reproducibility and stability of the dispersion.

Zeta potential is used as a measurement of the nanodispersion of SLN. Negative Z values are essential to ensure low toxicity of the nanoparticles. The other important parameter to optimize is DL, since, as mentioned previously, a disadvantage of SLN is partitioning of the drug in the aqueous phase producing low loading efficiency.

Yang and Zhu(Yang & Zhu, 2002) prepared SLN loaded with camptothecin (CA), an antitumor drug, by high-pressure emulsification. CA was dissolved in pH 8.0 absolute ethanol adjusted with ammonium hydroxide; stearic acid and soybean lecithin were added. The mixture was heated to above melting point and ultrasonicated to produce a clear melted lipid phase. Poloxamer 188 was dissolved in the aqueous phase containing glycerol as an isotonic agent. A predispersion of the lipid in this aqueous phase, which was previously heated to approximately the same temperature as the melted lipid, was prepared by magnetic stirring and ultrasonication. This premix was passed through a preheated high-pressure homogenizer for five cycles at different pressures.

A Taguchi orthogonal experimental design was used to study the influence of four different variables: emulsifiers, concentration of CA, concentration of emulsifiers, and homogenized pressure. Each variable had three value levels on nanoparticle size. Analysis of variance has been used to evaluate the preparation of CA-SLN and perform product optimization. The optimized CA-SLN suspension was lyophilized using mannitol and glucose as cryoprotectants. The physicochemical characteristics of CA-SLN were evaluated using TEM, electrophoresis, and DSC. The release of CA from CA-SLN in various media was evaluated using a high-performance liquid chromatography (HPLC) method. The results showed that the concentration of emulsifier and the homogenization pressure had a significant influence on the particle size. The optimized CA-SLN had an average diameter of about 200 nm, exhibited monodispersity with Dw/Dn of 1.06 (were Dn =Particle size and Dw/Dn =Size distribution), and carried a negative charge. The optimal cryoprotectants consisted of 10% mannitol and 5% glucose in nanoparticle suspension. Lyophilized product was reconstituted in distilled water within 0.5 min without change of nanoparticle size. CA might exist in an amorphous state in SLN. In vitro results showed that drug release was achieved for up to one week. The results demonstrated that SLN are good sustained-release delivery vehicles for lipophilic drugs.

Zhang et al. (Zhang, Fan, & Smith, 2009) performed a systematic investigation on the simultaneous influence of multiple formulations variables on the SLN properties. The SLN were prepared by emulsion-diffusion method. The lipid monostearin was dissolved in 5mL of a water-miscible solvent in a water bath at 70°C. Three solvents were used for screening purposes: Ethanol, isopropanol, and acetone. The resultant organic solution was injected into 50 mL of an aqueous phase containing surfactant at different temperatures, under mechanical agitation for 30 min. The nanosuspensions or nanoemulsions formed were then cooled to RT, resulting in SLN formations. The specific amount of ingredients added during the preparation strictly followed the experimental design generated by a central composite design. This method uses a relatively small number of empirical evaluations to determine mathematical trends that allow the prediction of final process parameters needed for a specific, optimized outcome. The results showed that lipid concentration and temperature are the critical parameters for the particle size of the nanoparticles.

Shah et. al in a study using simvastatin as the model drug, a 2^3 full-factorial experimental design was used to optimize SLN, prepared by solvent injection technique (Shah & Pathak, 2010). The amount of glycerol monostearate, concentration of poloxamer 407 and volume of isopropyl alcohol were selected as independent variables. Each factor was set at a high level and low level. Eight formulations of SLN were prepared according to the design. The particle size, percent EE, and percent cumulative drug release (CDR) were taken as response parameters. The design was validated by extra design checkpoint formulation, and the possible interactions between independent variables were studied. The responses of the design were analyzed using Design Expert 7.1.6. (Stat-Ease, Inc, USA), and the analytical tools of software were used to draw Pareto charts and response surface plots. Based on software analysis, optimized formulation with a desirability factor of 0.611 was evaluated for the independent parameters. Optimized formulation showed a particle size of 258.5 nm, % EE of 75.81%, with 82.67% CDR after 55 h. The release kinetics of the optimized formulation best fitted the Higuchi model and the recrystallization index of optimized formulation was found to be 65.51%.

Risperidone was formulated as SLN formulation using response surface methodology of DoE to evaluate compositional variations and their interaction (Rahman, Zidan, & Khan, 2010). The SLN were characterized by non-destructive methods of analysis. Box–Behnken DoE was constructed using drug (X1), lipid (X2) and surfactant (X3) level as independent factors. Compritol 888 ATO and sodium lauryl sulfate were used as lipid and surfactant, respectively. The SLN was prepared by solvent evaporation method and characterized by TEM, DSC, XRD, FTIR, NIR and NIR-chemical imaging (NIR-CI). Responses measured were EE (Y1), diameter (Y2), zeta potential (Y3), burst effect (Y4) and CDR in 8 h (Y5). Statistically significant ($p < 0.05$) effect of X1 on the Y1, Y2, Y3 and Y4 were seen. FTIR revealed no interaction between risperidone and compritol 888 ATO. TEM showed spherical and smooth surface SLN. Compritol retained its crystalline nature in the SLN formulation revealed by DSC and XRD studies. Homogenous distribution of risperidone and compritol 888 ATO was revealed by NIR-CI. Principal component analysis and partial least square (PLS) were carried out on NIR data of SLN formulation. PLS showed a correlation coefficient > 0.996 for prediction and calibration model of both risperidone and compritol 888 ATO. The accuracy of models in predicting risperidone and compritol 888 ATO were 1.60% and 11.27%, respectively. In conclusion, the DoE reveals a significant effect of DL on SLN characteristics, and chemometric models based on NIR and NIR-CI data provided a non-destructive method of estimation of components of SLN.

Vitorino et al. in another work prepared SLN by a modification of the emulsification-solvent evaporation to study the effect of variables (Vitorino, Carvalho, Almeida, Sousa, & Pais, 2011). The lipid (glyceryl tripalmitate, glyceryl behenate or glyceryl palmitostearate) was dissolved in dichloromethane and then added dropwise to a 30 ml of emulsifier solution in a high shear homogenizer. Ultrasound was applied to some samples. The dispersion obtained was magnetically stirred to allow the solvent evaporation. The stabilizer used was PVAL or polysorbate 80. Simvastatin was used as a model drug under the same conditions. An experimental design with a two level, three variable, 2k full factorial planning was performed for the optimization of the particle composition. The solvent: lipid ratio constituted the main factor influencing particle size. Increase the amount of solvent induce a decrease in the particle size. This was a general trend, essentially independent from solvent and lipid type. The amount of emulsifier had a non-trivial impact on the size, depending on whether systems were located below, above or close to the optimal surface coverage. The amount of lipid had a limited influence upon particle size, being more relevant for lower lipid concentrations. Sonication reduced both particle size and PDI. These particles were also tested as drug carriers using simvastatin as a model of a lipophilic drug. SLN were able to entrap a high amount of simvastatin, with little effect upon the size and zeta potential, constituting a promising carrier for lipophilic drugs.

SLN of quercetin, a natural flavonoid with established antioxidant activity, were developed for intravenous administration in order to improve its permeation across the blood-brain barrier into the central nervous system, and eventually to improve the therapeutic efficacy of this molecule in Alzheimer's disease (Dhawan, Kapil, & Singh, 2011). SLN of quercetin were formulated using Compritol as the lipid and Tween 80 as the surfactant through a microemulsification technique and optimized employing a 3^2 central composite design. Selection of the optimized SLN formulation, using brute-force methodology and overlay plots, was based on its efficiency of entrapping quercetin inside the lipophilic core, particle size, surface charge potential and ability of the SLN to release the entrapped drug completely. The optimized formulation was subjected to various *in-vivo* behavioral and biochemical studies in Wistar rats. The optimized formulation exhibited a particle size of less than 200 nm, 85.73% drug EE, and zeta potential of 21.05 mV. In all the *in-vivo* behavioral and biochemical experiments, the rats treated with

SLN-encapsulated quercetin showed markedly better memory-retention compared to pure quercetin-treated rats. The studies demonstrated successful targeting of the potent natural antioxidant, quercetin, to the brain as a novel strategy having significant therapeutic potential to treat Alzheimer's disease.

Hao et al. in a study optimized chloramphenicol-loaded SLN by investigating the relationship between design factors and experimental data using response surface methodology (Hao et al., 2011). A Box-Behnken design was constructed using solid lipid, surfactant, and drug/lipid ratio level as independent factors. SLN was successfully prepared by a modified method of melt-emulsion ultrasonication and low temperature solidification technique using glyceryl monostearate as the solid lipid, and poloxamer 188 as the surfactant. The dependent variables or responses were EE, DL, and turbidity. Properties of SLN such as the morphology, particle size, zeta potential, EE, DL, and drug release behavior were investigated, respectively. As a result, the nanoparticle designed showed nearly spherical particles with a mean particle size of 248 nm. The PDI of particle size was 0.277 ± 0.058 and zeta potential was −8.74 mV. The EE (%) and DL (%) could reach up to 83.29% ± 1.23% and 10.11% ± 2.02%, respectively. In vitro release studies showed a burst release at the initial stage followed by a prolonged release of chloramphenicol from SLN up to 48 hours. The release kinetics of the optimized formulation best fitted the Peppas–Korsmeyer model. These results indicated that the chloramphenicol-loaded SLN could potentially be exploited as a delivery system with improved drug EE and controlled drug release.

Hao et al. (Hao et al., 2012) also prepared SLN containing bacialin, an anxiolytic drug. The carrier system was composed of a stearic acid alkaline salt as lipid matrix and prepared as per the coacervation method in which fatty acids precipitated from their sodium salt micelles in the presence of polymeric nonionic surfactants comparing with a reference solution (baicalin dispersed in 1% hydroxypropyl-methylcellulose aqueous solution). A two-factor five-level central composite design was introduced to perform the experiments. A quadratic polynomial model was generated to predict and evaluate the independent variables with respect to the dependent variables. The composition of optimal formulation was determined as 0.69% (w/v) lipid and 26.64% (w/w) drug/lipid ratio. The results showed that the optimal formulation of baicalin-loaded SLN had %EE of 88.29%, the particle size of 347.3 nm and PDI of 0.169. The morphology of nanoparticles was found to be nearly spherical in shape by SEM observation. The DSC analysis indicated that the drug incorporated into SLN was not in an amorphous form but in a crystalline state. Using a rat model, the maximum plasma concentration (C_{max}), Medium Residence Time, area under the first moment curve from zero to infinite ($AUMC_{0\text{-}\infty}$) and area under the plasma concentration-time curve from zero to infinite ($AUC_{0\text{-}\infty}$) values of SLN were approximately 1.6-fold, 1.9-fold, 5.0-fold and 2.6-fold greater than that of reference preparation, respectively.

Cationic SLN were synthesized by the phase-inversion temperature technique, physicochemical characterization and statistical optimization of SLN was studied, as potential carrier for gene therapy (Carbone, Tomasello, Ruozi, Renis, & Puglisi, 2012). The preliminary screening from a physicochemical point of view on three cationic lipids (CTAB, DDAB and DOTAP), selected on the basis of their different chemical structure and increasing lipophilicity, allowed us to select SLN with DOTAP, due to its higher zeta potential and smaller particle size. Afterward, a 2^2 full factorial experimental design was developed in order to study the effects of two independent variables (amount of DOTAP and concentration of lipid matrix) and their interaction on mean particle size and zeta potential values. The factorial planning was validated by ANOVA analysis; the correspondence between the predicted values of size and Zeta potential and those measured experimentally confirmed the validity of the design and the equation applied for its resolution. The factorial design showed a significant influence of the independent variables on the selected parameters; in particular, a higher effect of DOTAP was observed on zeta potential value. Different dilutions of the optimized SLN containing 7% w/w of cetyl palmitate and 1% w/w of DOTAP, with size and zeta potential values respectively of 462.9 nm and 50.8 mV, were in vitro examined to evaluate the possible cytotoxicity on two models of cell cultures: human prostate cancer androgen-non-responsive DU-145 cells and primary cultures of rat astrocytes.

SLN is a very well tolerated carrier systems for dermal application due to the employment of physiological and/or biodegradable lipids. The effects of five factors, two categorical and three quantitative factors, were studied on the mean diameter and EE of the produced SLN using response surface method, D-optimal design (Ghadiri et al., 2012). Two methods of microemulsion and solvent diffusion and two types of lipid, cetyl palmitate and stearic acid, were examined comparatively. The quantitative variables were studied in three levels; amount of original Paromomycin (60, 90 and 120 mg), fraction of surfactant (0.5, 0.75 and 1 w/v %) and drug to lipid ratio (2, 4 and 6). Mean particle size and EE of the loaded Paromomycin were modeled statistically and the optimal condition was determined to approach to the maximum EE. The drug release profile of the optimal formulated material was examined in aqueous media and 64% of the Paromomycin loaded in SLN was gradually released during 24 h, which reveals efficient prolonged release of the drug.

Isotretinoin (ITR) is a drug for the treatment of all types of acne, including recalcitrant, severe and nodulocystic. Oral intake of the drug is reported to be associated with severe side-effects including teratogenicity, skin dryness and psychological disorders. Topical delivery, though advised for ITR, is marked with several hiccups like irritation, erythema and peeling of the skin. Raza et al. (Raza, Singh, Singal, Wadhwa, & Katare, 2013) optimized SLN of ITR employing formulation by design (FbD) approach. Microemulsification method was employed for the preparation of SLN. In brief, drug and the lipophilic antioxidant (BHT;2% of the total lipids) were dissolved in a portion of ethanol. The phosphatidylcholine was dispersed in a portion of water along with Tween 80 and hydrophilic antioxidant (sodium metabisufite). The solid lipid Compritol 888 ATO was melted at 70 oC. The lipid phase, aqueous phase and drug solution were mixed isothermally to obtain a clear microemulsion. The hot microemulsion formed was poured into the remaining portion of water (about 80% of the total), previously cooled and maintained at 4 oC. The mixture was stirred continuously at 3000 rpm for 20 min. The developed system was characterized and evaluated for skin compliance, skin transport characteristics and anti-acne potential against testosterone-induced acne in male Laca mice. The SLN were able to transport the drug to various skin layers effectively while formed drug micro-reservoirs. The nano-colloidal systems showed marked anti-acne potential and tolerability on the mouse skin vis-à-vis the marketed product. The optimized SLN exhibited drug entrapment of 89.49 ± 4.1%, while the size was found to be in the nano-range (i.e., 75.3 ± 2.4 nm).

Other types of nanoparticles, beside SLN, have been optimized for the loading of capsaicin. For example: De Freitas et. al, (De Freitas et al., 2018) optimized the preparation process of capsaicin-loaded albumin nanoparticles applicating a factorial design. Statistical analysis of the data was conducted by one-way analysis of variance. Factors as glutaraldehyde percent, pH value of the solution, rate of the added ethanolic capsaicin solution and bovine serum albumin (BSA)/ capsaicin ratio were analyzed to determinate the influence of each parameter on particle diameter and encapsulation efficiency, they present a diameter of about 200 nm, quasi-spherical morphology, EE of 98.3 ± 7.4% and negative zeta potential.

No significant difference in diameter and PDI was observed between the formulations achieved at different pH values, velocity of ethanol addition and proportion of BSA and capsaicin. The lower rate of ethanol addition produced significantly higher EE% than the faster addition, indicating that the process of drug encapsulation might require lower velocity in the desolvation step.

Case of Study

As described above, capsaicin is a metabolite with high therapeutic potential and low water solubility (Zhu et al., 2015). A case of study is presented for the formulation of capsaicin-loaded SLN, intended as a transdermal delivery system. The lipophilic characteristic of the drug was taken as an advantage to prepared SLN using the adapted solvent emulsification-diffusion method. The size of the SLN was optimized to obtain the diameter in the scale of 100 nm; since it is know that nanoparticles of such size have good skin penetration. Experimental design can be used to optimize every independent factor to obtain a specific response, in this case, to generate the nanoparticles of the target size (or smaller), using the minimum number of experiments.

Based on previously information showed in Table 1, the alternative hypothesis is to prepare nanoparticles with different drug concentration, solvent type, lipid concentration, and surfactant concentration, so the design can find an optimized formulation to obtain smaller drug-loaded nanoparticles.

Previously, a drug-free formulation was development by the solvent emulsion-diffusion method with 400 mg of solid lipid and 5% w/v of surfactant (PVAL); besides, a drug-loaded SLN formulation was evaluated from 5 to 50% w/w with respect to the lipid phase (Quintanar-Guerrero et al., 2005; Urbán-Morlán et al., 2010). In accordance with the authors these concentrations were taken as mean, and the lower and upper factor levels, as shown on Table 2, to guarantee the complete dissolution of reagents.

In this study the fractional factorial design $2^{(k-p)}$ considered four factors or independent variables (k = 4); % of capsaicin, solvent type, % of lipid Gelucire® 44/14 and % of PVAL, and one response or depend variable such as particle size. Using a fractional factorial design of half fraction with k = 4 factors, and p = 1 $^{(24-1)}$ with eight runs, IV resolution, plus two central points per block. Block shows ten different formulations, equal repeating per block (n = 3); the design used 30 random runs for this study showing on blocks.

The lower and higher concentration of drug was estimated at 10% and 30%, respectably, based on the lipid content. The solvents ethyl acetate (EtOAc) and methyl acetate (MeOAc), encoded with -1 and +1 respectively, were selected due to different aqueous miscibilities, but knowing the qualitative solubility of the lipid Gelucire® 44/14, and because their low toxic potential (Quintanar-Guerrero et al., 2005). Gelucire ® 44/14 was select on 150 mg and 250 mg for the lower and upper contents, respectively, in order to guarantee the solubility of the lipid in the solvents, PVAL was select on 3% and 7% respect to the aqueous phase, for lower and upper level, respectively.

Capsaicin-loaded SLN were successfully prepared using emulsion diffusion method. This method consist in dissolving the lipid and drug in water-saturated solvent which was emulsified with solvent-saturated aqueous solution containing PVAL (dispersion medium), using a Polytron homogenizer (PT 1600E, Kinematica, Lucerne, Switzerland), followed the formation of an O/W emulsion, finally simple water (dilution medium) were added to the system to allow the diffusion of solvent into the external phase, allowing the formation of SLN.

Formulation variables were optimized to get SLN of minimum (~100nm) particle size. The standard error based in total error with 25 degrees of freedom (d.f.), shows an orthogonal design, the largest variance inflation factor (V.I.F) for all factors are equal to 1.0. A perfectly orthogonal design will show a diagonal matrix with 1's on the diagonal and 0's off the diagonal.

Any non-zero terms of the diagonal imply that the estimates of the effects corresponding to that row and column will be correlated. In this case, there is no correlation amongst any of the effects. This means that this design will get clear estimates of all those effects.

Table 1. Summary of the reported information which applied a DoE, showing independent factors and principal responses to optimize

Author	Drug	Method	Experimental Design	Independent Factors	Significance	Response
Yang et al., 2002	Camptothecin	High-pressure emulsification	Taguchi orthogonal	Emulsifiers Concentration of CA Concentration of emulsifiers Homogenized pressure	- - $p<0.05$ $p<0.05$	Particle size
Zhang et al., 2009	Free drug	Emulsion-diffusion	Central composite design	Lipid concentration Surfactant % Stirring speed Temperature	$p<0.05$* - - $p<0.05$*	*Particle size Zeta potential
Shah et al., 2010	Simvastatin	Solvent injection technique	2^3 full-factorial	Amount of glycerol monostearate Concentration of poloxamer 407 Volume of isopropyl alcohol	- - -	Particle size % EE % CDR
Rahman et al., 2010	Risperidone	Solvent evaporation	Response surface	Drug Lipid Surfactant	- $p < 0.05$[1,2,3,4] -	[1]EE [2]Diameter [3]Zeta potential [4]Burst effect [5]CDR in 8 h
Vitorino et al., 2011	Simvastatin	Emulsification-solvent evaporation	2^3 full factorial	Lipid concentration Solvent:lipid ratio Emulsifier concentration	- - -	Particle size
Dhawan et al., 2011	Quercetin	Microemulsification	3^2 Central composite design	Lipid Surfactant	- -	Particle size EE Amount of drug release in 20 h Zeta potential
Hao et al., 2011	Chloramphenicol	Melt-emulsion ultrasonication and low temperature solidification	Response surface Box-Behnken	solid lipid, surfactant, Drug/lipid ratio	$p<0.05$[1] $p<0.05$[1,2,3,] $p<0.05$[2,3]	[1]EE [2]DL [3]Turbidity
Haoet al., 2012	Baicalin	Coacervation	Central composite factorial	Lipid amount Drug/lipid ratio	$p<0.05$[1,2,3] $p<0.05$[2,3]	[1]EE [2]Particle size [3]PDI
Carbone et al., 2012	Free drug	Hot high-pressure homogenization	2^2 full factorial	Amount of DOTAP Concentration of the wax cutina CP	- -	Zeta potential Particle size
Ghadiri et al., 2012	Free drug	Microemulsion and solvent diffusion	Two level half fractional factorial (2^{5-1}) followed by D-optimal design	Lipid drug ratio Surfactant % Amount of drug Lipid type Method	$p<0.05$[1,2] $p<0.05$[1,2] $p<0.05$[1] $p<0.05$[1,2] $p<0.05$[1,2]	[1]EE % [2]Particle size
Raza et al., 2013	Isotretinoin	Microemulsificatio	Face centred cubic design (3^2 factorial)	Amount of solid lipid Amount of Phospholipon® 90G	- -	Particle size EE Skin permeation flux Skin retention
De Freitas et al., 2018	Capsaicin	Desolvation-coacervacion	Factorial	pH value Velocity of ethanol addition Proportion of BSA:CAP	- - -	Mean size

Table 2. Factor level applied in the optimization

Independent Factor	Lower level	Upper level	Units
Capsaicin	10.0	30.0	%
Solvent	-1.0	+1.0	
Gelucire ® 44/14	150.0	250.0	mg
PVAL	3.0	7.0	%

Table 3. Analysis of variance for particle size of SLN loaded with CAP

Source	Sum of Squares	Df	Mean Square	F-Ratio	p-Value
A:Capsaicin	263509.0	1	263509.0	40.24	0.0000
B:Solvent	36624.1	1	36624.1	5.59	0.0261
C:Gelucire ® 44/14	22878.4	1	22878.4	3.49	0.0733
D:PVAL	13547.0	1	13547.0	2.07	0.1627
Total error	163694.0	25	6547.76		
Total (corr.)	500253.0	29			

R-squared = 67.27 percent
R-squared (adjusted for d.f.) = 62.04 percent
Standard Error of Est. = 80.9182
Mean absolute error = 64.0238
Durbin-Watson statistic = 2.00421 (P=0.5307)
Lag 1 residual autocorrelation = -0.0469237

Analysis of Variance (ANOVA)

Analysis of variance for particle size divides the particle size variability into separate pieces for each of the effects. It then tests the statistical significance of each effect by comparing the mean square against an estimate of the experimental error. In this case, two effects (capsaicin concentration and solvent employed) have a p-value of less than 0.05, indicating that they are significantly different from zero at 95.0% confidence level as shown on Table 3. In other words, particle size is significatively different when using 10%, 20, or 30% of drug to prepare nanoparticles, and in the same way, particle size will be different depending of the solvent used (MeOAc or EtOAc).

The R-Squared statistic indicates that the model as fitted explains 67.27% of the variability in particle size. The adjusted R-squared statistic, which is more suitable for comparing models with different numbers of independent variables, is 62.04%. The standard error of the estimate shows the standard deviation of the residuals to be 80.9182. The mean absolute error (MAE) of 64.0238 is the average value of the residuals. The Durbin-Watson (DW) statistic tests the residuals to determine if there is any significant correlation based on the order in which they occur in the data file. Since the p-value is greater than 5.0%, there is no indication of serial autocorrelation in the residuals at the 5.0% significance level. This model explains 67.27% of the variability in particle size.

Figure 3. Standardized Pareto chart for particle size of SLN loaded with CAP

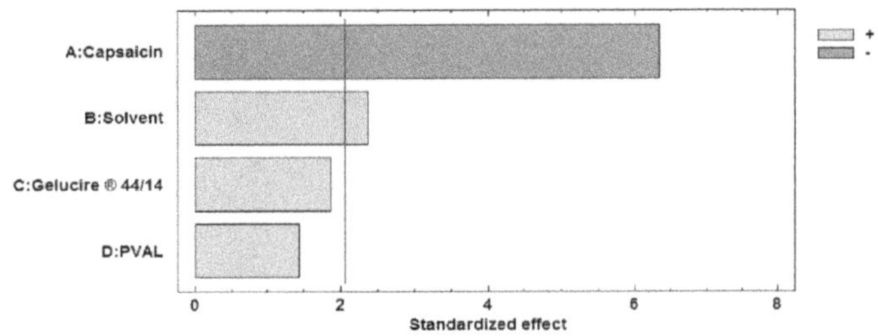

Standardized Pareto chart for particle size and the main effect plot for particle shows the effects statistically significative and predictions to increase or decrease the particle size for every factor, Figure 3 shows the influence of capsaicin to decrease the mean particle size, and the solvent to increase it. The other factors such as Gelucire ® 44/14 and PVAL the positive influence is not statically significative. The effect of surfactant percentage is not a statically significant (p=0.1627); Kwon et al. explained "above each concentration, only a small quantity of stabilizer is adsorbed at the interface, the excess remains in the continuous phase, and does not play any significative role in the emulsion" (Kwon, Lee, Choi, Jang, & Kim, 2001). According to the literature, the particle size decreased when PVAL concentration was increased from 1% to 3%, and only a few nanometer decrease was seen from 2.5 to 3% (Derman, 2015; Shi, Ahmad Khan, Wang, & Schneider, 2015). In fact, concentrations as low as 0.1% w/w of PVAL were enough to stabilize nanoparticles (Mante, Heider, Zlomke, & Mäder, 2016).

The effect of lipid percentage is not statically significative (p=0.0733). Gelucire ® 44/14 is approved by the US Food and Drug Administration (FDA). It has been used in many formulations as wetting, emulsifying, and stabilizing agent resulted in stable nanoparticles without the aid of surfactant, due to the amphiphilic properties, facilitated the emulsification and resulted in low particle size and PDI (El Assasy, Younes, & Makhlouf, 2019; Wehrung, Geldenhuys, & Oyewumi, 2012).

This design shows the regression equation which has been fitted to the data. The equation of the fitted model is:

Particle size = 231.224 - 10.4783*Capsaicin + 34.94*Solvent

+ 0.6175*Gelucire ® 44/14 + 11.8792*PVAL

The factor % capsaicin shows the influence to decrease the particle size, meanwhile the other factors has the influence to increase the particle size. The values of the variables are specified in their original units, except for the categorical factors which take the values -1 for the low level or EtOAc and +1 for the high level or MeOAc.

Applicating this DoE to evaluate the effects between levels and compared the effects statistically significative by ANOVA multifactorial for particle size the results are as following.

The capsaicin factor effect on the mean particle size in this case have a p-value less than 0.05, indicating that they are significantly different from zero at the 95.0% confidence level (Table 3). This independent variable with tree level groups (10, 20 and 30 percent) decreases the particle size when the concentration of drug is higher as shown on Figure 4., percent of capsaicin has the ability of decrease the particle size with high concentrations of drug, that mean nanoparticles prepared with 30% of cap-

Figure 4. Main effects plot for particle size of SLN loaded with CAP

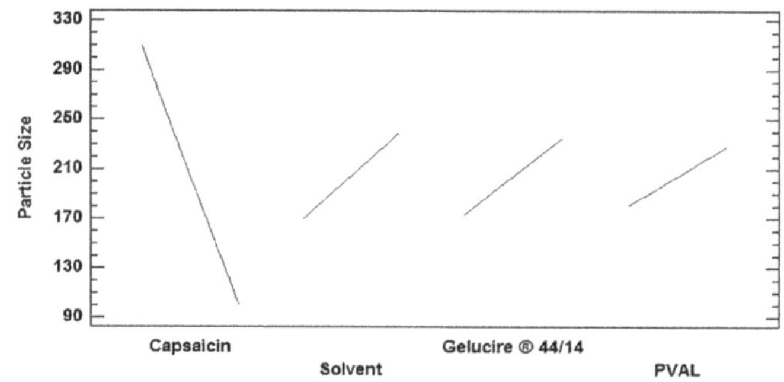

Figure 5. Mean plot of particle size and 95% LSD intervals for the different concentrations of capsaicin in loaded SLN

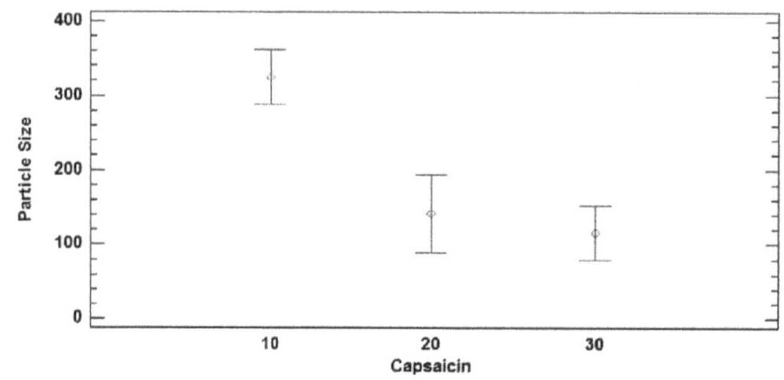

saicin show a mean particle size of 115.5 nm, prepared with 20% of capsaicin a mean of 141.6 nm and prepared with 10% of capsaicin a mean of 325.0 nm. The smallest nanoparticles can be prepared with 30% of capsaicin on a range of lower limit on 67.4 nm and upper limit at 163.5 nm as shown on Table 4. At the top of the levels 2 homogenous groups are identified (20,30), indicating that these pairs show statistically significant differences comparing with level 10 at the 95.0% confidence level as describe in Figure 5. Formulations with lower drug contain resulted in higher particle size, which indicates that capsaicin has an emulsification effect.

The method used to discriminate among the means is Fisher's least significant difference (LSD) procedure. With this method, there is a 5.0% risk of calling each pair of means significantly different when the actual difference equals 0. It applies a multiple comparison procedure to determine which means are significantly different from which others.

Table 4. Shows the mean particle size for each level of the factors. It also shows the standard error of each mean, which is a measure of its sampling variability. The rightmost two columns show 95.0% confidence intervals for each of the means.

The factor solvent has two discontinuous levels .The low level -1 was set up as EtOAc and the high level +1 was set up as MeOAc. The test of statistical significance on the mean particle size in this case have a p-value less than 0.05, indicating that they are significantly different from zero at the 95.0% confidence level (Table 3), with a predisposition to decrease the particle size with EtOAc and increase

Table 4. Table of least squares means for particle size with 95.0% confidence intervals

Level	Count	Mean	Stnd. Error	Lower Limit	Upper Limit
GRAND MEAN	30	194.061			
Capsaicin					
10	12	325.075	23.3697	277.038	373.112
20	6	141.6	33.0497	73.6653	209.535
30	12	115.508	23.3697	67.4712	163.545
Solvent					
EtOAc	15	159.121	21.4752	114.978	203.264
MeOAc	15	229.001	21.4752	184.858	273.144

Figure 6. Mean plot of particle size and 95% LSD intervals for solvent type in SLN loaded with CAP

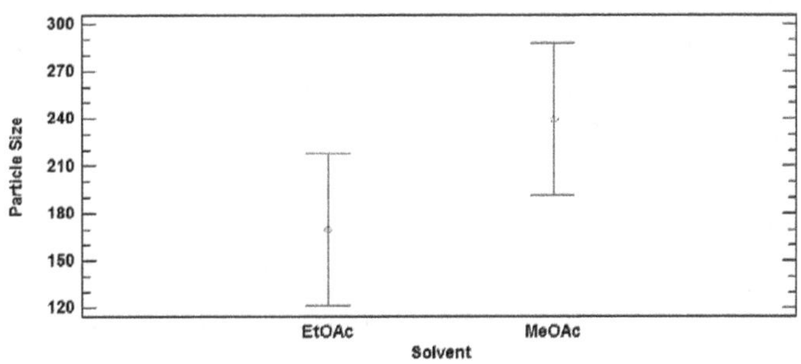

with MeOAc (Figure 4). Nanoparticles prepared using EtOAc as solvent are smaller than using MeOAc at 159.1 nm and 229.0 respectably as shown on Table 4 with 95.0% confidence intervals. The smallest nanoparticles can be prepared with EtOAc on a range of lower limit on 114.9 nm and upper limit at 203.2 nm and lower limit on 184.8 nm and upper limit at 273.1 nm with MeOAc as shown on Table 4. The method LSD indicating this pair show statistically significant differences (Figure 6).

This has a relationship with the miscibility of each solvent, the water miscibility of MeOAc (33, referred to 100 parts of water) is higher than EtOAc (8.5, referred to 100 parts of water), this is a determinant factor on the particle size by a decrease in the interfacial turbulence during the diffusion step (Quintanar-Guerrero et al., 2005). The lower water miscibility of EtOAc can dissolve a high content of hydrophobic molecules (Kocbek, Baumgartner, & Kristl, 2006), such as capsaicin. In general, solvents with high water miscibility and stabilizers able to form stable emulsions are preferred (Quintanar-Guerrero et al., 2005).

Optimized Formulation

The main effects plot for particle size (Figure 4) shows the inclination of each one factor for particle size by level. The factor % of capsaicin on the lower level (10%) has a tendency to produce higher size particles, the upper level (30%) has a bearish tendency of particle size. For the factor solvent smallest particles size are obtained using EtOAc, for the factor lipid concentration the particle size increase when the concentration of lipid is raised up, that mean that using a low concentration generate smaller particles; and finally, with the factor surfactant on the lower level (3%) produce smallest particles, which are increasing when the concentration of PVAL is raised up.

Optimized formulation was obtained with 28% of drug, 153 mg of lipid content and 3% of PVAL using MeOAc as a solvent. These numbers show the combination of factor levels which maintains particle size at 100 nm over the indicated region.

As described in Table 3, Gelucire concentration is not statically significative. However, the prediction to optimize the formulation is to use a lower concentration (150 mg). With this concentration it is possible to make particles smaller than using higher concentration (250 mg). On the same way, the surfactant PVAL is a not statically significative factor, within a prediction that using the lower concentration (3%), smaller particles are possible; opposite to this, high concentration will reflect on higher particle size.

PVAL acts as a viscosity increasing agent and stabilizer under continuous stirring due to its amphipathic character. Its exhibit surface activity at the O/W interface and produce effective stearic stabilization against strong flocculation and coalescense. Tadros (2015) showed that SLN present a clear transition from predominantly viscous to predominantly elastic response, as the oil volume fraction was increased. In this case a higher concentration of oil, such that the inter-droplet distance becomes smaller than twice the adsorbed layer thickness of surfactant results in interpenetration, and possible compression of the long dangling tails of PVAL. Already, small amounts of the surfactant lead to a significant decrease in the drop size (Cocke & Maaß, 2017; Jeevanandam, Chan, & Danquah, 2016; Tadros, 2015).

Figure 7 shows the estimated response surface first order plot for particle size of SLN loaded with capsaicin. In the operability region the independent variables capsaicin and solvent type are delimited by lower level until the uper level, considering all the possible combinations with the factor levels, describing the influence of this factors above the response on particle size. Keeping constant Gelucire ® 44/14=200 mg and PVAL=5.0%, this plot finds target conditions on the process. The height of the surface represents the predicted value of the performance in the space of capsaicin and solvent, with the other two factors constant in their average values.

The best treatment, while the experimental region is near to 100 nm particles, was obtained with the highest amount of capsaicin, (upper level), and MeOAc as a solvent. In this case, the surface doesn't have a curvature, which means it is correctly described by the first order model. This model is available to move over the target direction until detect a tendency to change, employing another design with the aim of being more precise.

SOLUTIONS AND RECOMMENDATIONS

Using DoE the optimization of particle size of SLN loaded with capsaicin was possible where the authors found two significative variables: drug concentration and solvent type. The other independent variables such as stabilizer concentration was not significative. DoE is a very convenient tool for the development of SLN since multiple variables are involved in the process of nanoparticles synthesis. Reduction of the number of experiments which reduces the time of formulation development and the cost of testing is very appreciated. It is highly recommended that all new formulation get evaluated for the effect of the drug concentration on the size of the nanoparticles.

Figure 7. Estimated response surface plot for particle size of SLN loaded with CAP showing of the influence of the independent variables capsaicin and solvent type maintaining constant Gelucire ® 44/14=200 mg and PVAL=5.0%

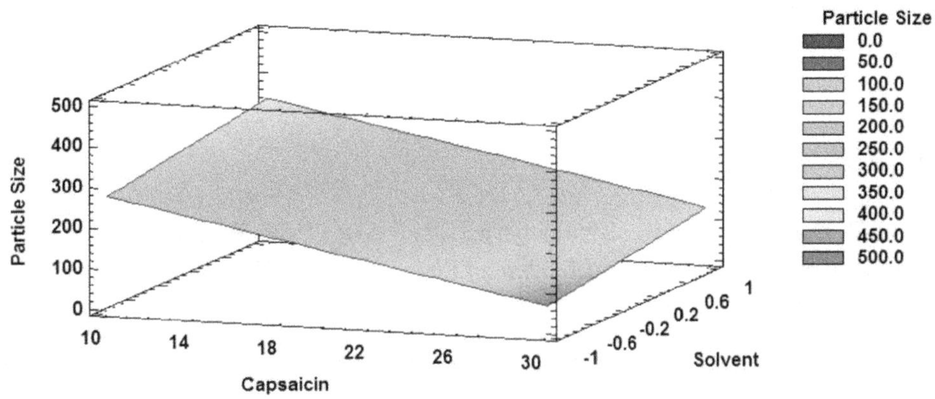

FUTURE RESEARCH DIRECTIONS

The complete development of a capsaicin formulation of SLN require optimization of other parameters such as DL. Also, it will be important to study drug permeation *in-vitro* and *in-vivo* to asses if the optimum formulation actually correlates with the size of the nanoparticles. As mentioned previously, DoE is a tool that should be implemented for each drug formulated in a nanoparticulated system.

CONCLUSION

The chapter gives a general overview of the importance of SLN in the field of drug delivery. Information on the most common methods of preparation of SLN, as well as the variables involved in the preparation of nanoparticles, including drug concentration, drug/lipid ratio, stabilizer concentration, etc. It also includes a critical survey of studies regarding the uses of DoE in the optimization of SLN. From the review of applications of DoE on the optimization of nanoparticles was possible to conclude that the variables that optimized more frequently are particle size and EE. The chapter include a case study about the optimization on the particle size of SLN loaded with capsaicin in accordance with the principal optimized variables. It was observed that drug concentration and solvent type were the variables that significatively affect the size of the prepared nanoparticle. These factors were also significative in previous researches. However, factors such as surfactant and drug/lipid ratio are frequent evaluated for the influence over the response. The chapter illustrate the importance of DoE in the optimization of SLN reducing the number of experiments needed for a successful formulation which translate in a reduction of experimental work.

ACKNOWLEDGMENT

Work supported by 20th Internal Call for Research Project Funding UABC (Grant 3900), SEP-PRODEP Call for Support for the Strengthening of Academic Bodies 2018 (UABC-CA-287). Capsaicin (98%) was provided by Victory Enterprises, Tijuana Mexico. Gelucire ® 44/14 was a kind gift of Quimicos Lyontec.

REFERENCES

Adhikari, P., Pal, P., Das, A. K., Ray, S., Bhattacharjee, A., & Mazumder, B. (2017). Nano lipid-drug conjugate: An integrated review. *International Journal of Pharmaceutics*, *529*(1–2), 629–641. doi:10.1016/j.ijpharm.2017.07.039 PMID:28723407

Agrawal, U., Gupta, M., & Vyas, S. P. (2015). Capsaicin delivery into the skin with lipidic nanoparticles for the treatment of psoriasis. *Artificial Cells, Nanomedicine, and Biotechnology*, *43*(1), 33–39. doi:10.3109/21691401.2013.832683 PMID:24040836

Al-Kassas, R., Wen, J., Cheng, A. E. M., Kim, A. M. J., Liu, S. S. M., & Yu, J. (2016). Transdermal delivery of propranolol hydrochloride through chitosan nanoparticles dispersed in mucoadhesive gel. *Carbohydrate Polymers*, *153*, 176–186. doi:10.1016/j.carbpol.2016.06.096 PMID:27561485

Albano, J. M. R., Morais Ribeiro, L., Muñiz Couto, V., Barbosa Messias, M., Rodrigues da Silva, G. H., Breitkreitz, M. C., ... Pickholz, M. (2019). Rational desing of polymer-lipid nanoparticles for docetaxel delivery. *Colloids and Surfaces. B, Biointerfaces*, *175*(1), 56–64. doi:10.1016/j.colsurfb.2018.11.077 PMID:30517905

Anand, P., & Bley, K. (2011). Topical capsaicin for pain management: Therapeutic potential and mechanisms of action of the new high-concentration capsaicin 8% patch. *British Journal of Anaesthesia*, *107*(4), 490–502. doi:10.1093/bja/aer260 PMID:21852280

Arana, L., Salado, C., Vega, S., Aizpurua-Olaizola, O., Arada, I., de la, & Suarez, T. (2015). Solid lipid nanoparticles for delivery of Calendula officinalis extract. *Colloids and Surfaces. B, Biointerfaces*, *135*, 18–26. doi:10.1016/j.colsurfb.2015.07.020 PMID:26231862

Carbone, C., Tomasello, B., Ruozi, B., Renis, M., & Puglisi, G. (2012). Preparation and optimization of PIT solid lipid nanoparticles via statistical factorial design. *European Journal of Medicinal Chemistry*, *49*, 110–117. doi:10.1016/j.ejmech.2012.01.001 PMID:22244589

Cavalli, R., Caputo, O., & Gasco, M. R. (2000). Preparation and characterization of solid lipid nanospheres containing paclitaxel. *European Journal of Pharmaceutical Sciences*, *10*(4), 305–309. doi:10.1016/S0928-0987(00)00081-6 PMID:10838020

Cocke, J., & Maaß, S. (2017). Cross linking between the baffling effect and phase inversion during liquid–liquid monomer mixing. *Macromolecular Reaction Engineering*, *11*(4), 1–17. doi:10.1002/mren.201700015

Czajkowska-Kośnik, A., Szekalska, M., & Winnicka, K. (2019). Nanostructured lipid carriers: A potential use for skin drug delivery systems. *Pharmacological Reports*, *71*(1), 156–166. doi:10.1016/j.pharep.2018.10.008 PMID:30550996

Das, S., Ng, W. K., Kanaujia, P., Kim, S., & Tan, R. B. H. (2011). Formulation design, preparation and physicochemical characterizations of solid lipid nanoparticles containing a hydrophobic drug: Effects of process variables. *Colloids and Surfaces. B, Biointerfaces*, *88*(1), 483–489. doi:10.1016/j.colsurfb.2011.07.036 PMID:21831615

Derman, S. (2015). Caffeic Acid Phenethyl Ester Loaded PLGA nanoparticles: Effect of various process parameters on reaction yield, encapsulation efficiency, and particle size. *Journal of Nanomaterials*, *2015*, 1–12. doi:10.1155/2015/341848

Dhawan, S., Kapil, R., & Singh, B. (2011). Formulation development and systematic optimization of solid lipid nanoparticles of quercetin for improved brain delivery. *The Journal of Pharmacy and Pharmacology*, *63*(3), 342–351. doi:10.1111/j.2042-7158.2010.01225.x PMID:21749381

Din, F., Mustapha, O., Kim, D. W., Rashid, R., Park, J. H., Choi, J. Y., ... Choi, H.-G. (2015). Novel dual-reverse thermosensitive solid lipid nanoparticle-loaded hydrogel for rectal administration of flurbiprofen with improved bioavailability and reduced initial burst effect. *European Journal of Pharmaceutics and Biopharmaceutics*, *94*, 64–72. doi:10.1016/j.ejpb.2015.04.019 PMID:25979136

Ding, Y., Pyo, S. M., & Muller, R. H. (2017). smartLipids ® as third solid lipid nanoparticle generation – stabilization of retinol for dermal application. *Original Articles*, *72*, 728–735. doi:10.1691/ph.2017.7016 PMID:29441957

Ebrahimi, H. A., Javadzadeh, Y., Hamidi, M., & Barzegar Jalali, M. (2016). Development and characterization of a novel lipohydrogel nanocarrier: Repaglinide as a lipophilic model drug. *The Journal of Pharmacy and Pharmacology*, *68*(4), 450–458. doi:10.1111/jphp.12537 PMID:27114047

Ekambaram, P., & Abdul, H. S. A. (2011). Formulation and evaluation of solid lipid nanoparticles of ramipril. *Journal of Young Pharmacists: JYP*, *3*(3), 216–220. doi:10.4103/0975-1483.83765 PMID:21897661

El Assasy, A. E.-H. I., Younes, N. F., & Makhlouf, A. I. A. (2019). Enhanced oral absorption of Amisulpride via a nanostructured lipid carrier-based capsules: development, optimization applying the desirability function approach and in vivo pharmacokinetic study. *AAPS PharmSciTech*, *20*(2), 82. doi:10.120812249-018-1283-x PMID:30652198

Eldem, T., Speiser, P., & Hincal, A. (1991). *Optimization of spray-dried and -congealed lipid micropellets and characterization of their surface morphology by scanning electron microscopy. Pharmaceutical Research.* An Official Journal of the American Association of Pharmaceutical Scientists; doi:10.1023/A:1015874121860

Ferreira, M., Silva, E., Barreiros, L., Segundo, M. A., Costa Lima, S. A., & Reis, S. (2016). Methotrexate loaded lipid nanoparticles for topical management of skin-related diseases: Design, characterization and skin permeation potential. *International Journal of Pharmaceutics*, *512*(1), 14–21. doi:10.1016/j.ijpharm.2016.08.008 PMID:27530292

Gaur, P. K., Mishra, S., & Purohit, S. (2013). Solid lipid nanoparticles of guggul lipid as drug carrier for transdermal drug delivery. *BioMed Research International*, *2013*, 1–10. doi:10.1155/2013/750690 PMID:24058913

Ghadiri, M., Fatemi, S., Vatanara, A., Doroud, D., Najafabadi, A. R., Darabi, M., & Rahimi, A. A. (2012). Loading hydrophilic drug in solid lipid media as nanoparticles: Statistical modeling of entrapment efficiency and particle size. *International Journal of Pharmaceutics*, *424*(1–2), 128–137. doi:10.1016/j.ijpharm.2011.12.037 PMID:22227603

Godolphin, J. D. (2019). Two-level factorial and fractional factorial replicates in blocks of size two. *Computational Statistics & Data Analysis*, *133*(1), 120–137. doi:10.1016/j.csda.2018.09.006

Hao, J., Fang, X., Zhou, Y., Wang, J., Guo, F., Li, F., & Peng, X. (2011). Development and optimization of solid lipid nanoparticle formulation for ophthalmic delivery of chloramphenicol using a Box-Behnken design. *International Journal of Nanomedicine*, *6*, 683–692. doi:10.2147/IJN.S17386 PMID:21556343

Hao, J., Wang, F., Wang, X., Zhang, D., Bi, Y., Gao, Y., ... Zhang, Q. (2012). Development and optimization of baicalin-loaded solid lipid nanoparticles prepared by coacervation method using central composite design. *European Journal of Pharmaceutical Sciences*, *47*(2), 497–505. doi:10.1016/j.ejps.2012.07.006 PMID:22820033

Hayman, M., & Kam, P. C. A. (2008). Capsaicin: A review of its pharmacology and clinical applications. *Current Anaesthesia and Critical Care*, *19*(5–6), 338–343. doi:10.1016/j.cacc.2008.07.003

Hu, L., Jia, Y., & Ding, W. (2010). Preparation and characterization of solid lipid nanoparticles loaded with epirubicin for pulmonary delivery. *Die Pharmazie*, *65*(8), 585–587. doi:10.1691/ph.2010.0023 PMID:20824958

Jeevanandam, J., Chan, Y. S., & Danquah, M. K. (2016). Nano-formulations of drugs: Recent developments, impact and challenges. *Biochimie*, *128–129*, 99–112. doi:10.1016/j.biochi.2016.07.008 PMID:27436182

Jiang, S., Zhu, R., He, X., Wang, J., Wang, M., Quian, Y., & Wang, S. (2017). Enhanced photocytotoxicity of curcumin delivered by solid lipid nanoparticles. *International Journal of Nanomedicine*, *12*, 167–178. doi:10.2147/IJN.S123107 PMID:28053531

Joshi, M. D., & Müller, R. H. (2009). Lipid nanoparticles for parenteral delivery of actives. *European Journal of Pharmaceutics and Biopharmaceutics*, *71*(2), 161–172. doi:10.1016/j.ejpb.2008.09.003 PMID:18824097

Kelidari, H. R., Saeedi, M., Akbari, J., Morteza-Semnani, K., Gill, P., Valizadeh, H., & Nokhodchi, A. (2015). Formulation optimization and in vitro skin penetration of spironolactone loaded solid lipid nanoparticles. *Colloids and Surfaces. B, Biointerfaces*, *128*, 473–479. doi:10.1016/j.colsurfb.2015.02.046 PMID:25797482

Kim, J. H., Ko, J. A., Kim, J. T., Cha, D. S., Cho, J. H., Park, H. J., & Shin, G. H. (2014). Preparation of a capsaicin-loaded nanoemulsion for improving skin penetration. *Journal of Agricultural and Food Chemistry*, *62*(3), 725–732. doi:10.1021/jf404220n PMID:24417234

Kim, S., Diab, R., Joubert, O., Canilho, N., & Pasc, A. (2016). Core–shell microcapsules of solid lipid nanoparticles and mesoporous silica for enhanced oral delivery of curcumin. *Colloids and Surfaces. B, Biointerfaces, 140*, 161–168. doi:10.1016/j.colsurfb.2015.12.040 PMID:26752213

Kocbek, P., Baumgartner, S., & Kristl, J. (2006). Preparation and evaluation of nanosuspensions for enhancing the dissolution of poorly soluble drugs. *International Journal of Pharmaceutics, 312*(1–2), 179–186. doi:10.1016/j.ijpharm.2006.01.008 PMID:16469459

Kwon, H. Y., Lee, J. Y., Choi, S. W., Jang, Y., & Kim, J. H. (2001). Preparation of PLGA nanoparticles containing estrogen by emulsification-diffusion method. *Colloids and Surfaces. A, Physicochemical and Engineering Aspects, 182*(1–3), 123–130. doi:10.1016/S0927-7757(00)00825-6

Lander, R., Manger, W., Scouloudis, M., Ku, A., Davis, C., & Lee, A. (2000). Gaulin homogenization: A mechanistic study. *Biotechnology Progress, 16*(1), 80–85. doi:10.1021/bp990135c PMID:10662494

Lasa-Saracibar, B., Estella-Hermoso De Mendoza, A., Guada, M., Dios-Vieitez, C., & Blanco-Prieto, M. J. (2012). Lipid nanoparticles for cancer therapy: State of the art and future prospects. *Expert Opinion on Drug Delivery, 9*(10), 1245–1261. doi:10.1517/17425247.2012.717928 PMID:22950878

Lemos-Senna, E., Wouessidjewe, D., Lesieur, S., & Duche, D. (1998). Preparation of amphiphilic cyclodextrin nanospheres using the emulsification solvent evaporation method. Influence of the surfactant on preparation and hydrophobic drug loading, *170*, 119–128.

Levy, D. M., Abraham, R. R., & Tomlinson, D. R. (1991). Topical capsaicin in the treatment of painful diabetic neuropathy. *The New England Journal of Medicine, 324*(11), 776–777. doi:10.1056/NEJM199103143241116 PMID:1997848

Mahant, S., Rao, R., & Nanda, S. (2018). *Nanostructured lipid carriers: Revolutionizing skin care and topical therapeutics. Design of nanostructures for versatile therapeutic applications*. Elsevier; doi:10.1016/B978-0-12-813667-6.00003-6

Malam, Y., Loizidou, M., & Seifalian, A. M. (2009). Liposomes and nanoparticles: Nanosized vehicles for drug delivery in cancer. *Trends in Pharmacological Sciences, 30*(11), 592–599. doi:10.1016/j.tips.2009.08.004 PMID:19837467

Mante, A., Heider, M., Zlomke, C., & Mäder, K. (2016). PLGA nanoparticles for peroral delivery: How important is pancreatic digestion and can we control it? *European Journal of Pharmaceutics and Biopharmaceutics, 108*, 32–40. doi:10.1016/j.ejpb.2016.08.009 PMID:27553262

Mehnert, W., & Mäder, K. (2012). Solid lipid nanoparticles: Production, characterization and applications. *Advanced Drug Delivery Reviews, 64*(SUPPL.), 83–101. doi:10.1016/j.addr.2012.09.021 PMID:11311991

Mukherjee, S., Ray, S., & Thakur, R. S. (2009). Solid lipid nanoparticles: A modern formulation approach in drug delivery system. *Indian Journal of Pharmaceutical Sciences, 71*(4), 349–358. doi:10.4103/0250-474X.57282 PMID:20502539

Müller, R. H., Mder, K., & Gohla, S. (2000). Solid lipid nanoparticles (SLN) for controlled drug delivery - A review of the state of the art. *European Journal of Pharmaceutics and Biopharmaceutics, 50*(1), 161–177. doi:10.1016/S0939-6411(00)00087-4 PMID:10840199

Nakamura, Y., Mochida, A., Choyke, P. L., & Kobayashi, H. (2016). Nanodrug delivery: Is the enhanced permeability and retention effect sufficient for curing cancer? *Bioconjugate Chemistry*, *27*(10), 2225–2238. doi:10.1021/acs.bioconjchem.6b00437 PMID:27547843

Priyanka, K., & Sathali, A. A. H. (2012). Preparation and evaluation of montelukast sodium loaded solid lipid nanoparticles. *Journal of Young Pharmacists: JYP*, *4*(3), 129–137. doi:10.4103/0975-1483.100016 PMID:23112531

Prow, T. W., Grice, J. E., Lin, L. L., Faye, R., Butler, M., Becker, W., ... Roberts, M. S. (2011). Nanoparticles and microparticles for skin drug delivery. *Advanced Drug Delivery Reviews*, *63*(6), 470–491. doi:10.1016/j.addr.2011.01.012 PMID:21315122

Quintanar-Guerrero, D., Tamayo-Esquivel, D., Ganem-Quintanar, A., Allémann, E., & Doelker, E. (2005). Adaptation and optimization of the emulsification-diffusion technique to prepare lipidic nanospheres. *European Journal of Pharmaceutical Sciences*, *26*(2), 211–218. doi:10.1016/j.ejps.2005.06.001 PMID:16046105

Rahman, Z., Zidan, A. S., & Khan, M. A. (2010). Non-destructive methods of characterization of risperidone solid lipid nanoparticles. *European Journal of Pharmaceutics and Biopharmaceutics*, *76*(1), 127–137. doi:10.1016/j.ejpb.2010.05.003 PMID:20470882

Rawat, M. K., Jain, A., & Singh, S. (2011). Studies on binary lipid matrix based solid lipid nanoparticles of repaglinide: in vitro and in vivo evaluation. *Journal of Pharmaceutical Sciences*, *100*(6), 2366–2378. doi:10.1002/jps.22435 PMID:21491449

Raza, K., Singh, B., Singal, P., Wadhwa, S., & Katare, O. P. (2013). Systematically optimized biocompatible isotretinoin-loaded solid lipid nanoparticles (SLNs) for topical treatment of acne. *Colloids and Surfaces. B, Biointerfaces*, *105*(January), 67–74. doi:10.1016/j.colsurfb.2012.12.043 PMID:23357735

Saeidpour, S., Lohan, S. B., Solik, A., Paul, V., Bodmeier, R., Zoubari, G., ... Teutloff, C. (2017). Drug distribution in nanostructured lipid particles. *European Journal of Pharmaceutics and Biopharmaceutics*, *110*, 19–23. doi:10.1016/j.ejpb.2016.10.008 PMID:27789357

Shah, M., & Pathak, K. (2010). Development and statistical optimization of solid lipid nanoparticles of simvastatin by using 23 full-factorial design. *AAPS PharmSciTech*, *11*(2), 489–496. doi:10.120812249-010-9414-z PMID:20309652

Sharif Makhmal Zadeh, B., Niro, H., Rahim, F., & Esfahani, G. (2018). Ocular delivery system for propranolol hydrochloride based on nanostructured lipid carrier. *Scientia Pharmaceutica*, *86*(2), 16. doi:10.3390cipharm86020016 PMID:29677103

Sharma, A., Jindal, M., Aggarwal, G., & Jain, S. (2010). Development of a novel method for fabrication of solid lipid nanoparticles: Using high shear homogenization and ultrasonication. *Research Journal of Pharmaceutical, Biological and Chemical Sciences*, *1*(2), 265–274.

Shi, C., Ahmad Khan, S., Wang, K., & Schneider, M. (2015). Improved delivery of the natural anticancer drug tetrandrine. *International Journal of Pharmaceutics*, *479*(1), 41–51. doi:10.1016/j.ijpharm.2014.12.022 PMID:25510598

Siafaka, P. I., Üstündağ Okur, N., Karavas, E., & Bikiaris, D. N. (2016). Surface modified multifunctional and stimuli responsive nanoparticles for drug targeting: Current status and uses. *International Journal of Molecular Sciences*, *17*(1440). doi:10.3390/ijms17091440 PMID:27589733

Singh, I., Swami, R., Jeengar, M. K., Khan, W., & Sistla, R. (2015). p-Aminophenyl-a-d-mannopyranoside engineered lipidic nanoparticles for effective delivery of docetaxel to brain. *Chemistry and Physics of Lipids*, *188*, 1–9. doi:10.1016/j.chemphyslip.2015.03.003 PMID:25819559

Sjöström, B., & Bergenståhl, B. (1992). Preparation of submicron drug particles in lecithin-stabilized o w emulsions: I. Model studies of the precipitation of cholesteryl acetate. *International Journal of Pharmaceutics*, *84*(2), 107–116. doi:10.1016/0378-5173(92)90051-3

Tadros, T. (2015). Viscoelastic properties of sterically stabilised emulsions and their stability. *Advances in Colloid and Interface Science*, *222*, 692–708. doi:10.1016/j.cis.2015.03.001 PMID:25900262

Taveira, S. F., Araújo, L. M. P. de C., de Santana, D. C. A. S., Nomizo, A., de Freitas, L. A. P., & Lopez, R. F. V. (2012). Development of cationic solid lipid nanoparticles with factorial design-based studies for topical administration of doxorubicin. *Journal of Biomedical Nanotechnology*, *8*(2), 219–228. doi:10.1166/jbn.2012.1383 PMID:22515073

Urbán-Morlán, Z., Ganem-Rondero, A., Melgoza-Contreras, L. M., Escobar-Chávez, J. J., Nava-Arzaluz, M. G., & Quintanar-Guerrero, D. (2010). Preparation and characterization of solid lipid nanoparticles containing cyclosporine by the emulsification-diffusion method. *International Journal of Nanomedicine*, *5*(1), 611–620. doi:10.2147/IJN.S12125 PMID:20856836

Vitorino, C., Carvalho, F. A., Almeida, A. J., Sousa, J. J., & Pais, A. A. C. C. (2011). The size of solid lipid nanoparticles: An interpretation from experimental design. *Colloids and Surfaces. B, Biointerfaces*, *84*(1), 117–130. doi:10.1016/j.colsurfb.2010.12.024 PMID:21242064

Wang, X. R., Gao, S. Q., Niu, X. Q., Li, L. J., Ying, X. Y., Hu, Z. J., & Gao, J. Q. (2017). Capsaicin-loaded nanolipoidal carriers for topical application: Design, characterization, and in vitro/in vivo evaluation. *International Journal of Nanomedicine*, *12*, 3881–3898. doi:10.2147/IJN.S131901 PMID:28579775

Wehrung, D., Geldenhuys, W. J., & Oyewumi, M. O. (2012). Effects of gelucire content on stability, macrophage interaction and blood circulation of nanoparticles engineered from nanoemulsions. *Colloids and Surfaces. B, Biointerfaces*, *94*, 259–265. doi:10.1016/j.colsurfb.2012.02.005 PMID:22386863

Wissing, S., Kayser, O., & Müller, R. (2004). Solid lipid nanoparticles for parenteral drug delivery. *Advanced Drug Delivery Reviews*, *56*(9), 1257–1272. doi:10.1016/j.addr.2003.12.002 PMID:15109768

Wong, H. L., Bendayan, R., Rauth, A. M., & Wu, X. Y. (2004). Development of solid lipid nanoparticles containing ionically complexed chemotherapeutic drugs and chemosensitizers. *Journal of Pharmaceutical Sciences*, *93*(8), 1993–2008. doi:10.1002/jps.20100 PMID:15236449

Yang, S. C., & Zhu, J. B. (2002). Preparation and characterization of Camptothecin solid lipid nanoparticles. *Drug Development and Industrial Pharmacy*, *28*(3), 265–274. doi:10.1081/DDC-120002842 PMID:12026219

You, P., Yuan, R., & Chen, C. (2017). Design and evaluation of lidocaine- and prilocaine-coloaded nanoparticulate drug delivery systems for topical anesthetic analgesic therapy: A comparison between solid lipid nanoparticles and nanostructured lipid carriers. *Drug Design, Development, and Therapy, 11*, 2743–2752. doi:10.2147/DDDT.S141031 PMID:29075099

Zhang, J., Fan, Y., & Smith, E. (2009). Experimental design for the optimization of lipid nanoparticles. *Journal of Pharmaceutical Sciences, 98*(5), 1813–1819. doi:10.1002/jps.21549 PMID:18781636

Zhu, Y., Wang, M., Zhang, J., Peng, W., Firempong, C. K., Deng, W., ... Zhang, W. (2015). Improved oral bioavailability of capsaicin via liposomal nanoformulation: Preparation, in vitro drug release and pharmacokinetics in rats. *Archives of Pharmacal Research, 38*(4), 512–521. doi:10.100712272-014-0481-7 PMID:25231341

ADDITIONAL READING

Abdel-Mottaleb, M. M. A., Neumann, D., & Lamprecht, A. (2010). In vitro drug release mechanism from lipid nanocapsules (LNC). *International Journal of Pharmaceutics, 390*(2), 208–213. doi:10.1016/j.ijpharm.2010.02.001 PMID:20149853

Abdel-Mottaleb, M. M. A., Neumann, D., & Lamprecht, A. (2011). Lipid nanocapsules for dermal application: A comparative study of lipid-based versus polymer-based nanocarriers. *European Journal of Pharmaceutics and Biopharmaceutics, 79*(1), 36–42. doi:10.1016/j.ejpb.2011.04.009 PMID:21558002

Balzus, B., Colombo, M., Sahle, F. F., Zoubari, G., Staufenbiel, S., & Bodmeier, R. (2016). Comparison of different in vitro release methods used to investigate nanocarriers intended for dermal application. *International Journal of Pharmaceutics, 513*(1–2), 247–254. doi:10.1016/j.ijpharm.2016.09.033 PMID:27628784

Gaur, P. K., Mishra, S., & Purohit, S. (2013). Solid Lipid Nanoparticles of Guggul Lipid as Drug Carrier for Transdermal Drug Delivery. *BioMed Research International, 2013*, 1–10. doi:10.1155/2013/750690 PMID:24058913

González-Zamora, A., Sierra-Campos, E., Pérez-Morales, R., Vázquez-Vázquez, C., Gallegos-Robles, M. A., López-Martínez, J. D., & García-Hernández, J. L. (2015). Measurement of capsaicinoids in chiltepin hot pepper: A comparison study between spectrophotometric method and high performance liquid chromatography analysis. *Journal of Chemistry, 2015*, 1–10. doi:10.1155/2015/709150

Nnamani, P. O., Hansen, S., Windbergs, M., & Lehr, C. M. (2014). Development of artemether-loaded nanostructured lipid carrier (NLC) formulation for topical application. *International Journal of Pharmaceutics, 477*(1–2), 208–217. doi:10.1016/j.ijpharm.2014.10.004 PMID:25290810

Shang, L., Nienhaus, K., & Nienhaus, G. U. (2014). Engineered nanoparticles interacting with cells: Size matters. *Journal of Nanobiotechnology, 12*(1), 1–11. doi:10.1186/1477-3155-12-5 PMID:24491160

Silva, C. O., Rijo, P., Molpeceres, J., Figueiredo, I. V., Ascensão, L., Fernandes, A. S., ... Reis, C. P. (2015). Polymeric nanoparticles modified with fatty acids encapsulating betamethasone for anti-inflammatory treatment. *International Journal of Pharmaceutics*, *493*(1–2), 271–284. doi:10.1016/j.ijpharm.2015.07.044 PMID:26222745

Yadav, N., Khatak, S., Vir, U., & Sara, S. (2013). Solid lipid nanoparticles - a review. *International Journal of Applied Pharmaceutics*, *5*(2), 8–18.

KEY TERMS AND DEFINITIONS

Diffusion: A movement of a solvent from an area of high concentration to an area of low concentration.

Drug Delivery: Is the process of the system to achieve a drug on the target site.

Lipid: A class of organic compounds as fatty acid or their derivates, soluble in organic solvents.

Optimization: Is a process to get target particle size, finding best concentrations of reagents with a percent of confidence.

Particle Size: The hydrodynamic diameter of a solid particle.

Permeation: The activity of the molecule to cross de stratum cornea and the epidermis.

Solid Lipid Nanoparticles: Particles with nanometric size elaborated with lipid and stabilized with surfactant.

Surfactant: Is a class of water-soluble polymer used to make stable emulsions oil/water.

Topical: Site on the skin to applicate a drug formulation as a enter drug point on the body.

Chapter 7
Advantages, Disadvantages, and Future Trends on the Use of Design of Experiments in Cross-Over Trials in Nutritional Clinical Investigation

Jose Manuel Miranda
Universidade de Santiago de Compostela, Spain

Alicia del Carmen Mondragon Portocarrero
Universidade de Santiago de Compostela, Spain

Alexandre Lamas Freire
Universidade de Santiago de Compostela, Spain

Carlos Manuel Franco Abuin
Universidade de Santiago de Compostela, Spain

Alberto Cepeda Saez
Universidade de Santiago de Compostela, Spain

ABSTRACT

The use of clinical trials to demonstrate effect of foods consumption on human health has increased significantly in recent years at the global level. As in other areas of human health, some authors choose to use parallel trial designs, while others prefer to use crossover designs for these trials. Because crossover trials have the advantage of reducing the number of subjects needed and the economic cost to be performed, they have many advocates within the scientific community. However, these types of tests also have numerous drawbacks, due to the difficulty of carrying out adequate statistical analyses, the lack of reliable standards adapted to them or confounding factors. In this chapter, the advantages and disadvantages of crossover designs and whether they are a recommended option for human nutrition research are shown. The usefulness of design of experiments coupled to crossover trials, especially when comparing various levels of the dependent variable, are also discussed.

DOI: 10.4018/978-1-7998-1518-1.ch007

INTRODUCTION

Human clinical trials are often designed to assess the effectiveness of an intervention or treatment on human health. Typically, these trials are used to test the efficacy of different therapeutic options in medicine, pharmacology or nutrition. However, in the last decades the use of such trials is increasingly common for the investigation of the effect of other types of habits and interventions on human health, as may be the case with nutrition baseline research. In addition to basic research, one of the most common purposes for conducting nutrition trials is to demonstrate the beneficial effects of a specific food component in order to use this information in the in food labeling. Food products that are labeled with messages related to the promotion of human health are marketed at a price 30–50% higher than their counterparts that cannot use these health claims, and in some cases can reach even more (Miranda et al., 2018). In order to obtain authorization for the use of one of these health claims, it is usually necessary to present a meta-analysis demonstrating the beneficial effects of the food on human health. As an example, in the European Union it is compulsory to include human trials to obtain authorization for the use of health claims in marketing food products (European Food Safety Authority, EFSA, 2011). Following this obligation imposed by the European Union, the number of nutrition-related clinical trials published worldwide increased significantly. As it can be seen in Figure 1, clinical trials related to human nutrition showed gradual growth between 2000 and 2010, which accelerated substantially from that date onwards for both cross-clinical trials and for other types of clinical trials.

Randomized clinical trials are considered the "gold standard" to evaluate therapeutic effectiveness due to its ability to avoid or minimize bias associated with imbalance in potentially confounding variables (Leonard, Lafrenaye & Goffaux, 2012). The most employed randomized clinical trials are the so-called parallel-group trial or design (Harris & Raynor, 2017), in which subjects are randomized to a unique intervention or control group during the entire trial, that occur simultaneously in time. Identical dependent outcome variables are measured in all groups included in the trial, with outcomes compared between groups, or between subjects, with the aim to determine the intervention effectiveness. In contrast, in a so-called crossover design, all subjects receive all levels of the independent variable at some point in the study, but subjects do not receive all levels at the same time (Figure 2). To determine intervention effectiveness, dependent outcome variables are measured for the two levels of the intervention and afterwards, then they are compared within the same subject (Harris & Raynor, 2017).

Although cluster-randomized crossover trials have become a popular tool in nutritional clinical investigation, some important characteristics of this kind of studies have not been clearly standardized for these emergent trials. In many cases, the principles from cluster-randomized trials with parallel design are not easily applied to a crossover setting (Reich, Myers, Obeng, Milstone & Perl, 2012). Additionally, on many occasions, researchers found complications to include crossover trials in systematic reviews and the meta-analysis, required to obtain an authorization for obtaining a health claim (Li, Tsung, Hawkins & Dickersin, 2015). In many published meta-analyses, the authors choose to include the results from crossover trials unfolded as if they become from a parallel design (Reich et al., 2012). A clear example of this difficulty can be seem in a recent meta-analysis (Yan, Guan, Gao, & Peng, 2018) related to omega-3 fatty acids effects and their effect in non-alcoholic fatty liver disease, in with textually excluded "uncontrolled, crossover, cross-sectional, and not reporting outcomes of interest or primary data."

Another crucial point in nutritional trials is that clinical trials are usually designed with the implicit assumption that data analysis will occur only after the trial is completed. While this affirmation is correct in pharmacological trials (volunteers received a treatment in order to treat a pathology), in the case

Figure 1. Total clinical trials and crossover trials related to human nutrition published from 2000 to 2018 (source: ISI web of knowledge®). Search criteria: total clinical trials were search by "clinical trial" in the field "title" and "nutrition" in the field "topic". Crossover trials were search by "crossover trial" in the field "title" and "nutrition" in the field "topic".

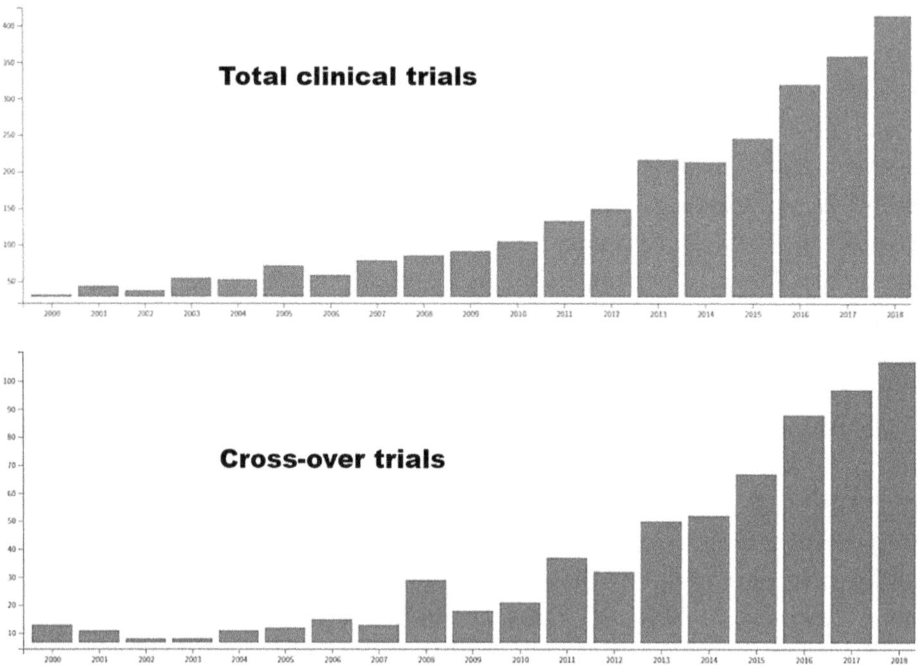

of nutritional trials, this is not always the case (Reich et al., 2012). Nutritional clinical trials with foods, food ingredients, or bioactive compounds are usually conducted over long periods, requiring changes in eating habits, and participants do not usually need the food tested to combat a pathology. All these factors make subject drop-outs much more frequent than in the case of pharmacological trials (Harris & Raynor, 2017).

In the present chapter, the appropriateness of the use of cross-clinical trials for research in the field of nutrition will be discussed, reviewing their advantages, drawbacks, and prospects for future use. The advantages of coupling cross-over trials with design of experiments to reduce the number of subjects and time employed is also discussed.

Parallel and Crossover Design in Nutrition Research

In a classical parallel clinical trial, each subject is randomly assigned to one, and only one of the nutritional interventions, or control groups. Identical dependent outcome variables are measured in all groups in the trial, with outcomes compared between groups, or between volunteers, to determine the effectiveness of nutritional intervention (Harris & Raynor, 2017). Another option in the design of clinical trials is always allow each patient to receive both the study nutritional intervention and the control intervention, establishing a certain order, and for a set period (Lui, 2016). Thus, all subjects receive all levels of the independent variable at some point in the study, but subjects do not receive all levels at the same

Figure 2. Basic (2x2) crossover trial design

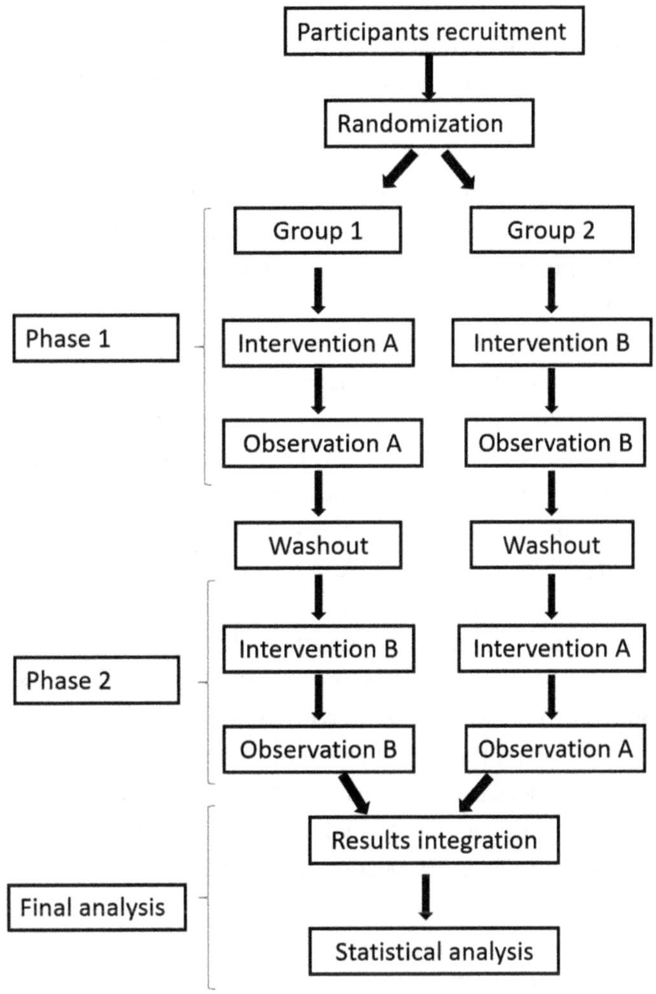

time. In a crossover design, subjects can be divided to one of more groups. One group receives dietary intervention while the other group acts as control, without any nutritional intervention, or receiving a placebo (Harris & Raynor, 2017). Once this phase is complete, the groups switch or cross over to the next phase, with the group having received the dietary intervention acting as control group, and the other group receiving the dietary intervention (Figure 2). To determine nutritional intervention effectiveness, dependent outcome variables are measured for the two levels of the independent variable and are compared within the same subject. Thus, each subject act as their own control, experiencing both interventions in a sequence of periods established randomly and both periods separated by a stabilization or washing period. Considering that all subjects are their own controls, the influence of seasonal, ambient, or other factors external to subjects have much less influence on the results of the trial (Harris & Raynor, 2017).

Crossover designs used in nutrition research can be employed to determinate the effects of a dietary intervention over a brief period on the amounts and types of food consumed in a large variety of physiological responses by humans (Alleleyn et al., 2018). These responses can include large consequences of the food intake such as sensation of hunger, satiation, satiety, anthropometric and laboratory data, nutrition-focused physical or psychic findings, and a large variety of health parameters (Alleleyn et al., 2018, Zhang et al., 2018).

In these studies, the food, nutrient or bioactive component being examined in the dietary intervention is provided to subjects in their usual daily life. Other way is that they may be instructed on the types and amounts of food and nutrients to be consumed, with the instruction being such that a permanent change in intake is not expected. The dietary intervention is often over a longer period, often significantly higher than those employed for the cases of medicines. Additionally, the dependent variables measured may need a longer period to recover to baseline levels due to the relationship between the diet and the health parameter. Thus, the overall length of the study for a volunteer is usually between large weeks to even various months (Harris & Raynor, 2017).

In nutrition trials using crossover designs, it is usual that the independent variable investigated has more than two levels. In this situation, the different number of sequences in which the levels of the independent variable can be implemented in volunteers increases (Harris & Raynor, 2017). For example, when are investigated three levels of the independent variable, this translates into the fact that there are six different sequences in which the levels of the independent variable can be implemented (Figure 3). Using traditional methods, to examine the potential for sequence effects in a crossover design, subjects need to be randomized to all potential sequences. As a result, the number of subjects needed in the study may increase substantially (Harris & Raynor, 2017).

When researchers want not to use all possible sequences due to the number of When researchers want not to use all possible sequences due to the number of volunteers required, design of experiments design is a common strategy applied in a crossover design with more than two sequences used to determine the sequences to use in the trial (Galvain, Barolo, Macchietto & Bezzo, 2010). Design of experiments allow to minimize the number of sequences used in the study, maintaining an appropriate degree of statistical significance. Consequently, the number of potential subjects required decrease substantially, reducing the time, technical and economic resources needed to perform a cross-over trial correctly (Harris & Raynor, 2017).

Eating Behavior Characteristics

In several trials aimed to investigate dietary habits or the effects of food on human composition or health, often behavioral factors are usually not included in the variables of the study. In behavioral terms, the intake of a food is the result of a large variety of a complex array of eating behaviors, including bites, chews, and pauses (Langlet, Bach, Odegi, Fagerberg & Ioakimidis, 2018). However, most studies regarding food effects ignore this complex behavior, that makes the people eat differently according to the type of food eaten and according to interpersonal differences (Schork & Goetz, 2017). For example, liquid foods, foods with high carbohydrate content, and palatable foods appear to increase food intake when compared to solid foods, foods rich in protein, and bland foods, respectively (Langlet et al., 2018). This also occurs with the environmental conditions, because a warm environment predisposes less to eat than a cooler environment, resulting in potential changes in behavior at the group level (McCrickerd & Forde, 2017).

Reliability studies, using both solids and semi-solids foods, have shown a high correlation of most eating behavior parameters under identical conditions, which indicates high relative reliability. However, the interindividual variability in these studies is usually large, with the standard deviation often exceeding half the size of the mean group value for most of the quantified eating behavior characteristics (Langlet

Figure 3. Crossover trial design with three or more degrees of the independent variable

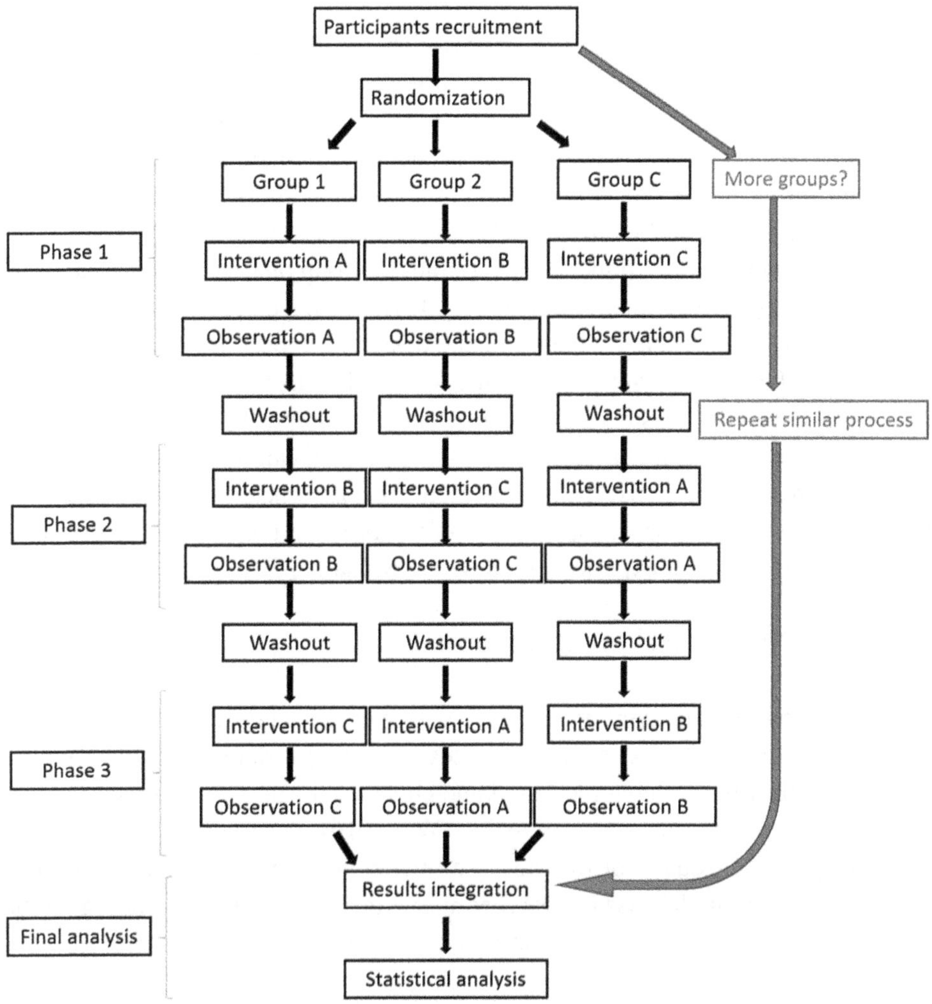

et al., 2018). On a group level, manipulating food properties, the environment, and even eating behavior characteristics of the consumer, have been shown to cause changes in food intake, as well as the timing of food intake during the day (Schork & Goetz, 2017).

Another parameter that have a potential effect in meal progression is the size of the tested food, likely through the modification of the oral exposure time to the food and the increased chewing. For example, an increase in the size of the food unit can induce an increase in chewing for proper bolus formation, resulting in a subsequent reduction of food intake. On the other hand, manipulating food unit segmentation can also induce the so-called "unit bias" leading subjects to consume a set number of food units (Geier, Rozin & Doros, 2006).

In trials with enough subjects, randomization process is essential to ensure that the different groups included in the study will have similar profiles of all measured and unmeasured characteristics (Reich et al., 2012). However, in cluster-randomized trials derived from design of experiments, achieving adequate balance often requires more attention (Reich & Milstone, 2014). In these cases, matching and stratification are two common techniques used. These techniques rely on the grouping of the clusters together based on a small number of possible confounding variables and then ensuring that some from each group are assigned to each sequence of the trial (Reich & Milstone, 2014). The cluster-randomized crossover design retains the advantages of a cluster-randomized design while leveraging its unique design to minimize imbalance. In a cluster-randomized crossover design, researchers get that a cluster will often be its own best control (Reich & Milstone, 2014). Consequently, in many cases, using a crossover design it can be used fewer clusters in the trial, while maintaining adequate statistical power (Reich et al., 2012).

Unlike with matching and stratification, to achieve balance in this setting, the researcher would not need to pre-specify the two or three "important" variables to group on, as most covariates will naturally be controlled for by the crossover (Reich & Milstone, 2014). Thus, it can be obtained important benefits derived from adding a crossover that by means of the design of experiments enables to include fewer clusters in the study (Harris & Raynor, 2017). In a standard analysis of cluster-randomized studies, comparisons are made between clusters in the intervention and control groups, assuming an even distribution of characteristics between clusters. The crossover design can limit the impact of imbalance of characteristics between clusters because the intervention effect is compared within a given cluster during the two periods. A crossover can be an especially useful tool when the number of available clusters is limited, as in the case of design of experiments (Reich & Milstone, 2014).

Measurements in Nutritional Investigation

A critical first step in any measurement process is to define clearly and specifically what variables will be measured. A measurement process can only be assessed as appropriate or inappropriate depending if researchers know exactly what is being measured and choose indicators that measure the variable correctly, with as little external interference as possible (Wu, 2016). A process that is appropriate for one type of measure may be inadequate for measuring a related but slightly different concept. This point may be especially important in the case of anthropometric measurements. Depending on the measured individual's anthropomorphic characteristics and physical exercise habits, the recommended measurement points may be different. Even these measurement points vary depending on the authors, as some recommend taking measurements on the right side of the subjects and others on the left side (Moreno et al., 2006). Additionally, even in the case of experienced professionals, anthropometric measurements on the same subject can yield variable results due to differences in criteria regarding anatomical references of the subject (Moreno et al., 2006).

Self-report methods are the most commonly methods used in nutrition studies to examine an individual`s usual nutrient intake. These shelf-reports methods can be either recording (prospective) or recall (retrospective). Obviously, recording methods are more reliable than recall methods, because they are based on recording food consumption later and do not influence the memory of the subject or the interview ability of the researcher (Ortega, Pérez-Rodrigo & López-Sobaler, 2015). A single 24-hour dietary recall (the most basic nutritional recording method), may provide a valid and reliable measure of a single individual`s target day intake of a specific nutrient, but will likely be a poor measure of that individual´s usual nutrient intake. At the same time, studies often include measures made by the researcher or laboratories, such as anthropometric, laboratory, and clinical measurements (Gleason et al., 2010). Dietetic practitioners diagnose nutrition-related pathologies using data from various sources and employing statistical tools to help with this process. Two well-known statistical tools are sensitivity and specificity, which are related to validity and reliability, in that they capture how well particular indicators measure what they are intended to measure (Gleason et al., 2010).

With the aim to assess and adequate measuring quality, two questions should be addresses. First, does the process used to perform the measure produce consistent results, when repeated under similar circumstances? Second, does the measurement resulting from this process reflect what it is intended to reflect? While the answer to the first question stablish the reliability of the measure, the answer to the second questions gives information about the measurement`s validity (Gleason et al., 2010).

Statistical power, the most important factor to address the statistical significance of a trial, can be defined as the probability of having conclusive evidence for one hypothesis over another given the existing study design. Obtaining an accurate estimate of the necessary sample size to answer a given scientific question in a setting ensures that researchers do not include too few of too many volunteers in a study (Reich et al., 2012), which would have very adverse economic consequences. For complex study designs, such as the cluster-randomized crossover, simple formulas to calculate statistical power may not adequately capture the expected variability from observed data. In these cases, the estimation of statistical power via simulation methods could be needed (Reich et al., 2012).

Because of the conflation of self-reported estimative data (self-reported measured data, data manually measured by the researchers, and data obtained in laboratory measures) several confounding effects could be present in a single nutritional trial (Moreno et al., 2006). For instance, in most nutrition studies involving volunteers, multivariate techniques must be employed to adjust or control for the effects of basic demographic factors, such as age, sex, race, or body mass index on the outcome of interest (Gleason et al., 2010). Essentially multivariate techniques allow the analysis of the relationship between more than one independent variable and one or more dependent variables without spurious results and inappropriate interferences (Sheean et al., 2011). Thus, multivariate analysis or modeling is an efficient analytical tool used to control for confounding effects, to assess effect modification, and to summarize the association of several predictor variables with some outcome variable of interest. More simplistically, multivariate analytical techniques represent a variety of mathematical models often used to classically measure the exposure-outcome association, taking in account important factors that can influence this relationship (Sheean et al., 2011).

Bias Risk in Randomized Trials

In clinical trials it is very important to minimize the potential biases. Especially, is basic to avoid potential biases related to recruiting methods of volunteers, recording the outcome measurements, the loss of volunteers to follow-up, and volunteers not providing outcome measurements (Sedgwick, 2015). Between the different types of bias that can occur in a clinical trial, most habitual are the selection bias, allocation bias, and ascertainment bias.

Selection bias is a generic term used to describe any type of biases and effects that can result in obtaining a sample significantly different from the population that is intended to represent. Non-response bias and volunteer bias can also be allocated to this group (Sedgwich, 2014). Non-response bias describes the potential differences between subjects who accepted the invitation to be included in a trial and those who did not accept. Non-response bias would have occurred if subjects who accepted the invitation to be included in a clinical trial are systematically different from those who did not accept. The volunteers would be expected to differ from the population because of their sociodemographic, behavior, attitudes, and health status. It has been reported that, in global terms, volunteers who participate in studies are more educated, come from a higher social class, and are more sociable than those subjects who do not accept to participate in a trial. Contrariwise, volunteer bias describes the potential differences between those who volunteered to be in the sample and the population which it is intended to represent (Sedgwick, 2015).

Ascertainment bias could occur for a wide variety of reasons. One example is the case of researchers that may favor one of the interventions because they think that it is the most effective intervention. If the volunteers perceive this preference, they might be conditioned to report their answers in a way that they perceive would please the researchers (Sedgwick, 2015). If the researchers measure the outcomes, they could then record their responses inaccurately or encourage the volunteers to respond so that their experiences favor an intervention. Volunteers who receive their preferred intervention could be more motivated and they would show greater adherence to their intervention. By contrast, subjects who do not receive the intervention preferred by researchers, might exhibit demoralization, whereby they comply poorly the intervention and possibly withdraw from the trial. Resentful demoralization of the volunteers participating in a trial can be prevented with an adequate double- or even triple-blinding (Sedgwick, 2015). Consequently, blinding of the volunteers, researchers and even lead researcher is necessary to prevent ascertainment bias. However, in nutritional trials it is not always possible to blind volunteers to their intervention allocation, because they are asked to intake a known food. Only in the cases that nutrients are applied to subjects through supplements or encapsulate nutrients, then an effective placebo effect can be achieved (Alleleyn et al., 2018).

Allocation bias would occur if there is a systematic difference between subjects in how they were allocated to intervention groups. For example, researchers may favor the intervention, wishing to show that it is more effective than the control, and allocating some volunteers to the intervention whom they believe that subjects will obtain the greatest benefit from the intervention (Sedgwick, 2015).

To minimize any risk of external biases, both external and internal validity are essential components in the design, analysis, and inference of clinical trials. External validity is the extent to which the study results can be generalized to the population. This will largely depend on the characteristics of the sample members and the extent to which they represent the population. Internal validity is the extent to which the observed intervention effects can be ascribed to differences in intervention and not interferences, thereby allowing the inference of causality to be ascribed to a group (Sedgwick, 2014).

In clinical trials, volunteers are normally recruited using random sampling method, and consequently, each population member has a known and typically equal probability of being selected (Sedgwick, 2014). However, in some cases, volunteers can be recruited by other methods, such as convenience sampling based on the degree of easiness to recruit volunteers for a trial. Thus, it is usual that the volunteers are recruited from consumers that have a previous pathology, from patients from a hospital or nutritionist clinic, or even from the habitual consumers of a food group (Ortega et al., 2015). All these circumstances resulted in a better accessibility and predisposition of the potential volunteers to participate in the trial. Therefore, not all population members had an equal probability of being selected (Sedgwick, 2014), and this is an additional potential source of bias.

Missing Data

Clinical trials are usually designed with the implicit assumption that data analysis will occur only after the trial is completed. Although in pharmacological trials this is usually the intention of the participants (patients need treatment in order to treat a pathology), in the case of nutritional trials, this is not always the case. Clinical trials with foods, food ingredients, or bioactive compounds are usually conducted over long periods, requiring changes in eating habits, and volunteers do not usually need the food tested to combat a pathology (Wu, 2016). All these factors make subject drop-outs much more frequent than in the case of pharmacological trials.

It is common that diverse problems arise during the trial for volunteers. For example, non-compliance with the exclusion criteria in a surreptitious manner, voluntary abandonment or force majeure forcing the volunteer to leave the trial prematurely. How missing data are dealt with in the analysis of a trial can have a major effect on the results and therefore, on the conclusions obtained from the trial (Marston & Sedgwick, 2014). The aim of randomization, if the sample size is large enough, is to produce groups of volunteers similar in baseline characteristics. Therefore, any differences between groups in outcome would be due to differences in intervention received, not differences in baseline characteristics, and the trial would have internal validity. However, the exclusion of subjects with missing data from the analysis would introduce confounds because the balance in baseline characteristics would disappear. As a result, internal validity would be threatened. Furthermore, the exclusion of subjects would ultimately lead to reduced statistical power (Marston & Sedgwick, 2014).

To prevent the loss of statistical power of a trial by the exclusion of volunteers, imputation is an alternative approach to deleting subjects with missing data. This strategy consists in replacing a missing observation with a plausible data value, which may be an existing value or one that is predicted from the subject's available data. Imputation alternative ensures that all randomized subjects are included in the analyses, thereby maintaining comparability between groups at baseline and minimizing confounding factors (Marston & Sedgwick, 2014).

Although the methods of baseline observation carried forward and the last observation carried forward have some advantages, they are generally discouraged as methods of imputation. Recently, more sophisticated and adequate methods, including regression mean imputation and multiple imputations, have been developed that are thought to give more plausible and unbiased values (Marston & Sedgwick, 2014).

Advantages of Crossover Design With Respect to Parallel Design

The crossover design has several advantages that researchers may want to use for early stage trials. The higher strength of crossover design is that the interventions under investigation are evaluated within the same volunteers, and consequently, avoid between-subject variability. Further, crossover design allows opportunities of head-to-head trials, and volunteers receiving multiple interventions can express preferences for or against particular interventions (Mills et al., 2009). Thus, the results obtained by crossover trials provide a more efficient statistical comparison than in parallel tests, because no confounds can occur due to subtle differences between the comparison groups that may be seen in a randomized clinical trial (Harris & Raynor, 2017; Sedwick, 2014).

Other important advantages of the crossover design are that the number of volunteers needed for each level of the independent variable is substantially smaller than those needed for a parallel clinical trial. In addition, even in cases where there was no difference in the number of volunteers needed for each of the levels of the independent variable, the overall sample size needed for the study is lower, because there are not randomization into unique groups representing the levels of the independent variable (Reich & Milstone, 2014). For a parallel clinical trial design with the same variables, it is necessary five times the number of subjects in each group to yield a similar study power and level of significance (Harris & Raynor, 2017). In example, in a recent work (Wijnhoven, van der Meij & Visser, 2015) investigated the effectiveness of strategies to increase intake in older women with low appetite. In this work, authors anticipated needing 17 volunteers to achieve 80% statistical power and significance level <0.05 using a cross-over design. In the case that the same work had be designed by a parallel clinical trial, the minimal number of volunteers would be over 70.

The main difference in determining sample size in both parallel and crossover designs is determining the expected variances (Golkowski, Friede & Kieser, 2014). The ability to calculate an appropriate sample size for both designs needs preliminary data that capture means and variances, as well as minimal meaningful clinical outcomes. These data may come from previous experiences of researchers, or previously literature data published by other authors but using similar samples and manipulations (Golkowski et al., 2014).

This reduction in the sample size involves other advantages, especially logistical and economic, arising from the handling of fewer samples or cheaper liability insurance that are necessary for a bioethics committee to authorize a clinical trial (Harris & Raynor, 2017). This is a very important task because nutrition clinical trials often do not investigate imminent risks to human life, but risk factors that predispose to chronic conditions. For this reason, they are not usually approached from a cost-effective perspective, but from a cost-benefit perspective, and therefore, their funding tends to be more modest than in other areas of human health, which investigates more immediate risks (Lui, 2018).

Another important advantage of cross-designing nutritional trials is that there is no ethical risk. Since no research is normally done on imminent risks to human life, there is no ethical risk in subjecting volunteers to a period in which they will act as controls (without potentially beneficial intervention). This ethical dilemma is an important barrier to using crossover trials in specialties where being assigned to the control group can result in a significant deterioration in the health of the subjects, as is the case with oncology trials.

Potential Confounding Variables and Disadvantages in the Crossover Design

Although a crossover design has many advantages, it cannot be used for all type of trials (Reich & Milstone, 2014). The main disadvantage of crossover designs for volunteers is that they are given so much intervention and over longer periods. This is a very important limitation not only because of the consideration towards the subjects but also because it increases the risk of dropouts during the study. Due to the initial number of subjects in crossover design is lower than that in a parallel trial, the losses of subjects participating in a crossover trial are much more problematic than in parallel trials from a statistical point of view (Mills et al., 2009). Additionally, there may be ethical concerns for the use of crossover designs in cases where the population of interest in the study has serious health concerns that are being investigated in the trial (Prasad & Grady, 2014). These concerns are moving volunteers to a level that provides less intervention than the earlier level, as well as removing all benefits derived from the intervention during the washout period (Prasad & Grady, 2014).

Another limitation is that it is important that subjects must maintain their initial status at the beginning of each period, so these studies are only useful if they are very stable in symptoms or subject's physiology. In the cases of important variations in symptoms or physiology, the small sample is not only not essential and may also be also inadequate.

Additionally, crossover trials can also have disadvantages due to the so-called residual effect, sequence effect, and period effect (Wu, 2016). The residual or carryover effect is produced when in a period of the study, the effect of the previous period intervention persists. This is especially important in the trials in which there is no "washout" between periods. Therefore, if an intervention effect cannot be "washed out", a crossover design cannot be effectively implemented (Reich & Milstone, 2014). Thus, an effective washout period between phases of a crossover design need to be used to prevent carryover effects. During this period, all subjects are taken off all manipulations to let the effects of the nutritional intervention on volunteer's health and theoretically allow the dependent variable to return to baseline levels (Harris & Raynor, 2017).

In this sense, the washing interval must guarantee the complete elimination of the active substance previously administered, prolonged dose for a period, not less than five times the terminal half-life of active substances. However, it is very difficult to determine the carryover effects, unless there is a large washout period or if the carryover effects due to the two groups are equal (Geneus, Li, Weerahandi, Whalen & Ray Yu, 2018). Special care is required if the active substance or its active metabolites have long half-lives, as in the case of liposoluble nutrients, that can be stored in different allocations of the human body for a long period. In cases of high variability in the rates of elimination between the subjects, the washing period is calculated as a function of the slower elimination rate. However, some interventions may cause permanent changes on physiological, behavioral, or psychosocial outcomes. In these cases, the crossover design of trials cannot be applied.

A sequence effect occurs when effect of the nutritional intervention is influenced by the order in which the levels of the independent variable are implemented in a volunteer. It can happen that subject characteristics tend to change throughout the study, modifying their response to the two interventions. A period effect can occur when the intervention effect is not constant over time resulting in intervention by period interaction. The period effect is more likely to occur when the intervention periods are long (such in the case of crossover trials), and when the underlying physiological or health condition is not stable.

A phase effect occurs when variables different than the independent variable, potentially influence the dependent outcome variables vary in the differing phases. Amongst these effects, the seasonal differences in eating patterns could especially be cited, because in most geographical areas in the world the diet in the summer months is widely different to winter months. When all these situations occur, a crossover design is inappropriate (Gleason et al., 2010).

Finally, important effect issues are often underreported in randomized crossover trials. Given their popularity, few reported important methodological issues such as potential confounding variables, that are mainly the carryover, phase, or sequence effects (Harris & Raynor, 2017). These potentially confounding variables may influence the dependent variable investigated in a different way, even changing the conclusions obtained about the independent variable in the trial. To avoid this concern, crossover designs should be designed taking the necessary precautions to minimize these potential confounders (Harris & Raynor, 2017). Globally, the advantages and disadvantages of parallel design, with respect to cross-over design and cross-over designs coupled to design of experiments are stated in Table 1.

CONCLUSION AND FUTURE PERSPECTIVES

In order to obtain more reliable results from crossover trials, it is mandatory to improve their transparency and results interpretation. It can be achieved by creating standards specifically adapted to crossover trial designs, similarly to CONSORT reporting guidelines, but adapted specifically for crossover trials. Another major concern is related to the use of inappropriate statistical analysis often applied to obtain conclusions from crossover designs. Because volunteers act as their own controls in crossover trials, the analysis could be based on paired data and the within-subject variability in outcomes could be consid-

Table 1. Advantages, disadvantages of parallel design with respect to cross-over trials and cross-over trials coupled to design of experiments

	Parallel Design	**Crossover Design**	**Crossover Design Plus Design of Experiments**
Advantages	• CONSOT standards guidelines ensures good standardization and facilitates the inclusion of the results obtained in systematic reviews and meta-analysis • The effects of dropouts during the study do not seriously distort the results • Requires shorter periods than cross-over designs • No occurrence of residual, sequence, period or phase effects	• Low effect of external factors in the results • Minimize imbalance during randomization process • Avoid between-subject variability • More efficient statistical comparison that parallel trials • Requires low number of subjects than parallel trials (5 times lower) • Low consumption of logistical and economic resources	• The advantages of cross-over designs are generally maintained and enhanced • Allow to minimize the number of sequences in the study, while maintaining adequate statistical power • Higher statistical significance even with few subjects • Low time, and economic resources needed • Better estimation of statistical power and other validity indicators • Better adjust or control for basis demographic factors • Decrease risk of unbiased values • Ability to isolate confounding factors
Disadvantages	• High number of subjects required • High risk of selection, allocation and ascertainments bias • High risk of bias in randomization process • Need of external and internal validity • Between-subject variability can seriously disturb the results • Poor statistical power • Requires high number of volunteers, that in many cases are difficult to achieve • High consumption of logistical and economic resources due to the high number of samples to process	• Absence of published standards guidelines • Difficulties to include results in systematic reviews and meta-analysis • High effect of subject dropouts during the study • Necessary very long periods • Dependent variables may need a long period during washing period to recover baseline levels • Exclusion of subjects that do not follow the intervention cause imbalance in the baseline • Possibility of occurrence of residual effect, sequence effect and period effects • Possibility of occurrence of phase effect • Effect issues and often underreported	• Necessary longer time of planification prior to begin the trial • Necessary trained personnel in biostatistics, design of experiments and bioinformatics

ered in sample size calculations. The use of a paired data-based analysis is much more efficient than a parallel group design when researchers expect a high correlation between volunteers` responses to the different interventions.

Additionally, even when properly applied, crossover trials may have certain concerns. The number of subjects who leave the trial prematurely after the first period of intervention and do not perform the second is much larger than in other types of trials. This makes within-subject comparison impossible and is particularly important if withdrawal is related to side-effects. Also, there may be a residual or carry-over of the effect or interventions across the study period, but randomization typically does not occur in the second period. Additionally, carry-over effect could potentially distort the results obtained during the second intervention or subsequent periods. Thus, the observed intervention effects will depend upon the order in which they were received.

Yet due to the absence of a long washout period or a four sequence, the crossover design has done more harm than good, when analyzed by classical methods. Considering the advantages and disadvantages of cross-clinical trials, it is not clear at present that these represent clear improvements over more classical designs for nutrition trials.

REFERENCES

Alleleyn, A. M. E., van Avesaat, M., Ripken, D., Bleiel, S. B., Keszthelyi, D., Wilms, E., ... Masclee, A. A. M. (2018). The effect of encapsulated nutrients mixture on food intake and satiety: A double blind randomized cross-over proof of concept study. *Nutrients*, *10*(11), 1787. doi:10.3390/nu10111787 PMID:30453597

European Food Safety Authority (EFSA). (2011). Scientific and technical guidance for the preparation and presentation of an application for authorization of a health claim. *EFSA Journal*, *9*(5), 2170. doi:10.2903/j.efsa.2011.2170

Galvain, F., Barolo, M., Macchietto, S., & Bezzo, F. (2010). Identification of physiological models of type 1 diabetes mellitus by model-based design of experiments. Process Systems Engineering, 7, 545-581. https://doi.org doi:10.1002/9783527631209.ch76

Geier, A. B., Rozin, P., & Doros, G. (2006). Unit bias: A new heuristic that helps explain the effect of portion size on food intake. *Psychological Science*, *17*(6), 578–582. doi:10.1111/j.1467-9280.2006.01738.x PMID:16771803

Geneus, V., Li, C., Weerahandi, S., Whalen, E., & Ray Yu, C. (2018). Interferences on treatment and carryover effects in two-period two-sequence crossover designs. *Current Trends in Biomedical Engineering & Biosciences*, *11*(5). doi:10.19080/CTBEB.2018.11.555821

Gleason, P. M., Harris, J., Sheean, P. M., Boushey, C. J., & Bruemmer, B. (2010). Publishing nutrition research: Validity, reliability, and diagnostic test assessment in nutrition-related research. *Journal of the American Dietetic Association*, *110*(3), 409–419. doi:10.1016/j.jada.2009.11.022 PMID:20184991

Golskowski, D., Friede, T., & Kieser, M. (2014). Blinded sample re-estimation in crossover bioequivalence trials. *Pharmaceutical Statistics*, *13*(3), 157–162. doi:10.1002/pst.1617 PMID:24715672

Harris, J. E., & Raynor, H. A. (2017). Crossover designs in nutrition and dietetics research. *Journal of the Academy of Nutrition and Dietetics*, *117*(7), 1023–1030. doi:10.1016/j.jand.2017.03.017 PMID:28479137

Langlet, B., Bach, M. T., Odegi, D., Fagerberg, P., & Ioakimidis, I. (2018). The effect of food unit sizes and meal serving occasions on eating behavior characteristics: Within person randomized crossover studies on healthy women. *Nutrients*, *10*(7), 880. doi:10.3390/nu10070880 PMID:29986529

Leonard, G., Lafrenaye, S., & Goffaux, P. (2012). Randomized placebo-controlled cross-over designs in clinical trials: A gold standard to be reassessed. *Current Medical Research and Opinion*, *28*(2), 245–248. doi:10.1185/03007995.2011.653558 PMID:22208674

Li, T., Tsung, Y., Hawkins, B. S., & Dickersin, K. (2015). Design, analysis and reporting of crossover trials for inclusion in a meta-analysis. *PLoS One*. doi:10.1371/journal.pone.0133023 PMID:26284684

Lui, K. J. (2016). *Crossover designs: testing, estimation, and sample size*. Hoboken, N.J.: John Wiley & Sons; doi:10.1002/9781119114710

Lui, K. J. (2018). Sample size determination for testing equality in frequency data under an incomplete block crossover design. *Journal of Applied Statistics*, *45*(8), 1517–1529. doi:10.1080/02664763.2017.1380788

Marston, L., & Sedgwich, P. (2014). Randomised controlled trials: Missing data. *British Medical Journal*, *349*(jul18 2), g4656. doi:10.1136/bmj.g4656 PMID:25038075

McCrickerd, K., & Forde, C. G. (2017). Consistency of eating rate, oral processing behaviors and energy intake across meals. *Nutrients*, *9*(8), 891. doi:10.3390/nu9080891 PMID:28817066

Mills, E. J., An-Wen, C., Wu, P., Vail, A., Guyatt, G. H., & Altman, D. G. (2009). Design, analysis, and presentation of crossover trials. *Trials*, *10*(1), 27. doi:10.1186/1745-6215-10-27 PMID:19405975

Miranda, J. M., Anton, X., Redondo-Valbuena, C., Roca-Saavedra, P., Rodriguez, J. A., Lamas, A., ... Cepeda, A. (2015). Egg and egg-derived foods: Effects of human health and use as functional foods. *Nutrients*, *7*(1), 706–729. doi:10.3390/nu7010706 PMID:25608941

Moreno, L. A., Joyanes, M., Mesana, M. I., González-Gross, M., Gil, C. M., Gutiérrez, A., ... Marcos, A. (2006). Harmonization of anthropometric measurements for a multicenter nutrition survey in Spanish adolescents. *Nutrition (Burbank, Los Angeles County, Calif.)*, *19*(6), 481–486. doi:10.1016/S0899-9007(03)00040-6 PMID:12781845

Ortega, M., Pérez-Rodrigo, C., & López-Sobaler, A. M. (2015). Dietary assessment methods: Dietary reports. *Nutrición Hospitalaria*, *31*, 38–45. doi:10.3305/nh.2015.31.sup3.8749 PMID:25719769

Prasad, V., & Grady, C. (2014). The misguided ethics of crossover trials. *Contemporary Clinical Trials*, *37*(2), 167–169. doi:10.1016/j.cct.2013.12.003 PMID:24365533

Reich, N. G., & Milstone, A. M. (2014). Improving efficiency in cluster-randomized study design and implementation: Taking advantage of a crossover. *Open Access Journal of Clinical Trials*, *14*(6), 11–15. doi:10.2147/oajct.S56730

Reich, N. G., Myers, J. A., Obeng, D., Milstone, A. M., & Perl, T. M. (2012). Empirical power and sample size calculations for cluster-randomized and cluster-randomized crossover studies. *PLoS One*, *7*(4), e35564. doi:10.1371/journal.pone.0035564 PMID:22558168

Schork, N. J., & Goetz, L. H. (2017). Single-subject studies in translational nutrition research. *Annual Review of Nutrition*, *37*(1), 395–422. doi:10.1146/annurev-nutr-071816-064717 PMID:28715990

Sedwick, P. (2014). Randomised controlled trials: "Within subject" versus "between subject" designs. *British Medical Journal*, *349*(aug06 2), g6435. doi:10.1136/bmj.g6435 PMID:25954990

Sedwick, P. (2015). Bias in randomized controlled trials: Comparison of crossover group and parallel group designs. *British Medical Journal*, *351*, h4283. doi:10.1136/bmj.h4283 PMID:26253465

Sheean, P. M., Bruemmen, B., Gleason, P., Harris, J., Boushey, C., & Van Horn, L. (2011). Publishing nutrition research: A review of multivariate techniques- Part 1. *Journal of the American Dietetic Association, 111*(1), 103–110. doi:10.1016/j.jada.2010.10.010 PMID:21185971

Wijnhoven, H. A. H., van der Meij, B. S., & Viser, M. (2015). Variety within a cooked meal increases meal energy intake in older women with a poor appetite. *Appetite, 95*, 571–576. doi:10.1016/j.appet.2015.08.029 PMID:26321418

Wu, C. H. (2016). Randomized response model versus mixture model in the crossover design for clinical trials. *Communication in Statistics-Theory and Methods, 45*(15), 4611-4627. doi:. doi:10.1080/03610926.2014.927490

Yan, J. H., Guan, B. J., Gao, H. Y., & Peng, X. E. (2018). Omega-3 polyunsaturated fatty acid supplementation and non-alcoholic fatty liver disease. *Medicine, 97*(37). doi:10.1097/MD.0000000000012271 PMID:30212963

Zhang, C. X., Long, W. Q., Ye, Y. B., Lu, M. S., Zhang, N. Q., Xu, M., ... Su, Y. X. (2018). Effects of chocolate-based products intake on blood glucose, insulin and ghrelin levels on satiety in young people: A cross-over experimental study. *International Journal of Food Sciences and Nutrition, 69*(7), 882–891. doi:10.1080/09637486.2018.1426737 PMID:29457516

KEY TERMS AND DEFINITIONS

Carry-Over: Effect that persist from one experimental condition to another. Whenever subjects perform in more than one condition (as they do in within-subject designs) there is a possibility of carryover effects.

Clinical Assay: Medical research conducted on people who voluntarily participate in these studies and who help discover better ways to treat, prevent, diagnose and understand diseases that affect humans.

Confusing Factor: Variable that influences both the dependent variable and independent variable, causing a mistaken association.

Cross-Over: Observational epidemiological design to assess whether a given intermittent or unusual exposure may have triggered an immediate short-term, acute event.

Design on Experiments: Set of active techniques that manipulate a process to induce it to provide the information required to improve it through changes in its variables and their interaction or sequence of execution.

Nutrition: Food intake in relation to the dietary needs of the human body.

Randomization: Method based on chance alone by which study participants are assigned to a treatment group.

Sequence Effects: Potential confounding influences in experiments where subjects are exposed to multiple conditions.

Section 3
Food

Chapter 8
Optimization of a Spectrophotometric Flow Injection:
Method for Determination Copper and Manganese in Wines by Design Experiments

Germán Adán Sánchez Gallegos
Universidad Autónoma de Baja California, México

Maria del Pilar Haro Vazquez
Universidad Autónoma de Baja California, Mexico

José Manuel Cornejo-Bravo
https://orcid.org/0000-0002-0013-8937
Universidad Autónoma de Baja California, Mexico

Eugenia Gabriela Carrillo-Cedillo
https://orcid.org/0000-0003-4239-5904
Universidad Autónoma de Baja California, Mexico

ABSTRACT

This chapter presents the optimization of the hydrodynamic and chemical parameters of the FIA system in the determination of copper and manganese in wine samples by VIS spectrophotometry. This technique has been based on the injection of liquid samples in the non-segmented movement, within a continuous carrier current of a suitable liquid. The injected sample forms a zone that disperses on its way to a detector. The later continuously records the absorbance or other physical parameters, since it continuously passes the sample material through the flow cell, using the factorial designs Plackett-Burman, Box-Behnken, and the factorial design 2^4. The methods have the advantages of low-cost, easy availability of chemicals, and instrumentation and straightforward application to real samples.

DOI: 10.4018/978-1-7998-1518-1.ch008

INTRODUCTION

Wine has been defined as an alcoholic beverage obtained from the alcoholic fermentation of only fresh grape musts (Vitis vinifera), with or without pomace, or the mixture of concentrated grape musts and water. The alcohol content in wine is from 8% to 16% Alc. Vol. The alcohol content may be up to 18% Alc. Vol. for products that are regulated by a denomination of origin. In the case of wines from late harvests, the alcohol content could reach 18% Alc. Vol by natural fermentation. Wines are classified by their content of total reducing sugars. A dry wine has less than 4 g/L; semi-dry: between 4.1 and 12 g/L; semi-sweet: between 12.1 to 50 g/L; and sweet: more than 50 g/L. Mexican wine is wine produced with grapes of 100% Mexican origin, and also is completely fermented and packed in the national territory (NOM-199-SCFI-2017, 2017).

The macromolecules present in wine also have the potential to interact and bind to metals, and may include tannins, polysaccharides, proteins, and combinations of complexes or aggregates of these macromolecules (Kontoudakis, Guo, Scollary, & Clark, 2017). Taking into account the elemental composition, the wine contains macro-elements with concentrations higher than 10 mg/L (Na, K, Mg, Ca), for the case of the micro-elements between 10 µg/L and 10 mg/L (Fe, Cu, Zn, Mn, Pb), and ultra-microelements less than 10 µg/L (Cr, As, Cd, Ni). Data on the mineral content in wines have been studied and reported due to their implications for organoleptic, hygienic and nutritional characteristics, as well as their toxicological implications (Guriérrez, Rubio, Moreno, & González, 2017).

The mineral content of the wine depends on several factors, including the absorption of metals from the soil, contamination by emissions to the atmosphere, type of grape, herbicide treatment, and the process of winemaking. The content of minerals has a significant influence on the quality of a wine, and several studies have done on this subject (Ferreira, et al., 2008).

Most of the abundant mineral elements found in wine come from the grape itself through the absorption of the soils where the grapes are grown, reflecting the elemental profile of the ground. These factors included age, root depth, soil pH, rainfall, temperature and will vary widely from region to region (Orellana, Johansen, & Gazis, 2019). In China, the concentrations of these elements (Cu, Fe and Mn) was regulated on imported wine and the limits fixed are (Copper 1 mg/L, Iron 8 mg/L and Manganese 2 mg/L), these minerals can vary from one area to another and from one variety the wine to another due to the presence of nutrients in the soil. When the grapes were growing, the uptake of these nutrients by the vine itself and the process by which wine was producing. Due to this significant variability, there is no way to guarantee that a particular wine meets import specifications with outperforming analytical tests (Spivey, Thompson, Shelton, & Kavan, 2015).

Schut et al., reported the content of Cu in 72 wines showing an average of 0.18 mg/L with a maximum of 0.55 mg/L. The Organization International de la Vigne et du Vin (OIV) recommended the copper concentration limit in wines is equal to 1.0 mg/L. The Germany national regulations allow the presence of 2.0 mg/L Cu in drinking water and German wines (Schut, Zauner, Hampel, Konig, & Claus, 2011). The OIV recommended a Mn concentration at interval of 0.5 to 5 mg/L (Gomez-Miguel & Sotes, 2014).

During fermentation, Cu is mostly removed through its association and precipitation with yeast cells (Hsia, Plack, & Nagel, 1975). To repress sulfidic-off odors in the wine, the winemaker's add Cu, as copper (II) sulfate pentahydrate or copper citrate, (Kontoudakis, Schmidtke, Bekker, & Smith, 2019). In USA, Cu sulfate can be added up to 6 mg Cu (II) per litre, although the residual level in wine cannot be over 0.5 mg/L (Code of Federal Regulations 2014) (Clark, Wilkes, & Scollary, 2015).

Previous work has shown the ability of Cu to interact with proteins and induce the formation of haze in wines, particularly in wines with high concentrations of Cu, present low oxygen conditions and residual proteins (Kontoudakis, Guo, Scollary, & Clark, 2017). Ultimately, new wines contain only 0.3 to 0.4 mg/L of copper, which is insufficient to cause turbidity in wine (Ribereau-Gayon, Glories, Maujean, & Dubourdieu, 2006).

Cacho et al., studies the influence of iron, copper and manganese on the oxidation of wine, measuring the evolution of different compounds in this process, such as anthocyanin's, tannins, total phenol and acetaldehyde content. They conclude that the Cu does not seem to influence acetaldehyde formation during oxidation and the Mn clearly favors acetaldehyde formation during oxidation of wine, can be observed again that for extreme oxidations acetaldehyde concentration decreases even at high manganese concentrations. This fact also explains the influence of manganese in the disappearance of phenolic compounds and tannins: the higher the manganese concentration the higher the acetaldehyde formation and therefore, the higher the polymerization and later precipitation of phenolic compounds (Cacho, Castells, & Barroso, 2002).

Due to the essential effects that metal ions have on the organoleptic properties of the wine (characteristics of aging, final aroma, flavor, and color), the monitoring of its content during production is necessary to achieve the appropriate quality parameters of the final product. The constant tracking of copper, iron, and manganese in the wine has required due to its toxicity in case of excessive intake. Therefore, in almost all countries, there is national legislation on the metallic content of wine (Cepo, Pelajic, Vrcek, & Krivohlavek, 2018).

At present, the low concentration of Cu in wine is no longer easily measured by specific analytical procedures, such as flame atomic absorption spectroscopy (FAAS), without prior preconcentration of the sample (Kontoudakis, Schmidtke, Bekker, & Smith, 2019). The electrochemical stripping methods of anodic stripping voltammetry (ASV) and stripping potentiometry (PSA) are very useful for determination speciation of copper in wine (Pyrzynska, 2004).

There are different methods to analyze manganese, including X-ray fluorescence, voltammetry, atomic absorption spectrometry (AAS), and UV-VIS spectrophotometry. AAS is a technique widely used for quantification of manganese, but the apparatus used is more expensive than UV-VIS spectrophotometry (Liu, Zhan, Heng, & Ma, 2013).

There is a need to develop less complicated methods using conventional laboratory equipment less sophisticated, which is sensitive and selective for the quantification of copper and manganese to monitor them. For the determination of copper, we continued with the previous work of Carrillo et al. they determinates copper to the analysis by injection in flow (FIA) by forming a yellow complex, which was measured at 434 nm using a fractional factorial design (Carrillo, Cañizares, Ames, & Sepúlveda, 2016). This technique has been basing on the injection of liquid sample in the non-segmented movement, within a continuous carrier current of a suitable liquid. The usual working conditions of the FIA cause that the injected sample bolus not to reach the chemical or physical equilibrium and create dispersion in the carrier solution or the reagent. The laminar regimen produces a well-defined area of the injected sample bolus, creating a concentration gradient along with the entire bolus (Cañizares, 2002). In the determination of copper and manganese in wine samples it is required to optimize the hydrodynamic and chemical parameters of the FIA system, the factorial designs Plackett-Burman, factorial design 2^4, and Box-Behnken have been using.

The main effects have a complicated, confusing relationship with the interactions of two factors in the Plackett-Burman designs. These designs should be applied to the study of the main influences when they can have been thinking that bidirectional interactions are insignificant. Full or fractional two-level factorial designs, as well as Plackett-Burman designs, are often used to analyze the main factors that influence the output measures of the process or the quality of the product. These designs are use fitting first-order models (which detect lineal effects) and can to ease information on the existence of second-order effects (curvature) when the design includes center points (Carrillo, Haro, Díaz, & Cañizares, 2014).

The Factorial 2^4 design is perhaps the largest factorial that can still has been running with two replicas, which implies making 32 runs, but even with a replica (16 tests) of this design many times It is enough to study the 10 effects that they are usually of greater interest (Gutiérrez & Salazar, 2012). In the determination of copper a fractional factorial design was used Plackett-Burman 2^6*3/16 of resolution III, with three central points and 8 degrees of freedom. For the case of the determination of manganese two designs were a resolution V^+, with two central points and 7 degrees of freedom, and other factorial design Box-Behnken with three central points and 5 degrees of freedom.

BACKGROUND

Kontoudakis et al., propose a method allows the determination of total copper concentration in white wine. It utilizes the colorimetric reagent 2,2′-bicinchoninic acid dipotassium salt (BCA) to react with copper(I) and form a purpled coloured complex, which absorbs at 563 nm. During the analysis, ascorbic acid is added to aid the conversion of copper (II) to copper (I). In addition, silver (I) nitrate is added in order to induce dissociation of suspended copper (I) sulfide in wine, and thereby to provide non-sulfide bound copper ions and silver (I) sulfide (Kontoudakis, Smith, Smith, Wilkes, & Clark, 2018).

Moise G. proposed a method the stripping potentiometric analysis is a simple analytical procedure to monitor copper in white wine, the conditions were optimized for a supporting electrolyte having the following composition: 1 mol/L HCl and 0.5 mol /L $CaCl_2$. It was noted that mercury oxidizes more convenient than oxygen, and to determine labile copper concentration it was necessary to change the agent the detection limit for copper ion is 0.05 mg/L (Moise, 2015).

Ferreira S. L. C. et al., propose a method to correct matrix effects in the determination of manganese and iron in wine samples, using FSFAAS. Indium, nickel, silver, and cobalt have been testing as reference elements. The results they indicated that cobalt and indium at a concentration of 2 and 10 mg/L were efficient for quantification of iron and manganese, respectively. Under these conditions, iron and manganese could have been determining with quantification limits of 27 and 40 µg/L, respectively. The method has been applying to 16 wine samples in the determination of iron and manganese. The content of manganese change from 0.78 to 2.89 mg/L, and that of iron has been finding in 0.88 to 9.22 mg/L (Ferreira, et al., 2008).

Ou et al., propose a spectrophotometric method for determining manganese in cooking wine was established by oxidation of chromotrope 2B. In the presence of sodium bismuthate, all low-valent Mn ions in samples were oxidized to form to Mn (VII), which oxided chromotrope 2B at room temperature in sulfuric acid medium resulting in fading. The most considerable reduction in the absorbance at 515 nm was observed, which was directly proportional to Mn (VII) concentration. The absorbance obeyed Beer's law in the Mn (VII) concentration range of 0.05 to 5.0 µg/mL with an apparent molar absorption coefficient of 2.05×10^4 L/(mol•cm). The recovery of the developed method for spiked samples of commercial liquor and cooking wine ranged between 98.5% and 101.9% (Ou, Li, & Wei-Xing, 2013).

General guidelines have been suggested for the recommended total concentrations of metal ions in wine to limit their detrimental impacts. A variety of techniques are now available to measure, not only the total concentrations of the metal ions but also different forms of the metal ions (metal speciation). Such techniques include electrochemical, liquid extractions, HPLC, solid phase extraction, and colorimetric methods. In all cases, the measurement technique itself may have some impact on the distribution of the different metal forms in the wine and hence the analytical results are commonly termed 'operationally-defined' measures of metal speciation (Rousseva, Kontoudakis, Schmidtke, Scollary, & Clark, 2016).

METALS IN WINES

Wine has been a favorite and widely-consumed alcoholic beverage worldwide since early civilization. The quality, origin, aroma/flavor characteristics and health safety of wine consumption have been influencing by environmental and anthropogenic factors and can has been identifying by varying contents of inorganic and organic substances forming its chemical composition. The environmental factors are geography (e.g. region, orography, presence of water resources), climate (e.g. temperature, precipitation, humidity, wind, etc.), soil type, composition, and grape variety. The anthropogenic factors, principle impacts are associated with the pollution of vineyards (e.g. soil contamination and irrigation water quality) viticulture management practices such as use of seed preservatives, chemical sprays, fertilizers, grape-growing approaches and winemaking technology and storage (Vystavna, Rushenko, Diadin, Klymenko, & Klymenko, 2014).

The external impurities that reach the wine during the growth of the grapes or in different stages of the winemaking (from the harvest to the bottling and the cellar) are associated with the contribution of the metals of secondary origin. During the increase of the grapes in a vineyard, the contaminations can have been classifying as protection, geographical (originated in the soil) and cultivation practices or environmental pollution.

Accordingly, the vineyards near sea or ocean result in a higher Na in wines content contrast to wines from other regions. Wind from sea or ocean affects the coastal area and vineyards located therein by a marine spray. Differences in Cu, K, and Ca content can be due to fertilizers employees for cultivation. Application of fungicides, pesticides, and fertilizers containing Mn, Cu, Cd, Zn, and Pb compounds during the growing season of vines leads to increases in the amounts of these metals in wine. Wine from vineyards located in industrial areas or near traffic contain higher levels of Cd and Pb due to vehicle exhaust or other emissions to the air, water, and soil.

Pollution can occur at different stages of wine production, as there is an enological (winemaking) source of metals. The reason for this is the extended contact time of the wine with the materials (wood, glass, brass, stainless steel and aluminum) from which the pipes and machines have been making for winemaking, and casks and barrels used to manipulate and store wine. This is the usual source of Zn, Cd, **Cu**, Cr, Fe, and Al. Contamination with Ca, Al or Na can be associated with fining and clarifying substances (flocculants, such as bentonites) added to wine to remove suspended solids after fermentation and to reduce turbidity. Ca concentration can also be altering by adding $CaSO_4$ or $CaCO_3$ for de-acidification of must and wine or enhancement of acidity of grape juices, respectively (Pohl, 2007).

Metals affect the organoleptic characteristics of wine, incorporating flavor, aroma, freshness, taste, and color, mainly due to precipitates being forming (yeast, fining and filtration sediments) or clouding during wine fermentation, maturation, and storage. Most metals are essential for efficient alcoholic fermentation. Na, K, Ca, and Mg takes part in regulating the cellular metabolism of yeasts by maintaining adequate pH and ionic balance. The minor metals (Zn, Fe, Cu, and Mn) and some trace metals are also convenient for the yeasts since they are necessary for the prosthetic activation of metallo-enzyme. Precipitation of Ca and K tartrates changes pH, which enhances Cu and Fe oxidation in addition to forming Cu, Al and Fe clouding. Both oxidation and clouding affect wine conservation (Pohl, 2007). The analysis of certain metals in wines is of particular interest due to toxicity in the case of excessive intake, effect on organoleptic properties and use for the description and classification according to the geographical origin and assessment of authenticity (Vystavna, Rushenko, Diadin, Klymenko, & Klymenko, 2014).

Manganese and Copper in Wines

Most metals are essential for efficient alcoholic fermentation. Ca, K, Mg, and Na take part in regulating the cellular metabolism of yeasts by maintaining adequate pH and ionic balance. Minor metals (Cu, Fe, Mn and Zn) and some trace metals are also favorable for yeasts, as they have required for prosthetic metallo-enzyme activation. Precipitation of K and Ca tartrates changes pH. This enhances Cu and Fe oxidation in addition to forming Al, Cu and Fe clouding; affecting wine conservation. Cu, Fe, and Mn form stable complexes with amino acids, polyphenols, and melanoids. These occur during wine maturation and storage and determine ageing characteristics, final aroma, taste, and even the color of the wine. In addition, the association of Cu, Fe and Mn cations with organic chelating ligands is an important natural anti-oxidative mechanism that decreases the rate of formation of reactive oxygen species responsible for reactions causing staleness and spoilage of wine (Pohl, 2007).

The oxidation process of wine was found to depend directly on the concentration of iron, copper, and manganese in wine, and that manganese and iron play an important role in chemical processes with acetaldehyde: manganese favors their formation, and iron catalyzes acetaldehyde combination with phenolic compounds (Santos, Brandao, Portugal, David, & Ferreira, 2009).

Manganese (Mn) is an essential microelement for human, but excessive intake should have avoided due to its harmful effect. The Mn content in wine depends on several factors including grape type, vintages, soil contamination, treated fungicides, and wine-making process (Jung, Kang, Choi, Y. S., Lee D., Lee J. & Park, 2019).

Studies on manganese in wine products are rare, as it has been considering so far that its toxicity is very low and hardly causes risks to food safety: manganese is one of the least toxic elements for mammals and only exposure at extreme concentrations resulting from human activities has adverse effects. Cabrera-Vique et al. (2000) they determined the Mn content in a total of eighty samples of wines from different regions of France and different years of harvest and affirmed that for a given vineyard and winery, there are variations of the Mn content during the conservation period. Mn concentrations varied from 0.435 to 7.836 mg/L in red wine, 0.674-2.203 mg/L in white wine, 0.844 to 1.805 mg/L in rosé wine and 0.358 to 0.733 mg/L in champagne (Gomez-Miguel & Sotes, 2014).

In terms of oxidative influences, both Cu and Fe are known to accelerate the oxygen consumption within the wine, leading to a concomitant loss of sulphur dioxide, and in turn, the release of aroma compounds, such as phenylacetaldehyde and methional.

In wines with reductive aromas, Cu has shown the most obvious, but also paradoxical effects. Initially, Cu can substantially decrease the odor associated with reduced wines (i.e., hydrogen sulfide, methanethiol), but during bottle aging, under low oxygen conditions, it has been linked to increased production of these sulphur compounds in some wines (Rousseva, Kontoudakis, Schmidtke, Scollary, & Clark, 2016).

APPLICATION OF THE DESIGN OF EXPERIMENTS FOR THE DETERMINATION IN WINE USING FLOW INJECTION ANALYSIS

The concept of flow-injection analysis (FIA) was first introduced in 1975, it has had an intense impact on how modern analytical procedures have implemented. There are many scientific publications to FIA that have been generated in all corners of the world. Thus, by the middle of 2006, there had been more than 16,500 papers. Obviously, the reason for this momentum is that the FIA has allowed us to execute novel and unique procedures, which are complicated and, in many cases, not even possible through traditional batch procedures, whose use has dominated readers on the scene. analytical for years; or, perhaps explained more precisely, since generation after generation scientists have been teaching, or rather convinced to think that this was the only sensible way to perform such analytical tests.

At present, the FIA has been modernized and different analytical approaches have been generated for the development of applications in food, environment and metals. Familiar with the knowledge of the three pillars of FIA that are:

1. Insertion, or injection, of a discrete, well-defined volume of sample solution into a flowing carrier stream.
2. Accurate and reproducible moment of the manipulation to which the injected sample zone is subjected in the system, from the point of injection to the point of detection.
3. The creation of a concentration gradient of the injected sample, providing a transient, but strictly reproducible readout of the recorded signal.

The segmented flow analysis (SFA) is an analytical technique where the samples are aspirated sequentially and between them are located air bubbles that separate (segmented) the established flow, including a wash cycle to avoid contamination between samples; the air bubbles have eliminated before they reach the flow cell located in the detector. The FIA has its antecedent in the SFA; but in FIA, unlike the SFA, the flow is not segmented and the sample is injected instead of aspirated, at the time of detection, neither physical nor chemical equilibrium is reached (Figure 1) (Cañizares Macías, 2002).

An appropriate detection device must monitor the final reading, which is the result of two matching kinetic processes, the physical process of the dispersion zone and the superimposed chemical processes that result from the reaction between the analyte and the reactive species. The first generation of FIA, which utilizes continuous pumping of carrier and reagent solutions, was in 1990 supplemented by the second generation, termed sequential injection analysis (SIA). Then, in 2000, the third appears FIA generation, the so-called valve laboratory (LOV), with which an even greater reduction was achieved and the concept of microsphere injection (BI) was introduced, which involves microsphere renewal approaches, which also offers new ways for testing chemists (Hansen & Mir6, 2007).

Figure 1. Example for flow injection analysis system (FIA).

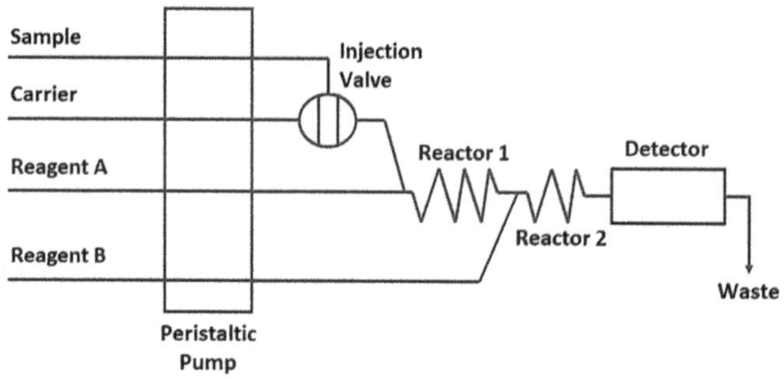

Figure 2. Stages for DoE (Armstrong, 2006).

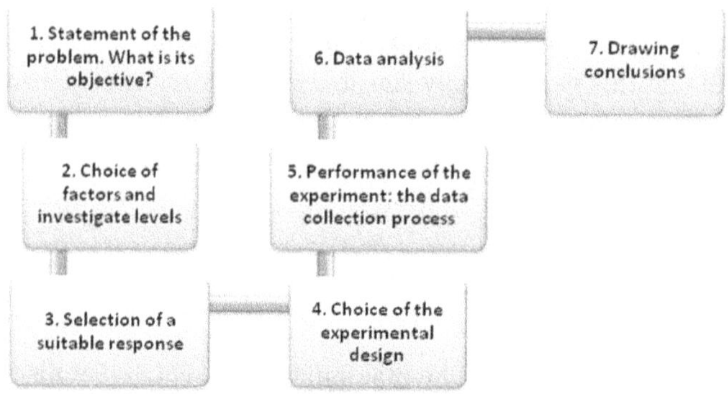

Experiments are one of the most important research activities and have used in many areas of science. As part of the conducted experiments, the influence of certain factors (independent variables) on selected answers (dependent variables) is determined, and that is why it is possible to check the accuracy of the work. Therefore, the goal of any project is to obtain reliable results that will lead us to proper conclusions. To get them, we need to follow many guidelines aimed at preventing from making a mistake that could affect results. Mistakes at an early stage of the research usually result in significant errors, since their consequences are being intensified at each subsequent phase of the experiment, ultimately leading to biased or incorrect answers (Jacyna, Kordalewska, & Markuszewski, 2019).

DESIGN OF EXPERIMENTS

The design of experiments (DoE) is the arrangement of the experimental units used to control the experimental error, while accommodating the treatments (Kuehl, 2001). The DoE approach is a tool for the systematic examination and documentation of the impact of each input factor on the responses studied and for finding the optimal parameters to obtain the desired values. DoE methods use statistical models and analysis of the measurement data obtained to obtain as much information about the process as possible (Jacyna, Kordalewska, & Markuszewski, 2019).

Experimentation may be defined as the investigation of a defined area with a firm objective, using appropriate tools and drawing conclusions that are justified by the experimental data so obtained. Most experiments consist of measuring the effect that one or more factors have on the outcome of the experiment. The factors are the independent variables, and the outcome is the response or dependent variable (Armstrong, 2006). The overall experimental process may be divided into the following stages:

Factorial design is a technique introduced by Fisher in 1926, and provides a means by which factors that influence on a reaction or a process can be evaluated simultaneously and their relative importance assessed. It is thus a means of separating those factors that are important from those that are not. The technique can be applied to many chemical, environmental, industrial processes and pharmaceutical, and it forms the basis for many tests that seek to find an optimum solution (Armstrong, 2006).

There are different types of experimental design we will focus on three types: The Factorial 2^k, Plackett-Burman and Box-Behnken.

Factorial designs are widely used in experiments involving several factors where it is necessary to study the joint effect of the factors on a response. The most important of these special cases is that of *k* factors, each at only two levels. These levels may be quantitative, such as two values of reactor length, injection volume, reagent concentration or speed; or they may be qualitative, such as two machines, two operators, the ¨high¨ and ¨low¨ levels of a factor, or perhaps the presence and absence of a factor. A complete replicate of this design requires 2 x 2 x … x 2 = 2^k observations and is called a 2^k factorial design (Montgomery 2005).

Two-level full factorial designs are the most powerful screening designs, once they allow estimating the main effects of input factors and their interactions on output responses. The main limitations of two-level full factorial designs rely on the large number of experiments required, when compared to Plackett-Burman designs. The number of experiments required for two-level full factorial designs may be calculated as 2^k, where *k* is the number of input factors to be studied (Fukuda, et al. 2018).

In the 1946, Plackett–Burman introduced statistical method offers a design where n variables are studied in n + 1 experimental runs. These experimental designs are available in multiples of four runs and hence they are excellent screening methods, as the number of experimental runs required are very few, leading to saving of time, chemicals, glassware and manpower (Naveena, Altaf, Bhadriah, & Reddy, 2005).

In the 1960, Box and Behnken introduced a simple plan that was based on factorial design and, nevertheless, met the assumptions of Response Surface Methodology. The newly created plan was a modification of a three-level factorial plan, where one factor has to have kept in their 'zero' value (central value, medium value) in every experiment, plus at least one experiment in 'zero point' – with all the factors tested in their central settings (Jacyna, Kordalewska, & Markuszewski, 2019). Box-Behnken design method is used to examine the relationship between one or more response variables and a set of quantitative experimental variables or factors. In order to describe the nature of response surface in the experimental region and elucidate the optimal concentrations of the most significant independent variables Box-Behnken design was used (Gopinadh, et al. 2015).

Figure 3. FIA Configuration for the Determination of Copper

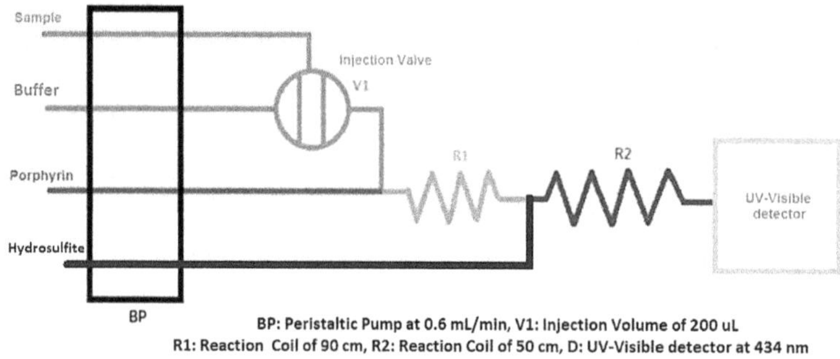

Determination Copper and Manganese in Wines by Design of Experiments

Copper Determination

Materials and Methods

Reagents. Copper standard solution ampule 75 mg/L as Cu, Sodium Hydrosulfite ($Na_2S_2O_4$), Ascorbic Acid ($C_6H_8O_6$), 5,10,15,20-Tetrakis (1-methyl-4-pyridinio) porphyrin tetra p-toluenesulfonate ($C_{72}H_{66}N_8O_{12}S_4$) (H_4tmpyp). All reagents were of analytical-reagent grade, unless stated otherwise. Deionized water was used in all experiments.

Apparatus. Spectrophotometer DR5000 with Sipper Module HACH, Peristaltic pump (Watson Marlow 400 sci), injection valve (VICI. Vaco Instruments Co. Inc). Analytical Balance (PB 303-5 Mettler Toledo), Ultrasonic cleaner (Branson 2510), quartz cell 160 uL. Computer and Statgraphics Centurion XVI.I software.

Experimental design. The useful analytical conditions were obtained by optimizing different physical and chemical variables of the FIA system using Plackett-Burman fractional factorial design 2^6 *3/16. The variables included were two reaction coil length (R1 and R2), sample loop size, flow rate, and the concentrations of 5,10,15,20-Tetrakis (1-methyl-4-pyridinio) porphyrin tetra p-toluenesulfonate ($C_{72}H_{66}N_8O_{12}S_4$) (H_4tmpyp), and Sodium Hydrosulfite ($Na_2S_2O_4$). The design was performed and optimization graphs were constructed using Statgraphics Centurion XVI.I software.

Procedure for flow injection analysis (FIA). The continuous flow analysis system used for the spectrophotometric determination of the copper is illustrated in Figure 3. The injection valve (VICI. Vaco Instruments Co. Inc) had two position microelectric actuators: sampling and injection. In the sampling position, the sample loop (100 cm length, 200 μL internal volume) was filled with the standard or sample solution, while the porphyrin reagent ($3.71E^{-5}$ mol/L) was pumped directly to the confluence through Tygon tubing (0.76 mm i.d.) at a flow rate of 0.6 mL/min. Simultaneously, the carrier solution of 0.015 mol/L ascorbic acid was also pumped to the confluence through Tygon tubing (0.76 mm i.d.) at a flow rate of 0.6 mL/min, hence providing the FIAgram baseline signal (the detector zero). In the injection position, the carrier solution was then diverted to the sample loop, transferring a 200 μL aliquot of standard or sample to the confluence, where it mixed with the porphyrin reagent. The mixture then passed through the reaction coil (R1, 90 cm), later solution of 0.034 mol/L sodium hydrosulfite was also pumped to the confluence through Tygon tubing (0.76 mm i.d.) at a flow rate of 0.6 mL/min. The

Figure 4. The Spectrum of Absorption of the Copper-Porphyrin Complex.

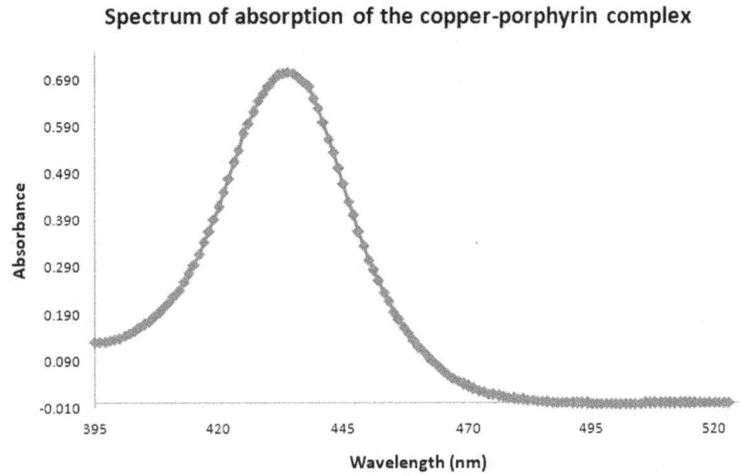

mixture then passed through the reaction coil (R2, 50 cm), kept at room temperature, and the yellow colored product formed was carried to the detector flow cell, where the transient absorbance signal was monitored at 434 nm is illustrated in Figure 4. After recording the FIAgram peak, the valve was reset to the initial (sampling) position to start another cycle. Peak height was used as the analytical signal, and its magnitude (expressed in units of absorbance) was proportional to the copper concentration.

In figure 3, you can see the configuration used for the flow analysis proposed by Carrillo et al. (Carrillo, Cañizares, Ames, & Sepúlveda, 2016), for the spectrophotometric determination of the copper.

Linearity, limits of detection and quantification. The linearity of the proposed FIA method for determination of copper was evaluated by constructing calibration curves over the concentration range from 40 to 280 µg/L, under the optimized operating conditions. The analytical curves considered a linear dependence between peak height (absorbance units) and analyte concentration. Linear regression analysis by the least squares method was used to determine the slope, intercept, and linear correlation coefficient. The limits of detection (LOD) and quantification (LOQ) were determined according to IUPAC recommendations and Mexico QFB association (García, et al. 2002): LOD = 3s/b and LOQ = 10s/b, where s is the standard deviation of the analytical response and b is the slope of the dynamic linear range.

Results and Discussion

Formation of complex. Experiments confirmed the formation of yellow complex, with absorbance maxima at 434 nm, from reaction between copper and porphyrin, in acid medium and at room temperature. The absorbance spectra of the complex are illustrated in Figure 4. It was observed that the maximum absorbance (at 434 nm) of the reaction final product was free from spectral interferences.

The probable reactions between copper and porphyrin (Tabata, M., & Morita, H., 1997) are shown schematically in equations 1 and 2. The complex formation can be divided into two stages. In the first, at ambient temperature, copper reacts with phorphyrin in the presence of ascorbic acid to form intermediary $[Cu(tmpyp)]^{+4}$. In the second stage, under acid conditions, the intermediary formed reacts with the hydrosulfite, producing a complex that can be measured by spectrophotometry.

*Table 1. Analysis of Variance of Fractional Design Plackett-Burman 2^6*3/16.*

Source	Sum of Squares	Df	Mean Square	F-Ratio	p-Value
A:Reaction coil R1	4.98E-07	1	4.98E-07	0.29	0.6026
B:Reaction coil R2	2.45E-04	1	2.45E-04	144.49	0.0000
C:Loop size	1.08E-03	1	1.08E-03	639.25	0.0000
D:Porphyrin	8.53E-05	1	8.53E-05	50.37	0.0001
E:Hydosulfite	2.09E-05	1	2.09E-05	12.31	0.0080
F:Flow rate	3.00E-06	1	3.00E-06	1.77	0.2200
Total error	1.36E-05	8	1.69E-06		
Total (corr.)	1.45E-03	14			

R-squared = 99.0 percent
R-squared (adjusted for d.f.) = 98.3 percent
Standard Error of Est. =1.3016E-03
Mean absolute error = 7.72902E-04
Durbin-Watson statistic = 2.8267 (p=0.9608)

Figure 5. Pareto Chart by Copper-Porphyrin.

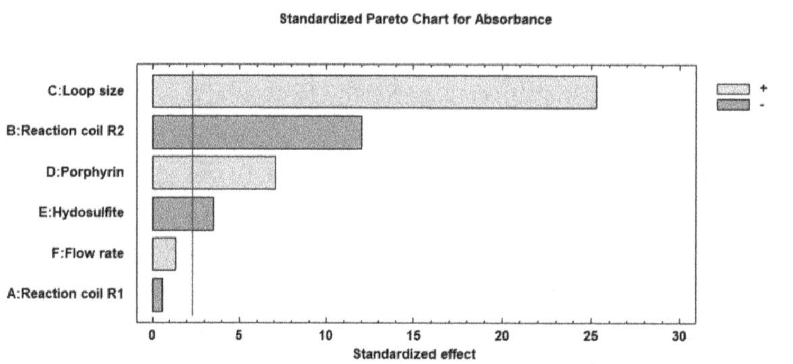

$$Cu^{+2} + \left[H_2tmpyp\right]^{+4} \rightleftarrows_{K_{-1}}^{K_1} \left[Cu\left(tmpyp\right)\right]^{+4} \tag{1}$$

$$\left[Cu\left(tmpyp\right)\right]^{+4} + \left[S_4O_4\right]^{-2} \rightarrow_{K_2} \left[Cu\left(tmpyp\right)\right]^{+5} + 2\left[HSO_3\right]^{-1} \tag{2}$$

Optimization of variables. Plackett-Burman fractional factorial design 2^6 *3/16 of resolution III, with three central points and 8 degrees of freedom was performed to establish the best conditions for the flow analysis, enabling statistical evaluation of factors that significantly affected the formation complex [Cu(tmpyp]$^{+5}$ reaction. Preliminary results obtained after monitoring the maximum absorbance at 434 nm revealed uniform peaks and a stable baseline. Fifteen experiments were performed randomly, in triplicate, using a constant 160 μg/L concentration of copper.

Figure 6. The Plot of Response Surface by Copper-Porphyrin Complex.

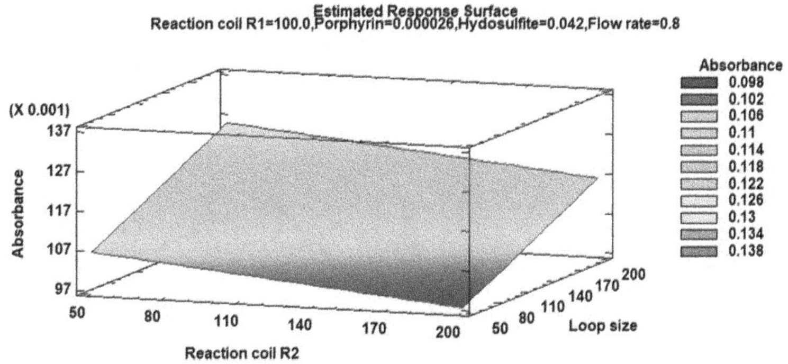

The ANOVA shown in Table 1 illustrates the effects of the individual parameters. This effects were standardized (each effect was divided by its standard error). In this case, four effects have p-values less than 0.05, indicating that they are significantly different from zero at the 95.0% confidence level.

The Pareto chart analysis (Figure 5) showed that the variables with the greatest influence were the loop size and concentration of porphyrin, but the reaction coil length R2 and concentration of hydrosulfite has a negative influence on the absorbance measurement. The values of these variables were therefore fixed at levels that were most appropriate for the methodology and that helped to reduce the consumption of reagents.

The three-dimensional response surface graph obtained by fitting the experimental data is shown in Figure 6, where the variables influence are the loop size and the reaction coil length R2. The green area reflects maximum absorbance and indicates the concentration levels of the factors to which the analysis was most sensitive.

The R-Squared statistic indicates that the model, as fitted, explains 99.0% of the variability in absorbance. The adjusted R-squared statistic, which is more suitable for comparing models with different numbers of independent variables, is 98.3%. The standard error of the estimate shows the standard deviation of the residuals to be 0.0013016. The mean absolute error (MAE) of 0.000772902 is the average value of the residuals. The Durbin-Watson (DW) statistic tests the residuals to determine if there is any significant correlation based on the order in which they occur in your data file. Since the p-value is greater than 5.0%, there is no indication of serial autocorrelation in the residuals at the 5.0% significance level

This panel shows the regression equation, which has been adjusting to the data. In equation 3, reaction coil R1 was omitted, there is not a statistically significant difference according to the ANOVA Table 1. The equation (3) of the fitted model is:

$$
\begin{aligned}
&Absorbance = 1.01E^{-1} - 6.02E^{-5} * R_2 + 1.27E^{-4} * \text{Loop size} \\
&+240.24 * \text{Porphyrin} - 9.41E^{-2} * \text{Hydrosulfite} + 6.67E^{-5} * \text{Flow rate}
\end{aligned}
\tag{3}
$$

In the following graph (Figure 7), the effects of the six variables can be observed, and by not showing curvature, it indicates that the values can continue to increase to improve the measurement of absorbance; the variable with the greatest influence is a loop size.

Figure 7. Main Effects Plot for Absorbance by Copper-Porphyrin Complex.

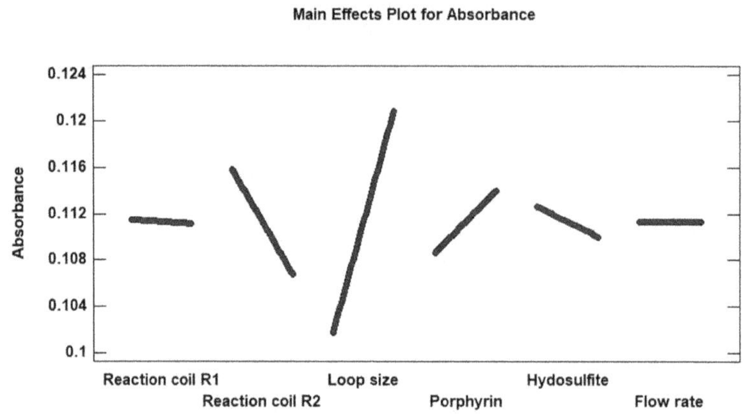

Table 2. Optimal Values of the Copper-Porphyrin.

Factor	Low	High	Optimum
A:Reaction coil R1 (cm)	50	150	90
B:Reaction coil R2 (cm)	50	200	50
C:Loop size (μL)	50	200	200
D:Porphyrin (mol/L)	1.49E-05	3.71E-05	3.71E-05
E:Hydosulfite (mol/L)	0.028	0.056	0.034
F:Flow rate (mL/min)	0.5	1.1	0.6

Figure 8. FIAgram of the Copper-Porphyrin.

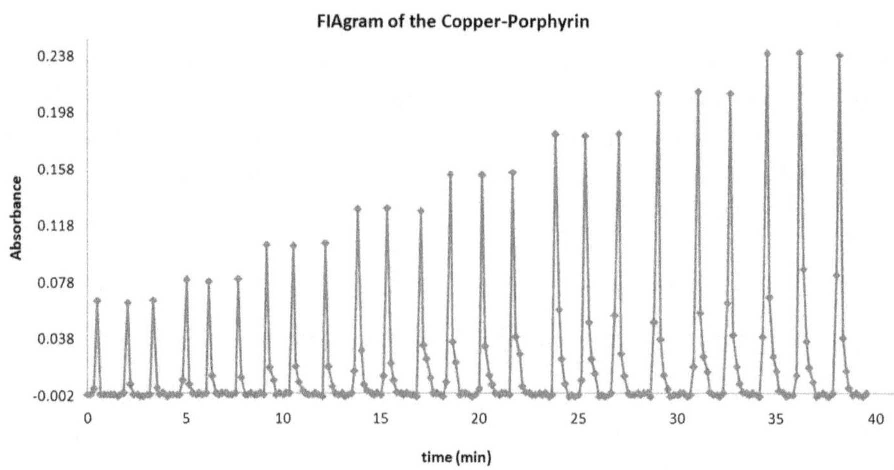

Table 3. Linearity of the Copper-Porphyrin.

Criteria	Value
Slope (b_1)	6.72 E^{-04}
Ordinate at the origin (b_0)	0.0506
Regression coefficient (r^2)	0.9993
Correlation coefficient (r)	0.9996
Confidence interval for the slope ($IC_{\beta 1}$)	6.62 E^{-04} – 6.82 E^{-04}
Confidence interval for the ordinate at the origin ($IC_{\beta 0}$)	0.0488 - 0.0524
Regression variation coefficient ($CV_{y/x}$)	1.0
Detection limit	7.7 µg/L
Quantification limit	23.4 µg/L

Figure 9. FIAgram of Accuracy as Repeatability of the Copper-Porphyrin.

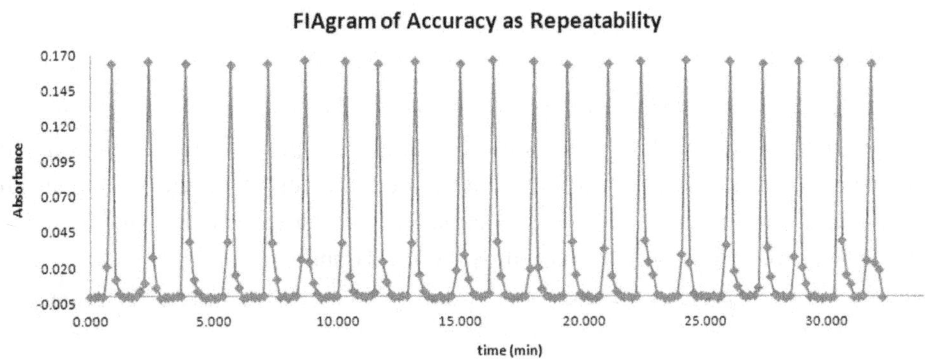

The optimized values of the operational variables for the formation of the Copper-Porphyrin complex have shown in Table 2.

Under the optimal experimental conditions, excellent linear relationships were obtained between the measured absorbance values and concentrations of Copper-Porphyrin complex with a range of 40 to 280 µg/L, and the FIAgram has been showing in Figure 8.

Analytical curves, LOD, and LOQ. Under the optimal experimental conditions, excellent linear relationships were obtained between the measured absorbance values and concentrations of copper in the range from 40 to 280 µg/L. The results indicated that the technique had sufficient sensitivity and linearity for the determination of copper in wines (Table 3).

Standard addition and recovery. Precision has been determining as reproducibility from the standard copper solution, a copper standard of 168 µg/L was injected seven times in triplicate, and the obtained FIAgram has been showing in Figure 9. The analytical variability was expressed in terms of relative standard deviation (%RSD). For precision, acceptable RSD values are considered to be less than or equal to 3%. The concentration percentages found for the fortified solutions were in the ranges 99.8 - 101.6%.

Table 4. Determination of copper in commercial wines

Sample	Without Standard (µg/L)	Added Standard (µg/L)	Found Value [a] (µg/L)	RSD (%)	Recovery (%)	t-Test[b]	F-Test[c]
Blanc of Zinfandel Mexico	64	100	162.3 ± 3.21	2.0	101.0	0.814	0.221
Merlot Central Valley ViñaEstefamyar Chilean Red Wine	40	100	141.3 ± 3.21	2.3	99.1	0.651	0.141
Dulzino White Moscato Sweet Wine	80	100	184.0 ± 2.00	1.1	97.8	2.771	2.560
Lambrusco Dulce Amore Rose Italy	62	100	167.0 ± 2.65	1.6	99.4	0.569	0.108

[a]Average ± standard deviation (SD), n= 3.
[b]Critical values of t at 95% confidence level (t = 4.303).
[c]Critical values of F at 95% confidence level (F=19).

The t values calculated to evaluate the accuracy of the proposed method were lower than the tabulated t values t calculated =0.828, df =40 ≤ t tabulated=2.021, df =40.

Analytical Application

The applicability of the proposed flow method was evaluated by analysis of several wines samples by a denomination of origin (Mexico, Chile and Italy) containing copper. The results obtained by the proposed method were statistically compared with those obtained using the addition standard method, as shown in Table 4. Considering the criteria of precision and accuracy, the calculated F and t values did not exceed the theoretical values (at a 95% confidence level), indicating that there was no significant difference between the added standard and found value. This confirmed the effectiveness of the new flow analysis technique.

Manganese Determination

Materials and Methods

Reagents. Manganese standard solution ampule 250 mg/L as Mn, Sodium citrate ($Na_3C_6H_5O_7$), Sodium periodate ($NaIO_4$). All reagents were of analytical-reagent grade, unless stated otherwise. Deionized water was used in all experiments.

Apparatus. Spectrophotometer DR5000 with Sipper Module HACH, Peristaltic pump (Watson Marlow 400 sci), injection valve (VICI. Vaco Instruments Co. Inc). Analytical Balance (PB 303-5 Mettler Toledo), Ultrasonic cleaner (Branson 2510), quartz cell 160 uL. Computer and Statgraphics Centurion XVI.I software.

Procedure for flow injection analysis (FIA). The continuous flow analysis system used for the spectrophotometric determination of the manganese is illustrated in Figure 10. The injection valve (VICI. Vaco Instruments Co. Inc) had two position microelectric actuators: sampling and injection. In the sampling position, the sample loop (500 µL internal volume) was filled with the standard or sample solution, then sodium citrate reagent (0.1 mol/L) was pumped directly to the confluence through Tygon

Figure 10. FIA Configuration for the Determination of Manganese.

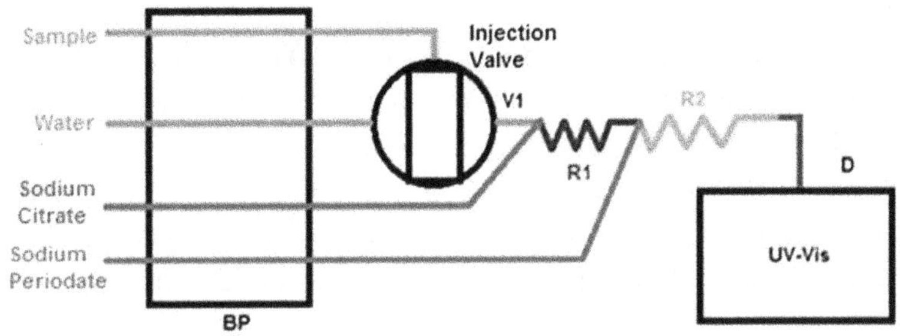

BP: Peristaltic pump, V1: Injection volume of 500 uL, R1: Reaction coil of 50 cm, R2: Reaction coil of 100 cm, D: UV-Vis detector at 524 nm

Figure 11. The Spectrum of Absorption of Permanganate

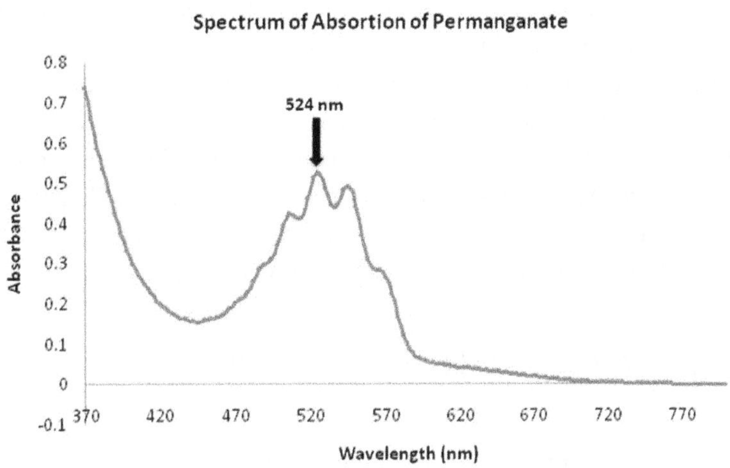

tubing (1.02 mm i.d.) at a flow rate of 0.8 mL/min. Simultaneously, the carrier solution water distilled was also pumped to the confluence through Tygon tubing (1.02 mm i.d.) at a flow rate of 1.1 mL/min, hence providing the FIAgram baseline signal (the detector zero). In the injection position, the carrier solution was then diverted to the sample loop, transferring a 500 μL aliquot of standard or sample to the confluence, where it mixed with the sodium citrate reagent. The mixture then passed through the reaction coil (R1, 50 cm), later solution of 0.1 mol/L sodium periodate was also pumped to the confluence through Tygon tubing (0.76 mm i.d.) at a flow rate of 0.6 mL/min. The mixture then passed through the reaction coil (R2, 100 cm), kept at room temperature, and the purple colored product formed was carried to the detector flow cell, where the transient absorbance signal was monitored at 524 nm is illustrated in Figure 11. After recording the FIAgram peak, the valve was reset to the initial (sampling) position to start another cycle. Peak height was used as the analytical signal, and its magnitude (expressed in units of absorbance) was proportional to the manganese concentration.

Linearity, limits of detection and quantification. The linearity of the proposed FIA method for determination of manganese was evaluated by constructing calibration curves over the concentration range from 0.5 to 50 mg/L, under the optimized operating conditions. The analytical curves considered a linear dependence between peak height (absorbance units) and analyte concentration. Linear regression analysis by the least squares method was used to determine the slope, intercept, and linear correlation coefficient. The limits of detection (LOD) and quantification (LOQ) were determined according to IUPAC recommendations and Mexico QFB association (García, et al. 2002): LOD = 3s/b and LOQ = 10s/b, where s is the standard deviation of the analytical response and b is the slope of the dynamic linear range.

Results and Discussion

Formation of permanganate. Experiments confirmed the formation of permanganate, with absorbance maxima at 524 nm, from reaction between Mn^{+2} and periodate, in presence of citrate and at room temperature. The absorbance spectra of the complex are illustrated in Figure 11. It was observed that the maximum absorbance (at 524 nm) of the reaction final product was free from spectral interferences.

A large number of early researchers studied the reaction between manganous ions and periodates with more or less contradictory results, but Willard and Greathouse were the first to use it as the basis of a spectrophotometric method for the quantitative determination of manganese. Its method depends on the oxidation in acid solution of manganese to permanganate salts by potassium periodate and affirm that it is free of all the faults of the other methods for manganese and produce results of a high degree of precision, recently it has been applied with success in water, animal and plant tissues (Mehlig, 1939).

The method most used in the spectrophotometric determination of manganese is based on the oxidation of the manganese ion (Mn^{2+}) to the permanganate $[MnO_4]^-$, as you can in the equation 4:

$$2Mn^{2+}_{(ac)} + 5\left[IO_4\right]^-_{(ac)} + 3H_2O \leftrightarrow 2\left[MnO_4\right]^-_{(ac)} + 5\left[IO_3\right]^- + 6H^+_{(ac)} \quad (4)$$

Optimization of variables. To optimize the parameters of the FIA system a first factorial design 2^4 of resolution V^+, with two central points and 7 degrees of freedom, and other factorial design Box-Behnken with three central points and 5 degrees of freedom.

Factorial Design 2^4, with two central points and 7 degrees of freedom was performed to establish the best conditions for the flow analysis, enabling statistical evaluation of factors that significantly affected the formation permangante reaction. Preliminary results obtained after monitoring the maximum absorbance at 524 nm revealed uniform peaks and a stable baseline. Eighteen experiments were performed randomly, in triplicate, using a constant 16 mg/L concentration of copper. The solution of citrates of total concentration 0.1 mol/L was used as working buffer, adjusted to pH = 6 with HCl 0.0925 mol/L. A phosphate solution with a total concentration of 0.2 mol/L was used as a working buffer, adjusted to pH = 8 and pH = 10 with 0.2 mol/L NaOH (FEUM, 2011).

The ANOVA Table 5 partitions the variability in Absorbance into separate pieces for each of the effects. It then tests the statistical significance of each effect by comparing the mean square against an estimate of the experimental error. In this case, five effects have p-values less than 0.05, indicating that they are significantly different from zero at the 95.0% confidence level.

Table 5. Analysis of Variance of Factorial Design 2^4

Source	Sum of Squares	Df	Mean Square	F-Ratio	p-Value
A: pH	2.12E-02	1	2.12E-02	376.49	0.0000
B: Loop size	2.67E-02	1	2.67E-02	475.41	0.0000
C: Reaction coil R2	2.07E-03	1	2.07E-03	36.82	0.0005
D: Periodate	1.37E-03	1	1.37E-03	24.35	0.0017
AB	2.60E-03	1	2.60E-03	46.26	0.0003
AC	1.60E-05	1	1.60E-05	0.28	0.6102
AD	2.50E-07	1	2.50E-07	0.00	0.9487
BC	2.56E-04	1	2.56E-04	4.55	0.0703
BD	2.50E-07	1	2.50E-07	0.00	0.9487
CD	2.72E-04	1	2.72E-04	4.84	0.0637
Total error	3.94E-04	7	5.62E-05		
Total (corr.)	5.49E-02	17			

R-squared = 99.3 percent
R-squared (adjusted for d.f.) = 98.3 percent
Standard Error of Est. = 0.00749868
Mean absolute error = 0.00375309
Durbin-Watson statistic = 2.0793 (p=0.2411)

Figure 12. Pareto Chart by the Oxidation of the Mn^{+2} of Factorial Design 2^4

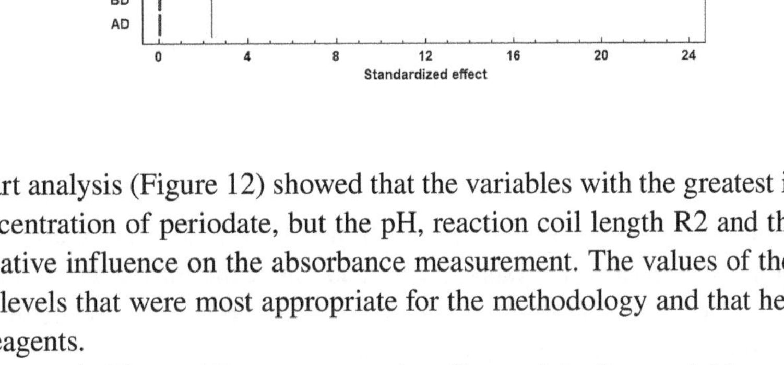

The Pareto chart analysis (Figure 12) showed that the variables with the greatest influence were the loop size and concentration of periodate, but the pH, reaction coil length R2 and the interaction loop size-pH has a negative influence on the absorbance measurement. The values of these variables were therefore fixed at levels that were most appropriate for the methodology and that helped to reduce the consumption of reagents.

In the following graph (Figure 13) you can see the effects of the four variables, and it has observed that non-curvature tells us where is the limit for each one of the variables with the work ranges that have proposed.

Figure 13. Main Effects Plot for Absorbance by Oxidation of the Mn^{+2} of Factorial Design 2^4.

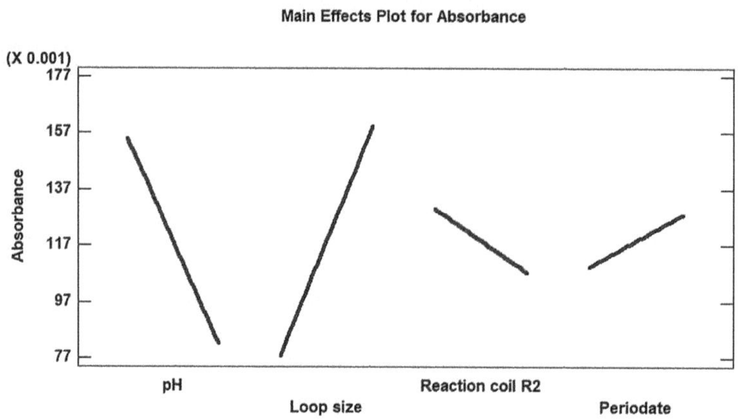

Table 6. Estimated effects for Absorbance of the Factorial Design Box-Behnken.

Source	Sum of Squares	Df	Mean Square	F-Ratio	p-Value
A:Periodate	5.62E-03	1	5.62E-03	123.38	0.0001
B:Loop size	1.25E-02	1	1.25E-02	274.13	0.0000
C:Flow rate	5.00E-07	1	5.00E-07	0.01	0.9206
AA	1.24E-03	1	1.24E-03	27.13	0.0034
AB	1.23E-05	1	1.23E-05	0.27	0.6261
AC	1.32E-04	1	1.32E-04	2.90	0.1491
BB	5.41E-03	1	5.41E-03	118.90	0.0001
BC	2.03E-05	1	2.03E-05	0.44	0.5344
CC	2.71E-05	1	2.71E-05	0.59	0.4754
Total error	2.28E-04	5	4.55E-05		
Total (corr.)	2.50E-02	14			

R-squared = 99.1 percent
R-squared (adjusted for d.f.) = 97.4 percent
Standard Error of Est. = 0.00674784
Mean absolute error = 0.00348889
Durbin-Watson statistic = 1.89943 (p=0.5661)

A 2^4 factorial design was performed to establish the pH=6 and reaction coil R2 = 100 cm, this are the best conditions for the flow analysis on the oxidation of the manganese ion (Mn^{2+}) to the permanganate [MnO_4]. The other variables, loop size and concentration of periodate they are not defined, since they present non-curvature in the main effects plot (Figure 13), so they were studied again with the Box-Benken factorial design to find their optimum value. Flow rate is included in this new data matrix due to the speed of reaction.

Figure 14. Pareto Chart by the Oxidation of the $Mn^{+2}$of the Factorial Design Box-Behnken.

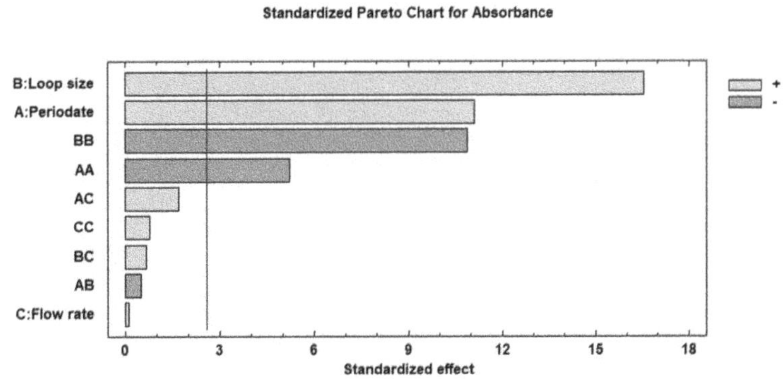

Figure 15. Main Effects Plot for Absorbance by the Oxidation of the $Mn^{+2}$of the factorial design Box-Behnken.

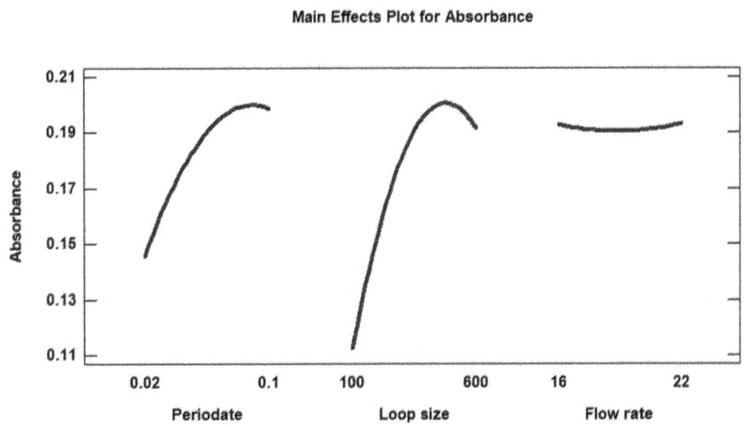

Factorial Design Box-Behnken, with three central points and 5 degrees of freedom was performed to establish the best conditions for the manganese flow analysis at pH=6 and reaction coil R2 = 100 cm, enabling statistical evaluation of factors: loop size, concentration of periodate and flow rate that significantly affected the formation permangante reaction.

The ANOVA Table 6 partitions the variability in Absorbance into separate pieces for each of the effects. It then tests the statistical significance of each effect by comparing the mean square against an estimate of the experimental error. In this case, 4 effects have p-values less than 0.05, indicating that they are significantly different from zero at the 95.0% confidence level.

The Pareto diagram shown in Figurem14 illustrates the effects of the individual parameters and their interactions. The length of each bar is proportional to the absolute value of the associated regression coefficient or estimated effect. The effects of all parameters and interactions were standardized (each effect was divided by its standard error). The Pareto chart analysis (Figure 14) showed that the variables with

Figure 16. Plot of Response Surface by the Oxidation of the Mn^{+2} of the Factorial Design Box-Behnken.

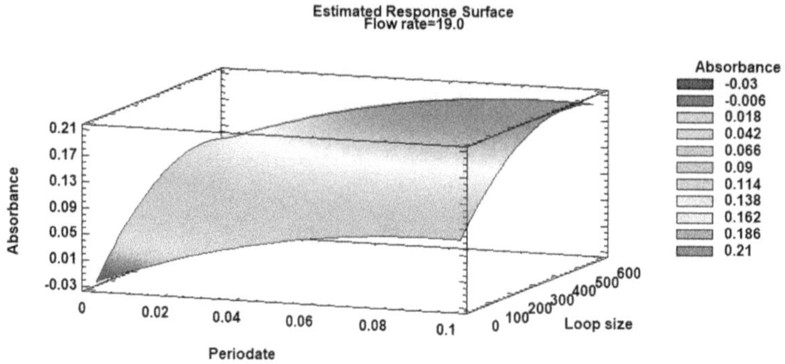

Table 7. Optimal Values by the Oxidation of the Mn^{+2} of the factorial design Box-Behnken.

Factor	Low	High	Optimum
A:Periodate (mol/L)	0.02	0.1	0.1
B:Loop size (μL)	100.0	600.0	500
C:Flow rate (rpm)*	16.0	22.0	22.0

*16 rpm equal 0.6 mL/min and 22 rpm equal 0.8 mL/min

the greatest influence were the loop size and concentrations of periodate, but the interaction loop size-loop size and interaction flow rate –flow rate have a negative influence on the absorbance measurement.

The R-Squared statistic indicates that the model, as fitted, explains 99.0876% of the variability in Absorbance. The adjusted R-squared statistic, which is more suitable for comparing models with different numbers of independent variables, is 97.4452%. The standard error of the estimate shows the standard deviation of the residuals to be 0.00674784. The mean absolute error (MAE) of 0.00348889 is the average value of the residuals. The Durbin-Watson (DW) statistic tests the residuals to determine if there is any significant correlation based on the order in which they occur in your data file. Since the P-value is greater than 5.0%, there is no indication of serial autocorrelation in the residuals at the 5.0% significance level.

This panel shows the regression equation, which has been adjusting to the data. In equation 5, flow rate, interactions: periodate-loop size (AB), periodate-flow rate (AC), loop size- flow rate (BC) and flow rate-flow rate (CC) were omitted, there is not a statistically significant difference according to the ANOVA Table 1. The equation (5) of the fitted model is:

$$Absorbance = -2.20E^{-2} + 2.05 * Periodate + 5.892E^{-4} * Loop\,size$$
$$-11.562 * Periodate^2 - 6.127E^{-7} * Loop\,size^2 \quad (5)$$

In the Pareto Diagram (figure 14), it has observed that loop and periodate has a positive influence on the oxidation of de Mn^{+2}.

Figure 17. FIAgram of the Oxidation of the Mn^{+2}.

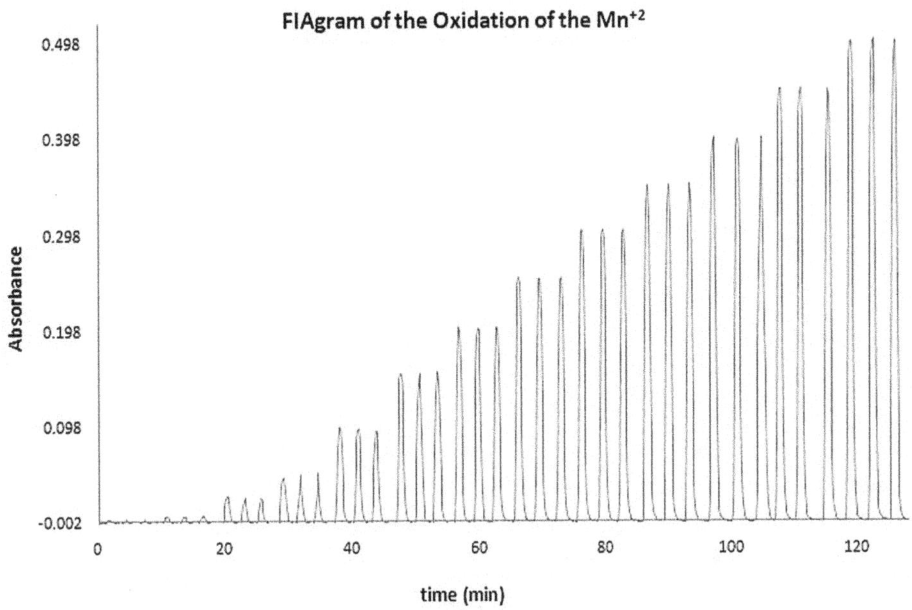

Table 8. Linearity of the Oxidation of the Mn^{2+}

Criteria	Value
Slope (b_1)	0.0101
Ordinate at the origin (b_0)	-3.33 E^{-04}
Regression coefficient (r^2)	0.9998
Correlation coefficient (r)	0.9999
Confidence interval for the slope ($IC_{\beta 1}$)	9.97 E^{-03} – 1.02 E^{-02}
Confidence interval for the ordinate at the origin ($IC_{\beta 0}$)	-0.0029 – 0.0023
Regression variation coefficient ($CV_{y/x}$)	1.2
Detection limit	0.9 mg/L
Quantification limit	2.7 mg/L

In the following graph (Figure 15), the effects of the three variables can be observed, and by showing curvature, it indicates that values can no longer increase to improve absorbance measurement; the variable with the greatest influence is a loop size.

The three-dimensional response surface graph obtained by fitting the experimental data is shown in Figure 16. The darkest area (red) reflects maximum absorbance and indicates the concentration levels of the factors to which the analysis was most sensitive for the oxidation of the Mn^{+2}.

Figure 18. FIAgram of Accuracy as Repeatability of Oxidation of the Mn^{+2}.

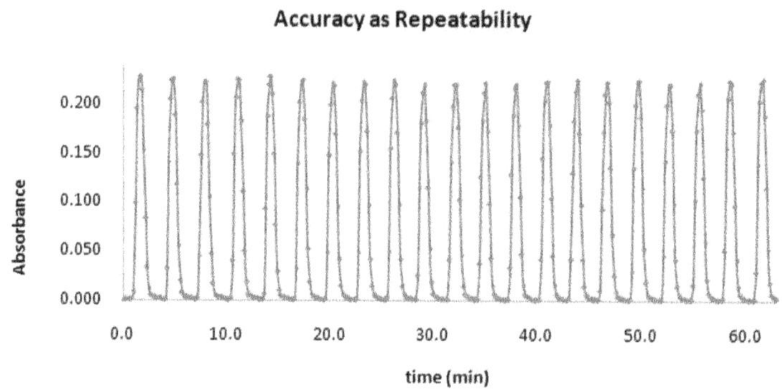

Table 9. Recovery of manganese added to wines.

Sample	Without Standard (mg/L)	Added Standard (mg/L)	Found Value [a] (mg/L)	RSD (%)	Recovery (%)	t-Test[b]	F-Test[c]
Red Wine Cabernet Sauvignon	2.04	15	17.25 ± 0.31	1.8	101.2	1.046	0.365
White Wine Blanc of Zinfandel	1.24	15	16.15 ± 0.31	1.9	99.4	0.469	0.073
RedWinePetiteSirah	3.13	15	17.81 ± 0.32	1.8	98.2	1.588	0.841
White Wine Blanc of Blancs	0.64	15	15.45 ± 0.39	2.5	98.8	0.805	0.216

[a]Average ± standard deviation (SD), n= 3.
[b]Critical values of t at 95% confidence level (t = 4.303).
[c]Critical values of F at 95% confidence level (F=19).

This Table 7 shows the combination of factor levels that maximize absorbance over the indicated region. Use the Analysis Options dialog box to indicate the region in which the optimization will take place. You can set the value of one or more factors in a constant by setting the high and low limits for that value as shown in the following table:

With the optimal values suggested by the design Box-Behnken, a calibration curve was made for Oxidation of the Mn^{+2} with a range of 0.5 to 50 mg/L, and the FIAgram is shown in figure 17.

Analytical curves, LOD, and LOQ. Under the optimal experimental conditions, excellent linear relationships were obtained between the measured absorbance values and concentrations of manganese in the range from 0.5 to 50 mg/L. The results indicated that the technique had sufficient sensitivity and linearity for the determination of manganese in wines (Table 8).

Standard addition and recovery. Precision has been determining as reproducibility from the standard manganese solution, a manganese standard of 22 mg/L was injected seven times in triplicate, and the obtained FIAgram has been showing in Figure 18. The analytical variability was expressed in terms of relative standard deviation (%RSD). For precision, acceptable RSD values are considered to be less than or equal to 3%. The concentration percentages found for the fortified solutions were in the ranges

99.2 - 100.6%. The t values calculated to evaluate the accuracy of the proposed method were lower than the tabulated t values t calculated =0.562, df =40 ≤ t tabulated=2.021, df =40.

Analytical Application

The applicability of the proposed flow method was evaluated by analysis of several wines samples from the Valle de Guadalupe region of Baja California, Mexico; containing manganese. The results obtained by the proposed method were statistically compared with those obtained using the addition standard method, as shown in Table 9. Considering the criteria of precision and accuracy, the calculated F and t values did not exceed the theoretical values (at a 95% confidence level), indicating that there was no significant difference between the added standard and found value. This confirmed the effectiveness of the new flow analysis technique.

SOLUTIONS AND RECOMMENDATIONS

The authors recommend that using DoE is very important in the area of analytical chemistry, because time is cut by optimizing several variables, this leads to a saving of reagents that positively impacts the economic part and is environmentally friendly.

CONCLUSION

The advantages of the methods developed and validated for the determination of copper and manganese in wines by continuous flow injection analysis using factorial design, include operational simplicity and substantially reduce the cost and time of analysis with a sampling frequency of 38 samples per hour for copper and 21 for manganese. The proposed flow injection procedure contributes to green analytical chemistry because reduce generation waste. The chapter illustrates that using factorial designs can reduce the number of experiments necessary to achieve successful quantification of copper and manganese in wines.

ACKNOWLEDGMENT

SEP-PRODEP Call for Support for the Strengthening of Academic Bodies (UABC-CA-287).

REFERENCES

Armstrong, N. A. (2006). Pharmaceutical Experimental Design and Interpretation (Second Edition). Boca Raton, FL: Taylor & Francis Group.

Cacho, J., Castells, J., & Barroso, C. (2002). *Analytica Chimica Acta*, *458*, 197–202. doi:10.1016/S0003-2670(01)01499-4

Cañizares, M. P. (2002). Análisis por inyección en flujo: Herramienta clave para la automatización analítica. *Revista de la Sociedad Química de México*, *46*(2), 167–174.

Carrillo, E. G., Cañizares, M. P., Ames, A. I., & Sepúlveda, R. G. (2016). Aplicación de Diseño Factorial Plackett-Burman en la Optimización de Parámetros para Determinar Cobre. *Aportaciones Científicas en Química Analítica*, *2016*, 464–469.

Carrillo, E. G., Haro, M. P., Díaz, G. C., & Cañizares, M. P. (2014). Plackett-Burman factorial design for the optimization of a spectrophotometric flow injection method for phenol determination in tap and bottled water using 4-aminoantipyrine. (S. Q. México, Ed.) J. Mex. Chem. Soc., 58(2), 99-105.

Cepo, D. V., Pelajic, M., Vrcek, I. V., & Krivohlavek, A. (2018). Differences in the levels of pesticides, metals, sulphites and ochratoxin A between organically and conventionally produced wines. *Food Chemistry*, (246), 349–403. PMID:29291864

Clark, A., Wilkes, E., & Scollary, G. (2015). Chemistry of copper in white wine: A review. *Australian Journal of Grape and Wine Research*, *21*(3), 339–350. doi:10.1111/ajgw.12159

Ferreira, S. L., Souza, A. S., Brandao, G. C., Ferreira, H. S., dos Santos, W. N., Pimentel, M. F., & Vale, M. G. R. (2008). Direct determination of iron and manganese in wine using the reference element technique and fast sequential multi-element flame atomic absorption spectrometry. *Science Direct*, (74), 699-702.

FEUM. C. P. (2011). Farmacopea de los Estados Unidos Mexicanos (Décima edición ed., Vol. I). México: Secretaría de Salud.

Fukuda, I. M., Fidelis, C. F., Moreira, C., Morais, A., & Rebello, F. (2018). Design of Experiments (DoE) applied to pharmaceutical and analytical Quality by Design (QbD). *Brazilian Journal of Pharmaceutical Sciences*, (54): 1–16.

García, M. A., Soberón, E., Cortés, M., Rodríguez, R., Herrera, J. L., & Alcántara, A. (2002). *Guía de Validación de Métodos Analíticos. C.D.Mx.* México: Colegio Nacional de Químicos Farmacéuticos Biólogos.

Gomez-Miguel, V., & Sotes, V. (2014). *El Manganeso y la Viticultura: Una revisión*. Madrid, España: Ministerio de Agricultura, Alimentación y Medio Ambiente.

Gopinadh, R., Ayyanna, C., Ramakrishna, Ch., Narayana, K. V., Kandisa, R. V., & Jagadhi, R. (2015). Optimization of chemical parameters for the production of citric acid using Box-Behnken design. *Journal of Bioprocessing & Biotechniques*, (5): 1–6.

Gutiérrez, A. J., Rubio, C., Moreno, I. M., González, A. G., Gonzalez-Weller, D., Bencharki, N., ... Revert, C. (2017). Estimation of dietary intake and target hazard quotients for metals by consumption of wines from the Canary Islands. *Food and Chemical Toxicology*, *108*, 10–18. doi:10.1016/j.fct.2017.07.033 PMID:28733233

Gutiérrez, H., & Salazar, R. (2012). Análisis y Diseños de Experimentos (3rd. ed.). D. F., México: McGraw-Hill.

Hansen, E., & Miró, M. (2007). How flow-injection analysis (FIA) over the past 25 years has changed our way of performing chemical analyses. *Trends in Analytical Chemistry*, *26*(1), 18–26. doi:10.1016/j.trac.2006.07.010

Hsia, C., Plack, R., & Nagel, C. (1975). Influence of must processing on iron and copper contents of experimental wines. *American Journal of Enology and Viticulture*, *26*(2), 57–61.

Jacyna, J., Kordalewska, M., & Markuszewski, M. (2019). Design of experiments in metabolomics-related studies: An overview. *Journal of Pharmaceutical and Biomedical Analysis*, *164*, 598–606. doi:10.1016/j.jpba.2018.11.027 PMID:30469109

Jung, M. Y., Kang, J. H., Choi, Y. S., Lee, D. Y., Lee, J. Y., & Park, J. S. (2019). Analytical features of microwave plasma-atomic emission spectrometry (MP-AES) for the quantitation of manganese (Mn) in wild grape (Vitis coignetiae) red wines: Comparison with inductively coupled plasma-optica emission spectrometry (ICP-OES) 1. *Food Chemistry*, *274*, 20–25. doi:10.1016/j.foodchem.2018.08.114 PMID:30372927

Kontoudakis, N., Guo, A., Scollary, G. R., & Clark, A. C. (2017). The impact of aging wine in high and low oxygen conditions on the fractionation of Cu and Fe in Chardonnay wine. *Food Chemistry*, *229*, 319–328. doi:10.1016/j.foodchem.2017.02.065 PMID:28372180

Kontoudakis, N., Schmidtke, L. M., Bekker, M., Smith, M., Smith, P. A., Scollary, G. R., ... Clark, A. C. (2019). Analytical strategies for the measurement of different forms of Cu and Fe in wine: Comparison between approaches in relation to wine composition. *Food Chemistry*, *274*, 89–99. doi:10.1016/j.foodchem.2018.08.084 PMID:30373024

Kontoudakis, N., Smith, M., Smith, P. A., Wilkes, E. N., & Clark, A. C. (2018). The determination of total Cu in white wine by BCA colorimetric analysis. *National Wine and Grape Industry Centre & Charles Sturt University*, (1), 1-4.

Kuehl, R. O. (2001). Diseño de Experimentos Principios Estadísticos de Diseño y Análisis de Investigación. (Second Edition), Arizona: Thomson Learning.

Liu, Y. H., Zhan, H. W., Heng, L. J., & Ma, W.-X. (2013). Spectrophotometric determination of manganese in wine by oxidizing acid fuchsin. *Asian Journal of Chemistry*, *25*(4), 2270–2272. doi:10.14233/ajchem.2013.13824

Mehlig, J. (1939). Colorimetric determination of manganese with periodate. *Industrial & Engineering Chemistry*, *11*(5), 274–278.

Moise, G. (2015). Method of analysis for the determination of heavy metals from wine. *Scientific Papers. Series Management, Economic, Engineering in Agriculture, and Rural Development, 15*(3), 201–204.

Montgomery, D. C. (2005). *Design and Analysis of Experiments.* Phoenix, AZ: John Wiley & Sons.

Naveena, B. J., Altaf, M., Bhadriah, K., & Reddy, G. (2005). Selection of medium components by Plackett–Burman design for production of L(+) lactic acid by Lactobacillus amylophilus GV6 in SSF using wheat bran. *Science Direct,* (96), 485-490.

NOM-199-SCFI-2017. (2017). Bebidas alcohólicas-Denominación, especificaciones fisicoquímicas, información comercial y métodos de prueba. *Norma Oficial Mexicana, Diario Oficial de la Federación,* 1-37.

Orellana, S., Johansen, A. M., & Gazis, C. (2019). Geographic classification of U.S. Washington State wines using elemental and water isotope composition. *Food Chemistry, X*(100007), 1–8. PMID:31432007

Ou, S., Li, C., & Wei-Xing, M. (2013). Spectrophotometric determination of manganese in cooking wine samples by oxidation of chromotrope 2B. *Food Science China, 34*(12), 156–158.

Pohl, P. (2007). What do metals tell us about wine? *Trends in Analytical Chemistry, 26*(9), 941–949. doi:10.1016/j.trac.2007.07.005

Pyrzynska, K. (2004). Analytical methods for the determination of trace metals in wine. *Critical Reviews in Analytical Chemistry, 34*(2), 69–83. doi:10.1080/10408340490475858

Ribereau-Gayon, P., Glories, Y., Maujean, A., & Dubourdieu, D. (2006). *Handbook of enology, the chemistry of wine stabilization and treatments (* (2nd ed., Vol. 2). Bordeaux, France: John Wiley & Sons. doi:10.1002/0470010398

Rousseva, M., Kontoudakis, N., Schmidtke, L., Scollary, G., & Clark, A. (2016). Impact of wine production on the fractionation of copper and iron in Chardonnay wine: Implications for oxygen consumption. *Food Chemistry, 203,* 440–447. doi:10.1016/j.foodchem.2016.02.081 PMID:26948636

Santos, W., Brandao, G., Portugal, L., David, J., & Ferreira, S. (2009). A photo-oxidation procedure using UV radiation/H_2O_2 for decomposition of wine samples - Determination of iron and manganese content by flame atomic absorption spectrometry. *Spectrochimica Acta. Part B, Atomic Spectroscopy, 64*(6), 601–604. doi:10.1016/j.sab.2009.04.012

Schut, S., Zauner, S., Hampel, G., Konig, H., & Claus, H. (2011). Biosorption of copper by wine-relevant lactobacilli. *International Journal of Food Microbiology, 145*(1), 126–131. doi:10.1016/j.ijfoodmicro.2010.11.039 PMID:21195499

Spivey, N., Thompson, P., Shelton, C., & Kavan, A. (2015). The analysis of copper, iron, and manganese in wine with FAST flame atomic absorption. *Perkin Elmer,* 1-4.

Tabata, M., & Morita, H. (1997). Spectrophotometric determination of a nanomolar amount of ascorbic acid using its catalytic effect on copper (II) porphyrin formation. *Talanta, 44*(2), 151–157. doi:10.1016/S0039-9140(96)02010-3 PMID:18966728

Vystavna, Y., Rushenko, L., Diadin, D., Klymenko, O., & Klymenko, M. (2014). Trace metals in wine and vineyard environment in southern Ukraine. *Food Chemistry*, *146*, 339–344. doi:10.1016/j.foodchem.2013.09.091 PMID:24176352

KEY TERMS AND DEFINITIONS

Absorbance: Is a measure of the quantity of light absorbed by a sample. is calculated based on either the amount of light reflected or scattered by a sample or by the amount transmitted through a sample. If all light passes through a sample, none was absorbed, so the absorbance would be zero and the transmission would be 100%. On the other hand, if no light passes through a sample, the absorbance is infinite and the percent transmission is zero.

Box-Behnken: Is an independent quadratic design in that it does not contain an embedded factorial or fractional factorial design. In this design, the treatment combinations are at the midpoints of edges of the process space and at the center. These designs are rotatable (or near rotatable) and require 3 levels of each factor. The designs have limited capability for orthogonal blocking compared to the central composite designs.

Copper: Cu is chemical element, a reddish, extremely ductile metal of Group 11 (Ib) of the periodic table that is an unusually good conductor of electricity and heat. Copper is found in the free metallic state in nature.

Factorial Design: Is one involving two or more factors in a single experiment. The number of levels of each factor and the number of factors classify such designs; is often used by scientists wishing to understand the effect of two or more independent variables upon a single dependent variable.

Flow Injection Analysis: FIA is based on the injection of a liquid sample into a moving, non- segmented continuous carrier stream of a suitable liquid. The injected sample forms a zone, which is then transported toward a detector that continuously records the changes in absorbance, electrode potential, or other physical parameter resulting from the passage of the sample material through the flow cell.

Manganese: Mn is chemical element, one of the silvery white, hard, brittle metals of Group 7 (VIIb) of the periodic table. It was recognized as an element in 1774 by the Swedish chemist Carl Wilhelm Scheele while working with the mineral pyrolusite and was isolated the same year by his associate, Johan Gottlieb Gahn. Although it is rarely used in pure form, manganese is essential to steelmaking.

Plackett-Burman: Used to identify the most important factors early in the experimentation phase when complete knowledge about the system is usually unavailable. Developed in 1946 by statisticians Robin L. Plackett and J.P. Burman, it is an efficient screening method to identify the active factors using as few experimental runs as possible.

VIS Spectrophotometry: Refers to absorption spectroscopy in the visible spectral region; is used to determine the absorption or transmission of VIS light (400 to 820 nm) by a sample. It can also be used to measure concentrations of absorbing materials based on developed calibration curves of the material.

Wine: An alcoholic drink that is usually made from grapes, but can also be made from other fruits or flowers. It is made by fermenting the fruit with water and sugar.

Chapter 9
An Overview of the Design of Experiment Workflow:
Applications in Food Production Systems

Marco A. Miranda-Ackerman
https://orcid.org/0000-0002-7041-7130
Universidad Autónoma de Baja California, Mexico

Alejandra García-Lechuga
Universidad Autónoma de Baja California, Mexico

ABSTRACT

Design of Experiments as a subfield of statistics has its origins trying to understand systems and making decisions. In the context of food supply chains where agricultural and farm systems, packaging and processing plants, cold chains and food logistics, product design, warehousing, retailing, among many other steps and actors, come together to produce safe, healthy, nutritious, reliable, affordable, and sustainable food products. At each stage, many problems have to be solved and decisions have to be made from agricultural practices, scheduling, land use, costing, and pricing, at the farm and orchard level. A general outline is presented on the general workflow; illustrated through examples and a current review of some of the applications of design of experiments in the context of food production help provide some directives and guidelines. Global trends and challenges that influence both the food industry and the practice of design of experiments are discussed.

INTRODUCTION

Design of Experiments (DOE) as a subfield of statistics has its origins trying to understand systems and making decisions based off of the results that are obtained after experiment execution. It has a clear process route: Problem definition, designing the experiment, running the experiment and collecting data, processing data, analyzing, and either continue experimenting or reaching conclusions. Yet, at each stage decisions have to made, some objectively others heuristically. In the context of food supply

DOI: 10.4018/978-1-7998-1518-1.ch009

chains where agricultural and farm systems, packaging and processing plants, cold chains and food logistics, product design, warehousing, retailing, among many other steps and actors, come together to produce safe, healthy, nutritious, reliable, affordable and sustainable food products. At each stage many problems have to be solved and decisions have to be made from agricultural practices, scheduling, land use, costing and pricing, at the farm and orchard level; through the processing of fresh fruit such as cleaning, sorting, inspecting, and through the transformation of processed food, such as concentrating, pasteurizing, stabilizing, mixing, cooking; each step having problems and opportunities to maintain or exceed the market and customer demand.

Yet, the Design of Experiments is not always used to this end. This chapter will provide a practical overview of the Design of Experiments work flow. Provide some basic heuristics for each stage of the work flow; and illustrated in the context of the many stages of the food supply chain and its production systems.

At the problem definition stage issues related to identifying the problems and opportunities for improvement will be described. Heuristics on how to identify a problem that can be framed through the use of Design of Experiments will be provided. It is illustrated through case studies and examples in agriculture, food processing and food logistics. Common pitfalls will be described. Helping the Experimenter avoid costly mistakes, such as providing the basic insight to variable definition where for example, an experimenter takes 3 factors that influence an output, yet two are not of interest in the decision model, when the tools of blocking could provide a means to manage this arrangement.

For the designing of experiment stage a general decision algorithm will be presented in order to help define the type of problem in relation to the possible experimental design one can use. Knowing what techniques one can use, understanding the economics in terms of time, money and resources, and the reliability and validity of a design strategy will all be detailed and illustrated in this section. These issues are important in any DoE context, but hold special importance if food production given that products are meant for human and animal consumption, and because in many developing and underdeveloped countries small profit margins and unfavorable environmental and social conditions may create a restrictive environment for the application of experiments.

Basic considerations during the running of experiments and data collection will be described and illustrated. Issues on randomization, sample size, replication, measurement, experimentation scheduling, experimental execution logistics issues, data collection pitfalls, data table design, and other practical issues will be developed. Some examples of best practices in food production context will be explained. An example that could be developed is the measurement requirements, where for example chicken feed could be measured in a volume cup or weighed, and what reasoning could be used to choose on over the other. Some information on data sharing and data maintenance will be briefly reviewed. This will be done with an emphasis of knowledge management for reproduction and replication in future studies. Some useful information on open source and proprietary software will be discussed where cloud based services can be helpful.

Strategies for analyzing results will be developed using examples and illustrations provided in the previous sections. A generalized description of the hypothesis testing results during the development of the experiment will be clarified. An overview of conventional results presentation and how to generally interpret or use information resulting from a DoE calculation will be described. And a decision route on what actions to take depending on the results will be expanded on.

Finally, some useful elements to consider in the conclusion stage of the experimental design will be seen. With a special emphasis on practical logic that can be summoned to make the best choice given the accessible information and the limited resources available. Some tips on how to present findings taking into account the readers and decision makers that will use the information provided. Best practices on how to share information and data used for the conclusions will be considered.

Indeed, general considerations and some rules of thumb and heuristics will be cautiously provided. And risk considerations explained in relation to excess cost, use of resources, health and safety, among other issued that interface with ethics and professional responsibility in the use of Design of Experiments in the context of food systems. Providing some examples of how to avoid some basic mistakes one can make during de design, execution and analysis. Some of the topics that will be touched are: bias during the problem definition, execution and analysis; overfitting of models and abuse of degrees of freedom; outliers and other omissions; repeatability and reproducibility of the experiment and the importance of transparency; the risk in the art of analysis among other subjects. The idea is to have basic knowledge of the ethical and professional responsibility issues in the context of Design of Experiments used in food production, where as said before has its unique quality and risk because of the interactions between abiotic and biotic systems and the industrial ecology that interacts with nature.

The objective of the chapter is not then to be exhaustive, rather to provide an overview of the logic and mindset one requires to have to identify opportunities to apply DOE. By providing specific examples in the context of food production systems, readers in the field may gain a sense of how to observe through the lens of a Design of Experiments practitioner. And by doing so, gaining the confidence to implement the tools provided by the field in their everyday work environment.

BACKGROUND

Design of Experiments as a field has developed over time. It has its roots in the agricultural setting with the direction given by Sir Ronald A. Fisher during the 1920s and 30s Fisher alongside his colleges the the Rothamsted agricultural experimental station in England worked on developing some of the basic elements of design of experiments. According to (Montgomery, 2017) these where randomization, replication and blocking. The seminal work was used mainly on topics from the agricultural and life sciences. The principals of factorial design and analysis of variance were then adapted to an industrial setting. Yet some main differences from the agricultural and life sciences perspectives in contrast to the controlled and high turnover of industrial processes required new strategies. Montgomery points out that Box and Wilson responded to this need with the advent of Response Surface Methodology (RSM). Focusing on the fact that there is an immediate resonance and possible adaptation of the experimentation given the nature of the systems being studied. With the evolution of industry and global markets new forces placed tension in industrial systems to find better and more efficient ways of working. Driving the era of industrial design of experiments used to improve quality and performance. Leading to smaller more efficient and practical applications of the techniques in order to find opportunities for improvement. Currently there is a new era underway, the data science era. It capitalizes on the fact that so much information can now be obtained given the exponential growth and inexpensiveness of computing and sensor use. This has given rise to new strategies and opportunities for the use of the principals in experimental design and other related techniques.

Figure 1. General workflow in design of experiments.

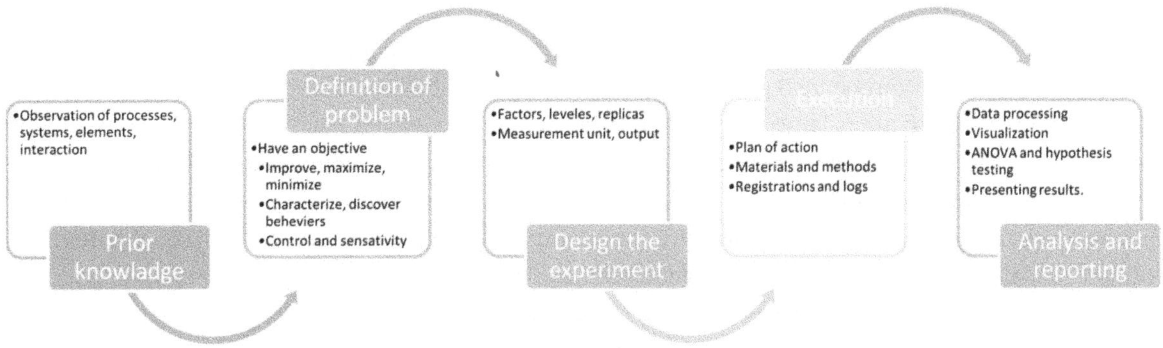

It is interesting to take notice that design of experiments as a field and a body of knowledge started in the agricultural sector, passed through the industrial processing industry and know is used system wide. This is the scope of DoE in the context of food production systems. Nevertheless the chapter will limit the reach of the application of DoE in production systems, while not excluding the connecting links in the full food value chain (i.e. agriculture, distribution, etc).

General Outline and Workflow in DOE

Experiments in the context of statistical design of experiments will be defined in this chapter as an investigation of a set of configurations and outcomes a system may have under a given set of rules and instructions. Further on in this chapter a detailed classification and examples of different uses and design of experiments will be presented. The general idea one should have to this point is that systems have inputs, processes and outputs. Experiments evaluate and compare the behavior of the system by changing control variables that have influence on the outcome of the system or output variables (aka results). DOEs in this sense are systematic comparative experiments. Systematic is due to the fact that they are not executed by trial and error alone, randomly exploring the control variable settings against their results. Rather principals such as randomization, statistical parameters, and other tools are used to order and aggregate information.

The initial stages require the Experimenter to have some previews knowledge of the system under study (see Figure 1). This is key given that the first step is to define and frame the hypothesis of the experiment. This definition is based on knowing a priori what controllable and uncontrollable variables may be of interest given that there is suspicion that they may be important to the behavior of the output variable. Uncontrollable variables have just been introduced. These are important given that they may influence the output yet cannot be easily controlled. Uncontrollable variables may exist given that there are limited resources to a given system, so take for example environmental fluctuation of temperature and rain on a farm. The farm could be changed to a controlled environment such as a greenhouse, however given that materials, energy, knowledge and other resources have to be invested, natural variability has to be considered part of the system. Related to this, are uncontrollable variables due to the unknown uncertain nature of elements in a system. Say that a large investment was made to protect the farm through a closed system green house, yet decease or volcanic eruption may strike. These are to side of a same coin.

So let us take an avocado processing plant as a case study in order to illustrate these basic part of Design of Experiments. The plant may have many unit operations, let us focus on avocado oil extraction process and more specifically a cold-press mechanical extraction (Costagli & Betti, 2015). In the kneading /malaxing process a recommended temperature of 30ªc is prescribed for a period of 30-45 minutes. The result is an emulsion that has large drops of oil that have coalesced and can be easier to separate from the fruit pulp. These guidelines have been developed through experience and academic studies related to oil extraction from olives (Angerosa, Mostallino, Basti, & Vito, 2001). So say, one would like to see if these temperature and time parameters are the best for avocado. What are the inputs? What are the outputs? What can be controlled? What cant? What do we want to measure? Why?

The initial stage is to know the basics of a system, and try to find affirmations or assumptions in order to define hypothesis that can be experimented on to accept or reject.

Let us define a simple experiment for the case we have just developed. We could state that oil drop size is the same when processes for 30 minutes and 45 minutes. These could be proven by a simple statistical hypothesis test of means, for example. But say we want to find the best setting, what could be done then? We could evaluate 30, 35, 40 and 45 minutes of the unit operation. And say it is difficult for the plant to maintain a fixed environmental temperature, and the factory manager would like to know how much effect does 2ªC up or down has on the oil drop size. This next level of complexity is where Design of Experiment, compared to simple hypothesis testing, has its advantages. We have defined the control variables temperature and processing time. We have defined the levels that the variables can take (e.g. 18,30 and 32ªC) .We have defined the output variable of interest: oil drop size in the emulsion after kneading and malaxing is finished.

Some other basic Design of Experiment setting have to be fixed. How many samples should be taken? How many times should a given treatment be performed? Are there any known sources of variability that can be blocked to limits their influence on the outcome (e.g. Species of avocado, avocado batch, night- and day-time processing). We have introduced the concepts of treatments and blocking. The former is the configuration at which an experiment will run; while the latter relates to the possibility to avoid the influence of a factor on the experimental result given that you cannot control it or it is not of interest. An experiment will have to have many runs, at least one for each treatment, if more than one for each treatment is performed, the second or the ones that follow are replicas. Replicas are important given that they provide more degrees of freedom, that translate in capturing with finer detail the natural variability of a given treatment, and allow us to calculate the experimental error and thus many other elements in what will be known as the Analysis of Variance or ANOVA Table.

The results from an experiment are, by convention, presented in an ANOVA Table. Many commercial and open source data processing and statistical tools that perform hypothesis testing and design of experiments present the results in such tables.

The most comment why to use in a very basic way is to use the p-value on the far left of the table. P value is the probability of getting the observed value of the test statistic, or a greater value, given that the null hypothesis is actually true. Until now, it is applied in practice as a "passing grade" usually set at 0.5 related to the alfa value. If the P-value is smaller than the alfa value, say 0.5 than the null hypothesis is rejected.

Once we have run are experiments and have are results, we are required to verify that some assumptions have been maintained given that general linear models are used in ANOVA testing. These assumptions are: independence of errors, normality, homogeneity of variance, linearity.

It all assumptions are met one can search for the factor, level or combination that makes a difference on the outcome. To do this other testing may be performed. This is to say that Design of Experiments is an iterative and sequential process.

APLICATION OF DESIGN OF EXPERIMENTS IN FOOD PRODUCTION SYSTEMS

Purpose

It may seem at first glass a simple strategy to explore "what if" combinations for a given system. Yet, DOE and hypothesis testing in general are powerful tools in engineering practice. In food production they are used continuously. The most important uses are:

- **Characterization:** It used to find the values and parameters from a given outcome or system behavior that are physically possible. This is to say, the feasible and desirable values for a given system without knowing what is the best setting or attributes.(Dominguez, Aurell, Gullett, Eninger, & Yamamoto, 2018; Therdthai & Zhou, 2009)
- **Optimization:** The objective is to find the minimal or maximal value for the response variable or output. This objective in the Design of Experiments is very common given that it is a simple, thus far scientifically valid mean to find optimal values. (S. Wang et al., 2007; Y. Wang, Yang, Bao, Xun, & Zhang, 2011)
- **Design:** The objective as is the case for the other two objectives is to make a choice. Yet in this classification of Design of Experiments we can evaluate design alternatives to the system. (Jonkman, Bloemhof, van der Vorst, & van der Padt, 2017; Miranda-Ackerman, Azzaro-Pantel, Montastruc, Domenech, & Lasserre, 2013, p.)

Scope and Scale

The application of Design of Experiments in food production can be at many different scales. At the smallest scale we have product system. It starts at the molecular level, take for example enzymes to clarify fruit juice, beer production, meat tenderization, cheese flavoring, food preservation, among many applications (Raveendran et al., 2018); and scales upward towards processing systems, these can be physicochemical changes such as those due to frying, microwaving, micro and nanoencapulating; biological such as bacteriological stabilization through Pulse Electric Field Preprocessing, High-Pressure Processing, pasteurization (Devahastin, 2010), they can also be mechanical such as mixing, separating, homogenizing, cutting, mincing, shredding, squeezing, crumbling, hashing, grinding, chopping, kneading, many of these operations can be designed, optimized and characterized to become the food products we know from our daily lives.

Further up in the scale of food production are process system, this is to say equipment or sets of equipment acting together to transform raw materials into product goods. Different to the product system scale, the inputs, process and output is focused on the hardware used not the materials itself, although they cannot be separated or neglected. So, food can be minced, but with this scale we do not focus on, for example, the texture or physical structure of the product, rather the speed it is minced, the mean time between blade changes, the energy consumption during processing, this is to say, inputs to the process system not the product output alone. (Bendicho, Barbosa-Cánovas, & Martín, 2002; Charles-Rodríguez, Nevárez-Moorillón, Zhang, & Ortega-Rivas, 2007; Costagli & Betti, 2015)

Interactions between the products' design and the manufacturing and processing design are intertwined and make up the production system. Depending on the product design or strategy the manufacturing and even the distribution and sales will be greatly influenced. Take for example packaging material for a given fresh fruit product. Most packaging material used as barrier are based on conventional plastics, which have specific physicochemical properties, and that have consolidated tools, equipment and suppliers; nevertheless there is a social, political and economic imperative to find substitutes form these types of material uses. One strategy is developing biofilms. In order to find the physical properties to substitute the conventional material characterization of alternatives has to be done. For example in a study by (Cazón, Vázquez, & Velazquez, 2018) they studied through a Box-Behnken design, a second-order, incomplete, factorial deign, in order to evaluate toughness, burst strength, distance to burst, water absorption, light barrier properties and transparency and thermal properties of the film. By changing the percentages of cellulose, glycerol and polyvinyl alcohol they could change different properties. If a film like this would substitute a conventional plastic the raw materials, processing steps, shelf life and many other quality and safety issues would change in the production system as a whole.

Variable

In order design an experiment no matter the purpose or scope, variables have to be defined. These can be of different types. Quantitative variables are numerical variables that have a relationship to the values, they are measured with instruments and standards. These are the most comment types of variables, given that they objectively describe attributes, characteristics and behavior in numerical sense, which can then be treated through different statistical techniques. Qualitative variables can be subjectively collected and defined through non-numerical values. It is useful to note that many qualitative variables can be transformed into qualitative variables with the understanding that scales, measurements and norms will not hold universally and will have special and temporal resolution inconsistencies. In sensory analysis the use of sensation measurements such as intensity, complexity, thermal perception, and body have both objective measurements that can be related to the attribute of interest but also have human subjective measurements and perception. Some of the most important application is in wine tasting. Many studies such as those from (Alvino, van der Lubbe, & Constantinides, 2018; Coste, Sousa, & Malfeito-Ferreira, 2018; Frost, Blackman, Ebeler, & Heymann, 2018) use qualitative and quantitative variables in order to understand and characterize wine and its sensory attributes.

Basic Design Strategies

The simplest design is called the One-way or Simple ANOVA where only one factor is evaluated at different levels. It may be important to clarify what constitutes a factor. A system that has imputs a process and an output is the framework design of experiments operates in. The inputs are factor, this is to say, elements that influence the process and thus the results or output. These factors are in a more abstract sense variable that can take different values in different scales. They can be numeric or linguistic, ordered or nominal. The experiment assesses certain factor influence the outcome of a process, yet are not sure, nor do they know a priori to what extend do they influence the process or output. While the levels they can take are the value or setting of the variable (factor). An interesting use of the one-way ANOVA, to give an example, comes from a study by (Spada, Conte, & Del Nobile, 2018). They used the design in order to study shelf life in order to model its relationship to food waste. The factor in the one-way ANOVA was the Return-Good-Ratio which is a measurement of the unsold goods that return from the market to the supplier, while the response variable (output) of interest was the shelf life of the product.

If more than one factor is evaluated the design becomes a multifactorial one. Factorial designs have many variables, each of which may have different levels that can be set or are of interest to the experimenter. Some recent examples of factorial design used in food production are (Jribi et al., 2019; Pineau et al., 2019).This type of experiments is one of the most often used both in industry and academia. Yet, in many cases, especially where the execution of experiments is very costly in terms of resources, time and effort, other principals and strategies can be used in order to perform less runs. Two popular strategies are to reduce the number of replicas, while the second is to reduce the number of levels for each factor. The latter has given way to what is known as the 2^k factorial design. The basic principal is to evaluate only two extreme levels for each factor. By doing this the number of runs is reduced dramatically, however it can only be used as an exploratory or heuristic tool. It is useful as a base to continue experimenting with a posteriori knowledge. An illustrative and didactical example of the implementation of this design strategy is the study executed by (Elhalil et al., 2016). The authors evaluated how the degradation efficiency (in a catalytic degradation process) was affected by the concentration of malachite green, concentration of ferrous ions, concentration of hydrogen peroxide and temperatures. Each of these factors was evaluated at two extreme levels, and a third central point in order to observe if non-leaner behavior was present. By using this strategy, insight was found on the behavior of the response variable and on the importance each factor has.

Another way to reduce the number of experiments to be performed is to take advantage of the use of aliases. These are treatments that can shed information on more than one factor level or combination of factors' effect. An example of the use of this technique can be studied in (Wong et al., 2015) used to study the significant contribution of ethanol concentration, temperature, liquid to solid ratio and time on the antioxidant capacity, total phenolic an flavonoid contents in palm oil.

One of the most used experimental design strategies in the food industry - to optimize and improve a system's attributes- is the so called Response Surface Methodology (RSM). It is a group of statistical techniques to develop empirical models. The objective is to find the optimal response given a combination of settings of the input variables. The main idea is to develop tree dimensional response surfaces, and sequentially steps towards regions where the optimal value lays. Depending on the model needs and characteristics of the system being studies and the behavior of the preliminary results different design strategies can be used. A sample of such techniques could include: multiple response design, central composite design and Box-Behnken design among others. An interesting example of the palication of RSM comes from (Elhalil et al., 2016) that was previously mention. They used 2^k factorial design for one part of the experiment, yet the main experiment consisted in an RSM design. The focus was to optimize the catalytic degradation of malachite green dye in aqueous solution. Other applications of RSM are for example (Noorbakhsh-Soltani, Zerafat, & Sabbaghi, 2018) a comparative study of gelatin and starch based nanocomposite films for food packaging; (Zhou, Fu, & Li, 2015) an optimization study to recover active monacolins and removing toxic citrinin from red yeast rice; (S. Wang et al., 2007) another study to find optimal approach to pectin extraction from apple pomace; (Ahmadi, Vahabzadeh, Bonakdarpour, Mofarrah, & Mehranian, 2005) a study with the goal of understanding the effect of the Fenton's peroxidation on the removal of organic pollutants in olive oil production process wastewater.

Solution and Recommendations

It is clear that this is not an extensive review of the possible design of experiments, however it is a basic initiation of the workflow and compendium of some used of the technique in food production setting. There are many points that are not covered in this overview, however it could be useful to enumerate some of the most important: number of samples for each replica, center points and curvature, computational statistics, statistical validity, reproducibility, repeatability, open data principals and transparency, ethics of statistical information and inference (impact of use of statistics), decision making and inference, incomplete or missing information, unbalanced designs, types and characteristics of errors, human factor, open source and commercial software, among many other important issues one should take into consideration.

It may also be important to point out the some of the most experiences researches, teachers and practitioner have found that there are some practical principals that could be essential to follow.

Understand the system and use previews knowledge: Understanding the system is important to both experimenters how are familiar with the system as well as for those who are new. This is because an experimental design forces us to as questions we would not otherwise make (e.g. when will the next preventive maintenance be executed for the equipment understudy?). Assumptions of the system must be validated. External factors should be identified that are not always present in daily operations (e.g. the state of the technician assisting in the experiment).

- **Use common sense and simplicity as the basis of the design**: Using common sense may seem an obvious approach to experimental design, nevertheless it is quite common to find oneself complicating some elements of experimentations for the sake of rigor. Nonetheless, it is very useful, to avoid mistakes to use common sense when in doubt or faces with a dilemma. It is also related to looking for simplicity. Complexity makes understanding what is going on and what should be interpreted a very hard task. In order to have a straightforward conclusion, a good strategy is to prefer simplicity over complexity despite the innate instinct during experimentation towards completeness and thoroughness.
- **Contrast and identify the statistical and practical significance of findings**: It is common to read in textbooks and scientific journals that a stated alpha level was used to test a hypothesis. This is important, nevertheless one should not discard hypothesis or reach conclusions and inferences based exclusively on statistical significance (Nature Editorial, 2019; Wasserstein, Schirm, & Lazar, 2019). Practical significance of the finding in the course of an experiment's results analysis is valid if it is useful and trustworthy in context of the objective of the study and the understanding of the results by the parties involved.
- **Experimental design usually is an iterative and sometimes sequential other times not**: Exploratory statistical experimentation can be a very useful precursor to a full statistical experimental design. Taking all of the principles previously stated, it is consistent to understand a system by applying Design of Experiments to filter ideas on the important of factor, the sensitivity of the output to specific parameters, the feasibility and cost (time, resource, etc.) of executing runs in a full experiment, among other knowledge acquisition. Changes mid experiments are very difficult to justify. It is preferable to start a new experiment than it is to change one, nevertheless, one should be alert to what is happening and what can be seen without calculations or data processing necessarily. In characterization and optimization design of experiments, equipment, tools, human

resources, materials are put in stressful extremes that are not known. It is thus important to know when to change setting and design parameters (number of replicas, factor levels, principal factors and interaction exclusions, etc.) in order to have a safe and successful study.

FUTURE RESEARCH DIRECTION AND TRENDS

Resent advancement in the production of information technology infrastructure and the ubiquitous acceptance of portable digital devises has change all landscapes, including food industry. The capacity to collect large sets of data from measurement and analytical equipment, manufacturing equipment, production plant facilities, suppliers, customers, distributers, and other food supply chain links and stakeholders has made access to information much easier. Alongside with the development of cloud computing and the cheapening of data processing have made using the vast sums of information possible.

Recent trends and the realization of the usefulness of data science tools have laid the foundation for innovation in food processing. **Data science** tools draw from computational statistics and design of experiments as a field can benefit from computer science. As pointed out in (Granato et al., 2018) there are many application of **chemometrics** (a discipline focused on using large datasets in chemical analysis) in food related disciplines and applications. Issues on authentication of geographic origin of food, agricultural systems, chemical composition of foods and chemical extracts from food based products are just some of the applications of this field in the realm of food production.

The same technological advancements that have allowed us to process so much information have also help with the development of **artificial intelligence** tools. One of the most useful aspects of AI in statistical experimentation is that AI can help overcome some important assumption requirements needed to apply validly many of the statistical models, through techniques such as Artificial Neural Networks. AI analytics can also help the experimenters and analyst thinking of applying design of experiments to look at developing models that can help with virtual experimentation. At the same time it is interesting to find that given the wealth of knowledge of the physicochemical world it is possible to use **simulation modeling** from molecular scales through to industrial processing scales. Given the complexity of the virtual environments, simulation can also benefit from design of experiments as an approach to perform "what if" analysis in a more structured and systematic way.

Stepping aside from technological and methodological trends one can also perceive a dramatic shift in social awareness of global challenges. The, now well known, **Sustainable Development Goals** laid out by the United Nations (Griggs et al., 2013) are a guideline on the importance and forms to measure the advancement to overcome global challenges. Among these are many that interact with food production, in the so called, **water-energy-food nexu**s (Azapagic, 2015; Namany, Al-Ansari, & Govindan, 2019). Water use in food production is very important, as well as natural resources and energy use. Design of experiments in the food production context is an important tool in the new set of objectives beyond economic benefits (Miranda-Ackerman, Azzaro-Pantel, & Aguilar-Lasserre, 2017).

One of the most important of the global challenges related to food production is what is known as **food security**, this is to say, the need to achieve enough production to supply the worlds' population. In order to overcome this challenge many small and large incremental improvement to the different steps in the food supply chain. **Six Sigma** is a quality and continues improvement tool developed in industry that can and is used in food production systems to improve productivity. It is based on the use of design of experiments, such as the **Taguchi methods** (Durmaz & Ozel, 2019; Go, Conag, & Bertumen, 2019;

Mahdi Jafari, Masoudi, & Bahrami, 2019). One of the most pressing issues within the sustainable development goals related to food security is the problem of **food loss and waste**. The power of design of experiments and its possible impact in efficiency improvement and new product and process systems design can provide a much-needed boost to improve food loss and waste.

CONCLUSION

In this chapter a brief overview to the workflow of design of experiments was introduced. With the aim at providing a general outline of how design of experiments can be used in the context of food production systems and their subcomponents. The general steps are described, as well as some classifications and applications. A description of some of the more important principals needed to apply design of experiments are presented. The importance of scale and scope of design of experiments studies was introduces, alongside some of the more trends and more pressing issues and global challenges faced in relation to food production in general. Through an illustrative set of examples in literature one can see the importance de design of experiments as a method and tool is to discover, characterize, improve, optimize and design many parts of the food supply system, including stages in research and development, through industrial application. Yet it is important to keep the use of the tool up to date with new capacities in infrastructure and technology, in order to continue improving systems, and finding new efficiencies needed to overcome the global challenges laid out in the previews section.

REFERENCES

Ahmadi, M., Vahabzadeh, F., Bonakdarpour, B., Mofarrah, E., & Mehranian, M. (2005). Application of the central composite design and response surface methodology to the advanced treatment of olive oil processing wastewater using Fenton's peroxidation. *Journal of Hazardous Materials*, *123*(1), 187–195. doi:10.1016/j.jhazmat.2005.03.042 PMID:15993298

Alvino, L., van der Lubbe, R. H., & Constantinides, E. (2018). Investigating individual preferences and brain activity in a wine tasting experience: a neuromarketing approach. *2018 Global Marketing Conference at Tokyo*, 559–564. 10.15444/GMC2018.05.03.04

Angerosa, F., Mostallino, R., Basti, C., & Vito, R. (2001). Influence of malaxation temperature and time on the quality of virgin olive oils. *Food Chemistry*, *72*(1), 19–28. doi:10.1016/S0308-8146(00)00194-1

Azapagic, A. (2015). Special issue: Sustainability issues in the food–energy–water nexus. *Sustainable Production and Consumption*, *2*, 1–2. doi:10.1016/j.spc.2015.08.002

Bendicho, S., Barbosa-Cánovas, G. V., & Martín, O. (2002). Milk processing by high intensity pulsed electric fields. *Trends in Food Science & Technology*, *13*(6–7), 195–204. doi:10.1016/S0924-2244(02)00132-2

Cazón, P., Vázquez, M., & Velazquez, G. (2018). Cellulose-glycerol-polyvinyl alcohol composite films for food packaging: Evaluation of water adsorption, mechanical properties, light-barrier properties and transparency. *Carbohydrate Polymers*, *195*, 432–443. doi:10.1016/j.carbpol.2018.04.120 PMID:29804996

Charles-Rodríguez, A. V., Nevárez-Moorillón, G. V., Zhang, Q. H., & Ortega-Rivas, E. (2007). Comparison of thermal processing and pulsed electric fields treatment in pasteurization of apple juice. *Food and Bioproducts Processing*, *85*(2), 93–97. doi:10.1205/fbp06045

Costagli, G., & Betti, M. (2015). Avocado oil extraction processes: Method for cold-pressed high-quality edible oil production versus traditional production. *Journal of Agricultural Engineering*, *46*(3), 115–122. doi:10.4081/jae.2015.467

Coste, A., Sousa, P., & Malfeito-Ferreira, M. (2018). Wine tasting based on emotional responses: An expedite approach to distinguish between warm and cool climate dry red wine styles. *Food Research International*, *106*, 11–21. doi:10.1016/j.foodres.2017.12.039 PMID:29579904

Devahastin, S. (2010). Physicochemical aspects of food engineering and processing. doi:10.1201/9781420082425

Dominguez, T., Aurell, J., Gullett, B., Eninger, R., & Yamamoto, D. (2018). Characterizing emissions from open burning of military food waste and ration packaging compositions. *Journal of Material Cycles and Waste Management*, *20*(2), 902–913. doi:10.100710163-017-0652-y

Durmaz, U., & Ozel, M. B. (2019). An experimental study on extraction of sugar from carob using with Taguchi method. *Sakarya University Journal of Science*, *23*(5), 916–923. doi:10.16984aufenbilder.541940

Editorial, N. (2019). It's time to talk about ditching statistical significance. *Nature*, *567*(7748), 283–283. doi:10.1038/d41586-019-00874-8 PMID:30894740

Elhalil, A., Tounsadi, H., Elmoubarki, R., Mahjoubi, F. Z., Farnane, M., Sadiq, M., ... Barka, N. (2016). Factorial experimental design for the optimization of catalytic degradation of malachite green dye in aqueous solution by Fenton process. *Water Resources and Industry*, *15*, 41–48. doi:10.1016/j.wri.2016.07.002

Frost, S. C., Blackman, J. W., Ebeler, S. E., & Heymann, H. (2018). Analysis of temporal dominance of sensation data using correspondence analysis on Merlot wine with differing maceration and cap management regimes. *Food Quality and Preference*, *64*, 245–252. doi:10.1016/j.foodqual.2016.11.011

Go, A. W., Conag, A. T., & Bertumen, M. M. N. (2019). Taguchi method to improve the production of sugar-rich hydrolysate from non-delipidated spent coffee grounds, and subsequent recovery of lipids and bioactive compounds. *Biofuels*, *10*(2), 193–205. doi:10.1080/17597269.2017.1309855

Granato, D., Putnik, P., Kovačević, D. B., Santos, J. S., Calado, V., Rocha, R. S., ... Pomerantsev, A. (2018). Trends in chemometrics: Food authentication, microbiology, and effects of processing. *Comprehensive Reviews in Food Science and Food Safety*, *17*(3), 663–677. doi:10.1111/1541-4337.12341

Griggs, D., Stafford-Smith, M., Gaffney, O., Rockström, J., Öhman, M. C., Shyamsundar, P., ... Noble, I. (2013). Policy: Sustainable development goals for people and planet. *Nature*, *495*(7441), 305–307. doi:10.1038/495305a PMID:23518546

Jonkman, J., Bloemhof, J. M., van der Vorst, J. G. A. J., & van der Padt, A. (2017). Selecting food process designs from a supply chain perspective. *Journal of Food Engineering*, *195*, 52–60. doi:10.1016/j.jfoodeng.2016.09.015

Jribi, S., Molnàr, H., Antal, O. T., Adànyi, N., Kheriji, O., Naàr, Z., & Debbabi, H. (2019). Zinc fortification as a tool for improving sprout hygienic and nutritional quality: A factorial design approach. *Journal of the Science of Food and Agriculture*, *0*(0). doi:10.1002/jsfa.9765 PMID:31032927

Mahdi Jafari, S., Masoudi, S., & Bahrami, A. (2019). A Taguchi approach production of spray-dried whey powder enriched with nanoencapsulated vitamin D3. *Drying Technology*, 1–13.

Miranda-Ackerman, M. A., Azzaro-Pantel, C., & Aguilar-Lasserre, A. A. (2017). A green supply chain network design framework for the processed food industry: Application to the orange juice agrofood cluster. *Computers & Industrial Engineering*, *109*, 369–389. doi:10.1016/j.cie.2017.04.031

Miranda-Ackerman, M. A., Azzaro-Pantel, C., Montastruc, L., Domenech, S., & Lasserre, A. A. A. (2013). Optimal design of a green supply chain: application to the agro-food industry supplier selection problem. Coll. Récents Progrès En Génie Des Procédés, N° 104 Les Sciences Du Génie Des Procédés Pour Une Industrie Durable XIVe Congrès SFGP 2013 8 Au 10 Octobre 2013, Lyon, 104. Société Française de Génie des Procédés.

Montgomery, D. C. (2017). *Design and analysis of experiments*. John Wiley & Sons.

Namany, S., Al-Ansari, T., & Govindan, R. (2019). Sustainable energy, water and food nexus systems: A focused review of decision-making tools for efficient resource management and governance. *Journal of Cleaner Production*, *225*, 610–626. doi:10.1016/j.jclepro.2019.03.304

Noorbakhsh-Soltani, S. M., Zerafat, M. M., & Sabbaghi, S. (2018). A comparative study of gelatin and starch-based nano-composite films modified by nano-cellulose and chitosan for food packaging applications. *Carbohydrate Polymers*, *189*, 48–55. doi:10.1016/j.carbpol.2018.02.012 PMID:29580425

Pineau, N., Moser, M., Rawyler, F., Lepage, M., Antille, N., & Rytz, A. (2019). Design of experiment with sensory data: A pragmatic data analysis approach. *Journal of Sensory Studies*, *34*(2). doi:10.1111/joss.12489

Raveendran, S., Parameswaran, B., Ummalyma, S. B., Abraham, A., Mathew, A. K., Madhavan, A., ... Pandey, A. (2018). Applications of microbial enzymes in food industry. *Food Technology and Biotechnology*, *56*(1), 16–30. doi:10.17113/ftb.56.01.18.5491 PMID:29795993

Spada, A., Conte, A., & Del Nobile, M. A. (2018). The influence of shelf life on food waste: A model-based approach by empirical market evidence. *Journal of Cleaner Production*, *172*, 3410–3414. doi:10.1016/j.jclepro.2017.11.071

Therdthai, N., & Zhou, W. (2009). Characterization of microwave vacuum drying and hot air drying of mint leaves (Mentha cordifolia Opiz ex Fresen). *Journal of Food Engineering*, *91*(3), 482–489. doi:10.1016/j.jfoodeng.2008.09.031

Wang, S., Chen, F., Wu, J., Wang, Z., Liao, X., & Hu, X. (2007). Optimization of pectin extraction assisted by microwave from apple pomace using response surface methodology. *Journal of Food Engineering*, *78*(2), 693–700. doi:10.1016/j.jfoodeng.2005.11.008

Wang, Y., Yang, Q., Bao, G., Xun, Y., & Zhang, L. (2011). Optimization design and experiment of fruit and vegetable picking manipulator. Nongye Jixie Xuebao=. *Nongye Jixie Xuebao*, *42*(7), 191–195.

Wasserstein, R. L., Schirm, A. L., & Lazar, N. A. (2019). Moving to a world beyond "p < 0.05." *The American Statistician, 73*(sup1), 1–19. doi:10.1080/00031305.2019.1583913

Wong, W. H., Lee, W. X., Ramanan, R. N., Tee, L. H., Kong, K. W., Galanakis, C. M., ... Prasad, K. N. (2015). Two level half factorial design for the extraction of phenolics, flavonoids, and antioxidants recovery from palm kernel by-product. *Industrial Crops and Products*, *63*, 238–248. doi:10.1016/j.indcrop.2014.09.049

Zhou, G., Fu, L., & Li, X. (2015). Optimisation of ultrasound-assisted extraction conditions for maximal recovery of active monacolins and removal of toxic citrinin from red yeast rice by a full factorial design coupled with response surface methodology. *Food Chemistry*, *170*, 186–192. doi:10.1016/j.foodchem.2014.08.080 PMID:25306334

ADDITIONAL READING

Granato, D., & Ares, G. (2014). *Mathematical and Statistical Methods in Food Science and Technology*. John Wiley & Sons. doi:10.1002/9781118434635

Granato, D., de Araújo Calado, V. M., & Jarvis, B. (2014). Observations on the use of statistical methods in Food Science and Technology. *Food Research International*, *55*, 137–149. doi:10.1016/j.foodres.2013.10.024

Jean-Vasile, A. (2016). Food Science, Production, and Engineering in Contemporary Economies. Retrieved from https://www.igi-global.com/book/food-science-production-engineering-contemporary/145605

Nunes, C. A., Alvarenga, V. O., de Souza Sant'Ana, A., Santos, J. S., & Granato, D. (2015). The use of statistical software in food science and technology: Advantages, limitations and misuses. *Food Research International*, *75*, 270–280. doi:10.1016/j.foodres.2015.06.011 PMID:28454957

Papajorgji, P., & Pinet, F. (2018). Innovations and Trends in Environmental and Agricultural Informatics. Retrieved from https://www.igi-global.com/book/innovations-trends-environmental-agricultural-informatics/195041

Raphaely, T., & Marinova, D. (2001). Impact of Meat Consumption on Health and Environmental Sustainability. Retrieved from https://www.igi-global.com/book/impact-meat-consumption-health-environmental/134810

Yu, P., Low, M. Y., & Zhou, W. (2018). Design of experiments and regression modelling in food flavour and sensory analysis: A review. *Trends in Food Science & Technology*, *71*, 202–215. doi:10.1016/j.tifs.2017.11.013

Chapter 10
Temperature Effects on the Physicochemical Properties and Composition of Fatty Acids in the Oil of Chia Seeds

María Gabriela Vargas Martínez
Universidad Nacional Autónoma de México, Mexico

Selene Pascual-Bustamante
Universidad Nacional Autónoma de México, Mexico

María del Carmen Beltrán-Orozco
Instituto Politécnico Nacional, Mexico

ABSTRACT

Chia seeds have a high content of omega 3, 6, and 9 fatty acids, so their consumption has become popular, often added to products subjected to heat treatments. This chapter evaluated the effect of temperature and time on the physicochemical properties and fatty acid composition of chia seed oil. The seeds, either whole or ground, were subjected to several treatments using a full factorial experimental design 3^2 where the factors were: the temperature (150, 187, and 225 °C) and treatment time (10, 25 and 40 min), while the studied responses were: iodine, acid, peroxide, and saponification values, Kreis test, and the content of omega 3, 6, and 9, the latter evaluated by gas chromatography. The use of moderate temperatures and short baking times is recommended (150 to 155 °C, less than 14 min) to avoid the loss in essential fatty acids and thus preserve the nutritional value of chia added in functional foods.

DOI: 10.4018/978-1-7998-1518-1.ch010

INTRODUCTION

Salvia hispanica L. is a herbaceous plant of the family of the *lamiaceae*, native to southwestern Mexico and Guatemala. The seed is rich in mucilage, starch and oil. The commercial chia seed is approximately 2 mm long and 1.5 mm wide. They have a small oval or circular shape, which often varies in color from dark brown to black gray. It is also possible to find white chia seeds (cultivar) or very small brown seeds that are little-known Mexican varieties of local consumption (Ghill, 2003).

In the pre-Columbian cultures of Mesoamerica, the specie is known as "Chía" (Cahill, 2003). At the time of the Aztecs, chia was part of the basic foods in the civilizations of Central America, after corn and beans and before amaranth (De Meester & Watson, 2008). Tenochtitlan, the capital of the Aztec Empire, received between 5,000 and 15,000 tons of chia as an annual tribute from conquered nations (Codex Mendoza, 1542). Chia seed was not used just as a food, but was also offered to Aztec gods (Sahagún, 1579).

In was reported (Ayerza & Coates, 2001) that the oil content of chia seeds is 32-39% which provides the highest known percentage of α-linolenic acid extracted from a natural source since it represents 60% to 63% in the profile of fatty acids (FAs) in the oil. The oil contained in the chia has high oxidation stability, so the seed can be stored under environmental conditions without deterioration. Their stability is attributed to the antioxidants contain, specifically as polyphenols as quercetin, kaempferol, chlorogenic acid, and caffeic acid (Reyes et al., 2008, Vargas-Martínez & Beltrán-Orozco, 2011).

Traditionally in Mexico, chia seed is consumed in fresh water and several products have been added with these seeds, such as the study carried out by Morales (2007), in which a crowbar based on chia seed, amaranth and oats was elaborated for the population with a normal diet, as well as for people with a special diet. Currently, the United States Food and Drug Administration (FDA) considers chia a food, not a food additive, therefore exempts it for being regulated (Ricker, 2007). On the other hand, the European Food Safety Authority deems chia suitable for use in bread and cereal products (whole seeds and ground) and in the European Union the use of chia oil as an additive was authorized on December 8, 2014 (Besana Portalagrario, 2014), but it was considered a novel food ingredient (SINC Science News, 2013) since 2009.

Nowadays, experimental design is applied in many areas of science and in industrial applications (Ilzarbe et al., 2008, Muthukumar & Mohan, 2005, Deiab & Elbestawi, 2005, Schonning, 2005). In traditional experiments, only one factor is varied at a time, while all others remain constant. In multivariate designs, all the factors of interest can be investigated in a single trial, minimizing the number of experiments required and providing information on the key process interactions. In the areas of food analysis (Tonon et al., 2008, Tiwary et al., 2012, Jing et al., 2011), food processing and development (Wadikar et al., 2010, Dutcosky et al., 2006), as well as food sensory evaluation (Deshpande et al., 2008), many applications of the use of experimental design have been reported.

In the development of products added with chia seeds or flour, Luna (2013) reported the effects of adding whole chia flour on the technological, nutritional and sensory qualities of cakes based on a 2^2 central composite rotational design. Silveira-Coelho and Salas-Mellado (2015) using the same experimental design concluded that it was possible to reduce the levels of saturated fat and increase the levels of polyunsaturated fat (PUFA), mainly omega-3 FAs, including chia seeds or flour in breads, in addition to increasing the level of fiber, yielding products with the features of functional foods.

Omega-3 FAs are basic units of polyunsaturated fats and are essential. They are polyunsaturated because they have more than two hydrogen bonds in their molecule between their carbon atoms and are essential because mammals lack the enzymes to synthesize them, so it is necessary to ingest them in the diet. Alpha linolenic acid (omega 3) belongs to the so-called PUFA family (Calder, 2008).

Due to the thermal processes to which different foods are subjected they can lose their physicochemical characteristics and also their nutritional value, specifically in their lipid compounds such as omega 3, 6 and 9 fatty acids (Biglar et al., 2012). Several studies using traditional designs (one factor at a time) have been carried out on seeds of sesame and sunflower, where changes are evaluated both in the composition of FAs and in some properties of the oils (formation of peroxides, iodine value, yield, etc.) when being heat treated, since these seeds regularly to be consumed require thermal processes, oscillating between 140 °C and 210 °C (Yen & Shyu, 1998, Lee et al., 2004, Akinoso et al., 2006, Lee et al., 2010 and Biglar et al., 2012). In natura, chia oil begins oxidation at approximately 145°C (Souza et al., 2017). Grampone et al. (2013) reported an oxidation temperature of 176° C using pure oxygen as oxidizing atmosphere at a higher rate in olive oil. Ixtaina et al. (2012) observed an oxidation temperature of 168.2 ± 2.8 °C monitoring peroxide values and using differential scanning calorimetry in studies on storage time of chia oil. Souza et al (2017) reported that the encapsulation of chia oil increased its oxidative stability since oxidation began at approximately 259° C. In the case of chia seeds due to their use in bakery products, cereal bars, among others, they go through a thermal process which can affect the physicochemical and nutritional characteristics of the oil they contain. However, so far there are no studies on the effect of temperature and exposure time on the composition of chia using similar temperatures to baking, nor the use of the multivariate experimental design to conduct these studies. For this reason, it is proposed to carry out the study of the effect of these factors on whole and ground chia seeds, using a full factorial experimental design 3^2.

BACKGROUND

At present, it has been demonstrated, historically and experimentally, how the consumption of fish fats decreases the prevalence of cardiovascular diseases, especially in coronary diseases. These virtues are attributed to a the PUFA family, specifically omega-3 fatty acids (Kris-Etherton, 2002).

Currently in the supermarket it is possible to find numerous foods enriched with omega-3. During the last decades there have been numerous epidemiological studies and trials on the effect of the consumption of omega-3 fatty acids on the health of human populations and clinical patients. They indicate a beneficial effect of omega-3 consumption (Flock et al., 2013).

Chia seed oil is very high in PUFA compounds, particularly in linolenic acid. Presenting a content of 19.8-20.8% of linoleic acid (omega-6) and 60.7-63.4% of linolenic acid (omega-3) (Ayerza, 1995).

Plants are good sources of essential fatty acids. Many of the vegetables contain basically gamma linolenic acid and alpha linolenic acid, so they provide other sources of consumption of these fatty acids, thus benefiting the health of consumers by increasing their consumption (Sergeant, et al., 2016). The oil from chia seeds is the richest known natural source of omega-3 FAs (α-linolenic acid), which ranges from 60.2-64.6 g/100g of FAs. An intake of 7.3 g of chia seed per day, provide 100% of the recommended intake of omega-3 FAs to prevent chronic diseases related to diet.

Fats and oils can undergo different transformations that, in addition to reducing the nutritional value of food, produce volatile compounds that transmit unpleasant odors and tastes. This is because the ester bond of acylglycerides is susceptible to chemical and enzymatic hydrolysis, and because unsaturated fatty acids are sensitive to oxidation reactions (Badui, 2006). The cause of these transformations is related to the deterioration of fats or oils, which are caused by factors such as: Type of storage, storage time, temperature, the presence of oxygen among others (Souza et al., 2017, Grampone et al., 2013, Ixtaina et al., 2012).

The challenge of incorporating polyunsaturated oils and omega-3 fatty acids (FAs) into foods is their susceptibility to oxidative deterioration (rancidity), resulting in changes in both negative alteration in food quality and nutrition (Decker et al., 2010, Frankel, 2005, Kubow, 1992). This is especially true of omega-3 FAs that not only oxidize quickly but also have oxidation products which humans can detect at concentrations as low at 1 μg/L (Frankel, 2005).

Several authors reported the inclusion of chia seeds in bread preparations to reduce the levels of saturated fat and increase the levels of polyunsaturated fat, mainly omega-3 FAs, in addition to increasing the level of fiber, yielding products with the features of functional foods (Luna, 2013). The technological quality of the loaves was affected by adding chia materials in the formulations, leading to a decrease in the specific volume and the total score values. In sensory evaluation, the breads with chia flour or seeds added obtained high levels of acceptability and purchase plans, demonstrating the commercial viability of these products, with the chia flour bread assigned a higher index of purchase intent than the chia seed bread (Silveira-Coelho & Salas-Mellado, 2015).

MAIN FOCUS OF THE CHAPTER

The general objective of this chapter was to evaluate the effect of temperature on the physicochemical properties and composition of the unsaturated fatty acids present in chia seeds (*Salvia hispanica* L.), subjecting them to thermal treatments at different temperatures and times. For this purpose, the heat treatment was carried out on whole and ground chia seeds, testing three temperatures (150, 187 and 225 °C) at three exposure times (10, 25 and 40 min), using a factorial design of 3^2 with 4 central points and performed in triplicate (39 experiments) for subsequent oil extraction (García et al., 2003). Temperature levels were proposed based on studies published by other authors (Yen and Shyu, 1998; Lee et al., 2004; Akinoso et al., 2006; Lee et al., 2010 and Biglar et al., 2012), which evaluated the different changes in the heat treated sesame and sunflower seed oil from 140 °C to 220 °C. On the other hand, for the baking time levels, what was reported in the manual of use of the electric oven used to cook traditional bread (25 to 30 min) was taken into account (Hamilton Beach 31199 User Manual, 2018). In order to observe differences in the effect of time, extreme intervals (10 to 40 °C) were planned.

The physicochemical parameters were evaluated in the extracted oil, such as: iodine, acid, peroxide, saponification values and kreis test. In addition, the composition of the fatty acids present in the chia seed oil was evaluated, as well as the content of omega 3, 6 and 9 fatty acids, before and after the treatment using gas chromatography. Finally, a statistical model was established to predict the responses (iodine, acid, peroxide, saponification values, Kreis test and fatty acid content) using the Desing Expert 7.0 program, in order to propose the temperature range and time that can predict the conditions where significant changes do not occur in the nutritional properties of the chia, guaranteeing the consumer, the preservation of the functional properties of the products baked with these seeds.

Table 1. Description of factors and levels studied in the experimental full factorial design 3^2 for thermal treatments carried out on chia seeds

Factors	Levels		
	-1	0	1
Temperature (°C)	150	187.5	225
Time (min)	10	25	40

1. Experimental

a. Biological Material

Chia seeds were obtained from the State of Colima, stored in plastic bags with hermetic closure, in batches of 250 g, for further treatment. The seeds that were subjected to the thermal treatments were in two presentations, which were whole and ground, for the subsequent extraction of the oil.

b. Planning of Thermal Treatments in Whole and Ground Chia Seeds Using Experimental Design

The whole and ground chia seeds were subjected to different thermal treatments, which consisted of placing them in a Hamilton Beach brand convection oven, model 31199, series A 4260BR. Temperature conditions were determined in the studies carried out by other authors (Yen & Shyu, 1998, Lee et al., 2004, Akinoso et al., 2006, Lee et al., 2010 and Biglar et al., 2012) reporting the changes in the sesame seed oil and sunflower that were heat treated (using 140 °C to 220 °C). Therefore, wider values were selected for chia seed experiments (150 to 250 °C, Table 1). To observe differences in the effect of baking time, extreme values (10 to 40 °C, Table 1) were planned. These values were decided considering the baking time suggested in the oven manual (25 to 30 min).

The experimental design for the heat treatment was a full factorial design 3^2 with 4 central points made in triplicate with a total of 39 experiments for each type of chia (whole or ground). It means that the entire design was done for 39 samples of whole chia seeds and for 39 samples of ground chia. Additionally, control samples (without heat treatment) of chia seeds were processed in triplicate.

The review of factors and levels are described in Table 1, while the series of experiments carried out can be seen in Table 2.

The experiments were planned using the Design Expert 7 software, (Stat Ease Inc.). The responses (output variables) were 8 and are defined in Table 3. Details of the procedure to be measured can be seen in sections d, e and f of this section and in Appendix 1 subchapter 1 at the end of the chapter.

The selection of multivariate optimization designs was preferred, in order to obtain models (equations) and response surfaces to predict the conditions in which the nutritional properties of chia seeds are still preserved. The response surfaces with their corresponding equation and the discussion of the effect of the factors on each measured response (output variable) will be carried out in its corresponding section, specifying whether the thermal treatments were performed on the whole chia seed or ground chia seed.

Table 2. Description of the thermal treatments carried out on chia seeds using experimental full factorial design 3^2 with 4 central points (13 treatments made in triplicate=39 experiments).

Experiments Number	Temperature °C	Time (min)	Temperature °C	Time (min)
1, 2, 3	150	10	-1	-1
4, 5, 6	187.5	10	0	-1
7, 8, 9	225	10	1	-1
10, 11, 12	150	25	-1	0
13, 14, 15	187.5	25	0	0
16, 17, 18	225	25	1	0
19, 20, 21	150	40	-1	1
22, 23, 24	187.5	40	0	1
25, 26, 27	225	40	1	1
28, 29, 30	187.5	25	0	0
31, 32, 33	187.5	25	0	0
34, 35, 36	187.5	25	0	0
37, 38, 39	187.5	25	0	0

Table 3. Description of the Responses (output variables) studied in the experimental full factorial design 3^2 for thermal treatments carried out on chia seeds (whole and ground).

Responses	Name	Units
Y1	Saponification Value	mg KOH/g Oil
Y2	Iodine value	g I_2/100 g Oil
Y3	Acid value	% Oleic acid
Y4	Peroxide value	meq O_2/kg Oil
Y5	Kreis test	545 nm/g Oil
Y6	Linolenic acid (Omega 3)	g/100 g sample
Y7	Linoleic acid (Omega 6)	g/100 g sample
Y8	Oleic acid (Omega 9)	g/100 g sample

c. Thermal Treatment of Chia Seeds

This was carried out in a Hamilton Beach oven (model 31199), using whole or previously ground chia seed, according to the following procedure:

- The oven rack was placed in the middle position.
- The temperature knob was adjusted to the desired temperature (confirming it is correct with the thermometer).
- After adjusting the baking function knob, the timer is set to preheat the oven for 10 minutes.
- In the 9.5 minute of preheating, 200 g of chia seed (whole or ground), previously homogenously placed in aluminum trays were introduced.
- The timer was set for the selected baking time.
- After the treatment time, the sample was taken out of the oven and allowed to cool at room temperature and it was stored in airtight polyethylene bags in a dark place.

d. Soxhlet Extraction of Chia Oil

Approximately 60 g of each sample of chia seeds were weighed (to extract the oil from the whole seeds it was necessary to grind them before the extraction) and by means of a Soxhlet equipment the extraction of the oil was carried out. Hexane (n-hexane) was used as solvent for 6 hours, at a temperature of 45 ± 2 °C and a solvent flow of 30 drops per minute. The hexane-oil mixture was distilled under reduced pressure in a rotary evaporator at 45 °C, to obtain the crude oil, which was packed in amber glass bottles and stored at 0 °C until further evaluation (García et al., 2003, Romero, 2016).

e. Evaluation of Physicochemical Parameters of Chia Seeds Oil

Different tests (iodine, peroxide, acid, saponification values and Kreis index) were measured in the oil, both for the one obtained from the thermally treated seeds (whole and ground) and for the oil obtained from the seed without any treatment. In order to evaluate the possible physicochemical changes presented in the samples subjected to the different treatments. The chemical fundamentals of each of the tests and their methodology can be seen in Appendix 1 subchapter 1.

f. Evaluation of the Fatty Acid Profile of Chia Seeds Oil by Gas Chromatography

i. Measurement of standards

The determination of the fatty acid profile present in the chia seed oil without heat treatment was carried out, for which a mixture of methyl ester standards was used (Supelco, F.A.M.E. MIX C14-C22, Cat. No. CRM18917).

Figure 1. Oven temperature ramp used for the determination of omegas 3, 6 and 9 in chia oil by GC

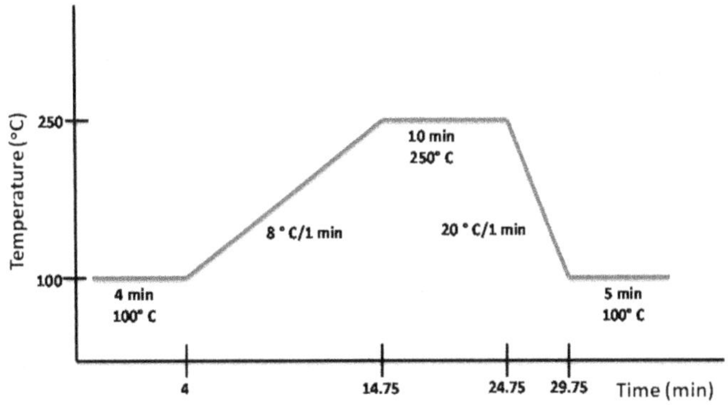

Table 4. Concentration levels of each PUFA compounds for the Calibration curves.

Methyl Oleate (mg/mL)	Methyl Linoleate (mg/mL)	Methyl Linoleate (mg/mL)	Methyl Heptadecanoate (mg/mL)
0.15	0.8	0.4315	1.4
0.3	1.6	2.1575	1.4
0.6	2.4	4.3150	1.4
0.9	3.2	6.4725	1.4
1.2	4.0	8.6300	1.4
1.5	4.8	12.9450	1.4

ii. Measurements of Omegas 3, 6 and 9 Present in the Chia Seed Oil of Samples Subjected to Thermal Treatments

The concentration of omega 3, 6 and 9 fatty acids (oleic, linoleic and linolenic acid respectively) was evaluated in the oil of seeds (milled and whole) heat treatment, a previous derivatization was necessary to convert fatty acids into fatty acid methyl esters that are volatile:

- Combine 100 mg of the sample and add 3 mL of BF_3-Methanol (Pierce Chemical Company, USA) in a 5 mL small reaction vials (Reacti-Vial™, Thermo Scientific)
- Cap the vial and heat at 60° C for 5-9 minutes.
- Cool and transfer to a separation funnel with 30 mL of hexane.
- Wash two times with a saturated NaCl solution.
- Discard the aqueous (bottom) layer after each wash.
- Dry the hexane extracts over anhydrous sodium sulfate and transfer to a clean, dry container.
- Analyze the hexane layer by gas chromatography (GC) directly or, if concentration is desirable, evaporate the hexane and analyze the residue.

Table 5. Characterization of chia seeds oil without heat treatment (control).

Test	Chia Seed Oil $X_{av} \pm S$ (n=3)
Peroxide value (meq O_2/kg of oil)	2.30 ± 0.20
Kreis test 545 nm /g of oil	2.611 ± 0.218
Acid value (mg NaOH/g of sample)	1.52 ± 0.09
Saponification value (mg KOH/g of oil)	191.53 ± 1.34
Íodine value (g I_2/100 g de aceite)	199.32±45.31

X_{av}=average, S= standard deviation, n= sample number

The prepared methyl esters were analyzed by GC under the conditions described by Chaves (2009). A Thermo Quest Trace GC 2000 Gas Chromatograph with a Flame ionization detector, injector temperature of 250 °C, a detection temperature of 270 °C and nitrogen as the gas flow was utilized for all the measurements. The temperature ramp of the oven is shown in Figure 1. A highly polarized Carbowax-20M 30 m x 0.25 mm column coated on the inside with a liquid phase of polyethylene glycol was used for all the measurements.

iii. Preparation of the PUFA Calibration Curves

The necessary milligrams of reagent were weighed to prepare the stock solutions of the PUFA methyl esters standards, diluting them in n-hexane. From the above solutions, aliquots were taken to prepare solutions at different concentration levels for the calibration curves shown in Table 4, using methyl heptadecanoate as the internal standard.

The solutions were prepared and analyzed by GC under the conditions described by Chaves, 2009. The data were processed using least squares and evaluating the validity of the models with the Statgraphics Centurion VI software (Statgraphics Technologies, Inc.).

2. Results and Discussion

The design of experiments is a statistical technique that allows identifying and quantifying the causes of an effect within an experimental study. In an experimental design, one or more variables are deliberately manipulated, linked to the causes, to measure the effect they have on another variable of interest; It prescribes a series of guidelines regarding which variables should be manipulated, in what way, how many times the experiment should be repeated and in what order (Massart, 1997).

Once the experimental results are available, the effects of the factors, as well as their interactions, can be calculated. The statistical tests allow verifying if the calculated effects are significant compared with the experimental error (Vargas, 1999). If a response surface model is constructed, as in this case, where a full factorial design 3^2 with 4 central points (made in triplicate) was used, the coefficients can be calculated by the least squares method and the model can be evaluated by replicating certain experiments and applying the test ANOVA. The model can be used to find the optimal area mathematically.

Figure 2. Response surface of saponification value of whole chia seeds oil with thermal treatments

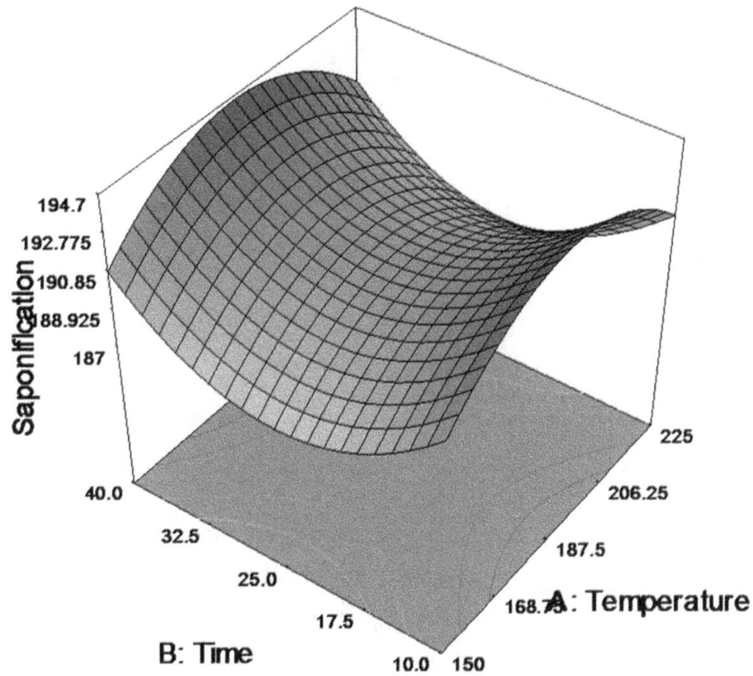

Using the response surfaces obtained for each design, the influence of the two factors: The temperature and exposure time in the nutritional quality of chia seeds will be discussed. The above for the 2 proposed cases: a) using whole chia seeds and b) using ground chia seeds.

a. Characterization of Chia Seed Oil Control

The physicochemical parameters were determined to the chia seed oil without any treatment (control) and can be seen in Table 5.

Regarding the data of peroxide value, Kreis test and % of free fatty acids (acid value), there are no specific references for the chia seed oil, however it is necessary to emphasize that according to the regulations for other types of oils as is the case of refined olive oil, canola or soybeans, the presence of peroxides in these oils should not be higher than 10.0 meq O_2 / kg oil (CODEX STAN, 1999). Ayerza & Coates, (2004) reported similar results in the peroxides values (from 2.7 to 3.8 meq O_2 / kg oil) of chia grown in different locations in South America.

Comparing the results of the saponification value obtained in the present work, with those reported by Cooper (2006), which was 192.53 mg KOH / g of fat. It can be seen that the result is very similar, indicating that the author worked with pure chia oil.

Figure 3. Response surface of the acid value of whole chia seed oil with thermal treatments

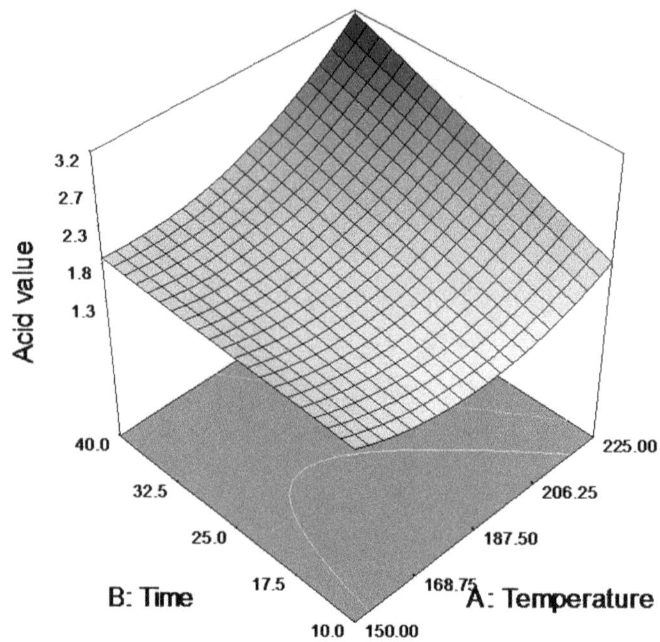

b. Physicochemical Characterization of Oils Obtained From Whole Chia Seeds Subjected to Heat Treatments

- **Saponification Value:** Represents the average molecular weight of the fatty acids in an inverse relationship. The response surface for the saponification value in the oil of whole chia seeds can be seen in Figure 2.

The model that fits is the quadratic, because it passes the lack of fit test. The effect of the factors that resulted significant were the temperature squared and the time squared (Eq. 1).

$$Saponification\ value = 190.56 - 3.56\left(Temperature\right)^2 + 3.95\left(Time\right)^2 \quad (1)$$

The saponification values obtained were found in a range of 185.21 to 195.34 mg KOH /g oil in whole seed without significant difference between them ($p \geq 0.05$). It is important to note that the data obtained are similar to those reported by the chia seed oil standard (NMX-F-592-SCFI-2017), where it is mentioned that the saponification value should be approximatly 192 mg KOH /g oil. However, there are no norms that indicate a limit for the saponification value of the oils.

- **Acid Value:** The acid value indicates the content of free fatty acids present in the oil, as well as being indicative of the degree of rancidity due to storage and heat (Figure 3), the data obtained for the different treatments adjust to a quadratic model.

Figure 4. Response surface of the peroxide value of whole chia seed oil with thermal treatments

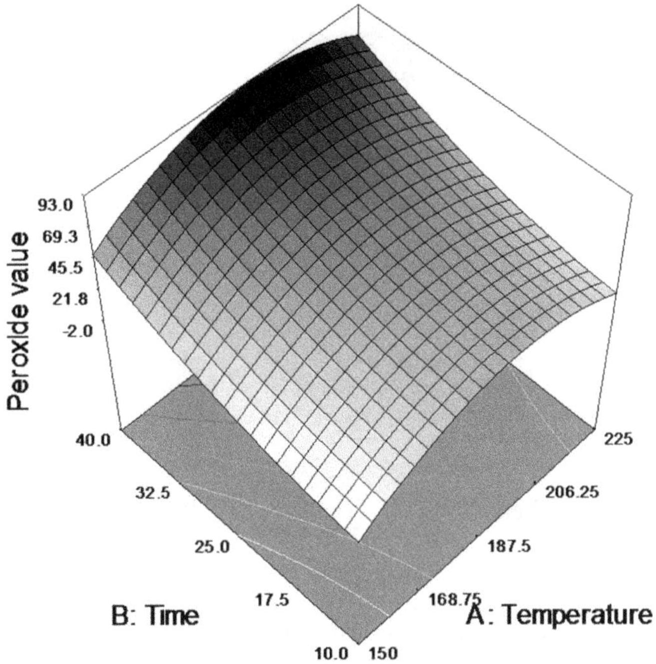

The model that fits is the quadratic, because it passes the lack of fit test. The effect of the factors that resulted significant were the main factors, the interaction of them and temperature squared (Eq. 2).

$$Acid\ value = 1.75 + 0.34Temperature + 0.36Time + 0.30Temperature * Time + 0.42\left(Temperature\right)^2 \tag{2}$$

The highest acidity values were for the oils obtained from chia seeds treated at 225 °C for 40 min, with 3.27 mg NaOH / g sample, which is due to the formation of free fatty acids generated by the increase in temperature. It can be said that the acid value of the oils presented a direct effect by the thermal treatments, without affecting the quality of the oil or the seed, as it does not exceed the limit specified in the CODEX of 4.0 mg NaOH /g oil.

- **Peroxide Value:** In the case of the peroxide values, the results obtained are presented on the response surface of Figure 4.

The model that fits is the quadratic, according to the lack of fit test. The effect of the factors that resulted significant were the main factors, the temperature squared and the time squared (Eq. 3).

$$Peroxide\ value = 55.69 + 10.80Temperature + 25.92Time - 30.47\left(Temperature\right)^2 + 9.78\left(Time\right)^2 \tag{3}$$

Figure 5. Response surface of the kreis test of whole chia seed oil with thermal treatments

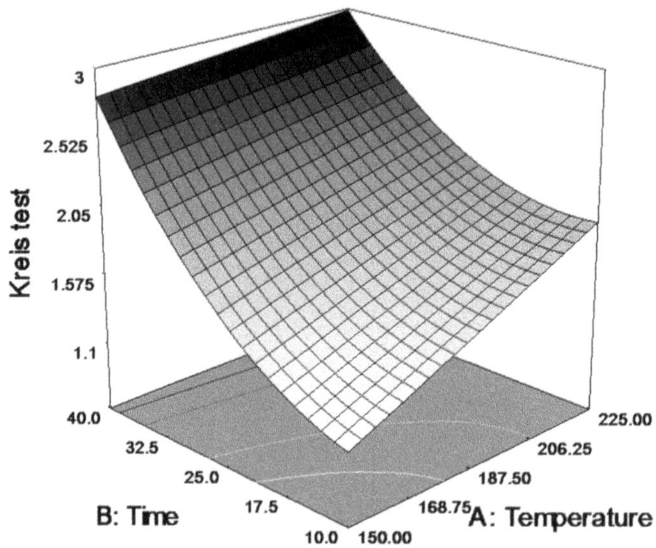

In the case of whole chia seeds, it can be observed that those treated at 187 °C had the highest peroxide value, reaching 93.27 meq O_2/ kg of oil, which shows a tendency to increase and then decreases when reaching 225 °C. The results obtained showed clear effects due to time and temperature. The considerable increase in peroxides values could catalyze the self-oxidation of the chia oil. As in all treatments at 187.5 and 225 °C exceed the limit values of 10 meq O_2/ kg of oil (CODEX). A thermal treatment of chia seeds of 155 °C is recommended for no more than 13 minutes to fulfill the peroxide value limit.

- **Kreis Test:** In the Kreis test in whole chia seed oil, the statistical analysis was carried out obtaining the response surface presented in Figure 5.

The model that fits is the quadratic, according to the lack of fit test. The effect of the factors that resulted significant were the main factors, their interaction and time squared (Eq. 4).

$$Kreis\,test = 1.91 + 0.26Temperature + 0.68Time + 0.17Temperature * Time + 0.31\left(Time\right)^2 \quad (4)$$

It can be observed that the treatment at 225 °C is the one with the highest Kreis test value, followed by the treatment of 187 °C and finally the treatment at 150 °C, observing the significant effect by treatment time and temperature as well as by the interaction of both factors. However, there are no norms that indicate a limit for the Kreis test of the oils.

- **Iodine Value:** Is one of the main characteristics of fats and oils, since it is used to determine the degree of unsaturation. However, no significant effect was observed due to temperature and time, nor when comparing values with chia control oil. In addition, no limit value has been reported in any norm.

Figure 6. Response surface of saponification value of ground chia seed oil with thermal treatments

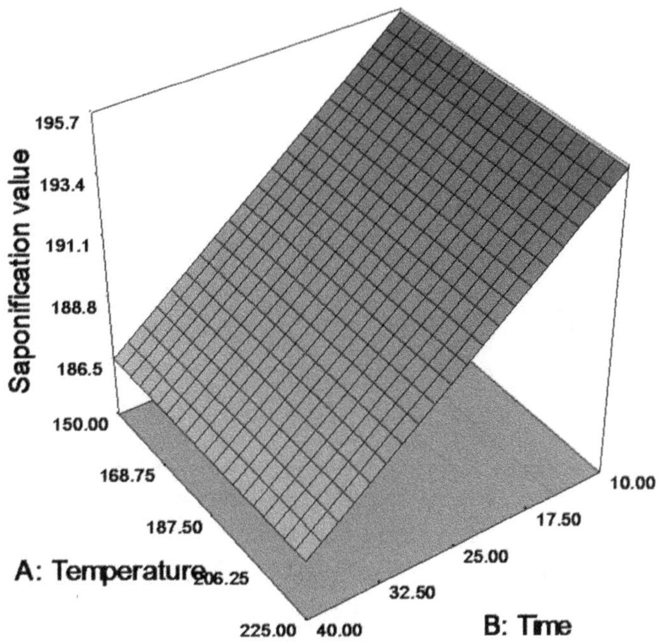

c. Physicochemical Characterization of the Oils Obtained From the Ground Chia Seed Subjected to Thermal Treatments

- **Saponification Value:** In the case of ground seed oil, it was observed that the saponification value decrease in the treatments carried out at 40 °C, obtaining data from 187.71 to 195.58 mg KOH / g of oil, these results are close to that reported by Cooper (2006) which reports 192.53 ± 0 .52 mg KOH / g of oil. Based on these results, the statistical analysis was performed, obtaining the response surface (Figure 6) for the saponification value in the oil of ground chia seeds.

The model that fits is the linear, because it confirmed by the lack of fit test. The effect of the factor that resulted significant was only the time (Eq. 5).

$$Saponification\, value = 191.05 - 4.54 Time \quad (5)$$

The saponification values obtained were found in a range of 188.0 to 207.3 mg KOH /g oil in ground chia seed. An increase in the dispersion of saponification values was observed with the ground seed subjected to thermal treatments. Although, there is no difference in values with the control sample. However, there are no norms that indicate a limit for the saponification value of the oils.

- **Acid Value:** It could be observed that in the ground seed oil (Figure 7), the acid value is approximately double compared to the whole seed oil, this is because the heat treatments directly affect the ground seed, while in the whole there may be a resistance for the mucilage that covers it, allowing a dissipation of the heat by the same one and generating smaller amount of free fatty acids.

Figure 7. Response surface of the acid value of ground chia seed oil with thermal treatments

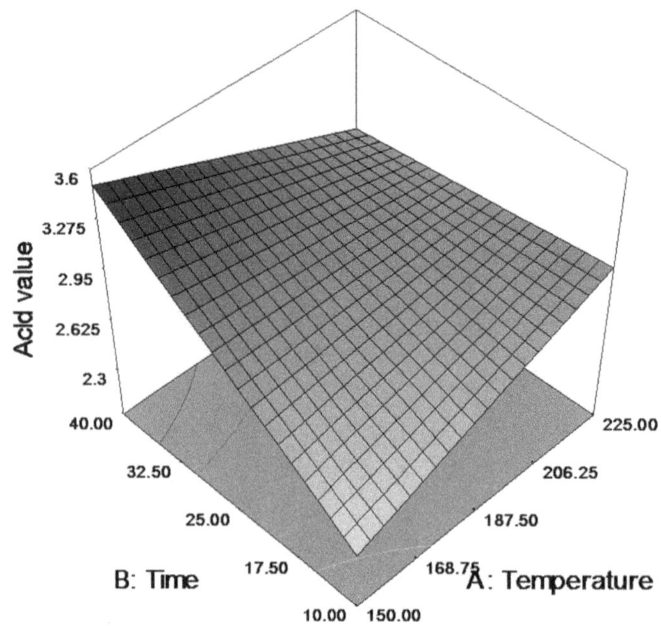

Figure 8. Response surface of the peroxide value of ground chia seed oil with thermal treatments

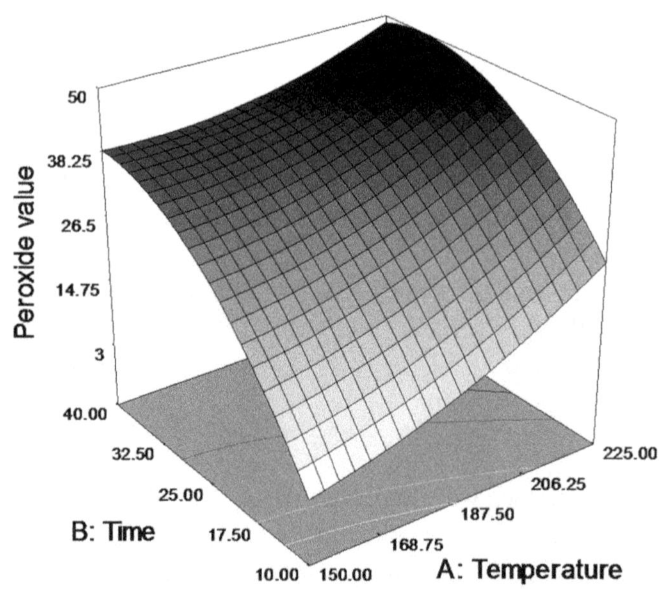

The model was linear, because it confirmed by the lack of fit test. The effect of the factors that resulted significant were the time and the factors interaction (Eq. 6).

Figure 9. Response surface of the Kreis test values of ground chia seed oil with thermal treatments

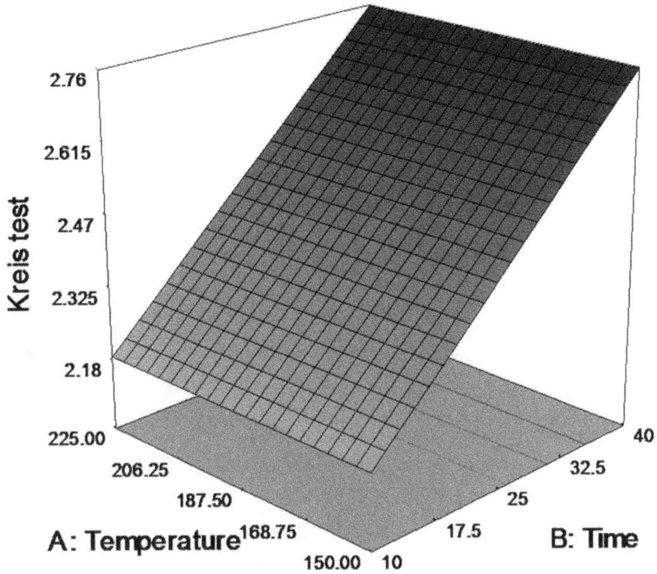

$$Acid\,value = 2.93 + 0.25Time + 0.32Temperature * Time \quad (6)$$

From de main effects only the time was significant and the interaction of both factors (time and temperature) had effect on the acid values in the oil of the ground seeds. However, the acid values do not exceed the limit specified in the CODEX of 4.0 mg NaOH /g oil in ground chia.

- **Peroxide value:** In Figure 8 can be observed that the oil obtained from the ground seed presents peroxide values of 3.7 to 48.58 meq of O_2/kg of oil. It behaves directly proportional to the temperature and the time of treatment, which indicates that the amount of peroxides formed goes proportionally with the increase of the temperature and time of treatment.

The model that fits is the quadratic, according to the lack of fit test. The effect of the factors that resulted significant were the main factors, their interaction, the temperature squared and the time squared (Eq. 7).

$$Peroxide\,value = 35.10 + 7.94Temperature + 14.85Time$$
$$-3.12Temperature * Time + 3.16(Temperature)^2 - 9.16(Time)^2 \quad (7)$$

It can be observed that the oil that presented the highest peroxide value is the one treated at 225 °C, reaching 48.58 meq of O_2/kg oil. The results obtained showed clear effects due to time and temperature. The considerable increase in peroxides values could catalyze the self-oxidation of the chia oil. All treatments at 187.5 and 225 °C exceed the limit values of 10 meq O_2/ kg of oil (CODEX). A thermal treatment of chia ground seeds of 155 °C is recommended for no more than 13 minutes to fulfill the peroxide value limit.

Figure 10. Conditions found using the desirability function to preserve the chemical values during heat treatment of whole and ground chia seed oil

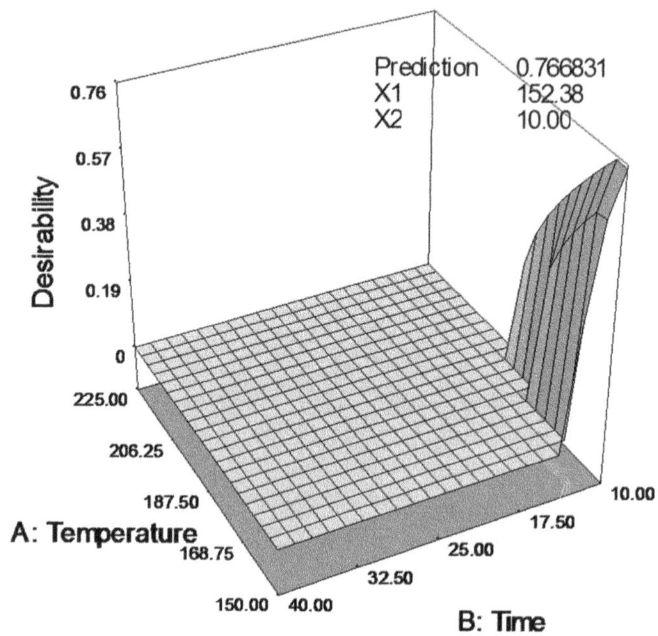

- **Kreis test:** Figure 9 shows the response surface obtained from the results of the kreis test for the oil of the ground chia seeds.

The model that fits is the linear, according to the lack of fit test. The effect of the unique factor that resulted significant was the time (Eq. 8).

$$Kreis\,test = 2.47 + 0.29Time \tag{8}$$

The oil of ground seeds presents a different trend, being the treatment at 187 °C for 40 min which presents the highest Kreis test value, followed by 225 °C at 40 min and ending at 150 °C at the same time. There are significant effects by time and the interaction of the factors (time and temperature). The Kreis test value is diminished in the ground seed compared to the whole seed, which may be due to the interaction of other compounds on the seed oil, since the milling allows the release of antioxidant compounds (such as phenols) due to the rupture of cells in the chia seed (Souza, 2017). However, there are no norms that indicate a limit for the Kreis test of the oils.

Optimization of the thermal condition (temperature and time values) using the desirability function for the treatment of whole and ground chia seed: Myers et al. (2016) describe a multiple response method called desirability. The method makes use of an objective function, D(X), called the desirability function. It reflects the desirable ranges for each response (di). The goal of optimization is to find a good set of conditions that meet all the objectives we want.

Table 6. Prediction of the chemical values of chia seed oil using the desirability function and conditions described in a) and b).

Chemical Values	Whole Seed	Ground Seed
Saponification Value (mg KOH/g of oil)	191.53	195.59
Iodine value (g I2/100 g de aceite)	197.61	195.00
Acid value (mg NaOH/g of sample)	1.64	2.38
Peroxide value (meq O2/kg of oil)	12.54	3.50
Kreis test (545 nm/g of oil)	1.23	2.19

The prediction of the temperature and baking time for preserving the chemical values of the chia seed was performed using the desirability function. Values in Table 5 of the sample chia seed control were introduced as the criteria to reach.

1. **For Whole Chia Seeds:** Using a temperature of 157.7 °C during no more than 11.5 min (Figure 10a), the desirability function predicted the chemical values of the Table 5.
2. **For Groud Chia Seeds:** Using a temperature of 152.38 °C during no more than 10 min (Figure 10b), the desirability function predicted the chemical values of the Table 6.

d. Determination of the Effect of Thermal Treatments on the Composition of Omega 3, 6 and 9 Fatty Acids in Chia Seed Oil Used as Control by Gas Chromatography

Control samples: From the identification of the fatty acids present in the chia seed oil without treatment, used as control (Figure 11), the fatty acid with the highest content in the oil was the linolenic acid, followed by the linoleic acid, also presenting a important content in oleic acid. These three fatty acids are known as omega 3 (linolenic acid), omega 6 (linoleic acid) and omega 9 (oleic acid) respectively, so the content of these essential FAs, present in the control chia seeds oil was evaluated. For the above, the response of each fatty acid measured by GC was interpolated in its corresponding model from the calibration curves (see Appendix 2) where the response was the relationship of areas. The internal standard (I.S.) used was heptadecanoate methyl ester. The analysis of the precision and details of the models can be seen in Appendix 2.

e. Effect of Thermal Treatments on the Content of Omegas 3, 6 and 9 in Whole Chia Seed Oil Measured by GC

- **Omega 3:** The content of linolenic acid in heat-treated chia seed oils was found in a range of 25.63 to 34.88%. Compared with the control it was observed that there is a decrease in the content of around 6% in the different treatments. Based on the results obtained, a statistical analysis was carried out where the response surface for the linolenic acid (omega-3) content is observed in Figure 12.

Figure 11. Fatty acid profile of chia seed oil without treatment (control) and 3 examples of thermal treatments measured by gas chromatography

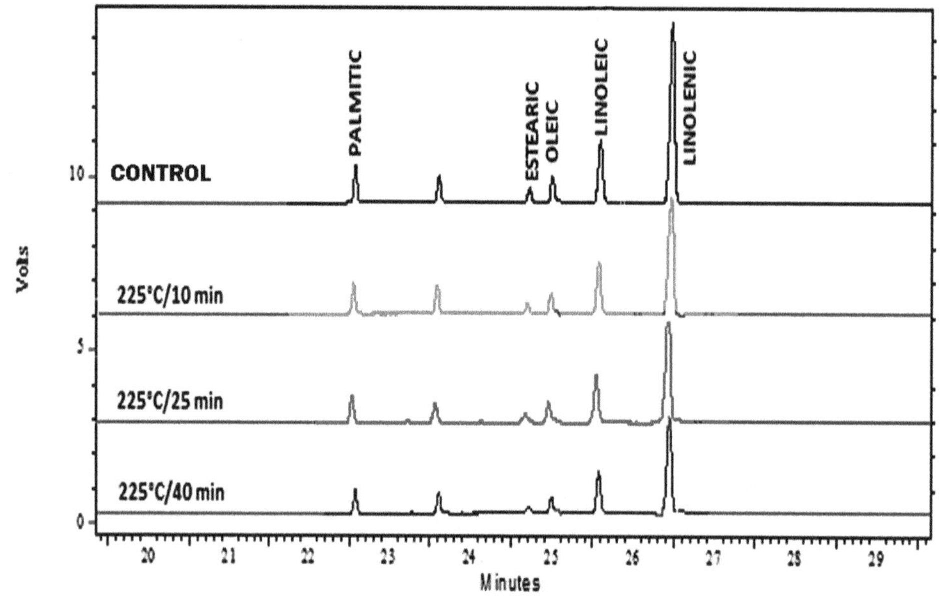

Figure 12. Response surface of the omega-3 content (g/100 g) of whole chia seed oil with thermal treatments

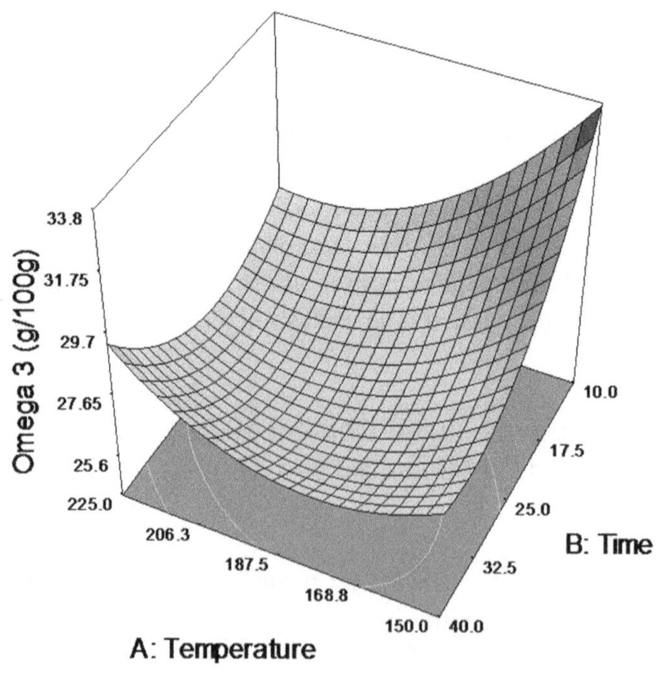

Figure 13. Response surface of the omega 6 content (g/100 g) of whole chia seed oil with thermal treatments

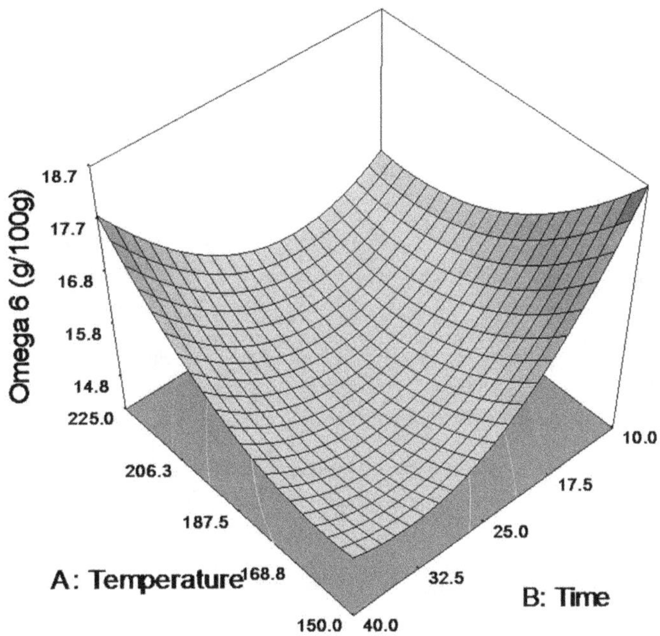

The model that fits is the quadratic, according to the lack of fit test. The effect of the factors that resulted significant were the main factors, their interaction, the temperature squared and the time squared (Eq. 9).

$$Omega\,3 = 25.81 - 1.05Temperature - 1.03Time + 1.86Temperature * Time + 1.87\left(Temperature\right)^2 - 2.10\left(Time\right)^2 \quad (9)$$

Omega-3 FA is the most abundant in chia oil and the most valuable because it is essential (it must be ingested in the diet). Its degradation in almost all thermal treatments of whole chia oil is considered an important loss. In the last section of this manuscript, the appropriate temperature and time conditions will be proposed to reduce the loss in this AF.

- **Omega 6:** The content of linoleic acid was lower (with significant effect) only in treatments at 187 °C (no matter the time). Its level decreasing approximately 2% compared to chia control seeds (17.29 g/100 g). The treatments at 150 and 225 °C showed similar contents to the control samples. Based on the results obtained, a statistical analysis was carried out where the response surface for the linoleic acid (omega-6) content is observed in Figure 13.

The model that fits is the quadratic, according to the lack of fit test. The effect of the factors that resulted significant were the main factors, their interaction, the temperature squared and the time squared (Eq. 10).

Figure 14. Response surface of the omega 9 content (g/100 g) of whole chia seed oil with thermal treatments

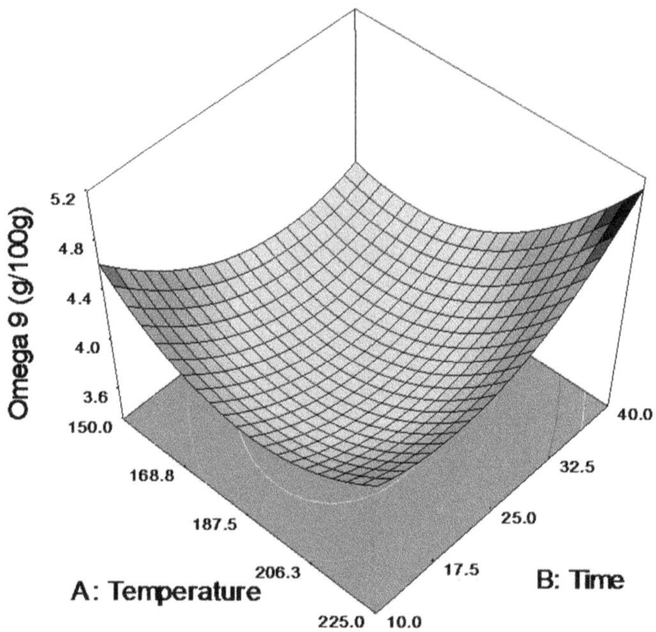

$$Omega\,6 = 14.87 - 0.41Time + 1.26Temperature * Time + 0.82\left(Temperature\right)^2 + 1.34\left(Time\right)^2 \quad (10)$$

In the last section of this manuscript, the appropriate temperature and time conditions will be proposed to reduce the loss in this compound.

- **Omega 9:** The content of oleic acid remained similar in the different treatments, with no differences observed with that quantified in the chia control seed oil (4.30 g / 100 g). Based on the results obtained, a statistical analysis was carried out where the response surface for the oleic acid (omega-9) content is observed in Figure 14.

The model that fits is the quadratic, according to the lack of fit test. The effect of the factors that resulted significant were the main factors, their interaction, the temperature squared and the time squared (Eq. 11).

$$Omega\,9 = 3.73 + 0.23Temperature + 0.32Temperature * Time + 0.47\left(Temperature\right)^2 + 0.37\left(Time\right)^2 \quad (11)$$

In the last section of this manuscript, the appropriate temperature and time conditions will be proposed to reduce the loss in this compound.

Figure 15. Response surface of the omega-3 content (g/100 g) of ground chia seed oil with thermal treatments

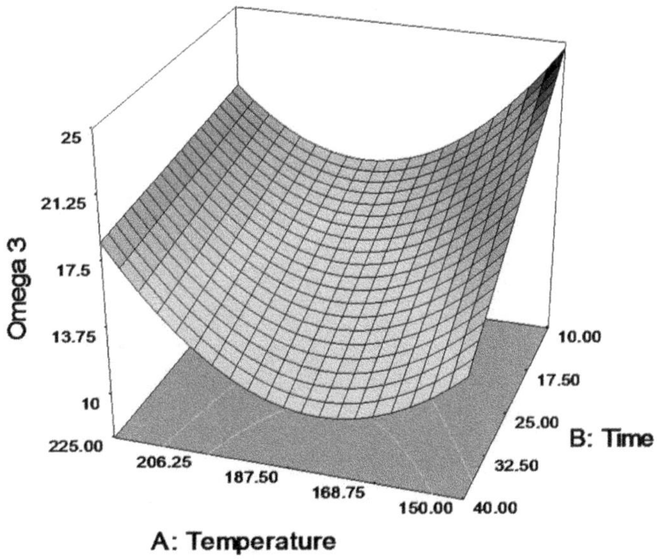

f. Effect of Thermal Treatments on the Content of Omegas 3, 6 and 9 in Oil of Ground Chia Measured by GC

- **Omega 3:** A considerable decrease in the content of linolenic acid was observed in ground chia seeds heat-treated compared to the control, however treatments at 150 ° C are those that maintained the content of linolenic acid to a greater extent. The higher temperature treatments decreased around 10% in the linolenic acid content compared to the control content (28.83 g / 100g). Based on the results obtained, a statistical analysis was carried out where the response surface for the linolenic acid (omega-3) content is observed in Figure 15.

The model that fits is the quadratic, according to the lack of fit test. The effect of the factors that resulted significant were the time, the main factors interaction and the temperature squared (Eq. 12).

$$Omega\,3 = 13.89 - 2.90Time + 2.19Temperature * Time + 5.69\left(Temperature\right)^2 \quad (12)$$

It is important to note that the nutritional contribution of omega 3 in the heat-treated ground seeds is reduced to 50%, being 3.62 g of linolenic acid for every 100 g of seeds approximately. Then ground chia seed is more susceptible to degradation of linolenic acid by the breakdown of cell walls and by its greater exposure to high temperatures. The appropriate temperature and time conditions will be proposed to reduce the loss in this compound that is essential in the human nutrition.

Figure 16. Response surface of the omega 6 content (g/100 g) of ground chia seed oil with thermal treatments

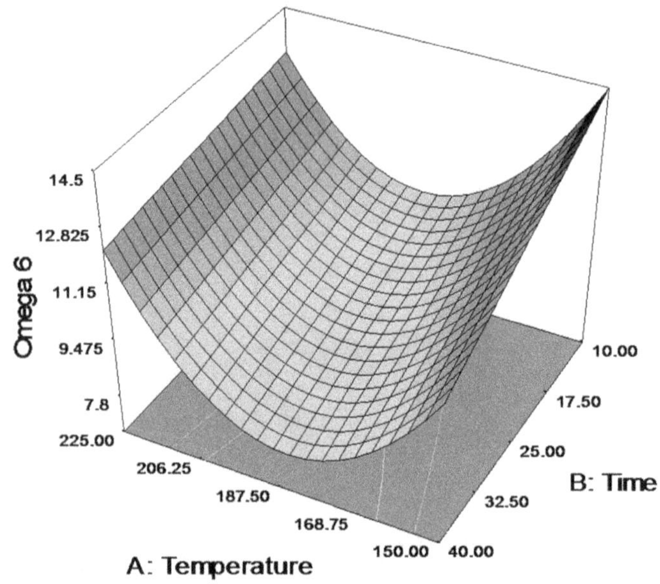

Figure 17. Response surface of the omega 9 content (g/100 g) of ground chia seed oil with thermal treatments

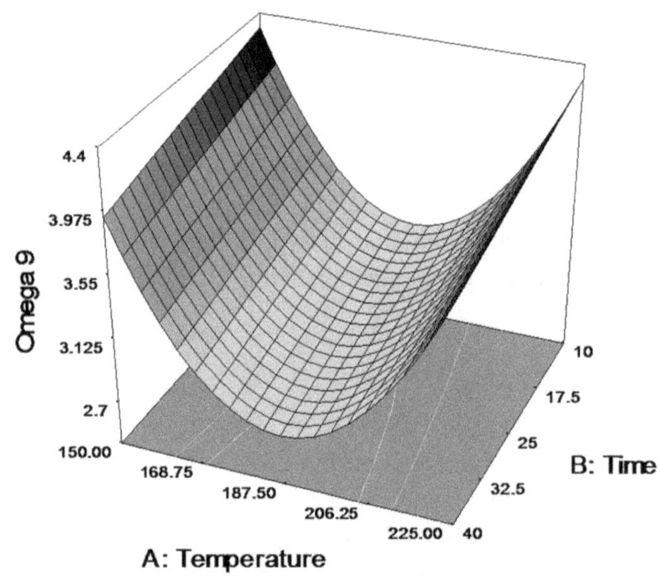

- **Omega 6:** It was reduced by treatment time and temperature. At 150 °C, ground chia seed oil had a linoleic acid content closer to this found for control seed oil. As the treatment time for the three temperatures (150, 187 and 225 °C) increased, the linoleic acid content decreased significantly ($p \leq 0.05$). Based on the results obtained, a statistical analysis was carried out where the response surface for the linoleic acid (omega-6) content is observed in Figure 16.

The model that fits is the quadratic, according to the lack of fit test. The effect of the factors that resulted significant were the time, the main factor interaction and the temperature squared (Eq. 13).

$$Omega\,6 = 8.93 - 1.10 Time + 0.65 Temperature * Time + 3.81 \left(Temperature\right)^2 \tag{13}$$

The ground chia seed oil had a lower content of linoleic acid compared to the oil of whole chia seeds, observing the same effect as the linolenic acid content. This would indicate that the heat treated ground seeds are more susceptible to degradation of these components.

In the last section of this manuscript, the appropriate temperature and time conditions will be proposed to reduce the loss in this compound that it is essential for human nutrition.

- **Omega 9:** Treatments at 150 °C presented similar contents in oleic acid to the control samples. While, at 187 °C its content decreases more than 2% compared with the content in control samples. Oleic acid concentration levels were found in a range of 2.78 to 4.38 g/100 g. Based on the results obtained, a statistical analysis was carried out where the response surface for the oleic acid (omega-9) content is observed in Figure 17.

The model that fits is the quadratic, according to the lack of fit test. The effect of the factors that resulted significant were the time, their interaction and the temperature squared (Eq. 14).

$$Omega\,9 = 2.91 - 0.17 Time + 1.23 \left(Temperature\right)^2 \tag{14}$$

In the last section of this manuscript, the appropriate temperature and time conditions will be proposed to reduce the loss in this compound.

Optimization of the thermal condition (temperature and time values) using the desirability function for the treatment of chia seed (whole and ground) to preserve omegas 3, 6 and 9 during heat treatment:

In this case, the composition of omegas 3, 6 and 9 of the control chia seed oil (31.10, 17.29 and 4.3 g/100 g oil, respectively) was utilized as the criteria to reach, in order (using the desirability function) to preserve the nutritional value of the samples. The predicted conditions were:

1. **Whole Chia Seeds:** Using a temperature of 155.1 °C during no more than 14.2 min (Figure 18a), the desirability function predicted the FAs composition of the Table 7.
2. **Groud Chia Seeds:** Using a temperature of 150.0 °C during no more than 10 min (Figure 18b), the desirability function predicted the FAs composition of the Table 7.

Figure 18. Conditions found using the desirability function to preserve omegas 3, 6 and 9 during heat treatment of whole and ground chia seed oil

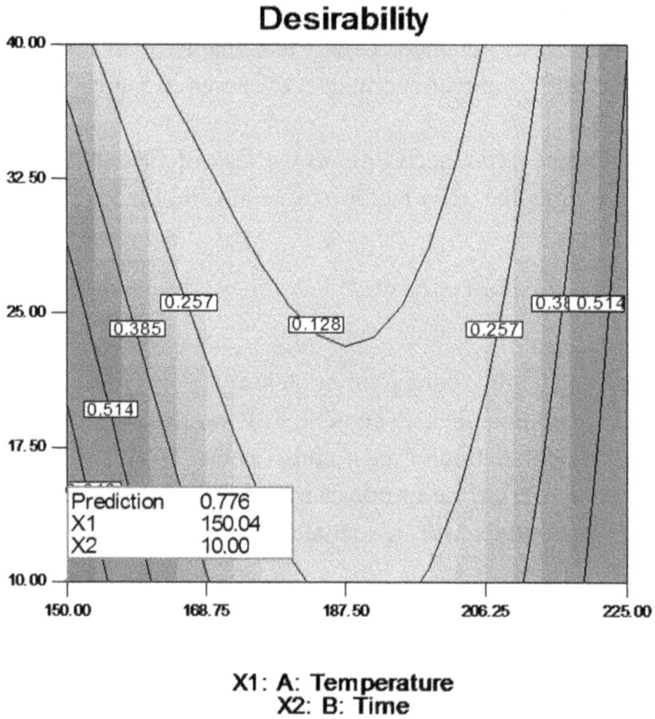

Table 7. Prediction of the content of omegas 3, 6 and 9 in chia seed oil using the desirability function and conditions described in a) and b).

FA's	Content (g/100g)	
	Whole Seed	Ground Seed
Omega 9	4.27	4.30
Omega 6	17.25	14.48
Omega 3	31.10	24.66

CONCLUSION

The results obtained from the measurement of different physicochemical parameters (iodine, acid, peroxides, saponification values and kreis test) indicate that the conditions optimized using the surface responses are 157.7 °C for 11.5 min for whole chia seeds and 152.38 °C for 10 min for ground chia seeds.

In the case of the composition of the omegas wants to be preserved 155.1 °C for 14.2 min as the ideal thermal treatment for whole chia seeds, and 150 °C for no more than 10 min for ground chia seeds, conditions that normally can be used in baking and roasting products.

Linolenic acid and linoleic acid were affected by thermal treatments in a meaningful way, while in the case of oleic acid this the effect was not so clear, because first it increase and then at very high temperatures decreases.

The ground chia seeds subjected to the various thermal treatments showed a decrease in the linolenic and linoleic acid content, but not oleic acid. In the optimization of the method it was found that these FAs are preserved to a greater extent at the temperature of 150 °C for 10 min. As a general conclusion, it was observed that the whole chia seeds maintained the physicochemical characteristics as well as the content of unsaturated fatty acids in comparison with the thermally treated ground chia seeds. Both in the whole chia seeds, and in the milled chia, when temperatures are close to 150 °C, they can maintain the physicochemical characteristics of the oil, as well as the nutritional value, in terms of the content of unsaturated fatty acids. Therefore, they can be used for baking and roasting products since these processes are carried out at similar temperatures.

ACKNOWLEDGMENTS

This research was supported by Dirección General de Asuntos del Personal Académico (DGAPA) with the project [PAPIIT IN221412] titled "Characterization and quantification of phenols, prolamines and fatty acids present in different varieties of chia" and by Secretaría de Investigación y Posgrado del Instituto Politécnico Nacional whit the projects [SIP: 20100619, and SIP: 20110350] "Caracterización de compuestos fenólicos presentes en la semilla y aceite de chía (Salvia hispánica L.), mediante electroforesis capilar", and "Capacidad antioxidante in vitro de la semilla de chía (Salvia Hispanica L.)".

REFERENCES

Akinoso R., Igbeka J., Olayanju T. (2006). Process optimization of oil expression from sesame seed (Sesamum indicum Linn). *Agricultural Engineering International,* (8), 1-7

Ayerza, R. (1995). Oil content and fatty acid composition of chia (Salvia hispanica L.) from five northwestern locations in Argentina. *Journal of the American Oil Chemists' Society, 72*(9), 1079–1081. doi:10.1007/BF02660727

Ayerza, R., & Coates, W. (2001). *Chia seeds: new source of omega-3 fatty acids, natural antioxidants, and dietetic fiber, Southwest center for natural products research & commercialization* (p. 3). Tucson, AZ: Office of Arid Lands Studies.

Ayerza, R., & Coates, W. (2004). Composition of chia (*Salvia hispanica*) grown in six tropical and subtropical ecosystems of South America. *Tropical Science, 44*(3), 131–135. doi:10.1002/ts.154

Badui, S. (2006). *Químca de los alimentos* (p. 285). México: Pearson Educación.

Biglar, M., Moghaddam, G., Sadeghi, N., Reza Oveisi, M., Jannat, B., Kaboli, A., ... Hajimahmoodi, M. (2012). Profiling of major fatty acids in different raw and roasted sesame sedes cultivars. *African Journal of Biotechnology, 11*(24), 6619–6623.

Cahill, J. P. (2003). Ethnobotany of chia, *Salvia hispanica* L. (Lamiaceae). *Economic Botany*, *57*(4), 604–618. doi:10.1663/0013-0001(2003)057[0604:EOCSHL]2.0.CO;2

Calder, P. C. (2008). Polyunsaturated fatty acids, inflammatory processes and inflammatory bowel diseases. *Molecular Nutrition & Food Research*, *52*(8), 885–897. doi:10.1002/mnfr.200700289 PMID:18504706

Chávez Torres, M. M. (2009). *Estudio de la composición de ácidos grasos insaturados (oleico, linoleico y linolénico) en aceites de oliva extra virgen de venta en el mercado mexicano, por cromatografía de gases. Tesis de Licenciatura en Ingeniería Química.* México: FES Cuautitlán, Universidad Nacional Autónoma de México.

Cooper, L. B. (2006). Obtención de un concentrado de ácidos grasos poliinsaturados a partir del aceite de chía (Salvia hispánica). Tesis de Licenciatura. Facultad de Química. UNAM. México. D.F.

De Meester, F., & Watson, R. R. (2008) Wild-type food in health promotion and disease prevention. In R. Ayerza (Ed.), Chapter 14, Chia as a new source of ω-3 fatty acids. Towa, NJ: Human Press Inc.

Decker, E. A., Chen, B., Panya, A., & Elias, R. J. (2010). Understanding mechanisms of oxidation and antioxidant activity. In E. A. Decker, B. Chen, A. Panya, & R. J. Elias (Eds.), *Oxidation in foods and beverages and antioxidant applications* (pp. 225–243). London, UK: Woodhead Publishing Series in Food Science, Technology, and Nutrition; doi:10.1533/9780857090447.2.225

Deiab, I. M., & Elbestawi, M. A. (2005). Experimental determination of the friction coefficient on the workpiece-fixture contact surface in workholding applications. *International Journal of Machine Tools & Manufacture*, *45*(6), 705–712. doi:10.1016/j.ijmachtools.2004.09.021

Deshpande, R. P., Chinnan, M. S., & McWatters, K. H. (2008). Optimization of a chocolate-flavored, peanut–soy beverage using response surface methodology (RSM) as applied to consumer acceptability data. *Lebensmittel-Wissenschaft + Technologie*, *41*(8), 1485–1492. doi:10.1016/j.lwt.2007.08.013

Design Expert 7 software, Stat Ease, Inc. (Minneapolis, MN). Retrieved from https://www.statease.com

Dutcosky, S. D., Grossmann, M. V. E., Silva, R. S., & Welsch, A. K. (2006). Combined sensory optimization of a prebiotic cereal product using multicomponent mixture experiments. *Food Chemistry*, *98*(4), 630–638. doi:10.1016/j.foodchem.2005.06.029

Eder, K. (1995). Gas chromatographic analysis of fatty acid methyl esters. *Journal of Chromatography. B, Biomedical Applications*, *671*(1-2), 113–131. doi:10.1016/0378-4347(95)00142-6 PMID:8520689

Flock, M. R., Harris, W. S., & Kris-Etherton, P. M. (2013). Long-chain omega-3 fatty acids: Time to establish a dietary reference intake. *Nutr. Rev. 71*, 692-707.

Frankel, E. N. (2005). *Lipid oxidation* (2nd ed.). Bridgewater, UK: The Oily Press. doi:10.1533/9780857097927

García, D., Viloria Matos, A., Belén, D., & Moreno-Álvarez, M. (2003). Características fisicoquímicas y composición de ácidos grasos del aceite crudo extraído de residuos de mora (*Rubus glaucus* Benth). *Grasas y Aceites*, *54*(3), 259–263.

Ghill, J. (2003). Ethnobotany of chia, *Salvia hispánica* L. (*Lamiaceae*). *Economic Botany*, *54*(4), 604–618.

Gow-Chin, Y., & Shyi-Liang, S. (1989). Oxidative Stability of sesame oil prepared from sesame seed with different roasting temperatures. *Food Chemistry*, *31*(3), 215–224. doi:10.1016/0308-8146(89)90059-9

Grompone, M. A., & Irigaray, B. (2013) Aceites de oliva: de la planta al consumidor. 2. Montevideo: INIA; Hemisferio Sur. 123-140.

Hamilton Beach 31199 User Manual, Glen Allen, VA, PN 840156300, p. 5.

Ilzarbe, L., Álvarez, M. J., Viles, E., & Tanco, M. (2008). Practical applications of Design of Experiments in the field of engineering. A bibliographical review. *Quality and Reliability Engineering International*, *24*(4), 417–428. doi:10.1002/qre.909

Ixtaina, V. Y., Nolasco, S. M., & Tomás, M. C. (2012). Oxidative stability of Chia (Salvia hispanica L.) Seed oil: Effect of antioxidants and storage conditions. *Journal of the American Oil Chemists' Society*, *89*(6), 1077–1090. doi:10.100711746-011-1990-x

Jing, P., Ruan, S.-Y., Dong, Y., Zhang, X.-G., Yue, J., Kan, J.-Q., ... Yu, L. L. (2011). Optimization of purification conditions of radish (Raphanus sativus L.) anthocyanin-rich extracts using chitosan. *Lebensmittel-Wissenschaft + Technologie*, *44*(10), 2097–2103. doi:10.1016/j.lwt.2011.06.003

Kira, R. S. (1991). Pearson's composition and analysis of foods. Ed. Longman Scientific & Technical. USA. 420-445.

Kris-Etherton, P. M., Harris, W. S., & Appel, L. J. (2002). Fish consumption, fish oil, omega-3 fatty acids, and cardiovascular disease. *Circulation*, *106*(21), 2747–2757. doi:10.1161/01.CIR.0000038493.65177.94 PMID:12438303

Kubow, S. (1992). Routes of formation and toxic consequences of lipid oxidation-products in foods. *Free Radical Biology & Medicine*, *12*(1), 63–81. doi:10.1016/0891-5849(92)90059-P PMID:1537572

Lee Seung, W., Jeung Min, K., Park Min, H., & Lee Soo, Y. (2010). Effects of roasting conditions of sesame seeds on the oxidative stability of pressed oil during thermal oxidation. *Food Chemistry*, *118*(3), 681–685. doi:10.1016/j.foodchem.2009.05.040

Luna-Pizarro, P., Lopes-Almeida, E., Sammán, N. C., & Chang, Y. K. (2013). Evaluation of whole chia (*Salvia hispanica* L.) flour and hydrogenated vegetable fat in pound cake. *Lebensmittel-Wissenschaft + Technologie*, *54*(1), 73–79. doi:10.1016/j.lwt.2013.04.017

Massart, D. L., Vandeginste, B. G. M., Buydens, L. M. C., De Jong, S., Lewi, P. J., & Smeyers-Verbeke. (1997). In Handbook of chemometrics and qualimetrics: part A. Amsterdam, The Netherlands: Elsevier.

Mendoza, C. (1542). *Edition of Francisco del Paso y Troncoso* (p. 1825). Museo Nacional de Arqueología, Historia y Etnología.

Morales, S. A. (2007). *Elaboración de un alimento funcional tipo palanqueta*. D.F., México: Tesis de Licenciatura. Escuela Nacional de Ciencias Biológicas. IPN.

Muthukumar, M., & Mohan, D. (2005). Studies on furan polymer concrete. *Journal of Polymer Research*, *12*(3), 231–241. doi:10.100710965-004-3206-7

Myers, R. H., Montgomery, C. D., & Anderson-Cook, C. M. (2016). *Response surface methodology: process and product optimization using designed experiments* (4th ed.). Wiley.

NMX-F-592-SCFI-2017. (2017). *Aceites y grasas vegetales o animales-aceite de semilla de chia-especificaciones.* Secretaría de Economía, Subsecretaría de Competitividad y Normatividad, Dirección General de Normas.

Horwitz, W. (1980). Official methods of analysis of the association of official analytical chemists. AOAC. Washington.

Portalagrario, B. (2014, Dec. 10) *Comercialización de aceite de chía (Salvia hispanica) como ingrediente alimentario.* Retrieved from http://www.besana.es/es/web/201412/comercializacion-aceite-chia

Reyes, E., Tecante, A., & Valdivia, M. (2008). Dietary fibre content and antioxidant activity of phenolic compounds present in Mexican chia (Salvia hispánica L.) seeds. *Food Chemistry, 107*(2), 656–663. doi:10.1016/j.foodchem.2007.08.062

Ricker, K. (2007). *FDA letter describing regulatory status of Salvia hispanica (chia).* US Dept. of Health and Human Services, Food and Drug Administration, Public Health Services.

Romero Aguilar, M. A., & Vargas-Martínez, M. G. (2016). *Identificación y clasificación de la chía (Salvia hispanica L.) silvestre y comercial mediante el perfil de proteínas obtenido por electroforesis capilar* (Undergraduate dissertation). Universidad Nacional Autónoma de México.

Sahagún, B. (1579). *Historia general de las cosas de la Nueva España* (p. 1982). Santa Fe, NM: Reprinted by School of American Research.

Schonning, A., Nayfeh, J., & Zarda, R. (2005). An integrated design and optimization environment for industrial large scaled systems. *Research in Engineering Design, 16*(1-2), 86–95. doi:10.100700163-005-0006-y

Sergeant, S., Rahbar, E., & Chilton, F. (2016). Gamma-linolenic acid, Dihommo-gamma linolenic, Eicosanoids and Inflammatory Processes. *European Journal of Pharmacology, 785*, 77–86. doi:10.1016/j.ejphar.2016.04.020 PMID:27083549

Silveira-Coelho, M., & Salas-Mellado, M. (2015). Effects of substituting chia (*Salvia hispanica* L.) flour or seeds for wheat flour on the quality of the bread. *Lebensmittel-Wissenschaft + Technologie, 60*(2), 729–736. doi:10.1016/j.lwt.2014.10.033

SINC Science News and Information Service. The Aztec and Maya Seed Arrives in Europe. (2013, Dec. 2) Retrieved from http://www.agenciasinc.es/en/News/The-Aztec-and-Maya-seed-arrives-in-Europe

Souza, M. F., Francisco, R. L., Sanchez, J. L., Guimarães-Inácio, A., Valderrama, P., Bona, E., ... Gonçalves, O. H. (2017). Fatty acids profile of chia oil-loaded Lipid microparticles. *Brazilian Journal of Chemical Engineering, 34*(3), 659–669. doi:10.1590/0104-6632.20170343s20150669

Statgraphics Centurion, V. I. Statgraphics Technologies, Inc., The Plains, Virginia. Retrieved from http://www.statgraphics.com

Supelco, Sigma-Aldrich Corporation subsidiary, St. Louis, MO. Retrieved from https://www.sigmaaldrich.com/catalog/product/supelco/crm18917?lang=es&region=MX

Tiwary, A. K., Puri, M., Sharma, D., & Barrow, C. J. (2012). Optimisation of novel method for the extraction of steviosides from Stevia rebaudiana leaves. *Food Chemistry*, *132*(3), 1113–1120. doi:10.1016/j.foodchem.2011.11.063 PMID:29243590

Tonon, R. V., Brabet, C., & Hubinger, M. D. (2008). Influence of process conditions on the physicochemical properties of acai (S Euterpe oleraceae Mart.) powder produced by spray drying. *Journal of Food Engineering*, *88*(3), 411–418. doi:10.1016/j.jfoodeng.2008.02.029

Vargas, M. G., Vander Heyden, Y., Maftouh, M., & Massart, D. L. (1999). Rapid development of the enantiomeric separation of beta-blockers by capillary electrophoresis using an experimental design approach. *Journal of Chromatography. A*, *855*(2), 681–693. doi:10.1016/S0021-9673(99)00701-3 PMID:10519104

Vargas-Martínez, M. G., & Beltrán-Orozco, C. (2011) Capítulo 7: Pharmacological activity of the chia seeds (Salvia hispanica L.). In R. M. Perez Gutierrez (Ed.), Edible plants in traditional medicine: Special Review Books, Research Singpost, Trivandrum, India.

Wadikar, D. D., Nanjappa, C., Premavalli, K. S., & Bawa, A. S. (2010). Development of ginger based ready-to-eat appetizers by response surface methodology. *Appetite*, *55*(1), 76–83. doi:10.1016/j.appet.2010.04.004 PMID:20417239

Young-Chul, L., Se-Wook, O., Jaehyn, C., & In-Hwan, K. (2004). Chemical composition and oxidative stability of safflower oil prepared from safflower seed roasted with different temperaturas. *Food Chemistry*, *84*(1), 1–6. doi:10.1016/S0308-8146(03)00158-4

KEY TERMS AND DEFINITIONS

Cultivar: Is a plant or group of plants that have been selected from a naturally occurring species and bred to enhance or maintain a particular set of desirable characteristics. These plants almost always originate from human cultivation, propagated through cutting or grafting, and often cannot be grown from seeds from the original plant.

Variety: Means a plant grouping, within a single botanical taxon of the lowest known rank, defined by the reproducible expression of its distinguishing and other genetic characteristics.

APPENDIX 1

Details of the Test for the Evaluation of Physicochemical Parameters of Chia Seed Oil

Saponification Value

The saponification value is defined as the weight in milligrams of potassium hydroxide necessary to saponify 1 g of fat. If the fat is acceptably pure, the method constitutes a system of classification of oils and fats since the saponification value is inversely related to the length of the fatty acids constituent of the glycerides of the fat (AOAC 920.160, 1980 and Kira 1991).

Weighed 1 g of oil with the addition of 12.5 mL of 0.5 M sodium hydroxide solution in alcohol, once obtained the mixture was connected in a condenser and boiled for 1 h, then it was cooled and 1 mL of phenolphthalein was added. Finally, it was titrated with a 0.5 N HCL solution. Calculating the saponification value with the following equation expressing the results in mg KOH/g of fat:

Saponification value = ((B-M)x(N))/P x 56.1

where:

B: is the volume of mL 0.5 N HCl required to title the target
M: is the volume of mL 0.5 N HCl required to titrate the sample
N: is the normality of the HCl solution
P: is the weight of the sample in grams
56.1: is the equivalent of potassium hydroxide (molar mass).

Iodine Value (Wijs)

It is defined as the weight of iodine adsorbed by the sample, that is, the chemical determination of the degree of unsaturation of a matrix containing diene or trienic bonds by the addition of iodine in the presence of the catalyst contained in the reagent of Wijs. The composition of Wijs solution is approximately 0.1 N of iodine monochloride in glacial acetic acid. It is also non-specifically related to the degree of unsaturation of fatty acids (AOAC 993.20, 1980).

The method consists in weighing about 0.5 g of oil which is diluted with 15 mL of chloroform, 25 mL of Wijs solution was added, allowing it to stand for 1 h, after which time 20 mL of 15% potassium iodide solution was added and 100 mL of distilled water. This is titrated with a 0.1 N solution of $Na_2S_2O_3$ until the appearance of a yellow color, 1 mL of starch indicator solution was added and the titration was continued until the the blue color disappear. At the same time a blank control was also prepared. It is expressed in grams of iodine adsorbed per 100 g of sample, to obtain the iodine value the following equation was used (Kira 1991).

Iodine value = ((B-S) x N) / (Sample mass (g)) x 12.69

Where:

B: is the volume (mL) of titrant solution used for the blank
S: is the volume (mL) of titrant solution used for the sample
N: is the normality of the sodium thiosulfate solution

Acid Value

It is defined as the weight of KOH in mg needed to neutralize the organic acids present in 1g of oil and it is a measure of the free fatty acids present in the fat or oil (AOAC 972.28, 1980 and Kira 1991).

1 g of sample was weighed by adding 15 ml of neutralized alcohol and stirred until the sample was dissolved. Finally, 2 mL of phenolphthalein was added, then this solution was titrated with 0.1 N sodium hydroxide. The results were expressed as % oleic acid using the following expression.

% of free fatty acids as oleic = (mL of sodium hydroxide x N x 28.2) / (weight of the sample).

Peroxide Value

The peroxide value is the amount (expressed in milliequivalents of active oxygen per kg of fat) of peroxides in the sample that cause the oxidation of potassium iodide. The released iodine is titrated with sodium thiosulfate solution. For the development of the method, 1 g of oil was weighed and 10 mL of the 3:2 solution of acetic acid and chloroform were added, then 0.16 mL of saturated potassium iodide solution was added, allowing to stand for 1 min, finished the time 10 mL of distilled water and 1 mL of the starch indicator solution were added. It was titrated with the 0.01N sodium thiosulfate solution until the blue color disappeared. A blank solution should be tested the same time as the samples (AOAC 965.33, 1980, and Kira 1991). The expression used to obtain the peroxide index values was as follows.

P.V. = ((M-T) x N x 1000) / (sample weight in grams)

Where:

T: is the volume (mL) of sodium thiosulfate used for the blank control
M: is the volume (mL) of sodium thiosulfate for the sample titration
N: is the normality of the sodium thiosulfate solution

Kreis Test

It is based on the production of red color due to the extremely sensitive reaction between phloroglucin and some compounds present in stale fats or oils: aldehydes and ketones (Kira, 1991).

0.5 g of sample is placed in test tubes by adding 1 mL of 30% tricloacetic acid in glacial acetic acid and 0.25 mL of Ploroglycine by covering them with aluminum. The sample is mixed by bubbling air for 2 or 3 s, then it is heated in a water bath at 45 °C for 15 min and after leaving the bath it is shaken again and 0.8 mL of ethyl alcohol is added. The absorbance of the sample at 545 nm is measured against a blank of reagents. The Kreis test is calculated as absorbance at 545 nm / g of fat.

Fundamentals of the Determination of Fatty Acid Profile by Gas Chromatography

It is based on the principles of gas chromatography and consists in the introduction to the injection port of methyl esters of fatty acids present in oils and grease using a flame ionization detector, with an injection temperature of 250 °C, detector temperature of 270 °C, and an oven temperature range of 100 to 250 °C. These are vaporized and transported by an inert gas that in this case is nitrogen through a capillary column (fused silica column, 30 mx 0.25 mm coated with a liquid phase of polyglycol on the base Carbowax-20M), and interact differentially with a distribution fluid (immobilized in the walls) that shows selective solubility with the components of the sample causing its separation (Eder, 1995).

The components that elute from the column pass one by one through the detector at different times, which generates a proportional electrical signal, which is transformed by the integrator into a graph of the signal obtained against time called the chromatogram.

APPENDIX 2

Evaluation of the Precision in the FA Measurements by GC

Eight solutions of 2 mg/ml of a mixture of the methyl esters of the FAs and its internal standard (heptadecanoate methyl ester, I.S.) were prepared. They were measured by GC under the conditions described in Appendix 1, number 2 and with the temperature gradient described in Figure 1. The relationship of areas (Eq. 15) was the measured response for the calculation of the precision of the chromatographic results, as can be observed in Table 8. The precision observed was acceptable since it was less than 5%.

$$Relationship\,of\,areas = \frac{area\,FA}{area\,I.S.} \tag{15}$$

Calibration Curves of the FAs Measured by GC

Six levels of concentration of each methyl fatty acid were measured in triplicate, using heptadecanoate methyl ester as the internal standard (I.S.), as it was explained in Table 8.

Table 8. Precision in the FA measurements by GC

Solution	Relationship of Areas		
	Methyl Oleate	**Methyl Linoleate**	**Methyl Linoleate**
1	1.06	1.16	1.11
2	1.18	1.27	1.21
3	1.14	1.22	1.14
4	1.09	1.17	1.09
5	1.07	1.18	1.11
6	1.20	1.25	1.15
7	1.14	1.21	1.14
8	1.09	1.17	1.10
Mean	1.12	1.20	1.13
S	0.05	0.04	0.04
%RSD	**4.60**	**3.47**	**3.39**

S= standard deviation, %RSD= percentage of relative standard deviation

Table 9. Equations of the calibration curves of the FAs measurements by GC

FA	Response	Slope	Intercept	r^2	Homoscedasticity
Oleic	Rel.Areas	0.3281±0.0113	0.014±0.0102	0.9848	Complies
Linoleic	Rel.Areas	0.7417±0.0181	-0.042±0.056	0.9905	Complies
Linolenic	Rel.Areas	0.7292±0.0147	0.010±0.1055	0.9935	Complies

r^2= determination coefficient, Rel.Areas=Relationship of areas (Eq.15)

Section 4
Industrial Applications

Chapter 11
Optimization of Injection Molding Process Parameters via Design of Experiments:
Medical Devices Environment

Karina Cecilia Arredondo-Soto
https://orcid.org/0000-0002-8929-7319
Universidad Autonóma de Baja California, Mexico

Arturo Realyvasquez-Vargas
https://orcid.org/0000-0003-2825-2595
Instituto Tecnológico de Tijuana, Tecnológico Nacional de Mexico, Mexico

Alejandro Jiménez-Zaragoza
Universidad Autónoma de Baja California, Mexico

ABSTRACT

The validation of processes is an innovative methodology where a process has been submitted to scrutiny to guarantee the products comply with the specifications of the company and the norms of the country of consumption. In the medical industry, this process is considered as a regulatory requirement, however, this also helps to improve quality, eliminate waste, and reduce costs, among other things. This chapter applied the methodology of process validation in a medical device company; the engineering tests were used in the company's clean room, using an injection molding machine and tests of several parameters were used even without being validated to know which are the best run and the best parameters, for its daily use. This project addresses the validation of the blood filter process by injection molding. The Design of Experiments applied was a 2K factorial design with central points where the replicas consisted of five of 45 in total for the three dimensions. The pieces are shared for those three dimensions that are being evaluated.

DOI: 10.4018/978-1-7998-1518-1.ch011

INTRODUCTION

In recent decades the consumption of medical devices has generated one of the fastest growing industries worldwide. This has resulted in organizations developing strategic manufacturing operations in different regions of the world in order to compete in a globalized market. So companies seek to generate high quality products, as well as low cost and timely deliveries. The main market for the consumption of medical devices is located in the United States of America. In this area, Mexico is considered a strategic point for the installation of manufacturing plants for medical products that are capable of supplying the main market. That is why, in recent years, Mexico registers the largest growth in the installation of medical device manufacturing plants in Latin America.

The Baja California region is characterized by concentrating 50% of the national total of the medical industry, consisting of 66 manufacturing plants that generate 42,000 jobs (Cervantes, 2013). Among the strengths that investors have identified to locate the manufacturing plants in this area are the following: geographic location, the availability of qualified human resources, the cost of competitive operation, logistics infrastructure, technological development and innovation. The last one is identified as a key element to ensure the manufacture of quality products that meet the specifications established by the company and country standards (Acevedo, Saldarriaga, & Garcia, 2015). The validation of processes is considered as an innovative methodology in the manufacturing processes of medical devices; by means of this methodology it is established that a process has been submitted to scrutiny in order to guarantee that the products comply with the specifications of the company and the norms of the country of consumption. In the medical industry this process is considered as a regulatory requirement, however, this also helps to improve quality, eliminate waste, and reduce costs, among other things.

The present study applied the methodology of process validation in a medical device company; the engineering tests were used in the company's clean room, using an injection molding machine and tests of several parameters were used even without being validated to know which are the best run and the best parameters, for its daily use. This project addresses the validation of the blood filter process by injection molding. Validation is part of the procedure to introduce a new injection molding process where the qualified specifications have to be validated so that it is controlled and can be constantly reproduced. The operational qualification was applied, which is where the parameters were defined to obtain the blood filter as a result.

It is important to optimize the processes of medical devices, but for that you need qualified personnel to carry out a stable process and have a point of reference to where the process was improved and something very important to reduce costs using advanced statistics and reduction the scrap indexes. With the methodology of design of experiments will give the solution to define the parameters of injection molding and know what parameters are those that are going to work to have the desired result of the blood filter. The Quality Policy, indicate to provide medical products and devices that meet all requirements, in order to obtain customer satisfaction through a process of continuous improvement, aiming zero rejections of the customer and deliveries on time. That is why advanced statistics is the important part to achieve the improvement of the process through the design of experiments, addressing issues of factorial designs with central points.

In recent years, the Design of Experiments (DoE) has been increasingly recognized as an essential tool for the validation of medical manufacturing processes (Mark, 1999), but in the experience of the authors, the adoption of these methods by the medical device sector has been limited. Research on British companies in a wide range of industries reported that only about 30% apply DoE techniques. This is due in part to the inadequate teaching of statistical methods such as DoE in university engineering courses (Alexander, 2000). The widespread teaching of statistical techniques as part of the Six Sigma training can improve the situation.

The Design of Experiments applied was a 2K factorial design with central points where the replicas consisted of five of the total 45 runs for the three dimensions, the pieces are shared for those three dimensions that are being evaluated. It is the blood breathing filter that is the internal diameter, the burr, and the distance of the angle of the filter that will be shown in the advanced chapters. In order to carry out this, it was necessary to have measurement equipment with great precision such as a Micro-vu, Caliper in order to have the results of the DoE that were selected by the Quality Department.

In the present work the general and specific objectives are defined to have the optimal parameters of the injection molding process for a finished product, such as the Blood Filter. Subsequently, the theoretical framework, procedures and documentation required for the validation of the blood filter are described. The theoretical framework presents research typologies defining the tools that were used to make the validations. Subsequently, the theoretical framework is analyzed and the DoE is applied using ANOVA studies and Factorial Designs to determine the ideal parameters for the injection molding process. The results are presented, analyzing if the Null Hypothesis is accepted or rejected, which were proposed at the beginning of the validation and compliance with the Process Capacity Index with the results of the experimental design tests. Finishing with the conclusion and sources of information where the research for the design of parameters based on DoE for plastic injection process was developed. In addition, the conclusions, method and effectiveness of the DoE Project are briefly described, in a total of 45 runs, obtaining a coefficient of determination in the results obtained in the tests, showing that there is significance.

BACKGROUND

Process validation can be defined as the documentation presenting the evidence for a given process has a high degree of security to consistently produce a product that satisfies predetermined specifications (Tobin, 2015). The application of this method is developed through the stages of Operation Qualification (OQ) and Process Qualification (P.Q.) (Mitu, 2012).

Tobin (2015) states a process validation is a regulatory requirement for the manufacture of medical devices. Just as it is also a requirement to obtain certification for certain international quality standards. Even when certification does not guarantee an organization complies with regulatory requirements, this can be a key to compliance. The operational qualification (OQ) shows that the equipment works consistently according to specifications under normal conditions, including: alarm tests, software function, extremes of operating ranges and machine consistency (Dixon, Eatock, Meenan, & Morgan, 2006).

The performance rating of the process or Process Qualification (P.Q) Regardless of the way in which the validation is subdivided, a protocol is written for each stage indicating the tests that will be carried out and the acceptance criteria (based on the product specification). The tests are carried out and a report is written based on the protocol. If the acceptance criteria are not met, it may be necessary to make changes in the process and repeat the qualification. It is often necessary to analyze historical data or conduct a prequalification study of the equipment before writing the validation documentation, that it is only possible to write a qualification protocol for a well-understood and stable process (Dixon, Eatock, Meenan, & Morgan, 2006).

The Six Sigma methodology has been adopted by companies in the medical device sector as an approach to quality improvement, especially where producing high volume products (for example, disposable). In 2004 six of the fifteen largest medical device companies worldwide promoted in fact they have adopted Six Sigma on their respective company websites (Dixon, Eatock, Meenan, & Morgan, 2006). The main objective of the Six Sigma quality method is to improve the defect rate and effectively address customer needs. Six Sigma, which are twelve deviations within tolerance, whose process conventionally presents 3.4 or fewer defects per million of opportunities.

Six Sigma

The Six Sigma method uses a series of established statistical tools, such as the DoE and the Statistical Process Control (SPC). The prevalence of Six Sigma training has led to a greater industrial awareness of DoE techniques. However, it should be noted that there are no regulations on the level of statistical training necessary to obtain Six Sigma qualifications, such as the Yellow Belt, Green Belt or Black Belt and the Master Black Belt. Therefore, this can lead to an inconsistency in its application throughout the sector (Pande, P., Mewman, R. & Cavanagh, R., 2000).

The Black Belt is an expert in the implementation of projects applying Lean Manufacturing and the Six Sigma tools. In addition, it generates and identifies improvement opportunities in the organization using the tools of Lean Manufacturing and Six Sigma. The Black Belt has the ability to train Yellow Belts and Green Belts, The difference of these two belts is that the Green Belt uses descriptive statistics where a set of data is collected that are represented by histograms, Pareto Diagrams; and the Black Belt uses advanced statistics for the optimization of the industrial process, trying to get control the process and reach the 3.4 or fewer defects per million that the Six Sigma system requires. The Master Black Belt is a member of the Directorate. Among other attributions, it is he who avoids conflicts of interest between departments, participates in the choice of projects and is formally informed of the progress of the projects.

Design of Experiments (DoE)

Tanco et al (2009) states the DoE is understood as a methodology to systematically apply statistics in experimentation processes. There are different DoE approaches such as the Shainin method, the Taguchi method developed in the 80's in the last century, as well as the classical method that served as the basis for the development of the two mentioned above (Martin, Elisabeth, & Lourdes, 2009). At present the classical method is the most used with the application of different techniques such as: Analysis of Variance (ANOVA), Factorial Design, Response Surface Methodology (RSM) among others.

A DoE approach allows efficient use of resources (staff time, machine time, materials, etc.), provides detailed analysis, provides information on reproducibility and errors, and provides a predictive capability (Montgomery, 1996). The DoE application reduces the size and therefore the cost of the Validation process. It is a regulatory requirement to perform sufficient trials to demonstrate the statistical significance of the results, and the DoE can assist in this procedural aspect (Dixon, Eatock, Meenan, & Morgan, 2006). The DoE covers the whole topic of professional thesis because with these designs they will give the function of being able to apply the methodology of DoE applying one of the types of factorial design that where operated in the company and it can be applied in the processes for injection molding in order to get the proper performance for delivery the product with good quality to the customer that meet the specifications they request.

The term DoE, refers to a set of statistical approaches for the design and analysis of experiments. The DoE is one of the key tools used in Six Sigma, and staff (for example, the Black Belt of Six Sigma) trained in its use can also apply the methodology to efficiently meet regulatory requirements. DoE is used to define the significance of one or more process factors in the chosen outputs.

Analysis of Variance

The Analysis of Variance (ANOVA) is the first method formally used in DoE, this was developed by Sir Ronald Fisher around 1920 ((Martin, Elisabeth, & Lourdes, 2009). This is used as a mathematical method to test the hypothesis that there is no significant difference between two or more sample or population means (Harish, U, & R, 2011). Kim (2017) states currently ANOVA is one of the most frequent statistical techniques used in the medical industry. When performing the ANOVA, a statistical test control is obtained where they compare with the critical value F to obtain the conclusion of the factors that are being studied are significant to the answers. It is also important if the factors are significant when interacting with them either in duple, triplet, etc. That is why it is very important to use this tool and be able to test hypotheses in order to accept the alternative hypothesis. With these ANOVA studies carried away out to validate the processes where they are critical points that must be taken care of when making the adjustments to the injection molding machines, where the parameters that can alter the specifications of the process should not be moved. With this method, profits are generated for the company by having a more stable and reliable process, generating stability in the quality department who ensures that the parameters established for the injection molding process are met.

Factorial Designs

A factorial experiment is defined as a statistical method allowing to investigate all possible combinations of the levels of the factors or conditions in each complete test (Medina & López, 2011). The objective is to identify the combination of levels of the winning treatment which has a maximum or a minimum where it obtains the result with its respective combination. Factorial designs are widely used in experiments in which several factors intervene to study the individual effect and together these on a variable of interest. The factors can be of a qualitative nature (machines, types of material, operator, the presence or absence of a previous operation, etc.) or of a quantitative type (temperature, humidity, speed, pressure, etc.). The most important thing is that these factors are significant where it is studied if the factor influences the result obtained. To study the way in which each factor influences the response variable, it is necessary to choose at least two levels of proof for each of them. With the complete factorial design, all the possible combinations that can be formed with the levels of the factors to be investigated are randomly run (Gutierrez & De la Vara, 2008).

At the same time these factorial designs are the most recommended because it is carried out in the validations of the medical industry where an operational qualification better known as the OQ (Operational Qualification, OQ for its acronym in English: Operational Qualification) is required. In order to complete this task it is necessary to asks for the validation of the process under which parameters will be working in the process. This tool is the most indicated to be able to define the required parameters and be able to validate them in said document and form part of the procedure of the operation that is being carried out. When, in a factorial design 2 at k, the factors admit an intermediate test level, it is advisable to implement an additional treatment formed by the combination of the intermediate or average level of all the factors. This combination is known as the central point.

There are two reasons why it is desirable to run the center point with a certain number of replicas. The first is to obtain additional degrees of freedom for the error in the ANOVA table, without affecting the balance in the estimate or the effects of interest. The convenience of interpreting ANOVA with at least 8 degrees of freedom in error has already been mentioned, a condition that is sometimes difficult to meet, for example, when for economic reasons the experiment is run without sufficient replicas. An example is running the factorial design 23 without replicas: there are 7 total degrees of freedom, so it is impossible to construct an analysis of variance with sufficient degrees of freedom for the error. Four or five repetitions at the center would add that amount of degrees of freedom to the error, in addition to providing a pure estimator independent of the estimated effects of the variance at that point. It is more feasible to obtain four runs at the center, than to repeat the treatments of the entire experiment (Gutierrez & De la Vara, 2008).

Regulations

The regulatory framework is based on the Food and Drugs Administration (FDA) that adds the Procedures for mandatory medical devices and in ISO 13485. Another criterion are the requirements specified by the regional or national authority, considering the useful life of the device for the conservation of documentation related to the manufacture and testing of the device not less than two years. For each type or model of medical device, the organization must establish and maintain a file that contains or identifies the documents that define the specifications. The responsibilities and authorities are defined, documented and communicated within the organization. Those responsible for monitoring and reporting adverse events should be defined as well as all stages of design, production and post-production. From the complete manufacturing process and if applicable, it must include the installation and the service (processes, specifications of inputs, controls, tests, for example). The applicable requirements of the organization's Quality Assurance System (QAS) considering the size of the organization and activities carried out, complexity and interaction of the processes and the competence of the personnel.

PROJECT DEVELOPMENT

This project addressed the application of the methodology for the validation of a mold (with six cavities) is used in the manufacture of a medical device, specifically a blood filter. The medical area has 26 machines for injection molding of the brand Sumitomo, Toyo, Toshiba. The production area has 100 operators. The general objective of the project was to define and apply the validation plan and the required tests by means of DoE, executed by qualified personnel, to successfully perform an Operation Qualification, determining the operation parameters for the injection molding machine in the production of blood filters. The questions to answer were:

- What are the parameters established within the injection molding process?
- What methodology was used for the DoE?
- How are the factors defined to perform the DoE?

- What are the defects that the process can have?
- What factors were selected to define the result of the experiment?

All the data obtained was re-calibrated using calibrated and controlled measuring equipment, such as the Micro-vu and Caliper, defining critical dimensions which are of higher priority to validate the process. The statistical model presented new studies in which the client's specifications are fulfilled in the validation of the operational qualification based on the capture of the data (30 samples) in Table 1.

When the results of tables 1 and 2 were analyzed, an operational qualification chart was made where the specification process complies with the requested indicators, Cpk≥1.33.

Table 1. Results of Dimensions OQ Low Shots

	Dim 1	Dim 2	Dim 3	Internal Burr	External Burr	Cosmetic	Result
USL	0.925	2.191	0.863	0.015	0.03		
Nominal	0.920	2.187	0.858				
LSL	0.915	2.183	0.853	0	0		
Equipment	Micro-Vu	Micro-Vu	Caliper	Micro-Vu	Micro-Vu		
Method	MD01-054	MD01-054	MD01-054	MD01-054	MD01-003		
				Shot 1			
9	Closed						
10	0.919	2.188	0.859	0	0.006	Pass	Pass
11	0.919	2.186	0.858	0	0.012	Pass	Pass
12	Closed						
13	0.920	2.187	0.859	0	0.007	Pass	Pass
14	0.921	2.185	0.860	0	0.002	Pass	Pass
15	0.918	2.186	0.861	0	0.000	Pass	Pass
16	0.922	2.187	0.862	0	0.004	Pass	Pass
				Shot 2			
9	Closed						
10	0.920	2.188	0.859	0	0.006	Pass	Pass
11	0.919	2.186	0.857	0	0.010	Pass	Pass
12	Closed				0.008		
13	0.920	2.187	0.860	0	0.002	Pass	Pass
14	0.920	2.186	0.860	0	0.000	Pass	Pass
15	0.918	2.186	0.861	0	0.004	Pass	Pass
16	0.922	2.188	0.863	0		Pass	Pass
				Shot 3			
9	Closed						
10	0.920	2.188	0.858	0	0.006	Pass	Pass
11	0.919	2.185	0.857	0	0.0125	Pass	Pass
12	Closed						

continues on following page

Table 1. Continued

	Dim 1	Dim 2	Dim 3	Internal Burr	External Burr	Cosmetic	Result
13	0.920	2.188	0.859	0	0.008	Pass	Pass
14	0.921	2.187	0.861	0	0.002	Pass	Pass
15	0.918	2.187	0.862	0	0.000	Pass	Pass
16	0.924	2.188	0.863	0	0.004	Pass	Pass
Shot 4							
9	Closed						
10	0.919	2.187	0.859	0	0.010	Pass	Pass
11	0.919	2.186	0.857	0	0.003	Pass	Pass
12	Closed						
13	0.920	2.187	0.861	0	0.002	Pass	Pass
14	0.918	2.186	0.861	0	0.000	Pass	Pass
15	0.918	2.187	0.862	0	0.002	Pass	Pass
16	0.922	2.187	0.863	0	0.002	Pass	Pass
Shot 5							
9	Closed						
10	0.92	2.189	0.859	0	0.009	Pass	Pass
11	0.92	2.187	0.858	0	0.002	Pass	Pass
12	Closed						
13	0.921	2.187	0.860	0	0.009	Pass	Pass
14	0.921	2.186	0.861	0	0.003	Pass	Pass
15	0.920	2.187	0.860	0	0.000	Pass	Pass
16	0.923	2.188	0.863	0	0.003	Pass	Pass

Table 2. Results of Dimensions OQ High Shots

	Dim 1	Dim 2	Dim 3	Internal Burr	External Burr	Cosmetic	Result
USL	0.925	2.191	0.863	0.015	0.03		
Nominal	0.920	2.187	0.858				
LSL	0.915	2.183	0.853	0	0		
Equipment	Micro-Vu	Micro-Vu	Caliper	Micro-Vu	Micro-Vu		
Method	MD01-054	MD01-054	MD01-054	MD01-054	MD01-003		
Shot 1							
9	Closed						
10	0.922	2.186	0.860	0	0.010	Pass	Pass
11	0.920	2.187	0.857	0	0.011	Pass	Pass
12	Closed						
13	0.922	2.186	0.860	0	0.007	Pass	Pass
14	0.921	2.188	0.861	0	0.009	Pass	Pass

continues on following page

Table 2. Continued

	Dim 1	Dim 2	Dim 3	Internal Burr	External Burr	Cosmetic	Result
15	0.922	2.187	0.862	0	0.003	Pass	Pass
16	0.923	2.187	0.863	0	0.008	Pass	Pass
			Shot 2				
9	Closed						
10	0.920	2.188	0.859	0	0.006	Pass	Pass
11	0.919	2.186	0.857	0	0.010	Pass	Pass
12	Closed				0.008		
13	0.922	2.187	0.860	0	0.002	Pass	Pass
14	0.922	2.186	0.860	0	0.000	Pass	Pass
15	0.920	2.187	0.861	0	0.004	Pass	Pass
16	0.925	2.188	0.863	0		Pass	Pass
			Shot 3				
9	Closed						
10	0.920	2.185	0.860	0	0.009	Pass	Pass
11	0.920	2.184	0.858	0	0.009	Pass	Pass
12	Closed						
13	0.922	2.185	0.860	0	0.006	Pass	Pass
14	0.921	2.188	0.861	0	0.009	Pass	Pass
15	0.920	2.188	0.862	0	0.004	Pass	Pass
16	0.924	2.188	0.863	0	0.006	Pass	Pass
			Shot 4				
9	Closed						
10	0.920	2.188	0.860	0	0.009	Pass	Pass
11	0.920	2.186	0.858	0	0.008	Pass	Pass
12	Closed						
13	0.922	2.185	0.860	0	0.007	Pass	Pass
14	0.922	2.187	0.861	0	0.009	Pass	Pass
15	0.920	2.187	0.862	0	0.003	Pass	Pass
16	0.924	2.187	0.863	0	0.005	Pass	Pass
			Shot 5				
9	Closed						
10	0.920	2.188	0.860	0	0.011	Pass	Pass
11	0.921	2.186	0.858	0	0.014	Pass	Pass
12	Closed						
13	0.922	2.187	0.861	0	0.006	Pass	Pass
14	0.921	2.187	0.862	0	0.008	Pass	Pass
15	0.920	2.187	0.862	0	0.003	Pass	Pass
16	0.924	2.187	0.864	0	0.004	Pass	Pass

DoE 2^k With Central Points

The format for the design of experiments is presented, which consists of the following factors: Temperature of the mold measured in degrees Fahrenheit (° F); Plastification zone (H1), Hold Press (Psi) using nine different runs, using high, low and nominal shots, as shown in the table 3. The results of the DOE consisted in reviewing the six cavities and three dimensions were revised in order to comply with the specifications of the piece.

ANOVA Dimension 1, Cavity 1

The ANOVA study of dimension 1 (Cavity 1) is presented where it explains that the factors are significant when using the p-value that is less 0.05. In the interactions there are factors when interacting with another factor, they are not significant as is the case of Mold Temperature and the Plasticity Zone; also Plasticity Zone and Sustaining Pressure.

ANOVA Dimension 2, Cavity 1

The ANOVA study of dimension 2, Cavity 1 is presented where the three factors are significant when using the p-value, which is less than 0.05. In the interactions, the factors that when interacting with another factor are all significant according to figure 2.

ANOVA Dimension 3, Cavity 1

The ANOVA study of dimension 3, Cavity 1 is presented where the three factors are significant when using the p-value, which is less than 0.05. In the interactions, the factors that when interacting with another factor, they are all significant according to figure 3.

The six cavities were analyzed for the three different dimensions, but for practical considerations only the ANOVA of cavity 1 is presented for the three dimensions, however the same methodology was followed with the other cavities.

Table 3. Format A

TAG	Mold Set Temperature (°F Water Input Line)	Plastification Zone (H1)	Hold Press (Psi)
A	75	380	55
B	75	380	65
C	75	400	55
D	75	400	65
E	85	390	60
F	95	380	55
G	95	380	65
H	95	400	55
I	95	400	65

Figure 1. Factorial Regression: Dimension 1

Analysis of Variance

Source	DF	Adj SS	Adj MS	F-Value	P-Value
Model	8	0.000392	0.000049	31.62	0.000
Linear	3	0.000302	0.000101	64.89	0.000
T Mold	1	0.000089	0.000089	57.09	0.000
T(H1)	1	0.000174	0.000174	112.43	0.000
Hold Press	1	0.000039	0.000039	25.16	0.000
2-Way Interactions	3	0.000019	0.000006	3.98	0.015
T Mold *T(H1)	1	0.000003	0.000003	2.13	0.153
T Mold *Hold Press	1	0.000014	0.000014	8.91	0.005
T(H1) *Hold Press	1	0.000001	0.000001	0.91	0.347
3-Way Interactions	1	0.000047	0.000047	30.51	0.000
T Mold *T(H1) *Hold Press	1	0.000047	0.000047	30.51	0.000
Curvature	1	0.000024	0.000024	15.80	0.000
Error	36	0.000056	0.000002		
Total	44	0.000448			

Model Summary

S	R-sq.	R-sq.(adj)	R-sq.(prod)
0.0012451	87.54%	84.77%	80.53%

Figure 2. Factorial Regression: Dimension 2

Analysis of Variance

Source	DF	Adj SS	Adj MS	F-Value	P-Value
Model	8	0.000141	0.000018	23.27	0.000
Linear	3	0.000092	0.000031	40.29	0.000
T Mold	1	0.000025	0.000025	32.64	0.000
T(H1)	1	0.000050	0.000050	65.14	0.000
Hold Press	1	0.000018	0.000018	23.10	0.000
2-Way Interactions	3	0.000020	0.000007	8.98	0.000
T Mold *T(H1)	1	0.000007	0.000007	8.96	0.005
T Mold *Hold Press	1	0.000008	0.000008	10.07	0.003
T(H1) *Hold Press	1	0.000006	0.000006	7.90	0.008
3-Way Interactions	1	0.000025	0.000025	32.64	0.000
T Mold *T(H1) *Hold Press	1	0.000025	0.000025	32.64	0.000
Curvature	1	0.000004	0.000004	5.70	0.022
Error	36	0.000027	0.000001		
Total	44	0.000169			

Model Summary

S	R-sq.	R-sq.(adj)	R-sq. (prod)
0.0008718	83.80%	80.19%	74.68%

Figure 3. Factorial Regression: Dimension 3

Analysis of Variance

Source	DF	Adj SS	Adj MS	F-Value	P-Value
Model	8	0.000416	0.000052	60.84	0.000
Linear	3	0.000285	0.000095	111.25	0.000
T Mold	1	0.000117	0.000117	137.33	0.000
T(H1)	1	0.000079	0.000079	92.11	0.000
Hold Press	1	0.000089	0.000089	104.31	0.000
2-Way Interactions	3	0.000087	0.000029	33.83	0.000
T Mold*T(H1)	1	0.000053	0.000053	62.20	0.000
T Mold*Hold Press	1	0.000018	0.000018	20.55	0.000
T(H1) *Hold Press	1	0.000016	0.000016	18.73	0.000
3-Way Interactions	1	0.000039	0.000039	46.13	0.000
T Mold*T(H1) *Hold Press	1	0.000039	0.000039	46.13	0.000
Curvature	1	0.000005	0.000005	5.33	0.027
Error	36	0.000031	0.000001		
Total	44	0.000446			

Model Summary

S	R-sq.	R-sq.(adj)	R-sq.(prod)
0.0009242	93.11%	91.58%	89.24%

Mean of the Results in the Six Cavities

With all the data the average of the six cavities of the three dimensions of the blood filter was presented in order to solve the ANOVA study, and to know that the three factors are significant factors as shown in table 4.

Table 5 presents the factorial designs with codified values (-1,0,1), where the results for the three dimensions are presented.

ANOVA Dimension 1 Mean for Six Cavities

The ANOVA study of dimension 1 (mean) is presented, the three factors are significant when using the p-value which is less 0.05. In the interactions there are factors that when interacting with another factor are not significant is the case of Mold Temperature and the Plasticity Zone; Molding temperature and holding pressure; also Plasticity Zone and the Sustaining Pressure (see figure 4).

ANOVA Dimension 2 Mean for Six Cavities

The ANOVA study of dimension 2 (mean) is presented; the three factors are significant when using the p-value less than 0.05. In the interactions there are factors that when interacting with another factor are not significant, it´s the case of Mold Temperature and the Plasticity Zone; Molding temperature and holding pressure; also Plasticity Zone and the Sustaining Pressure (see figure 5).

Table 4. Results of the dimensions of the six cavities

No.	Factor 1	Factor 2	Factor 3	No.	Factor 1	Factor 2	Factor 3
1	0.924833	2.19017	0.862167	24	0.920000	2.18692	0.857817
2	0.924833	2.19033	0.861500	25	0.919917	2.18763	0.859100
3	0.924899	2.19033	0.862500	26	0.999833	2.18667	0.857400
4	0.924667	2.19017	0.861833	27	0.919667	2.18613	0.857500
5	0.924667	2.18983	0.862333	28	0.919667	2.18742	0.858033
6	0.924000	2.18500	0.859083	29	0.919600	2.18752	0.859133
7	0.923000	2.18400	0.859083	30	0.919583	2.18830	0.859467
8	0.922833	2.18533	0.859167	31	0.919433	2.18780	0.858767
9	0.922500	2.18700	0.861000	32	0.919100	2.18808	0.859083
10	0.922167	2.18600	0.860817	33	0.919100	2.18600	0.858333
11	0.921667	2.18742	0.857433	34	0.918833	2.18517	0.858167
12	0.921500	2.18600	0.860017	35	0.918833	2.18758	0.859500
13	0.921500	2.18700	0.862000	36	0.918667	2.18542	0.857683
14	0.921333	2.18717	0.858667	37	0.918500	2.18575	0.857933
15	0.921000	2.18700	0.857000	38	0.918167	2.18578	0.859350
16	0.921000	2.18600	0.860083	39	0.917667	2.18400	0.858833
17	0.920667	2.18808	0.859200	40	0.917667	2.18500	0.858167
18	0.920667	2.18717	0.859083	41	0.914833	2.18267	0.853167
19	0.920333	2.18700	0.858700	42	0.914500	2.18283	0.853667
20	0.920333	2.18517	0.858033	43	0.914333	2.18300	0.853333
21	0.920333	2.18633	0.856067	44	0.914167	2.18325	0.853333
22	0.920000	2.18667	0.856750	45	0.913917	2.18333	0.853167
23	0.920000	2.18700	0.856933				

ANOVA Dimension 3 Mean for Six Cavities

The ANOVA study of dimension 3 (mean) is presented; the three factors are significant when using the p-value which is less 0.05. In the interactions there are factors when interacting with another factor are not significant, it is the case of Mold Temperature and the Plasticity Zone; Molding temperature and holding pressure; also Plasticity Zone and the Sustaining Pressure (see figure 6).

In the cube graphs it is shown the best encoded run: Mold Temperature (-1), Plasticity Zone (1) and Sustaining Pressure (-1) for the three dimensions of the plane because it approaches the desired specification Dimension 1=0.92, Dimension 2=2.187, Dimension 3=0.858.

Result of the Selection of the Optimal Run

The run C was selected as the optimal to work in the injection molding process (-1, 1, -1) the results are presented in the table 6

Table 5. Results of the 45 complete runs

No.	A	B	C	Answer	No.	A	B	C	Answer
1	-1	-1	-1	0.91450000	24	1	1	1	0.92466667
2	1	-1	-1	0.91900000	25	-1	-1	-1	0.91391667
3	-1	1	-1	0.92066700	26	1	-1	-1	0.92033333
4	1	1	-1	0.92216667	27	-1	1	-1	0.92150000
5	-1	-1	1	0.91966667	28	1	1	-1	0.92150000
6	1	-1	1	0.92100000	29	-1	-1	1	0.91816667
7	-1	1	1	0.92100000	30	1	-1	1	0.92000000
8	1	1	1	0.92400000	31	-1	1	1	0.92066667
9	-1	-1	-1	0.91416667	32	1	1	1	0.92483333
10	1	-1	-1	0.91766667	33	-1	-1	-1	0.91483333
11	-1	1	-1	0.91966667	34	1	-1	-1	0.92250000
12	1	1	-1	0.92300000	35	-1	1	-1	0.92000000
13	-1	-1	1	0.91883333	36	1	1	-1	0.92150000
14	1	-1	1	0.92000000	37	-1	-1	1	0.91850000
15	-1	1	1	0.9203333	38	1	-1	1	0.9203333
16	1	1	1	0.9248333	39	-1	1	1	0.9216667
17	-1	-1	-1	0.9143333	40	0	0	0	0.9246667
18	1	-1	-1	0.9176667	41	0	0	0	0.9191000
19	-1	1	-1	0.9188333	42	0	0	0	0.9199166
20	1	1	-1	0.9228333	43	0	0	0	0.9196000
21	-1	-1	1	0.9186667	44	0	0	0	0.9194333
22	1	-1	1	0.9198333	45	0	0	0	0.9195833
23	-1	1	1	0.9213333					

Table 6. Optimal run of the injection molding process.

TAG	Mold Set Temp (°F Water Input Line)	Plastification Zone (H1)	Hold Press (Psi)
A	75	380	55
B	75	380	65
C	75	400	55
D	75	400	65
E	85	390	60
F	95	380	55
G	95	380	65
H	95	400	55
I	95	400	65

Figure 4. Factorial Regression dim 1 six cavities

Factorial Regression: Mean 1 Versus T Mold, T(H1), Hold Press, Center Pt

Analysis of Variance

Source	DF	Adj SS	Adj MS	F-Value	P-Value
Model	8	0.000305	0.000038	50.35	0.000
Linear	3	0.000282	0.000094	123.90	0.000
T Mold	1	0.000093	0.000093	123.07	0.000
T(H1)	1	0.000143	0.000143	189.27	0.000
Hold Press	1	0.000045	0.000045	59.35	0.000
2-Way Interactions	3	0.000006	0.000002	2.45	0.079
T Mold *T(H1)	1	0.000000	0.000000	0.64	0.428
T Mold *Hold Press	1	0.000003	0.000003	3.58	0.067
T(H1) *Hold Press	1	0.000002	0.000002	3.14	0.085
3-Way Interactions	1	0.000017	0.000017	21.87	0.000
T Mold *T(H1) *Hold Press	1	0.000017	0.000017	21.87	0.000
Curvature	1	0.000001	0.000001	1.86	0.181
Error	36	0.000027	0.000001		
Total	44	0.000333			

Model Summary

S	R-sq	R-sq(adj)	R-sq(pred)
0.0008706	91.80%	89.97%	87.18%

Figure 5. Factorial Regression dim 2 six cavities

Factorial Regression: Mean 2 Versus T Mold, T(H1), Hold Press, Center Pt

Analysis of Variance

Source	DF	Adj SS	Adj MS	F-Value	P-Value
Model	8	0.000157	0.000020	50.73	0.000
Linear	3	0.000110	0.000037	94.80	0.000
T Mold	1	0.000014	0.000014	37.01	0.000
T(H1)	1	0.000055	0.000055	143.43	0.000
Hold Press	1	0.000040	0.000040	103.95	0.000
2-Way Interactions	3	0.000011	0.000004	9.72	0.000
T Mold *T(H1)	1	0.000001	0.000001	2.14	0.152
T Mold *Hold Press	1	0.000010	0.000010	25.67	0.000
T(H1) *Hold Press	1	0.000001	0.000001	1.36	0.251
3-Way Interactions	1	0.000025	0.000025	65.91	0.000
T Mold *T(H1) *Hold Press	1	0.000025	0.000025	65.91	0.000
Curvature	1	0.000010	0.000010	26.39	0.000
Error	36	0.000014	0.000000		
Total	44	0.000171			

Model Summary

S	R-sq	R-sq(adj)	R-sq(pred)
0.0006216	91.85%	90.04%	87.27%

Figure 6. Factorial Regression dim 3 six cavities

Factorial Regression: Mean 3 Versus T Mold, T(H1), Hold Press, Center Pt

Analysis of Variance

Source	DF	Adj SS	Adj MS	F-Value	P-Value
Model	8	0.000233	0.000029	57.42	0.000
Linear	3	0.000178	0.000059	116.57	0.000
T Mold	1	0.000043	0.000043	85.04	0.000
T(H1)	1	0.000112	0.000112	220.07	0.000
Hold Press	1	0.000023	0.000023	44.59	0.000
2-Way Interactions	3	0.000021	0.000007	13.54	0.000
T Mold *T(H1)	1	0.000000	0.000000	0.39	0.538
T Mold *Hold Press	1	0.000020	0.000020	40.17	0.000
T(H1) *Hold Press	1	0.000000	0.000000	0.06	0.814
3-Way Interactions	1	0.000033	0.000033	64.26	0.000
T Mold *T(H1) *Hold Press	1	0.000033	0.000033	64.26	0.000
Curvature	1	0.000002	0.000002	4.78	0.035
Error	36	0.000018	0.000001		
Total	44	0.000252			

Model Summary

S	R-sq.	R-sq.(adj)	R-sq.(pred)
0.0007127	92.73%	91.12%	88.64%

CONCLUSION

1. The method of Factorial Design 2^k with central points was very efficient because it performed fewer tests and nominal encodings were used.
2. We used a total of 45 runs with the factorial design 2^k with centered point that in other designs more tests could be used.
3. The three factors represent more than 90% (coefficient of determination) of the results obtained from the tests.
4. The three factors are significant; but the most significant of the three factors is the Plasticization Zone, with the Interaction Effects showing that there is significance.
5. The cube graph indicated the parameters approached the target or nominal value were with the run (-1, 1, -1).

REFERENCES

Alexander, K.Clarkson, P. J. (2000). Good design practice for medical devices and equipment. *Journal of Medical Engineering & Technology*, *24*(1), 5–13. doi:10.1080/030919000293987 PMID:10849882

Dixon, D., Eatock, J., Meenan, B., & Morgan, M. (2006). Application of Design of Experiment (DOE). Techniques to process validation in medical device manufacture. Newtownabbey, Irlanda del Norte, Irlanda: Journal of Validation Technology.

Gutierrez, H., & De la Vara, R. (2008). Análisis y diseño de experimentos. In *Análisis y diseño de experimentos (pág. 195)*. D.F., México: McGraw-Hill Interamericana.

Martin, T., Elisabeth, V., & Lourdes, P. (2009). Diferentes Enfoques de Diseño de Experimentos. Memoria de trabajos de difusión cientifica y técnica, 19 - 37.

Medina, P. D., & López, A. M. (2011). Análisis crítico de Diseño factorial 2k sobre Casos Aplicados. *Sciences et Techniques (Paris)*, 101–106.

Mitu, R. K. (2012). *Process validation SOP template*. ROFRI MED.

Montgomery, D. C. (1997). *Design and analysis of experiments*. New York: John Wiley & Sons.

Pande, P., Mewman, R., & Cavanagh, R. (2000). *The six sigma way*. McGraw-Hill.

Tobin, E. (2015). *A guide to process validation for medical devices*. Solo Validation Resources Limited.

KEY TERMS AND DEFINITIONS

ANOVA: Analysis Of Variance.

DOE: Design Of Experiments.

OQ: Operational Qualification.

Revalidation: It is the process validation of a device for a second time due to the product specification, the equipment and the parameters of the processor the materials have changed.

Chapter 12
The Design of Experiments as a Methodological Framework for the Improvement of Manufacturing Processes

Sukey Nakasima-López
Universidad Autónoma de Baja California, Mexico

Mydory Nakasima-López
Universidad Autónoma de Baja California, Mexico

Karla Frida Madrigal Estrada
Universidad Autónoma de Baja California, Mexico

Erika Beltrán Salomón
Universidad Autónoma de Baja California, Mexico

ABSTRACT

Industries seek changes in manufacturing processes by designing or redesigning them, to improve the quality of products, reduce costs and cycle times, change materials, modify methods, design innovative products, among others. Facing these demands requires a powerful methodological framework known as Design of Experiments. Most of literature focuses on the application of these techniques in the areas of statistics and quality. However, the variety of problems facing engineers in industry is wide and includes different levels of complexity, ranging from the design of new products, improvement of design, maintenance, control and improvement of manufacturing processes, maintenance and repair of products, among others. This chapter provides the reader different applications of this methodology in industry, to highlight the importance and benefits of knowing and applying these techniques. It will present the application of this methodology in a general way and finally, it will discuss different case studies that use this methodology in industry.

DOI: 10.4018/978-1-7998-1518-1.ch012

INTRODUCTION

The manufacturing industry faces constant challenges, mainly in the manufacture of innovative and high-quality products, due to the current conditions of the global market, and the high competitiveness of the organizations. Based on the concept of manufacturing, which is defined as: *"the transformation of materials and information into goods for the human needs satisfaction"* (Chryssolouris, 2006), the processes for product development are increasingly complex since the products become more versatile to meet the needs of massive customization (Ong, Yuan, & Nee, 2008). Therefore, it is necessary to make important changes to the manufacturing processes by designing or re-designing them, in order to improve the quality of products, reduce costs and cycle times, change materials, modify methods, design innovative products and highly competitive in the market, among others (Gutiérres Pulido & De la Vara Salazar, 2012).

Different manufacturing phases require experimentation in order to be able to face critical factors such as optimization of available resources, of the execution time and the process performance, as well as variability control, to assure the quality of the product and/or service produced (Tanco, Viles, Ilzarbe, & Álvarez, 2007). A powerful methodological framework that allows to create a multidimensional design space and its interactions between factors that might affect the performance and quality of the process is known as Design of Experiments (DoE) (Davim, 2012). This framework provides us with objective evidence that responds to diverse questions in the different stages of process development, clarifying uncertain aspects that lead us either to the improvements implementation or problem solutions with different complexity levels (Gutiérres Pulido & De la Vara Salazar, 2012). DoE can be understood as a practical methodology, a tool that is composed of a set of techniques with well-defined application phases, which are described below (Murray et al., 2016; Weissman & Anderson, 2015):

1. Planning, this phase begins with the recognition of a need and/or problem, as well as the definition of the objectives to be achieved, requires a good level of process knowledge, or an evaluation and comprehensive analysis that allows to identify the factors (variables) involved in the process, as well as, define the expected responses and list all possible interactions between them. This exercise contributes to the construction of an environment where the culture of decision-making is fostered based on experimental design techniques, rather than those based on trial and error, providing greater certainty to the decision making process.
2. Design, in this phase the most appropriate DoE is chosen, based mainly on the number of factors that must be studied, the different levels of detail, as well as the objectives to be achieved.
3. Conducting, refers to the experimentation, execution and the evaluation of the results obtained.
4. Analysis, in this stage the results are analyzed and interpreted, in order to obtain significant conclusions for the decision making process, to reach objectives such as parametric design or to identify the process variables that critically affect its performance, the influence or sensitivity due to variability of the external components, the identification of the different parametric levels that produce an optimal performance, and finally, to identify if the improvement or implementation can be executed in a real environment.

the planning and design phases are recognized as the most important in the DoE methodology application, in these phases we achieve a good level of knowledge about the operation of the process and the object of study, this helps to facilitate the conduction and analysis, providing a good detail of information to generate clear and well defined strategies to be applied in the real environment. It should be noted that in general terms, three crucial moments are identified in the process life where the application of DoE could be useful, these are (Murray et al., 2016; Weissman & Anderson, 2015) ;

1. The early stages of process design, where usually the conditions are still uncertain. DoE is utilized because it is a systematic process that offers a methodological framework for its application. The initial phase contains the definition of a context where the objectives are established, the variables are defined as well as the interaction between them, this initial evaluation and analysis aids to enhance the knowledge of the process to be implemented.
2. When the robustness of a production process is evaluated. Understanding as robustness the response to variation of uncontrolled parameters (noise), which allows knowing its stability or detecting the need to implement improvements in the process, through the measurement of precision and reproducibility, leading to significant results by experimentation.
3. Finally, to optimize the conditions of procedures already established, for instance, financial adjustments originated from the analysis of alternative materials and even the redesign of products and/or processes that reduce costs and time, and resource optimization.

Nowadays, the application of design of experiments in the industry is used in various fields such as Chemistry, Mechanics, Electronics, Computing, and Industrial Engineering, among others (Caldas, Tanco, Viles, & Sánchez-Arjona, 2007). However, this tool has not been explored completely to solve problems in the manufacturing industry (Tanco et al., 2007), in spite of its advantages in terms of cost and time. Most of the literature is focused on the application of these techniques in the areas of statistics and quality.

According to some authors Caldas, Tanco, Viles & Sánchez-Arjona (2007) and Tanco et al. (2007), this is due to the limited knowledge of the techniques and statistical tools by engineers, also, some barriers can be identified, such as education, the basic knowledge level in terms of probability and statistics, cultural barriers, limited amount of time to achieve results, modeling strategies based on trial/error observation, communication barriers, limited knowledge of the DoE methodology and its application, consequently, all of these causes make difficult to transmit or carry out the implementation of any decision made through DoE (Antony, 2003a).

The variety of problems faced by engineers in the industry its broad and includes different levels of complexity, ranging from the design of new products, design improvement, maintain, control and improve the manufacturing processes, the maintenance and repair of products, among others (Tanco et al., 2007).

In order to solve this type of problems, engineers perform experiments and data analysis; therefore, they need statistics to achieve results, thus the importance of the knowledge of advanced statistical techniques such as DoE, which allows solving problems in an effective way and at a low cost. This chapter focuses on providing the reader with different applications of this methodology in the industry, to highlight the importance and benefits of knowing and applying these techniques. In the first part, we will introduce the application of this methodology in a general way; lastly, different study cases using this methodology in the industry will be discussed.

DESIGN OF EXPERIMENT: BACKGROUND, RELEVANCE AND IMPACT

According to Tanco, M., Viles, E., & Pozueta (2009) the experimentation has been used as a common practice through time to understand different phenomenon of interest and also in the creative process to choose the best test that will lead us to find an optimal solution. Hernández Sampieri (2014) defines an experiment as "control situation in which one or more independent variables (causes) are intentionally manipulated to analyze the consequences of such manipulation on one or more dependent variables (effects), providing an ordering and analysis with high structuring of factors and variables (Niedz & Evens, 2016).

Experimentation is carried out when deliberate changes are made in the object, process or study phenomenon that allows identifying an impact on the answers. Anderson & Whitcomb, (2016) consider that the experiments are used to study the performance of processes and systems, whilst for Salazar & Baena (2009) a process or system has inputs and outputs, as well as factors which may or may not be controllable (see Figure 1). In this sense, the process can be visualized as a combination of operations, machines, methods, people and other resources that transform some input (often a material) into an output that has one or more observable response variables (Montgomery, 2017).

Czitrom, V. (1999); Montgomery D.C., (2005); Tanco et al. (2007) y Niedz & Evens (2016) mention that there are three types of experimentation strategies (Table 1), which are classified as: Best-guess approach, One-factor-at-time (OFAT) and DoE.

The trial and error approach (best-guess approach) consists of testing different scenarios, changing the levels of one or perhaps two factors between one test and another, continuing indefinitely. The result success is associated to the knowledge and experience of the individuals performing the experiment. According to Montgomery (2017), this experimentation strategy has at least two disadvantages; first, tests are made by making guesses about the correct combination of factor levels until the desired results

Figure 1. Classic model of a process or system

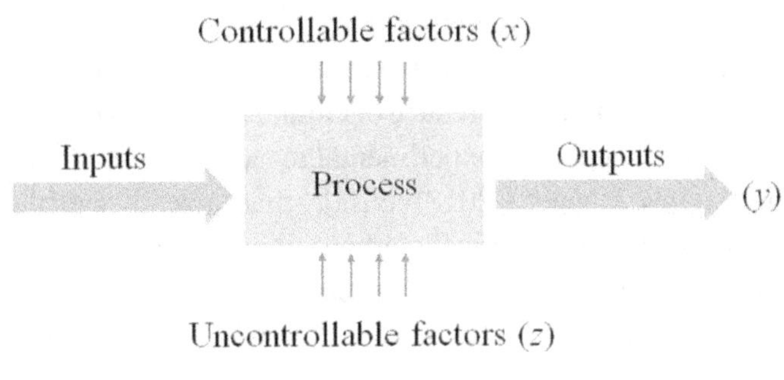

Table 1. Experiment Strategies

Best-Guess Approach	One-Factor-at-Time (OFAT)	DoE
Heuristic method to obtain knowledge, which consists in testing an alternative and verifying if it works.	Scientific method that modifies one variable at time in order to visualize the effect of each variable in the response.	Methodology to systematically apply the statistic to the experimentation process.

Source: (Montgomery 2005, 2009, 2017)

are found, this could take a long time and have no guarantee of success, and second, even if the initial test produces an acceptable result, there is no guarantee that the best solution has been found.

On the other hand, in engineering and sciences, one-factor-at-a-time (OFAT) experiments are often performed. In this traditional approach, the influence of the tested factors is measured by changing the level of a factor while keeping other factors at their nominal and/or basic levels (Molina, A. G., & Diego, 2016; Jacyna, Kordalewska, & Markuszewski, 2018). This kind of experiments are often inefficient and wasteful, with results that are often misleading, because they can only perform a small fraction of the total number of possible combinations due to resource limitations. This is a characteristic of this type of experiment (Niedz & Evens, 2016). Additionally, there is no opportunity to investigate the possibility and strength of the interactions between the studied factors. Therefore, its application should only be justified in the case of the optimization of simple processes, which consider one or two parameters. Manufacturing and industrial processes have changed over time, Industrial research cannot afford to experiment using the trial-and-error method, or by changing only one factor at a time, so the first two strategies, although they are still used, have become obsolete in many applications due to the shortcomings and little information they offer. For that reason, it was necessary to develop a better methodology to run tests where many factors and the combination of them affect the result (Astakhov, 2012). Thus, in the 20th century, DoE was born as a tool for experimentation in the 1920s and was developed by Sir Ronald Fisher who applied it for the first time in England in the agriculture area, where the interest was focused on the improvement of potato production. Finally, these results and experience led him to publish his book "Design of Experiments" in 1935 (Montgomery, 2017).

The DoE methodology was based on experimentation at the beginning and only dealt with physical experiments, hence, the results were affected by random errors (noise) which represented a clear disadvantage, this makes the data modeling complex (Yondo, Andrés, & Valero, 2018). On the other hand, when experiments are carried out using this methodology, replicas are generated and an estimate of the experimental error is obtained as suggested by Garza Villegas (2013) that sustain that the greater the number of these, the smaller the error, as long as the experiments are performed under the same conditions. In addition, this author considers that during the performance of the experiment, *randomization* is essential to avoid dependence between the samples and ensure that the results are actually caused by the dependent variables and not by the experimenter. While Montgomery (2017) argues that blocking is a design technique used to improve the accuracy with which comparisons are made between the factors of interest. And that, often the blockade is used to reduce or eliminate the variability transmitted by annoying factors, that is, factors that can influence the experimental response, but in those that are not of interest.

Authors such as Yondo, Andrés, & Valero (2018) consider that currently with the technological advance in computer systems and the advent of numerical methods, it is possible to design experiments through computer-based simulations that contrast with the physical experiments in which the analyzed process is stochastic, and the experiments based on simulation are deterministic, without the need for replication.

Exploring Definitions by Different Authors

DoE is a set of active techniques, in the sense that they do not expect the process send the expected signals, but that it is "manipulated" to provide the information that is required for its improvement. This methodology is considered useful, versatile and somewhat easy to execute in addition, it has been studied and applied by experts from different areas. Some of them have contributed their definition of it, as shown in the following table (Table 2).

Table 2. Definitions of DoE (by different authors)

DoE Definitions	Author (s)
It consists of a series of tests in which purposeful changes are made to the input variables (factors) of a product or process so that one may observe and identify the reasons for these changes in the output response.	Tanco et al. (2007)
Design of experiments is a set of methods that are used to manipulate a process in order to obtain information on how to improve.	Sanchez (2009)
The design of experiments is the most effective way to testing and consists of determining which tests should be performed and in what way, to obtain data that, when analyzed statistically, provide objective evidence to answer the questions raised, and thus clarify the uncertain process aspects, in order to solve a problem or achieve improvements.	Gutiérres Pulido & De la Vara Salazar (2012)
DoE is a statistical formal methodology allowing an experimentalist to establish statistical correlation between a set of input variables with a chosen outcome of the system/process under study under certain uncertainties, called uncontrolled inputs.	Astakhov, V. P. (2012)
DoE is a large and well-developed field for understanding and improving the performance of complex systems.	Niedz & Evens (2016)
Statistical design of experiments refers to the process of planning the experiment so that appropriate data that can be analyzed by statistical methods will be collected, resulting in valid and objective conclusions.	Mongomery (2017)
A well-defined and well-structured experimental design allows for the study of input parameters (predictor variables) and the generated output (response variables), as well as various interactions that may exist between the input variables. In general, experimental designs are denoted by a matrix, with columns representing the independent variables associated with the study, and rows representing samples or experimental runs.	Yu, Low, & Zhou (2018)
The DoE approach is a tool for systematical examination and documentation of the impact of each input factor on the studied responses and for finding the optimal parameter settings in order to obtain their desired values.	Jacyna et al. (2018)

In summary, DoE is a methodology used by entities from all sectors to know how a process works, study the variables that affect it and, using statistical tools, obtain the necessary information for its improvement. In addition, DoE offers an organized approach that connects experiments in a rational way, providing more accurate information with fewer experiments (Tye, 2004).

Methodological Framework

In order to achieve and adequately design an experiment it is necessary to have a methodology that clarifies the stages and steps on which it is built. Currently the procedure for the application of DoE is based on the contributions of Coleman & Montgomery (1993) who in their work "A systematic approach to planning for a designed industrial experiment" presented the general scheme to develop designs of experiments. This scheme has served as support for the development of other authors. Figure 2 shows the methodological steps to use it (Astakhov, V. P., 2012; Montgomery, 2017; Jacyna et al., 2018).

Following, each one of the stages is approached, as well as the steps that are developed in each of them (Farooq, Nóvoa, Araújo, & Tavares, 2016):

1. **Planning:** Stage where the goals and objectives to be pursued are defined, one must have full knowledge of the process, activity or object of study that needs some type of optimization or improvement, for which the design of an experiment is required.

Figure 2. General procedure for conducting experiments in accordance with the DoE methodology

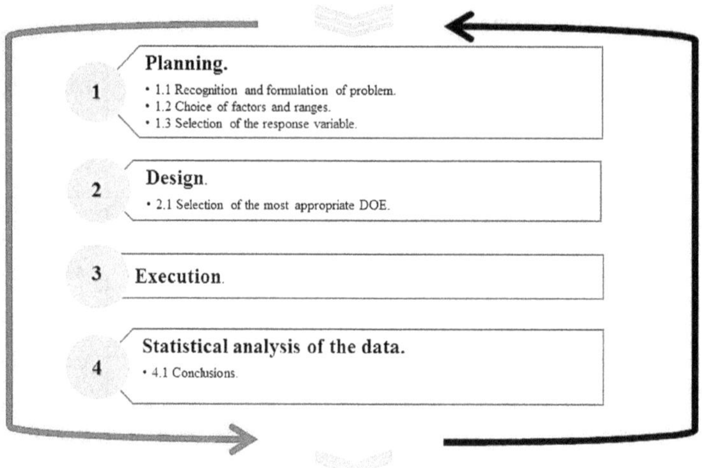

It should be emphasized that prior to using the statistical approach of the DoE, it is necessary that all those involved in the experiment have a clear idea in advance of what is going to be studied exactly, how the data should be collected and at least a brief understanding of the how will these be analyzed.

a. **Recognition and Formulation of Problem:** Initial step that allows the start of this first stage, for which it is necessary to develop all the ideas about the objectives of the experiment, that is, it is important to always keep in mind the general objective which should be clear and measurable. In this sense, it is useful to prepare a list of specific problems or questions that will be addressed in the experiment with the intention of obtaining a clear formulation, which contributes substantially to a better understanding of the phenomenon studied and the final solution of the problem.
b. **Choice of Factors and Ranges:** If the research focuses on a new or little known phenomenon, then the first stage should be the selection of factors and ranges, in the case of the factors, these should be easy to control, in addition, that the ranges should be determined in which they will be tested. Some techniques that can help in the selection of such factors are; experience and full knowledge of the process, brainstorming, historical data, cause-effect analysis, another frequent practice is to perform screening tests and, depending on the results, choose only the parameters that influence the process. This step is very important, because if you leave out some critical factors for the optimization or improvement that you want to carry out, the results will not be precise or simply the actions or strategies resulting from the experiments will not be useful for the improvement of the process.
c. Selection of the response variable (can be one or several), considered critical for the success of the experiment, it must be certain that said variable really provides useful information about the process or system of study, and that it can be measured accurately, so that a measurement system should be previously established in order to know what to measure?, where to measure?, and define the person in charge of making those measurements. Frequently, the response variable will be the average or standard deviation (or both) of the measured characteristic (for example: the average diameter of a particle).

2. **Design:** At this stage, according to the needs and conditions of the process or activity to be improved or optimized, the design of experiments that best suits the requirements and provides optimal solutions must be chosen.
    a. Selection of the most appropriate design of experiments is done according to the amount of information what is desired to get, the amount of resources available (time, materials, data, etc.), budget, time and if there are any other restrictions. Specifically, it will also depend on the number of factors to be processed, as well as the depth obtained from the definition of their levels, the nature of the problem and the objectives to be achieved. It is possible to choose from a variety of different schemes available that include logistic regression, factorial plans or response surface methodology.
3. **Execution of the Experiment:** In this step it is necessary to perform as planned, that is, to perform all the experiments following the previously defined procedure, with the intention of reducing the errors that could affect the experimental validity. Thereby, it is recommended to perform pilot tests that offer an idea of the possible results, in such a way that the decisions taken in the previous steps can be reconsidered.
4. **Statistical Analysis of the Data:** This phase implies that by using a statistical tool (usually a software) a treatment, representation and interpretation of the results obtained during the execution of the experiment (previous stage) is performed, in such way that can offer objective, verifiable, valid and solid conclusions. Currently, there is a lot of software package for the experimenter to choose, however, the usual is that when the design of experiments that will run is selected, it offers a direct interface with the most appropriate tools for statistical analysis. It must be considered that the correct selection of the statistical tool, added to a correct interpretation of results, and in turn, combined with a proper knowledge of object of study by the experimenters, generally leads to solid conclusions.
    a. **Conclusions:** once the analysis of the results was made it is possible to draw conclusions, as well as recommendations which indicate the course of action to follow. However, it is suggested to validate the results.

In practice, it is common that during the experimentation process changes, modifications, reconsiderations and/or new formulations are required. This result in a learning process that goes beyond the results obtained in the experiment itself and allows the acquisition of knowledge by the experimenter.

Application and Impact of DoE in Industry

The manufacturing industry is one of the engines that drives the growth and economic development of a country, it is identified as a secondary sector, and represents approximately 16% of the global gross domestic product (World Bank, 2017). The manufacturing industry faces constant challenges, mainly in the manufacture of innovative and high-quality products. This is due to the current conditions of the global market and the high competitiveness of organizations. The processes of product development are increasingly complex as products become more versatile to meet the needs of mass customization (Ong, 2008).

A key factor for the success of organizations is to use the maximum of their capacity for knowledge and learning, as well as their experience. In this sense, experimentation is one of the elements that can contribute most to learning and improving products and processes. Even experts associate the DoE as one of the most effective means to achieve an improvement and optimization of systems, processes and products, which has a lot of application in industry.

Some of DoE applications are listed below:

- Improve processes, whether improving their efficiency, reliability or performance.
- Comply with the specifications for a robust manufacturing.
- Formulate and design new products that are resistant and easy to produce.
- Establish specifications and tolerances for products and processes.
- Learn from the processes and their failures.
- Evaluate alternative materials.
- Improve quality, reduce times and costs.

Some of the advantages offered by the use of DoE in industry are the effectiveness and efficiency that they produce. DoE is effective when properly applied, reduces the predicted error and increases the probability that the results obtained represents the actual processes (Czitrom, 1999). It would be efficient that the information obtained will be consistent with the initial objective, and reduces the amount of resources used, it also maximizes the quality of the data (Niedz & Evens, 2016).

DoE can be applied in all types of industry and all kind of areas. However, Izquierdo, Tanco & Sanchez-Arjona (2007) mention that despite of its advantages and the opportunities, the application of DoE is not common in the industry, due to the limited knowledge of advanced statistical techniques. However, today there are software that facilitate the use of DoE (Salazar, J. C., & Baena Zapata, 2009), to support experimenters, creating a range of possibilities for the improvement of products and processes of all kinds (Silva, Filleti, Christoforo, Silva, & Ometto, 2015).

However, today there is a more important challenge to fulfill in manufacturing industry and corresponds to guarantee sustainable production and consumption patterns. This implies a big change in industrial, social and governmental sector about the way in which societies produce and consume goods and services (ONUDI, 2015). Because of this, in 1994 at the Oslo Symposium, the UNEP declared that sustainable consumption and production refers to "use of services and products that respond to basic needs and provide a better quality of life, while minimizing the use of natural resources and toxic materials, as well as emissions of pollutants and waste throughout life cycle of the service or product, so as not to endanger needs of future generations" (United Nations, 2015).

Therefore, it is necessary to make important changes to manufacturing processes by designing or re-designing them, in order to improve the quality of products, reduce costs and cycle times, replace materials, modify methods, design innovative products and high competition in market, among others (Gutiérrez & De La Vara, 2012). This is why it is very important to use the DoE as a statistical tool to solve problems and make decisions.

Main Uses of DoE in Manufacturing Industry

As mentioned above, the design of experiments is considered a multipurpose tool with a systematic and rigorous approach to solving engineering problems, that is, it is the most effective method to solve complex problems with many variables. Therefore, it is used under different scenarios.

In manufacturing industry, it is usually used to identify the most important input factors of process, commonly known as "input variables", and how these variables are related to the output, known as "response variables". It is very important to identify each and every one of the input variables to the process, since they are the ones that serve to control, to a certain extent, the state of system, and when being processed, will give the expected results and largely will help to know the functioning of system.

Under statistical concepts, the design of experiments is basically a regression analysis, which, as already mentioned, can be used under different scenarios. There are five areas of general engineering problems in which the DoE can be applied (see Figure 3) (Tanco et al., 2007; NIST/SEMATECH, 2012 & Durakovic, 2017):

Following explains in detail the five general engineering problem areas, seen in Figure 3:

1. **Comparison.** It refers that the engineer wants to evaluate if changing a single factor (variable) of process results in an improvement in system. In this area, multiple comparisons are made between factors to select the best option. In general, to bring out this comparison between factors, the t- test, Z-test or F-test is used.
2. **Selection.** After performing the design and analysis of process, the engineer's main objective is to "understand" how the process works in its entirety, in order to be able to select and prioritize the most important factors that may affect the performance of a system, product or process. Generally, for the selection of most important factors (variables), is performing through of two-level factorial designs.
3. **Modeling.** As the name implies, in this phase the engineer is interested in functionally modeling the process based on the most important input variables (identified in previous phases), the relationship between input variables and output variable, considering the output variable as a mathematical function that presents a good fit, that is, with a high predictive power and with a maximum precision of the coefficients in that function. The purpose of modeling is to be able to explore the performance of a system, product or process through the transfer function (mathematical modeling), before execution.
4. **Optimization.** In this phase, the engineer determines the optimal configuration of factors (variables) of process, with the main objective of improving the performance of system, product or process. This is achieved through the transfer function (mathematical modeling), moving the experiment to the optimal configuration of variables, that is, determining for each factor, the level of the factor that optimizes the response variable of process.

Figure 3. General engineering problems areas

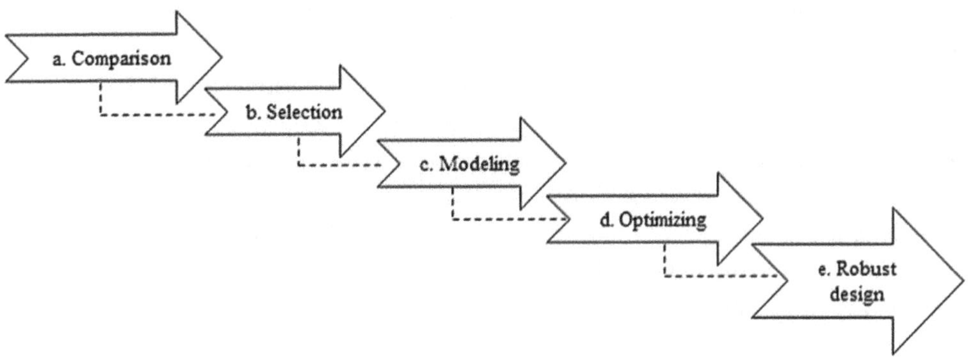

5. **Robust design.** The main objective of achieving a robust design is to reduce the variation of system, product or processes against noise, without eliminating its causes. Dr. Genichi Taguchi was the first Japanese engineer and statistician who made an important contribution to industrial statistics, improving the quality of manufactured products through the implementation of a robust system against noise, that is, against uncontrollable factors such as: environmental or external (for example: temperature, humidity, dust), internal (wear of a machine, fatigued workers, etc.) and variations in materials, processes and equipment, the latter known as "unit to unit variation".

Therefore, it is very important to mention the advantages that are had when applying DoE as a statistical tool for decision making in manufacturing industry, which are (Montgomery, 2005; Costa et al., 2006 & Jiju, 2014a):

- Reduction in process variability
- Improvement in process capacity
- Reduction in the design and development times of process
- Higher performance and stability of the process
- Reduction in manufacturing costs
- Higher profits and return on investment
- Greater understanding by engineers about the relationship of key inputs and outputs of the process
- Increased motivation of engineers to solve recurring problems successfully
- Increased in the profitability of the company by reducing the rate of waste and defects, rework, re-inspections, among others.

Challenges Faced by DoE Application in Manufacturing Industry

Despite all the efforts made by specialists in quality and statistics, there are still many obstacles to using DoE in the industrial sector. Studies have shown that less than 30% of people have knowledge about DoE (Jiju, 2014b), some of the most common obstacles for which DoE is not used in industrial sector, are listed below (Romeu, 2006; Tanco et al., 2008 & Tanco et al., 2009):

1. **Deficiency of knowledge.** Managers in industry fail to understand the importance of using DoE in their manufacturing processes to solve problems.
2. **Demanded quick solutions without following methodologies**. There are many organizations in which managers encourage engineers to solve problems related to processes and quality based on "experience", which commonly leads to many failures.
3. **Resistance to change.** There is still a high resistance to change to apply this statistical tool justifying that the results that can be obtained with the use of DoE shows "what is already known", "sounds good but is not applicable to the process that is being handled", "An additional effort is needed to demonstrate what is already known", among others.
4. **False beliefs.** Many managers in industries still think that DoE as a statistical tool does not add any value to management and decision making, since they believe that with the use of DoE a lot of time and resources are required.

5. **Limited communication.** There is still very little communication between the academic and industrial sectors, and this communication is considered vital since, on the one hand, the lack of statistical knowledge on the part of engineers leads to erroneous interpretation of the historical data collected as well as the interaction between the factors of experiment, and on the other hand, the lack of statistical knowledge on the part of academics generates problems such as the wrong selection of process variables, lack of precision, among others.

Managers usually seek quick solutions that provide short-term economic benefits for their organizations, without a statistical support that demonstrates that the decisions made are correct in short, medium and long term.

GENERAL STRUCTURE, BASIC CONCEPTS AND CLASSICAL TECHNIQUES

Basic Components of DoE

Changing operating conditions of inputs of a process is what will allow the execution of an experiment, as can be seen in Figure 4, the inputs are composed of factors, configuration levels and it is in the output where the effects are established through one or multiple responses, as well as the estimation of the random and experimental error.

Hereunder, some basic definitions of the elements that complement the general structure of a DoE are explored.

1. **Factors** are defined as inputs to process, on which different parameters can be configured, they provide the facility to be able to generate experiments, since it is these factors or variables that allow start-up and execution of a process under different conditions, These are highly configurable at different levels, providing important information for decision making after the experiment has been carried out, the factors are classified as follows (Giunta, Wojtkiewicz, & Eldred, 2013; Gutiérrez & De La Vara, 2012; Montgomery, 2009; Vanaja & Rani, 2007):
    a. **Controllable Factor or Input:** Also known as factor (x), which are defined as those input parameters that can be modified in an experiment or process, in order to optimize the output, some examples are; the temperature on a certain chemical process, materials to be used for the

Figure 4. General DoE Structure

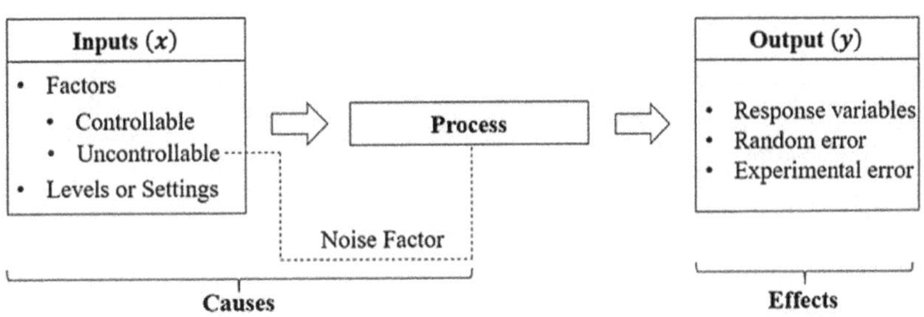

development of a product according to its mechanical properties or those that are friendly to the environment, suppliers that meet deliveries on time and greater availability of inventory, among others, these factors can be quantitative and qualitative.

b. **Non-Controllable Factor or Input:** Those parameters that cannot be changed, also known as noise factor, if there is no mechanism or technology that allows them to change their properties, then it is considered non-controllable, they can be external or internal, some examples of this type of factor are; the mood of the personnel that operates in an organization, weather conditions, social aspects, among others.

2. **Levels or Adjustment of Each Factor in the Study:** The values that could be established for the different controllable factors identified must be configured in the levels. And the combination of levels and factors is known as treatment or design point, which will allow observing the influence between factors under the different levels established, an example of which is shown in Table 3, where two factors with two level to each factor, and for each combination of the factors and their levels, a treatment is generated from which an output is expected.
3. **Response or output of the experiment** also known as variable (y) or dependent variable, it has an objective to establish the cause-effect relationship between the controllable factors and the dependent variable of interest.
4. **Random error** is the observed variability that cannot be explained through the factors studied.
5. **Experimental error** is identified as part of the random error, which reflects the errors of the experimenter when not choosing the controllable factors of greater impact for the experimental study, it is presented mainly in the stage of planning and execution of DoE.

Fundamental Principles

Due to DoE works with observable data which implicitly contain noise and therefore variability, given that a fact does not occur in the same way twice, mechanisms or strategies must be sought that, in this way, can guarantee robustness and precision of the experiment executed, for this reason it is important to have fundamental principles that provide the assurance that the experiment has been designed in an appropriate manner and can provide information or optimal indicators that serve for decision making, these principles can be observed in the Figure 5.

To guarantee the validity of the data collected and that are useful when used in an experiment, the fundamental principles will support this process during the planning stage, which are explained below (Antony, 2003b; Dean, Voss, & Draguljić, 2017; Griffith & Griffith, 2006; Joachim, 2000):

Table 3. Treatments or Design Points

Levels		Treatment	Response Variable (y)
Factor 1	Factor 2		
1	1	1	¿?
2	1	2	¿?
1	2	3	¿?
2	2	4	¿?

Figure 5. Fundamental principles of DoE

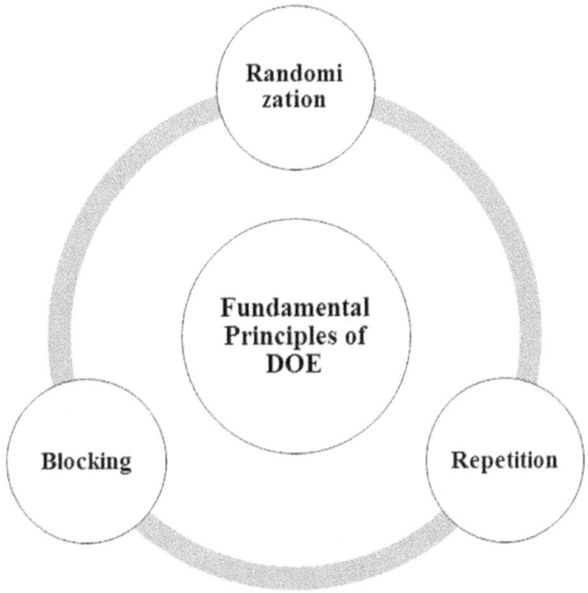

1. **Randomization:** Experimental executions are generated randomly, that is, haphazardly. This technique is useful to balance the effect that can influence the output responses from non-controllable external factors, such as materials, operators, equipment, among others, which in practice change without notice and highly affect the process, activity or study object. By randomizing the order of the experimental runs, as well as combinations of factors and levels, the tendency of bias in the results and other sources of atypical variations is reduced. During this process, the allocation of treatments to the experimental units is randomized, implying that each assignment of the treatment has the same probability of being performed.
2. **Repetition:** Refers to the repetition of the treatment defined in an experiment, this is executed in the order in which it was established when the principle of randomization was carried out, within the advantages offered, is that it allows to estimate the variability or random error, decreases the variance, improves the accuracy of estimating the effect and identify the factors that cause it, facilitating the construction of statistical analysis from the data.
3. **Blocking:** This principle allows the experimental units to be grouped into blocks, so it must result in more homogeneous data sets, for example, if the experimental block is carried out from the operated one regardless of the equipment it operates, it would be expected that measurements that this will perform will not vary much from one another. This technique is used to increase the precision with which comparisons are made between the factors of the experiment.

Experiment Techniques Commonly Applied

Design of experiments as a methodological tool in addition to the stage and steps to carry out its proper development and implementation, also offers a set of techniques that can be chosen depending on the needs and requirements of the problem that want to address. Table 4 shows the generic classification and its corresponding techniques (Narenderan, Meyyanathan, & Karri, 2019), which will be explained below.

Table 4. Design of Experiments Techniques

Design of Experiments Techniques	
Screening	**Optimization**
Factorial	**Response Surface**
• Full Factorial Design • Fractional Factorial Design • Taguchi Orthogonal Array Design	• Central Composite Design • Box-Behnken Design

1. **Screening:** The main objective of this type of design is to identify the critical factors that affect the flow and quality of a process, activity or object of study, its techniques are commonly used to examine a large number of variables with few experiments, focusing more on the significant effects than in their interactions (Ranga, Jaimini, Sharma, Chauhan, & Kumar, 2014; Woods & Lewis, 2017).
    a. **Factorial:** It allows to study the effects that several factors can have on a response, it helps to analyze the interactions between the factors and the changes of levels of said factors can be carried out at the same time. They can be designed from one factor to multiple factors. When the factorial design only contemplates one factor, this is known by the name of simple comparative experiment, in this case t-test and ANOVA are used for its analysis. When you use two factors, evaluate all possible combinations of factor levels. Finally, when the factorial design is carried out with multiple factors and at least two levels, the experiment can become very complex and represent a problem due to the number of combinations between the factors and levels that it has to perform, this type of design is not highly recommended because it has a high demand and consumption of resources, time and therefore it is very expensive to maintain it, besides that under very changing conditions the results could not be relevant, to face up with the problems presented by the multi factorial, the Full Factorial Design represent an ideal alternative, this technique going to be explained in the next point (Heiberger & Holland, 2015; Woods & Lewis, 2017).
        i. **Full Factorial Design:** Can be denoted as 2^k where k is the number of factors studied and 2 is the number of levels. The study levels are high (+1) and low (-1), this type of design is particularly used in the initial part of an experimental work, especially when the number of factors is less than or equal to 4. It is assumed that the response is approximately linear over the range of established factors. Here below, a generic example is shown, first, Table 5 on which its parameters or factors are defined, the labels that correspond to these parameters and the levels with their ranges and corresponding unit (for example, for label A they have established ranges in minutes and for label B ranges in degrees centigrade). Note, the data defined within the tables have no meaning, they have been put only to exemplify the type of design (Antony, 2014; Das & Dewanjee, 2018a).

In Table 6 the design layout of the experiment with respect to the response was defined, as it can observed in this experiment, the trial number was 4 since being a Full Factorial Design of level 2 with 2 factors, its calculation is done as follows $2^k = 2^2 = 4$, in based on the combination of the different levels, 5 replications were done in order to obtain an estimate of the reasonable experimental error.

Table 5. Definition of parameters, labels and levels (ranges)

Parameters (Factors)	Labels	Levels	
		Low	High
Factor 1	A	1 min	10 min
Factor 2	B	18°C	49°C

Table 6. Design layout of the experiment with respect to the response defined

Trial Number	A	B	Response Variable				
1	1	18	10.5	11.72	12.94	10.9	12.30
2	1	49	9.8	11.02	12.24	11	10.9
3	10	18	10	11.5	11.6	12.8	10.7
4	10	49	12.5	13.8	12.7	13	11.99

Table 7. Design matrix and signs for effect in full factorial design to 2^2 levels/factor with mean of response variable

A	B	AB	Mean Response Variable	Standard Deviation
-1	-1	+1	11.67	0.89
-1	+1	-1	10.99	0.77
+1	-1	-1	11.32	0.94
+1	+1	+1	12.80	0.60

Finally, Table 7 shows the design matrix and signs for effect in full factorial design to 2^2 levels/factors with the mean based on response variable.

As can be observed in Table 7, the signals are coded from the combination of the factors / levels established in Table 6, in the first record it is observed that A = -1 since the evaluated factor was performed with the minimum level and the factor with label B = -1 also corresponds to the minimum value of the range of the established level, AB = 1 is the result of the multiplication of A * B and finally the average obtained from the replicas generated by each test of the values is calculated of the response variable, based on these results, the level of significance, impact and sensitivity to the variability between the interactions of the evaluated factors can be determined.

ii. **Fractional Factorial Design:** Evaluates the effect of certain factors with a minimum trials number, that is why the experimental runs are made on selected or fractionated subsets of a complete factorial design, are suitable when resources are limited or when the number of identified factors are very large, reason whereby they are widely used in the industry. This design is generally represented as follows $2^{(k-p)}$ where 2 represents the level, k is the factors number and p is the full factorial design fraction 2^k (J, 2014; Ou, Zhang, & Qin, 2019; Rytz et al., 2017).

Table 8. Fractional factorial design matrix of a subset $2^{(4-1)}$ experimental trials

Trial Number	A	B	AB	C	AC	BC	D=ABC
1	-1	-1	1	-1	1	1	-1
2	1	-1	-1	-1	-1	1	1
3	-1	1	-1	-1	1	-1	1
4	1	1	1	-1	-1	-1	-1
5	-1	-1	1	1	-1	-1	1
6	1	-1	-1	1	1	-1	-1
7	-1	1	-1	1	-1	1	-1
8	1	1	1	1	1	1	1

As shown in Table 8, the construction of the fractional factorial design matrix of a subset $2^{(4-1)}$ is observed with 8 experimental tests, three factors (A, B, C and the generation of one more factor, from of the interaction of the three previous factors (D).

iii. **Taguchi orthogonal array design:** Considered as a type of generalized fractional factorial design, this type of design allows the study of multiple factors with different levels based on an orthogonal arrangement from which Dr. Genichi Taguchi proposed a design matrix, which allows to take into account a subset of selected multifactor and their combinations at multiple levels (Freddi & Salmon, 2019; Kul & Çetinkaya, 2010). This design is considered to be easy implementation and was thought to improve the quality of the manufacturing of products, at present it is widely used by engineer in several areas of the manufacturing sector. Taguchi orthogonal array design is denoted as $L_t\left(2^k\right)$ where L refers to the orthogonal array, t corresponds to the trials number, 2 refers to the levels (which could be greater than 2) and k corresponds to the factors number to be evaluated, the factor levels are weighted equally throughout the design, the columns of the matrix represent the control factors, the rows represent the executions (combination of factor / levels) and each cell of the table represents the factor level for that execution, an example is shown in the Table 9 (Silva, 2018).

b. **Optimization:** practice that allows to establish the optimal conditions of an experiment, for which an exact optimum can be obtained through the response surface methodology.

Table 9. Orthogonal array $L_4(2^2)$

Trial Number	Column Number		
	1	2	3
1	1	1	1
2	1	2	2
3	2	1	2
4	2	2	1

i. **Response Surface**: can predict the relationship between responses and independent variables, is an experimental strategy and analysis that allows to find the optimal operating conditions of a process, activity or object of study. Frequently it is used to refine the models generated by factorial designs from which the most important factors have been determined (Candioti, De Zan, Cámara, & Goicoechea, 2014). The equation of the factorial design with respect to the response surface equation is the addition of quadratic terms in the latter, which allows to generate a curved shape, unlike in the factorial design its shape is flat. Hereunder, two of the most commonly used techniques are shown:

ii. **Central composite design:** very used when a sequential experimental design plan is required, part of a factorial or fractional factorial design, but includes central points and star points also called axial points, which allow to estimate the curvature, efficiently calculates the terms of first and second order as shown in Figure 6 (Adeleke et al., 2018; Glyk, Solle, Scheper, & Beutel, 2015).

This technique is also used to determine the number of experiments to be evaluated for the optimization of the response variable, an experimental design matrix is generated for the optimization of the response factors, with the minimum, medium and high values, which are labeled with the values -1,0 y +1 respectively as shown in the Table 10. Note, the values within the table have no meaning, they have been put only as an example.

iii. **Box behnken design:** unlike the central composite design technique, this does not include a factorial or fractional factorial design. They are used to generate high order response surfaces using minimal experimental runs. This technique requires at least three continuous factors, which are found in the midpoints of the edges of the experimental space for which combinations of treatments are generated, as shown in Figure 7, the points on the edges represent the experimental executions carried out (Das & Dewanjee, 2018b; Okoro, Sun, & Birch, 2019).

Figure 6.

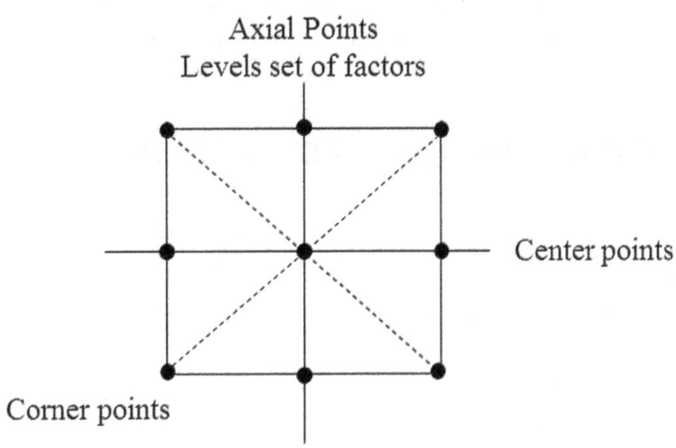

Table 10. Experimental matrix design to optimization of response factors

Parameters (Factors)	Labels	Levels		
		Low (-1)	Middle (0)	High (+1)
Factor 1	A	1 min	5 min	10 min
Factor 2	B	18°C	30°C	49°C

Figure 7. Experimental space to three continue factors and all of executions

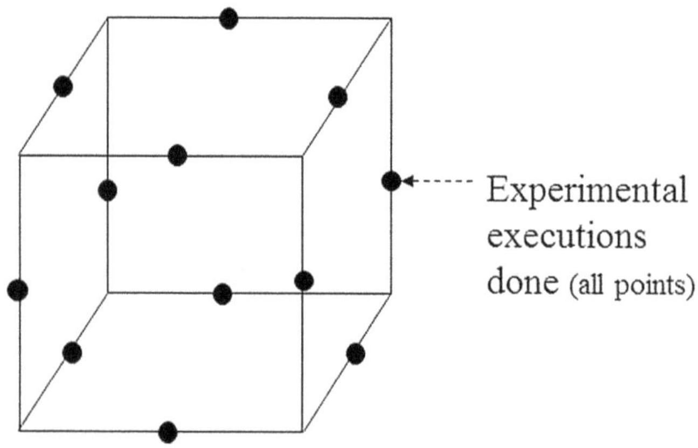

They are not suitable for sequential experiments, they are considered as efficient estimators of first and second order coefficients, they are usually less expensive than the central composite design because they handle less design points, they do not have axial points, this ensures the operation within a safe area of the process, unlike the central composite design which usually have axial points outside the cube, so they could be outside the region of interest, which would make it an impossible to process experiment because it is outside the marked limits as a safe operating area.

As can be seen, the classical techniques that were revised allows selection of main factors, whose effects on the response variable are critical for improvement and performance of the process, activity or object of study, as well as for generating models optimizing said factors.

DoE CASE OF STUDIES IN THE MANUFACTURING PROCESSES

In the last 20 years the application of DoE has grown rapidly both in the manufacturing industry and in other types of industries. It is a very popular tool in the scientific areas of medicine, engineering, biochemistry, physics, computer science, contributing with 50% of the applications, in comparison with all the other scientific areas (Durakovic, 2016).

It is considered of a great importance for this research to focus on case studies about the use of this methodology and oriented to manufacturing processes, covering a very broad spectrum ranging from the use and wear of tools, the production process, waste reduction, and process optimization.

In the following there will be a review of some of the most relevant case studies that were found in the literature, detailing the problems that were addressed when using DoE, the type of technique that was applied, as well as what were the results obtained.

In the first case study, the application of DoE is used in a problem to produce articles through the technologies of additive and digital manufacturing. Griffiths, Howarth, De Almeida-Rowbotham, Rees, & Kerton, (2016) mention that, due to these new disruptive technologies, in which anyone can produce their products, there will be a radical change in the consumption of materials and energy, which is why it is necessary to take into account both environmental and economic problems and the impact that these will have on our society due to the accelerated growth of this sector.

Under this approach the authors make an application of the design of experiments to reduce the waste of material generated in this process, as well as the reduction of production time, which translates into a reduction in energy expenditure.

In this work it is mentioned that in past years the focus was on the quality and functionality of the parts without considering the environmental impact of the manufacturing processes. Additive manufacturing was a way to offer flexibility to designers in terms of the types of materials to meet the mechanical, aesthetic and functional requirements. However, now the importance of creating sustainable production systems has made the approach take environmental considerations. It is mentioned that additive manufacturing is more efficient in terms of material consumption than traditional machining processes where material is subtracted to shape the part, but it is also observed that with regard to the energy use of these new technologies, more research is needed to determine how efficient they are in this regard compared to traditional machining methods.

The objective of these authors was to use an experimental design approach to obtain a balance between quality of parts and wasteful methods, and thus provide solutions to have a positive impact on sustainable development when these types of technologies are used. The authors sought to identify optimal construction adjustment parameters for the weight of the part and production time and explore different optimal routes that will help reduce energy consumption and the weight of waste. The response variables determined were; Weight of waste, weight of the part, energy consumption and production time, with the purpose of optimizing the properties of the parts produced with the FDM technology of additive manufacturing.

The Taguchi method was used in this study to determine how the process affects the performance of the material. The optimization was based on 4 process factors that were the following; orientation of the construction, infill, the number of layers and the height of the layers. And two levels were handled, therefore an orthogonal Taguchi L16 arrangement was used. A complete 2-level factorial model was used. A total of 320 tests were carried out. The results are summarized in Table 11, where the response variables are shown and the process factor that has the greatest influence on them.

Table 11. Process variables

Response Variable	Process Factors
Scrap weight	Slice orientation
Part weight	Infill in first place and second place number of shells
Energy consumption	Layer height following Slice orientation
Production time	Layer height following Slice orientation

As a result of these experiments it was possible to determine other important relationships which can help to optimize this type of processes. And it is the starting point for the investigation of these new manufacturing technologies and the environmental impact that these could have in the near future.

The following case study is about the application of DoE in a manufacturing process to increase productivity. This project was developed in an assembly line that is preceded by a cleaning machine. The takt time of the process and the takt time of the cleaning machine have a difference of 22 sec, the cleaning station fails to feed the parts on time to the assembly line or has to work extra time to solve this difference which translates in more hours of work and use of resources.

Adhikary (2014) mentions that there are certain processes such as in the aerospace and automotive industries, particularly where the cleaning of the parts before being assembled is of vital importance for the quality of the product, that the parts are free of dirt translates into a longer life time for the assembled parts, less service problems and therefore generate greater customer satisfaction. So, tests are done to measure that the cleaning process is appropriate, this measure is a millipore test where the amount of dirt and debris is measured once the part is cleaned. For this particular case where it is required to balance the takt time of the cleaning with that of the assembly line, this cannot be done arbitrarily, so an experiment design was carried out in order to determine if it is possible to reduce cleaning times and thus balance this process making it more productive, without affecting the results of this millipore test.

The objective of this DoE is to determine if modifying the cleaning cycles of the machine has any effect on the purity of the cleaning. In this design, 4 factors were considered, three of them corresponding to different washing cycles; rear butt stationary wash, rear, top, bottom, butting and top butt stationary wash, the last factor is the average temperature of the cleaning bath. Two levels of these factors were established, so we had a factorial design of 2 ^ 4. Sixteen sets of experiments were carried out with three runs for each experiment; therefore, 48 different results were obtained. The response variable was the millipore test.

As a result of this experiment it was found that one of the factors cannot be modified as indicated in the experiment, however, the other 2 cleaning cycles can be reduced, in 5 seconds each without affecting the response variable, so if the change is made the difference between the takt time of the two processes is reduced, which helps to improve the productivity of this process. These 10 seconds of difference translate into an increase per day of 16 parts in total.

In the following case study, the DoE is used as a tool to create the database for a neural network in order to be able to predict the lifetime of a milling tool. In this article, Khorasani, Yazdi and Safizadeh (2011), make reference to the fact that the milling process is the second most used method to cut metals, and due to the organizations that face global competition, they are always in search of optimize manufacturing processes to reduce costs and delivery times ensuring high quality. Therefore, the use of predictive models for machining processes and tool life can give them a competitive advantage, because they generate a lower cost than working trial and error. These models can provide a prediction of stable cutting conditions that help increase tool life.

The prediction of these models is done through neural networks and is currently being a highly explored field. To generate the model, we used an experimental design that considered 3 factors that were put in 5 levels, the factors were; spindle speed, feed rate, depth of cut, and the response variable was the tool lifetime measured in minutes. An orthogonal Taguchi L25 arrangement was used, this arrangement contains 25 repetitions. With these results, the database of the neural network that would generate the prediction model was fed, which turned out to be satisfactory.

In their conclusions, the authors mention as the first step of this process, the Taguchi method of DoE with different parameters should be used to generate this database. In this case, the DoE was used as an initial part of the experimentation, not as a result but as input to the network.

The following article talks about the application of DoE for the validation of processes in the medical device industry. In a research conducted by Dixon, Eatock, Meenan and Morgan (2006) two case studies that used DoE were presented, whereby validation process these are broken down in different stages.

The medical device industry is highly regulated, so all its processes must follow a procedure for its validation. This validation assures us that a process that goes through these stages of intense scrutiny can guarantee consistent results and according to specifications. In this type of industry, the Six Sigma approach has been adopted to improve quality.

This method uses statistical tools such as DoE and statistical process control, to control variability and improve the quality of its processes, for this reason the knowledge of DoE has been extended and in recent years it has been considered as an essential tool that is used for validation. The validation is divided into a series of stages; In general, there are 3; Installation qualification (IQ), operational qualification (OQ) and performance qualification (PQ). The first stage refers to that, all the main aspects of the installation of the equipment are according to the specifications of the manufacturer, the second stage shows that the equipment works consistently according to the specifications under normal conditions, and the third stage guarantees us that a product is produced within specification even when the conditions in which it operates are not the most optimal. For these validations it is necessary to write a protocol where the necessary tests are carried out and the results that must be within the specifications are analyzed in order to determine that the process is validated.

The first use of DoE referred in the article is given in the part of establishing the control limits of the process. Before any equipment can be validated it is necessary to establish the specifications or limits and ensure that the equipment works consistently within those limits. The effect produced by altering the parameters or specifications in the product quality can be established through the design of experiments. The test was conducted as part of the prequalification study to define the permitted operating conditions.

The process under study was the heat sealing of the product packaging. In this process a heated pleated tool is used to heat seal the two sides of a package. If this sealing fails and breaks during the sterilization process and the handling could result in the loss of the sterile product and if this reaches the customer, it would put their safety at risk, hence the importance of this process. The variables that intervene in this process and must be controlled are the temperature of the tool, the time of sealing and the pressure applied, to achieve a seal with quality and acceptable. The DoE was used to establish the limits of these variables because it is considered the most efficient method. With the traditional method, 150 runs would have to be done, using the DoE approach, only 20. The composite type of response surface model is used for this experiment.

This method studied each variable in 5 levels in the 20 runs. For the experiment the parameters were established in the equipment for each run and several samples were produced and tested, as the next step was data processing. In this case the answer variable was in terms of the peel strength of the bond formed during the sealing process, this measure had to comply according to the specifications of the data sheet. As a result, it was found that of the three variables that influence the process, the pressure had only a minimal effect on the resistance of the seal, so it was decided to set its operation level at 65 psi. On the other hand, longer times and higher temperatures result in a higher bond strength.

Therefore, the limits for this variable were established, ensuring the quality of the product. Thus, DoE works as a method to investigate the effect of these variables allowing to have evidence to establish the limits of these variables.

The following case mentioned in this article occurs in the validation stage of performance qualification. In this phase it is shown that an acceptable product is produced under all foreseeable operating conditions. The process that is mentioned is the one of reflow solder process used in the manufacture of Printed Circuits Boards (PCB) for medical devices application. Therefore, the procedure was to produce PCBs in the worst possible conditions by modifying the three possible sources of variation that are process star up, loading rate and paste height.

In order to know if the product was within specifications under the worst conditions of these variables, a two-level DoE was carried out. The experiment was carried out with the minimum number of runs and produced information about the interactions that could not be determined with other methods. They were tested with each combination of the three factors and the data were obtained. Once the plates were produced with this set of conditions they were tested visually and by flying test, with this it is possible to determine if the adverse conditions have effects on the quality of the product, if they did not comply with these tests they were considered defective. Once this data was calculated the average of defects per run if they did not exceed the upper limit it was concluded that the conditions did not affect.

In this case it was found that none of the proven variation sources has a negative effect on the quality of the product. The author of this article concludes that the advantage of using DoE allows to handle a greater number of variables at the same time in more efficient experiments than the traditional methods used in this industry.

The study cases shown were those that were most relevant to talk about the different uses of this tool. However, in the literature search that was conducted, other study cases were found that used DoE for different purposes.

In Lopez, Aisa, Martinez & Mercado (2016), the use of DoE in an injection molding process is shown, they used it to see the influence that process parameters have on the quality of complex molded parts. Finding results that allow them to conclude that for complex parts the behavior of the quality of the pieces according to the variations of the parameters of the process are different from that of the simple pieces.

In the Krishna, Sreeramulu & Venkatesh (2017), the DoE is used to optimize a turning process by analyzing the process parameters that will produce a higher quality product both in finish and in its mechanical properties. Finding as result the values that the 4 parameters of the process must have; speed, feed, depth of cut and material, to achieve the best results in the quality of the product.

In Jayasheree, Sharma, Shetty, Mahato & Gowrishankar (2017), DoE was used as in the previous case to optimize the TIG welding process for an aluminum alloy, resulting in the values of the parameters established for the process of such so that the quality of the product is optimal.

The design of experiments is a tool that has been used and will continue to be used to solve problems and optimize processes in the field of manufacturing, as can be seen in the aforementioned case studies, however, it is necessary to spread the advantages of using this tool in this sector.

CONCLUSION

Global competition means that the manufacturing industry requires powerful tools that allow it to generate strategies that lead to the improvement of its processes or activities, timely diagnosis, improvement of conditions and operations that guarantee the organization and its customers the delivery of products / services with quality, fulfilling the delivery deadlines promised. The DoE as a methodological tool, provides a robust framework, which from the different stages and steps of the methodology will allow the development and implementation of an organized, accurate solution that initially allows the representation of the phenomenon that is wants to study, and from the chosen technique or techniques to generate various experiments that lead to find solutions that are less susceptible and vulnerable to external sources that could put at risk the operation of the process, product or service.

DoE is an efficient tool for the solution of problems in the manufacturing industry and other scientific areas, since, as we observed in the cases of studies reviewed, this methodology and its techniques provides certainty for the decision making process due to the discovery of interacions that were not obvious and that can support the optimization and improvement of process.

The authors in the different cases of studies agreed that the use of DoE results in savings in time and efforts required to reach the response compared with traditional methods. Also, to increase in productivity. This methodology paired with techniques of computational intelligence, serves as a tool to evaluate innovative technology and detect opportunity niches and improvement.

However, the potential and advantages of DoE for process optimization are still unknown, for this reason, the contribution of this study regarding its methodology, techniques, and applications, is relevant and required as literature for engineers to learn about its importance, impact and its application, for process improvement, productivity, efficiency, energy management, and mainly in its decision making process compared to the traditionally used trial and error.

REFERENCES

Antony, J. (2014b). 4 - A systematic methodology for design of experiments. Design of experiments for engineers and scientists (2nd ed., pp. 33–50). Elsevier; doi:10.1016/B978-0-08-099417-8.00004-3

Adeleke, O. A., Latiff, A. A. A., Saphira, M. R., Daud, Z., Ismail, N., Ahsan, A., & Hijab, M. (2018). Locally derived activated carbon from domestic, agricultural and industrial wastes for the treatment of palm oil mill effluent. In *Nanotechnology in water and wastewater treatment* (pp. 35–62). Elsevier; doi:10.1016/b978-0-12-813902-8.00002-2

Adhikary, S. N. (2014). A case study in increasing productivity by DOE in manufacturing sector. *International journal of Advancement in Engineering Technology, Management, and Applied Science, 1*(7).

Anderson, M. J., Whitcomb, P. J., & Whitcomb, P. J. (2016). *DOE simplified.* Productivity Press; doi:10.1201/9781315382326

Antony, J., (2003a). Design of experiments for engineers and scientists. *Design of Experiments for Engineers and Scientists.* doi:10.1016/B978-0-7506-4709-0.X5000-5

Antony, J. (2003b). Fundamentals of design of experiments. In *Design of experiments for engineers and scientists* (pp. 6–16). Butterworth-Heinemann; doi:10.1016/B978-075064709-0/50003-X

Antony, J. (2014). *Full factorial designs*, doi:10.1016/B978-0-08-099417-8.00006-7

Astakhov, V. P. (2012). *Design of experiment methods in manufacturing: basics and practical applications*. Berlin, Germany: Springer.

Candioti, L. V., De Zan, M. M., Cámara, M. S., & Goicoechea, H. C., (2014). Experimental design and multiple response optimization. Using the desirability function in analytical methods development. doi:10.1016/j.talanta.2014.01.034

Chryssolouris, G. (2006). *Manufacturing systems: theory and practice* (2nd ed.). New York: Springer-Verlag.

Coleman, D. E., & Montgomery, D. C. (1993). A systematic approach to planning for a designed industrial experiment. *Technometrics, 35*(1), 1–12. doi:10.1080/00401706.1993.10484984

Costa, N. R. P., Pires, A. R., & Ribeiro, C. O. (2006). Guidelines to help practitioners of design of experiments. *The TQM Magazine, 18*(4), 386–399. doi:10.1108/09544780610671057

Czitrom, V. (1999). One-factor-at-a-time versus designed experiments. *The American Statistician, 53*(2). Retrieved from https://www.jstor.org/stable/2685731

Das, A. K., & Dewanjee, S. (2018a). Optimization of extraction using mathematical models and computation. In *Computational phytochemistry* (pp. 75–106). Elsevier; doi:10.1016/B978-0-12-812364-5.00003-1

Das, A. K., & Dewanjee, S. (2018b). Optimization of extraction using mathematical models and computation. In *Computational phytochemistry* (pp. 75–106). Elsevier; doi:10.1016/B978-0-12-812364-5.00003-1

Davim, J. P. (2012). *Statistical and computational techniques in manufacturing*. Statistical and Computational Techniques in Manufacturing; doi:10.1007/978-3-642-25859-6

Dean, A., Voss, D., & Draguljić, D. (2017). Principles and techniques. In *Design and analysis of experiments* (pp. 1–5). Cham, Switzerland: Springer International Publishing; doi:10.1007/978-3-319-52250-0_1

Dixon, D., Eatock, J., Meenan, B. J., & Morgan, M. (2006). Application of design of experiment (DOE) techniques to process validation in medical device manufacture. *Journal of Validation Technology, 12*(2), 92–100.

Durakovic, B. (2017). Design of experiments application, concepts, examples: state of the art. *Periodicals of engineering and natural sciences, 5(3).* pp. 421–439. Available at http://pen.ius.edu.ba

Farooq, M. A., Nóvoa, H., Araújo, A., & Tavares, S. M. O. (2016). An innovative approach for planning and execution of pre-experimental runs for Design of Experiments. *European Research on Management and Business Economics, 22*(3), 155–161. doi:10.1016/j.iedee.2014.12.003

Freddi, A., & Salmon, M. (2019). Introduction to the Taguchi Method. In *Design principles and methodologies: from conceptualization to first prototyping with examples and case studies* (pp. 159–180). Cham, Switzerland: Springer International Publishing; doi:10.1007/978-3-319-95342-7_7

Giunta, A., Wojtkiewicz, S., & Eldred, M. (2013). Overview of modern design of experiments methods for computational simulations. In *41st Aerospace Sciences Meeting and Exhibit*, Reston, VA. American Institute of Aeronautics and Astronautics. doi:10.2514/6.2003-649

Glyk, A., Solle, D., Scheper, T., & Beutel, S., (2015, Dec. 15). Optimization of PEG-salt aqueous two-phase systems by design of experiments. *Chemometrics and Intelligent Laboratory Systems*. Elsevier. doi:10.1016/j.chemolab.2015.09.014

Griffith, C. R. (2006). The principles of experimental method. In *Principles of systematic psychology* (pp. 436–470). Elsevier/Butterworth-Heinemann; doi:10.1037/11148-012

Griffiths, C. A., Howarth, J., De Almeida-Rowbotham, G., Rees, A., & Kerton, R. (2016). A design of experiments approach for the optimisation of energy and waste during the production of parts manufactured by 3D printing. *Journal of Cleaner Production, 139*, 74–85. doi:10.1016/j.jclepro.2016.07.182

Gutiérres Pulido, H., & De la Vara Salazar, R. (2012). Análisis y diseño de experimentos. *Journal of Chemical Information and Modeling*. doi:10.1017/CBO9781107415324.004

Gutiérrez, H., & De La Vara, R. (2012). *Análisis y diseño de experimentos*. México: McGraw-Hill.

Heiberger, R. M., & Holland, B. (2015). Design of experiments-factorial designs. In *Statistical analysis and data display: an intermediate course with examples in R* (pp. 427–478). New York, NY: Springer New York; doi:10.1007/978-1-4939-2122-5_13

Hernández Sampieri, R. (2014). *Metodología de la investigación. 6th Edición*. McGraw-Hill.

Antony, *J*. (2014). Fractional factorial designs. Elsevier; doi:10.1016/B978-0-08-099417-8.00007-9

AntonyJ, . (2014a). *1 - Introduction to industrial experimentation. Design of experiments for engineers and scientists* (2nd ed., pp. 1–6). Elsevier; doi:10.1016/B978-0-08-099417-8.00001-8

Jacyna, J., Kordalewska, M., & Markuszewski, M. J. (2018). Design of experiments in metabolomics-related studies: An overview. *Journal of Pharmaceutical and Biomedical Analysis, 164*, 598–606. doi:10.1016/j.jpba.2018.11.027 PMID:30469109

Jayashree, P. K., Sharma, S. S., Shetty, R., Mahato, A., & Gowrishankar, M. C. (2018). Optimization of TIG welding parameters for 6061Al alloy using Taguchi's design of experiments. *Materials Today: Proceedings, 5*(11), 23648–23655.

Joachim, K. (2000). *Experimental design - a handbook and dictionary for medical and behavioral research. Techniques in the behavioral and neural sciences* (Vol. 14). Elsevier; doi:10.1016/S0921-0709(00)80012-7

Khorasani, A. M., Yazdi, M. R. S., & Safizadeh, M. S. (2011). Tool life prediction in face milling machining of 7075 Al by using artificial neural networks (ANN) and Taguchi Design of Experiment (DOE). *IACSIT International Journal of Engineering and Technology, 3*(1), 30–35. doi:10.7763/IJET.2011.V3.196

Kul, M., & Çetinkaya, Ü. (2010). Application of the Taguchi orthogonal array design to Cr (VI) solvent extraction. *Mining, Metallurgy, & Exploration, 27*(2), 102–109. doi:10.1007/BF03402386

López, A., Aisa, J., Martinez, A., & Mercado, D. (2016). Injection moulding parameters influence on weight quality of complex parts by means of DOE application: Case study. *Measurement, 90*, 349–356. doi:10.1016/j.measurement.2016.04.072

Madhavi, S. K., Sreeramulu, D., & Venkatesh, M. (2017). Evaluation of optimum turning process of process parameters using DOE and PCA Taguchi method. *Materials today: Proceedings*, *4*(2), 1937–1946.

Molina, A. G., & Diego, S. M. M. (2016). Avanzada científica. Avanzada Científica, 19(1), 2016 (Ejemplar dedicado a: ENERO-FEBRERO-MARZO-ABRIL), pp. 1-11. Centro de Información y Gestión Tecnológica.

Montgomery, D. C. (2005). *Design and analysis of experiment* (5th ed.). John Wiley & Sons.

Montgomery, D. C. (2009). *Introduction to statistical quality control* (6th ed.). Jefferson City: John Wiley & Sons.

Montgomery, D. C. (2017). *Montgomery: design and analysis of experiments. John Wiley & Sons* (8th ed.)

Narenderan, S. T., Meyyanathan, S. N., & Karri, V. V. S. R. (2019). Experimental design in pesticide extraction methods: A review. *Food Chemistry*, *289*, 384–395. doi:10.1016/j.foodchem.2019.03.045 PMID:30955627

Niedz, R. P., & Evens, T. J. (2016). Design of experiments (DOE)—history, concepts, and relevance to in vitro culture. *In Vitro Cellular & Developmental Biology. Plant*, *52*(6), 547–562. doi:10.100711627-016-9786-1

NIST/SEMATECH. (2012). What is design of experiments (DOE)? Engineering Statistics Handbook. e-Handbook of Statistical Methods. Retrieved from https://www.itl.nist.gov/div898/handbook/pmd/section3/pmd31.htm

Okoro, O. V., Sun, Z., & Birch, J. (2019). Thermal depolymerization of biogas digestate as a viable digestate processing and resource recovery strategy. In *Advances in eco-fuels for a sustainable environment* (pp. 277–308). Woodhead Publishing; doi:10.1016/B978-0-08-102728-8.00010-3

Ong, S. K., Yuan, M. L., & Nee, A. Y. C. (2008). Augmented reality applications in manufacturing: a survey. *International Journal of Production Research*, *46*(10), 2707–2742. doi:10.1080/00207540601064773

ONUDI. (2015). Organización de las Naciones Unidas para el Desarrollo Industrial. Informe sobre el Desarrollo Industrial 2016. El rol de la tecnología y la innovación en el desarrollo industrial inclusivo y sostenible. Resumen. Viena. Número de identificación de ONUDI: 447. Retrieved from https://www.unido.org/sites/default/files/2015-12/EBOOK_IDR2016_OVERVIEW_SPANISH_0.pdf

Ou, Z., Zhang, M., & Qin, H. (2019). Tripling of fractional factorial designs. *Journal of Statistical Planning and Inference*, *199*, 151–159. doi:10.1016/j.jspi.2018.06.002

Ranga, S., Jaimini, M., Sharma, S. K., Chauhan, B. S., & Kumar, A. (2014). A review on design of experiments (DOE). *International Journal of Pharmaceutical and Chemical Sciences*, *3*(1), 216–224. Retrieved from www.ijpcsonline.com

Romeu, J. L. (2006). Teaching Engineering Statistics to Practicing Engineers, ICOTS-7, In *7th International Conference on Teaching Statistics*, Salvador, Brazil.

Rytz, A., Moser, M., Lepage, M., Mokdad, C., Perrot, M., Antille, N., & Pineau, N. (2017). Using fractional factorial designs with mixture constraints to improve nutritional value and sensory properties of processed food. *Food Quality and Preference*, *58*, 71–75. doi:10.1016/j.foodqual.2017.01.004

Salazar, J. C., & Baena-Zapata, A. (2009). Dyna. *Dyna (Bilbao)*, *76*(159), 249–257.

Sanchez, G. S. (2009). *Medición del ruido en la línea de transmisión eléctrica generado por electrodomésticos, para aplicaciones de PLC*. Universidad de las Américas Puebla.

Silva, D. A. L., Filleti, R. A. P., Christoforo, A. L., Silva, E. J., & Ometto, A. R. (2015). Application of Life Cycle Assessment (LCA) and Design of Experiments (DOE) to the monitoring and control of a grinding process. *Procedia CIRP*, *29*, 508–513. doi:10.1016/j.procir.2015.01.037

Silva, V., (2018). *Statistical approaches with emphasis on design of experiments applied to chemical processes*. BoD--Books on Demand.

Tanco, M., Viles, E., Ilzarbe, L., & Álvarez, M. J. (2007). Manufacturing industries need design of experiments (DoE). *Proceedings of the World Congress on Engineering, Vol. II*, WCE, London, UK.

Tanco, M., Viles, E., Ilzarbe, L., & Álvarez, M. J. (2008). Is design of experiments really used? A survey of Basque industries. *Journal of Engineering Design*, *19*(5), 447–460. doi:10.1080/09544820701749124

Tanco, M., Viles, E., Ilzarbe, L., & Álvarez, M. J. (2009). Barriers faced by engineers when applying design of experiments. *The TQM Journal*, *21*(6), 565–575. doi:10.1108/17542730910995846

Tanco, M., Viles, E., & Pozueta, L. (2009). Diferentes enfoques del diseño de experimentos (DOE), *7*, 29–37.

Tye, H. (2004). Application of statistical 'design of experiments' methods in drug discovery. *Drug Discovery Today*, *9*(11), 485–491. doi:10.1016/S1359-6446(04)03086-7 PMID:15149624

United Nations. (2015). Sustainable development goals, knowledge platform. Goal N.12 of the 2030 Agenda for Sustainable Development aims to ensure sustainable consumption and production patterns. Retrieved from https://sustainabledevelopment.un.org/topics/sustainableconsumptionandproduction

Vanaja, K., & Rani, R. H. S. (2007). *Design of experiments: Concept and applications of Plackett Burman design. Clinical Research and Regulatory Affairs*. Taylor & Francis; doi:10.1080/10601330701220520

Woods, D. C., & Lewis, S. M. (2017). Design of experiments for screening. Handbook of uncertainty quantification, 1143–1185.

World Bank. (2017). The World Bank Group, IBRID-IDA. Manufacturing, value added (% of GDP). World Bank national accounts data, and OECD National Accounts data files. License: CC BY-4.0. Retrieved from https://data.worldbank.org/indicator/NV.IND.MANF.ZS

Yondo, R., Andrés, E., & Valero, E. (2018). A review on design of experiments and surrogate models in aircraft real-time and many-query aerodynamic analyses. *Progress in Aerospace Sciences*, *96*, 23–61. doi:10.1016/j.paerosci.2017.11.003

Yu, P., Low, M. Y., & Zhou, W. (2018). Design of experiments and regression modelling in food flavour and sensory analysis: A review. *Trends in Food Science & Technology*, *71*, 202–215. doi:10.1016/j.tifs.2017.11.013

KEY TERMS AND DEFINITIONS

Critical Factors: The main factors that have an impact on the optimization or improvement in the outcome of an experiment.

Design of Experiments: Set of techniques, that do not expect the process to send the expected signals, but that it is "manipulated" to provide the information that is required for its improvement.

Experimentation: The action or process of performing a scientific procedure in order to determine something in and object, process or study phenomenon.

Manufacturing Industry: The economic sector based on the fabrication or processing of products from raw materials and commodities.

Optimization: Practice that allows establishing the right variables involved in a process or experiment to achieve the best results.

Process: Set of predetermined and successive phases or procedures to achieve an expected objective or result.

Statistical Tools: Set of techniques that allow the process improvement and the reduction of errors for troubleshooting.

Statistics: It is a science based on solid mathematical theorems that come from irrefutable logical laws. It consists of the collection, treatment, interpretation, and analysis of a set of data that allows the understanding of a phenomenon and the correct decision making.

Section 5
Diverse Applications

Chapter 13
Two-Level Factorial Designs

Enrique Barrado
https://orcid.org/0000-0002-9563-0376
Universidad de Valladolid, Spain

T. Alexandra Ferreira
https://orcid.org/0000-0001-8159-8611
Universidad Autonoma del Estado de Hidalgo, Mexico

ABSTRACT

The term design of experiments in analytical chemistry is associated to the establishment of adequate experimental conditions when working in the laboratory or process conditions used in industry to improve the instrumental conditions and/or extract the highest information from the experimental data. This chapter presents practical problem-solving strategies used to obtain a product or chemical process with desirable characteristics in an efficient mode, focused on the use of full and fractional (2-level) designs. The information is presented as a tutorial and the main advantages and disadvantages are presented and discussed, emphasizing the effect of reduction of experimentation in the data quality.

INTRODUCTION

Design of experiments in analytical chemistry is a term associated to the establishment of adequate process conditions in order to: improve the instrumental conditions and/or extract the highest information from the experimental data (Otto, 2017). Before apply a design of experiments strategy it is appropriate to consider the following questions (Miller & Miller, 2010): a) the initial knowledge of the system, what variables or factors (and their levels) are important to consider in our study? b) the weight of each variable in the system, what variable has a higher contribution? and are there interaction between variables? c) optimize the response, what is the response associated to the highest quality of the product/process? and d) evaluate the system robustness, what is the effect of uncontrolled variables?

DOI: 10.4018/978-1-7998-1518-1.ch013

A system can be considered as a conjunct of elements interdependent forming and integrating whole (Deming & Morgan, 1993). The system has an entrance (generally the analyte) expressed as a quantity or quality, the factors involved in the response of the system, an output variable (signal) influenced by the processes and chemical reactions which took place during the transformation of entrance into response.

The main drawback in all cases is the existence of factors so-called uncontrolled which are included in all the experiments, these factors cannot be eliminated and affect the response by adding a variability or variance to the output variable. The variance related to the response has then two contributions: variance of the signal and a second one known as uncertainty or noise. In this sense, the design of experiments can be applied to estimate and quantify the contributions of the controlled factors and minimize the uncontrolled (Figure 1).

The proposed chapter is planned to be a tutorial for application and resolution of experimental problems based on the use of *full and fractional (2-level) designs* of experiments. Taking into account the proposed aim, we must consider the following steps commonly employed when it is applied design of experiments (Goupy, 1993):

1. Problem statement
   a. Definition of the response or output variable which must describe adequately the process.
   b. Identification of the factors (continue or discrete)
   c. Selection of the levels and experimental domain
   d. Identification of uncontrolled factors (noise)
   e. Definition of signal/noise ratio
   f. Evaluation of the system robustness
   g. Selection of the experimental methodology (experimental arrangement)
2. Experimentation
   a. Randomization
   b. Blocks
   c. Replicates
3. Results and analysis (interpretation)
   a. Analysis of variance (ANOVA)
4. Optimization
   a. Selection of the experimental conditions which generate the adequate signal/noise ratio
5. Conclusions
   a. Feed-back

Figure 1. Scheme of a system

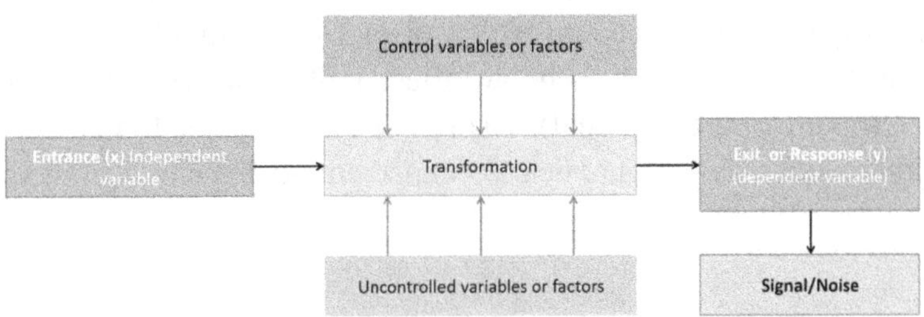

FULL FACTORIAL DESIGN

The first practical case is the evaluation of the instrumental conditions employed for analysis of metallic ions in water samples by differential pulse polarography at hanging mercury electrode. The evaluation of the factors was performed employing a complete factorial design (Masetto et al. 2011). The proposed example would be useful to demonstrate the importance of identification, selection and randomization of the experiments. It is proposed the inclusion a tutorial to determine the analysis of variance terms and the interpretation of the results obtained. Additionally, the possible of minimize the experimentation would be discussed.

In this sense, the proposed problem is achieving the best limit of detection possible which is directly associated to the highest peak height (or area). Once defined the output variable, it is possible to propose the chemical transformation related to our goal, in this case it can be the instrumental technique employed (i.e. spectroscopy, chromatography, electrochemical). The technique employed must be studied in order to identify the possible control factors involved. If metal ions are determined in water samples by polarography followed by anodic stripping voltammetry the main factors are the electrolyte composition (pH, presence of complexing agents, ionic strength) and the instrumental conditions (drop size, deposition time, pulse height, etc.) (Zinoubi, et al. 2017). All these numeric factors are continue, which means that they their variation in a scale is continue despite they are limited (i.e. pulse height adequate interval is from -10 to -100 mV). The values evaluated are so-called levels and the set selected is the experimental domain.

The factors involved in the problem must be evaluated deeply, they must be identified and separate them in control and noise factors with the corresponding experimental domain. This information is necessary to determine the methodology adequate to establish the factors that have a significant effect, the levels which generate the highest signals and in consequence the conditions with a minimal contribution of un-controlled factors.

The experimental tool useful for this purpose is the complete factorial design. This is a randomized block experimental design with replicates. The complete factorial design studies all the possible combinations of the levels and factors, randomizing the order and replicating each combination in order to have a degree of freedom useful to determine the residual variance (associated to noise). The use of random experimentation minimize the inclusion of systematic error (which invalidate the results) and the blocks are required when it is required continue experimentation. The factorial design is antithesis of the classical design that evaluates one variable at time while the others remain constants.

In order to explain the construction of a experimental matrix, it is presented the evaluation of pH and ionic strength (I) on the peak height. The following experimental matrix is presented in Figure 2, where X_{ijk}= mean of each pair of experiments.

According to Figure 2, it is required to be performed 16 randomized experiments (in duplicate: 16x2=32). The experimental conditions of each pair (1,2) (19,20), etc are equal, and the differences observed would be a consequence of the random error (noise). On the other hand, it is possible to estimate the variance employing the mean value of each row (effect of pH), each column (effect of ionic strength) and the uncertainty (random error or residual). The last one is attributed to the interaction between the control factors and this contribution cannot be determined employing classical methodologies such as: one variable at time.

Figure 2. Factorial design for two factors
**subscripts make reference to the evaluated levels of each factor*

		Ionic strength (I)			
		I_1	I_2	I_3	I_4
pH	pH_1	1,2	3,4	5,6	7,8
	pH_2		x_{ijk}		
	pH_3	17, 18	19,20	21,22	23,24
	pH_4	25,26	27,28	29,30	31,32

Determination of the contribution of each variable and the interaction is shown in Figure 3,; r=rows number; n=columns number; p=number of replicates in each cell; N=prn=number of total experiments, T_{ij}= sum of each cell=X_{ij}; T_{if}=sum of row i, T_{cj}=sum of column j; $T=\Sigma T_{if}=\Sigma T_{cj}$ and $C=T^2(rnp)^{-1}$=correction term. The result most interesting is the residual (noise).

The main disadvantage of the complete design of experiments is the high amount of experiments required; this fact limits its applicability (Kobilinsky et al. 2017). However, considering the potential of this statistical tool it was proposed the use of two-level factorial designs (2^{k+1}), where k is the number of factors (Tounsadi et al 2019). This type of design of experiments limits the levels evaluated. The reduction of the experimental domain produces two problems: the first assume a linear response of the out-variable *vs.* each factor (which cannot be true) and the second is the loss of a degree of freedom, then an algorithm must be used.

A practical example is described, considering the example mentioned above it is desired to evaluate the effect of the mercury dropping time (t) and the pulse potential (U) in the peak height of the response obtained during the analysis of copper in water samples. Considering that minimize of experimentation is required, it was selected a two-level factorial design (2^{2+1}), which means the evaluation of two factors at two levels with two replicates to determine the effect of the following factors:

Figure 3. Equations employed for the analysis of variance of a full factorial design

Source of variation	**Square Sum**	**Degrees of freedom**
Between rows	$\sum \frac{T_{if}^2}{pn} - C$	r - 1
Between columns	$\sum \frac{T_{cj}^2}{pr} - C$	n - 1
Interaction	**By subtraction**	**By subtraction**
Residual	$\sum x_{ijk}^2 - \left(\sum \frac{T_{ij}^2}{p}\right)$	r.n.(p - 1)
Total	$\sum x_{ijk}^2 - C$	p.r.n - 1

t (s) = 0.4 (low level, -1) and 0.8 (high level, +1)

U (mV) -20 (low level, -1) and -100 (high level, +1)

The number of experiments is: $2^{2+1} = 2^3 = 8$ ($2^2 = 4$ experiments with 2 repetitions = 8 experiments). The experimental matrix obtained is shown in Figure 4.

A graphical representation of the experimental data is presented in Figure 5. The variable that changes in experiments Y_1 and Y_2 is U (from low to high level) and the variable t remains constant (low level), while in experiments Y_3 and Y_4 U changes and t is at high level. This is represented in blue line, and the effect can be determined as the mean sum of the differences:

$$Effect\,of\,U = \frac{(Y_2 - Y_1) + (Y_4 - Y_3)}{2} = \frac{(39.25 - 23.95) + (41.20 - 25.40)}{2} = 15.85$$

Considering the associative property of the sum, the effect can be estimated considering the difference between the sum of the response at high (Y_2 and Y_4) and low (Y_1 and Y_3) levels:

$$Effect\,of\,U = \frac{(Y_2 + Y_4) - (Y_1 + Y_3)}{2} = \frac{(39.25 + 41.80) - (23.95 + 25.40)}{2} = 15.85$$

Figure 4. Experimental matrix 22+1

Factors and levels		Denomination	Individual responses (μA)	Mean response (μA)
t	U			
-1	-1	Y_1	24.0 23.9	23.95
-1	+1	Y_2	39.5 39.0	39.25
+1	-1	Y_3	25.9 24.9	25.40
+1	+1	Y_4	40.9 42.7	41.80

Figure 5. Graphical representation of the 22 two level factorial design

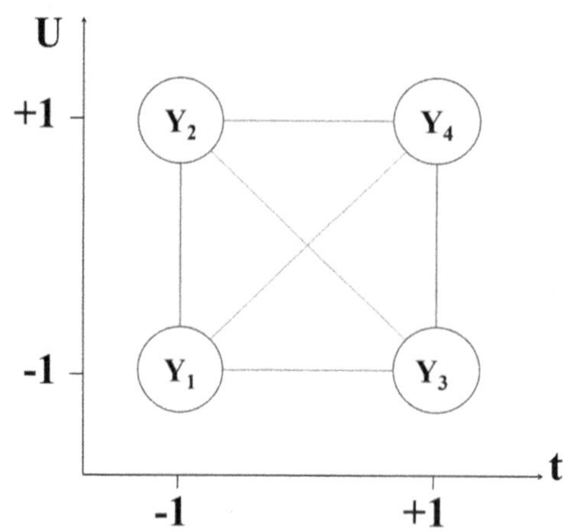

The effect of t can be obtained following the red line in Figure 5, the low levels (-1) are Y_1 and Y_2 while the high level (+1) are Y_3 and Y_4.

$$Effect\ of\ t = \frac{\left(Y_3 + Y_4\right) - \left(Y_1 + Y_2\right)}{2} = \frac{\left(25.40 + 41.80\right) - \left(23.95 + 39.25\right)}{2} = 2.00$$

The interaction is obtained from Figure 5 applying the rules of signs for multiplication, generating a new column tU which represents the interaction between these factors. The Figure 6 shows the complete experimental design. The effect of tU is represented in green (Figure 5) and its effect was calculated following the explanation described above: high levels (Y_1, Y_4) low levels (Y_2, Y_3).

$$Effect\ of\ tU = \frac{\left(Y_1 + Y_4\right) - \left(Y_2 + Y_3\right)}{2} = \frac{\left(23.95 + 41.80\right) - \left(39.25 + 25.40\right)}{2} = 0.55$$

The effect values are semi-quantitative are useful to evaluate the behavior of each evaluated factor in the response. In order to demonstrate their significance, it is required to determine the variance for each variable. It must be remembered that a negative consequence of the two-level factorial designs is the loss of one degree of freedom. Yates' algorithm has been applied to determine the square sum (SS) using the expression (Massart et al. 1997):

$$Square\ sum\left(SS_t\right) = \frac{N(Effect\ of\ t)^2}{4} = \frac{8(2.00)^2}{4} = 8.00$$

where: N is total number of experiments. SS estimation has one degree of freedom and the value obtained is equal to the mean squares (MS) or variance. Critical factors are selected using *F* test, the $SS_{residual}$ was estimated:

$$SS_{residual} = \Sigma x_{ijk}^2 - \frac{\Sigma\left(pT_{if}^2\right)}{p}$$

$$\Sigma x_{ijk}^2 = \left[24.0^2 + 23.9^2 + 39.5^2 + 39.0^2 + 25.9^2 + 24.9^2 + 40.9^2 + 42.7.0^2\right] = 9015.38$$

Figure 6. Complete experimental design (factors+ interaction)

Factors and levels	
t	U
-1	-1
-1	+1
+1	-1
+1	+1

⇨

Interaction	Denomination
tU	
+1	Y_1
-1	Y_2
-1	Y_3
+1	Y_4

$$\frac{\Sigma\left(pT_{if}^2\right)}{p} = \frac{(2*23.95)^2 + (2*39.25)^2 + (2*25.20)^2 + (2*41.80)^2}{2} = 9013.13$$

$$SS_{residual} = 9015.38 - 9013.13 = 2.25$$

The $MS_{residual}$ was obtained as:

$$MS_{residual} = \frac{SS_{residual}}{Degrees\ of\ freedom}$$

$$MS_{residual} = \frac{2.25}{7-3} = 0.5625$$

Once determined the variance of each factor, the experimental *F* (F_{exp}) are calculated as $MS_{variable}/MS_{residual}$. The results are contrasted to the critical *F* ($F_{1,4}$=7.71) and a variable is considered critical when $F_{exp}>F_{crit}$. The results obtained are collected in Figure 7, it can be concluded that in the proposed case both factors are critical, in addition the contribution of U is higher than t. It is important to observe that interaction tU is not critical and it can be estimated without additional experimentation.

The analysis of three factors involves the use of a 3 dimension graphic which is still suitable. In order to explain its applicability, it is presented the evaluation of a spectrophotometry methodology for analysis of Fe(III) based on liquid-liquid extraction of its complex with oxine (Chobot et al. 2018, Bahar, Zakerian 2012). The response variable (out-put or response) is the absorbance value and the control factors evaluated are: pH of the aqueous phase (Factor A), extraction time (Factor B) and volume of chloroform employed (Factor C). According to Figure 8 it is possible to observe the presence of different oxine species depending on the acidity of the media, the time is associated to the extraction yield and affects the analysis time while the solvent volume ratio (aqueous:organic) employed is related to the maximum concentration and in consequence affects the limit of detection (Hanson, 2013)..

Figure 9 shows the required experimental matrix, it contemplates to perform 2^{3+1}=16 experiments (eight combinations in duplicate). Each row shows the combination of the factors and the individual results, additionally it is included the mean values used for the analysis of the experimental data.

Figure 7. Analysis of variance of the analysis of a 22+1 factorial design
[a] $F_{crit\,(1,4)}$=7.71

Factor	Effect	Sum of Squares (SS)	Degree of freedom	Mean Squares (MS)	F_{exp} [a]
t	2.00	8.000	1	8.0000	14.22
U	15.85	502.445	1	502.4450	893.24
tU	0.55	0.605	1	0.6050	1.08
Residual		2.25	4	0.5625	

Figure 8. Schematic representation and reactions involved on separation of Fe(III) by liquid-liquid extraction using oxine

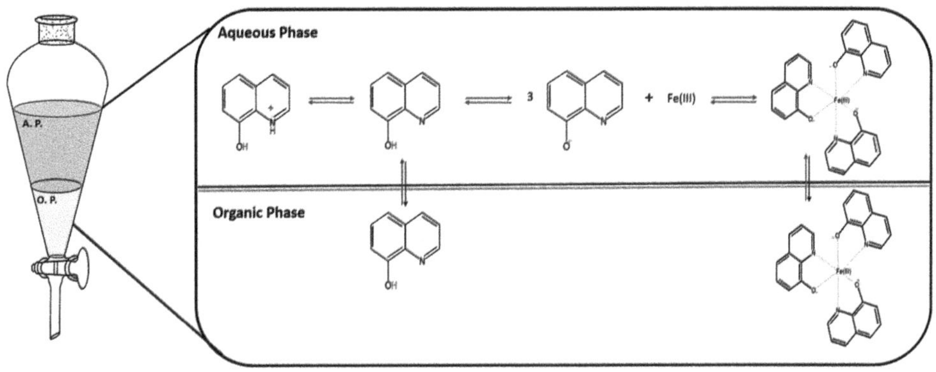

Figure 9. Experimental matrix 23+1 and 3D graphic used to evaluate 3 factors two level factorial designs

A	B	C	Denomination	Individual responses (A.U.)	Mean response (A.U.)
-1	-1	-1	Y_1	0.290 0.286	0.288
+1	-1	-1	Y_2	0.550 0.479	0.515
-1	+1	-1	Y_3	0.304 0.306	0.305
-1	-1	+1	Y_4	0.250 0.243	0.247
-1	+1	+1	Y_5	0.255 0.257	0.256
+1	-1	+1	Y_6	0.438 0.450	0.444
+1	+1	-1	Y_7	1.320 1.330	1.325
+1	+1	+1	Y_8	1.350 1.320	1.335

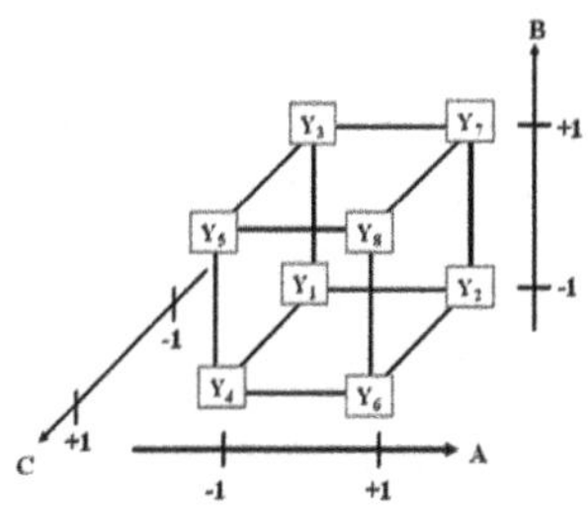

Considering the experimental data, the first step is the determination of the individual effects. Figure 10 includes the corresponding expression and the corresponding cubic part analysed. It must be remembered that it is used the Mean response to perform the determinations.

Once evaluated the individual contributions, it is determined the interactions. For three factors there are 2-way interactions (AB, AC and BC) and one 3-way interaction (ABC). The corresponding column was obtained applying the rules of signs for multiplication of the experimental matrix presented in Figure 9. The complete experimental matrix is composed of 7 columns (Figure 11) and it is useful to determine the effect of 2- and 3-way interactions, the graphical representation and the equations employed are presented in Figure 12.

Once determined the effect of all the factors included in the experimental design, it was determined the SS applying the Yates' algorithm, as example:

$$SS_A = \frac{16(Effect\,of\,A)^2}{4} = \frac{16(0.631)^2}{4} = 1.592$$

the effect of A=. The $SS_{residual}$ was determined employing the individual and mean data as mentioned above:

Figure 10. Determination of individual effects in a 23+1 experimental design

Factor	Effect	Representation
A	$\dfrac{(Y_2 + Y_6 + Y_7 + Y_8) - (Y_1 + Y_3 + Y_4 + Y_5)}{4}$	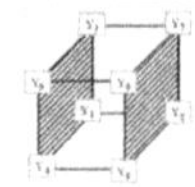
B	$\dfrac{(Y_3 + Y_5 + Y_7 + Y_8) - (Y_1 + Y_2 + Y_4 + Y_6)}{4}$	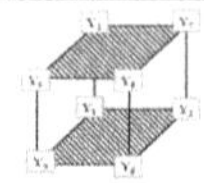
C	$\dfrac{(Y_4 + Y_5 + Y_6 + Y_8) - (Y_1 + Y_2 + Y_3 + Y_7)}{4}$	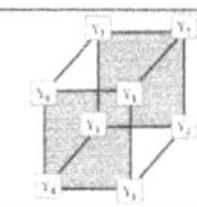

Figure 11. Complete experimental matrix of a 23+1 design

Factor			2-way interaction			3-way interaction	Denomination
A	B	C	AB	AC	BC	ABC	
-1	-1	-1	+1	+1	+1	-1	Y_1
+1	-1	-1	-1	-1	+1	+1	Y_2
-1	+1	-1	-1	+1	-1	+1	Y_3
-1	-1	+1	+1	-1	-1	+1	Y_4
-1	+1	+1	-1	-1	+1	-1	Y_5
+1	-1	+1	-1	+1	-1	-1	Y_6
+1	+1	-1	+1	-1	-1	-1	Y_7
+1	+1	+1	+1	+1	+1	+1	Y_8

Figure 12. Determination of 2 and 3 way effects in a 23+1 experimental design

Factor	Effect	Representation
AB	$\dfrac{(Y_1 + Y_4 + Y_7 + Y_8) - (Y_2 + Y_3 + Y_5 + Y_6)}{4}$	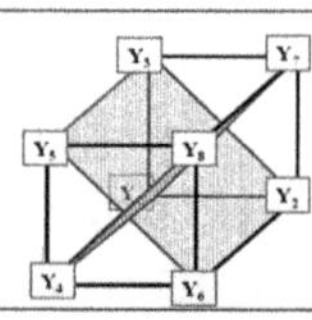
AC	$\dfrac{(Y_1 + Y_3 + Y_6 + Y_8) - (Y_2 + Y_4 + Y_5 + Y_7)}{4}$	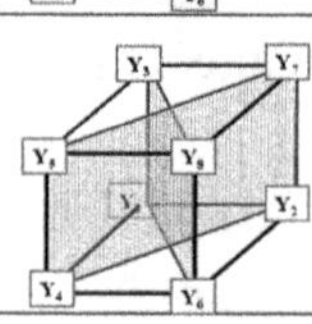
BC	$\dfrac{(Y_1 + Y_2 + Y_5 + Y_8) - (Y_3 + Y_4 + Y_6 + Y_7)}{4}$	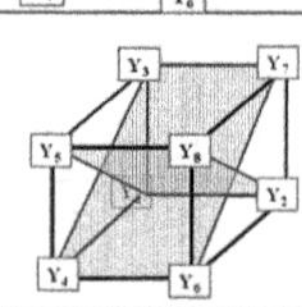
ABC	$\dfrac{(Y_2 + Y_3 + Y_4 + Y_8) - (Y_1 + Y_5 + Y_6 + Y_7)}{4}$	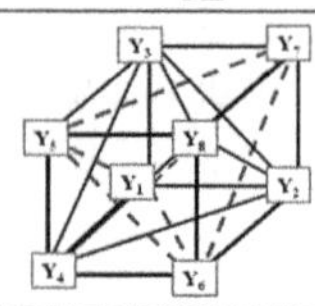

$$SS_{residual} = \sum x_{ijk}^2 - \frac{\sum\left(pT_{if}^2\right)}{p} = 8.607056 - \frac{17.207854}{2} = 0.003129\,.$$

The results obtained are collected in Figure 13. All the selected factors have a significant statistical contribution to the methodology, however A (pH), B(extraction time) and the 2-way interaction AB have a greater contribution. The AC, BC and ABC interactions are not significant, since their F_{exp} is not greater than the critical F_{value}= 5.32.

Duplicate experiments cannot be obtained in all cases, in these circumstances the use of non-parametric test are an interesting option (Fraser, 1956). The ranking test (Figure 14) is the statistical tool for evaluation of critical parameters (Conover & Iman, 1981). If it is supposed that results of mean values (Figure 9) corresponds to a unique experiment, it can be determined the effect of each value into the response (second column in Figure 13). The effects must be then ordered ascending (low to high) and identified. One ranks the effects from the most negative to the most positive and then proceeds to visualize the distribution (Massart, *et al.* 1997). The probability of each point was determined and it is represented the effect of each variable *vs.* probability.

A single line must be obtained when the experimental data are compatible with a normal distribution. If there are factors that do not belong to this distribution (critical variables) they would be outside the line. The graphic representation is shown in Figure 15, the factors AC, BC and ABC are align, the variable C is not clear enough. However, the factors A, B and AB are clearly critical factors which contribution is higher than random errors. The results obtained are congruent to previous analysis but the information is not accurate because of the loss of information in the experimental data (Cornell & Gorman, 1984).

Factorial designs are an interesting alternative to select better experimental conditions (Montgomery. 2017). The mean graphic can also be obtained from the experimental data and it can be used to select and evaluate qualitatively the contribution of each factor. Using the experimental data in Figure 9, it must be determined the contribution of each variable at low and high levels. The evaluation of factor A (pH) is explained in Figure 16. The contribution of A at low and high levels is the mean value of the variable at -1 and +1 levels.

The same scheme was followed to evaluate the contribution of each variable and the values are connected by a single line. The Figure 17 shows the mean graphic obtained, contribution of each factor can be estimated by the distance between low and high levels for each factor, which means that A and

Figure 13. Analysis of variance of the analysis of a 23+1 factorial design
[a] $F_{crit\,(1,8)}$=5.32

Factor	Effect	Sum of Squares (SS)	Degree of freedom	Mean Squares (MS)	F_{exp} [a]
A	0.631	1.592	1	1.592	4068.72
B	0.432	0.746	1	0.746	1908.59
C	-0.038	0.006	1	0.006	14.57
AB	0.419	0.701	1	0.701	1793.31
AC	0.007	0.000	1	0.000	0.58
BC	0.018	0.001	1	0.001	3.41
ABC	0.022	0.002	1	0.002	4.95
Residual		0.003129	8	$3.91x10^{-4}$	

Figure 14. Ranking for non-replicate two level factorial design (23)

Factor	Effect
A	0.631
B	0.432
C	-0.038
AB	0.419
AC	0.007
BC	0.018
ABC	0.022

Factor	Effect	Ranking	Probability
C	-0.038	1	$100\frac{1-0.5}{7}=7.1$
AC	0.007	2	$100\frac{2-0.5}{7}=21.4$
BC	0.018	3	$100\frac{3-0.5}{7}=35.7$
ABC	0.022	4	$100\frac{4-0.5}{7}=50.0$
AB	0.419	5	$100\frac{5-0.5}{7}=64.3$
B	0.432	6	$100\frac{6-0.5}{7}=78.6$
A	0.631	7	$100\frac{7-0.5}{7}=92.9$

Figure 15. Ranking test for analysis of two level factorial design

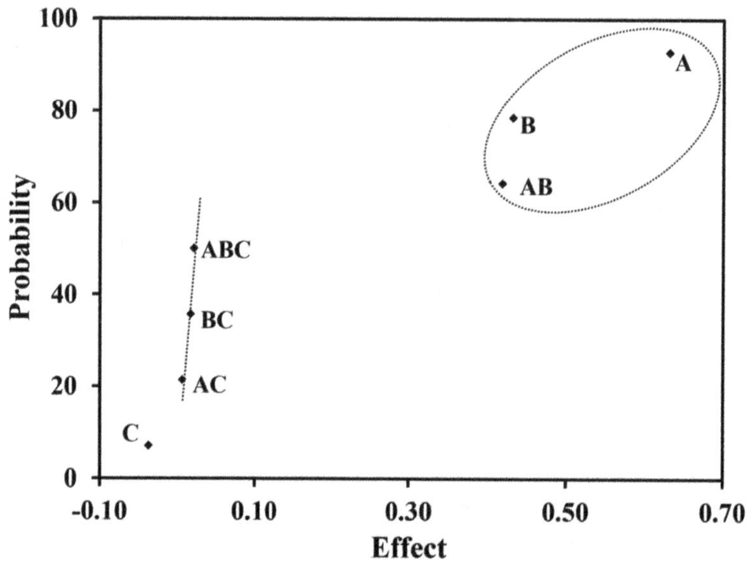

Figure 16. Determination of mean contribution at low (-) and high (+) levels for factor A

A	Denomination	Mean response (A.U.)	Mean low/high levels
-1	Y_1	0.288	$(A-)=\frac{Y_1+Y_3+Y_4+Y_5}{4}=\frac{0.288+0.305+0.247+0.256}{4}=0.274$
+1	Y_2	0.515	
-1	Y_3	0.305	
-1	Y_4	0.247	
-1	Y_5	0.256	$(A+)=\frac{Y_2+Y_6+Y_7+Y_8}{4}=\frac{0.515+0.444+1.325+1.335}{4}=0.90475$
+1	Y_6	0.444	
+1	Y_7	1.325	
+1	Y_8	1.335	

B contribute more to the response than C. The conditions selected to obtain the highest response are: A+, B+ and C-, which corresponds to experiment Y_7. From the experimental data it was inferred that experiment Y_8 has the highest response, nevertheless the contribution of C is not enough and there are no statistical differences between experiments Y_7 and Y_8.

The main problem is the analysis of more variables, the increment of the experiments increase exponentially. The analysis of 4 and 5 factors requires 32 and 64 experiments when considering replicates of the experiments, otherwise the number of experiments required would be 16 and 32, respectively. It is possible to minimize the amount of experiments by applying the fractional experimental designs.

FRACTIONAL DESIGN

The use of fractional factorial designs requires a deep knowledge of the studied system. In this section, it must be explained the advantages and disadvantages of the reduction of information. In the last example, it was concluded that factors A (pH) and B (extraction time) and their interaction AB were critical. In this case notation must be changed in order to minimize the mistake. The two level factorial design employed for 2 factors require performing 4 experiments in duplicate (8 in total) while the 3 factors involves 8 experiments in duplicate (16 experiments). The 2^{2+1} design generates 3 columns: two for factor and a two-way interaction. If the effect of interaction is not critical it can be neglected and we can evaluate an additional variable using this column. In the previous case, the AB interaction was critical and the column cannot be substituted (Robinson et al., 2004).

The interaction between pH (A) and organic solvent volume (C) was the value with less variance. These factors were selected to generate a 2^{2+1} matrix and the interaction was used to evaluate the effect

Figure 17. Means graphic for analysis of a two level factorial design

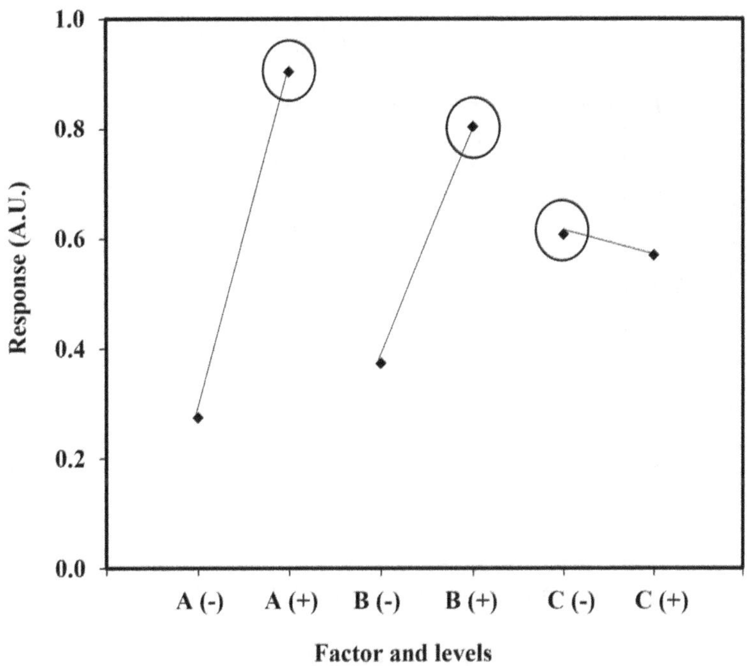

Figure 18. Generating of a Fractional factorial design

Factors		Interaction
A	C	AC
-1	-1	+1
+1	-1	-1
-1	+1	-1
+1	+1	-1

Factors		Factor added	Individual	Mean
A	C	B	Responses	Responses
-1	-1	+1	0.304 0.306	0.305
+1	-1	-1	0.550 0.479	0.515
-1	+1	-1	0.250 0.243	0.247
+1	+1	+1	1.350 1.320	1.335

of extraction time. The initial 2^{2+1} design and the fractional one are presented in Figure 18, it must be focused that it would be evaluated 3 factors with 4 experiments in duplicate (8 total), promoting a reduction of the experiments needed.

The mean graphic is presented in Figure 19, it can be observed that the optimal conditions corresponds to the experiment Y_8, however the contribution is similar for the 3 factors which is a different conclusion that the obtained with the complete design (2^{3+1}). The difference is attributed to the inclusion of the variance of AC and the variance of the variable. However, the three factors have significant contribution and this result is congruent with the former design but using the half amount of the experiments.

Fractional factorial designs allow the user to improve the experimental conditions reducing the number of experiments needed, in this way the analysis is more economical and the responses are obtained quickly, this is desirable especially at industrial level.

Figure 19. Means graphic for analysis of a fractional factorial design

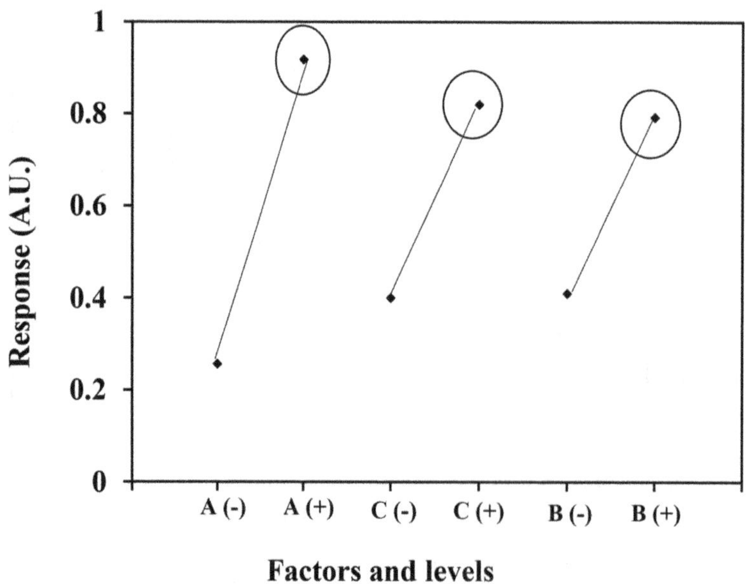

FUTURE RESEARCH DIRECTIONS

In perspective, evaluation and identification of the factors with significant effect must be a previous step before application of an optimization procedure. Therefore, one of the objectives of the 2-level designs is to find the factors that are significant. Once these factors have been identified, then optimization designs can be used (with at least 3 levels). It must be considered the knowledge of the system in order to minimize the experimentation required. Therefore, there are different designs of experiments options used to evaluate a process, however in chemistry it should be considered the economic, environmental and the time required for each test.

CONCLUSION

The use of factorial (2- level) design of experiments is an alternative to minimize experimentation and obtain the highest amount of information when experimental work is carried out. The use of two level factorial designs is a screening technique to evaluate a system; it can be applied not only for chemical purposes it can be extrapolated to industrial, pharmaceutical and food area. It is important to remember that any system can be evaluated without a previous knowledge of it. Before apply a design of experiments it must be considered the real experimental conditions than can be modified, with factorial designs the optimal conditions cannot be achieved due to practical consideration, since 2 level designs are just for monitoring, but it is possible to increase the response in order to find better experimental conditions. If an optimization is required, experimental designs with at least 3 levels can be used.

ACKNOWLEDGMENT

Authors thank the Consejería de Educación of the Junta de Castilla y León Project (Spain) VA171U14 for financial support. T. A. Ferreira, also thanks the Consejo Nacional de Ciencia y Tecnologia (Mexico) for the PhD. Grant.

REFERENCES

Bahar, S., & Zakerian, R. (2012). Speciation of Fe(II) and Fe(III) by using dispersive liquid-liquid microextraction and flame atomic absorption spectrometry. *Journal of the Brazilian Chemical Society*, *23*(5), 944–950. doi:10.1590/S0103-50532012000500021

Chobot, V., Hadacek, F., Bachmann, G., Weckwerth, W., & Kubicova, L. (2018). Antioxidant properties and the formation of iron coordination complexes of 8-hydroxyquinoline. *International Journal of Molecular Sciences*, *19*(12), 3917. doi:10.3390/ijms19123917 PMID:30544490

Conover, W. J., & Iman, R. L. (1981). Rank transformation as a bridge between parametric and nonparametric statistics. *The American Statistician*, *35*(3), 124–129.

Cornell, J. A., & Gorman, J. W. (1984). Fractional design plans for process variables in mixture experiments. *Journal of Quality Technology*, *16*(1), 20–38. doi:10.1080/00224065.1984.11978883

Deming, S. N., & Morgan, S. L. (1993). *Experimental design: A chemometric approach*. Amsterdam, The Netherlands: Elsevier.

Fraser, D. A. S. (1956). *Nonparametric methods in statistics*. Hoboken, NJ: John Wiley & Sons. doi:10.1037/13191-000

Goupy, J. L. (1993). *Methods for experimental design*. Amsterdam, The Netherlands: Elsevier.

Hanson, C. (2013). *Recent advances in liquid-liquid extraction*. Elsevier.

Kobilinsky, A., Monod, H., & Bailey, R. A. (2017). Automatic generation of generalised regular factorial designs. *Computational Statistics & Data Analysis*, *113*, 311–329. doi:10.1016/j.csda.2016.09.003

Masetto, A., Flores-Cotera, L. B., Díaz, C., Langley, E., & Sanchez, S. (2001). Application of a complete factorial design for the production of zeaxanthin by Flavobacterium sp. *Journal of Bioscience and Bioengineering*, *92*(1), 55–58. doi:10.1016/S1389-1723(01)80199-7 PMID:16233058

Massart, D. L., Vandegiste, B. G. M., Buydens, L. M. C., Lewi, S., de Jong, P. J., & Smeyers-Verbeke, J. (1997). Handbook of chemometrics and qualimetrics. Amsterdam, The Netherlands: Elsevier.

Miller, J. N., & Miller, J. C. (2010) *Statistics and chemometrics for analytical chemistry*, Pearson Education Limited, Exxes.

Montgomery, D. C. (2017). *Design and analysis of experiments*. John Wiley & Sons.

Otto, M. (2017). *CHEMOMETRICS: Statistics and computer application in analytical chemistry*. Weinheim, Germany: Wiley-VCH.

Robinson, T. J., Borror, C. M., & Myers, R. H. (2004). Robust parameter design: A review. *Quality and Reliability Engineering International*, *20*(1), 81–101. doi:10.1002/qre.602

Tounsadi, H., Khalidi, A., Farnane, M., Machrouhi, A., Elhalil, A., & Barka, N. (2019). Efficient removal of heavy metals by koh activated diplotaxis harra biomass: Experimental design optimization. *Environmental Engineering and Management Journal*, *18*(3).

Zinoubi, K., Majdoub, H., Barhoumi, H., Boufi, S., & Jaffrezic-Renault, N. (2017). Determination of trace heavy metal ions by anodic stripping voltammetry using nanofibrillated cellulose modified electrode. *Journal of Electroanalytical Chemistry*, *799*, 70–77. doi:10.1016/j.jelechem.2017.05.039

Chapter 14
Application of the Simultaneous Perturbation Stochastic Approximation Algorithm for Process Optimization: Case Study

Juan Carlos Castillo Garcia
https://orcid.org/0000-0002-4790-3245
Universidad Autónoma de Baja California, Mexico

Claudia Camargo Wilson
https://orcid.org/0000-0001-7803-6706
Universidad Autónoma de Baja California, Mexico

Jesús Everardo Olguín Tiznado
https://orcid.org/0000-0002-6205-0973
Universidad Autónoma de Baja California, Mexico

Juan Andrés López Barreras
https://orcid.org/0000-0002-5477-033X
Universidad Autónoma de Baja Calirfonia, Mexico

Rafael García Martínez
Instituto Tecnológico de Guaymas, Mexico

ABSTRACT

There are different techniques for the optimization of industrial processes that are widely used in industry, such as experimental design or surface response methodology to name a few. There are also alternative techniques for optimization, like the Simultaneous Perturbation Stochastic Approaches (SPSA) algorithm. This chapter compares the results that can be obtained with classical techniques against the results that alternative linear search techniques such as the Simultaneous Perturbation Stochastic Approaches (SPSA) algorithm can achieve. Authors start from the work reported by Gedi et al. 2015 to implement the SPSA algorithm. The experiments allow authors to affirm that for this case study, the SPSA is capable of equalizing, even improving the results reported by the authors.

DOI: 10.4018/978-1-7998-1518-1.ch014

BACKGROUND

When you are trying to improve a process, there are two basic ways to obtain the necessary information: one is to observe or monitor using statistical tools, until you obtain useful signals that allow you to improve it; it is said that this is a passive strategy. The other way is to experiment, making strategic and deliberate changes to the process to provoke these useful signals. Design of experiments (DOE) is a set of active techniques, in the sense that they do not expect the process to send useful signals, but rather that it is "manipulated" to provide the information required for its improvement (Pulido, H. G. et al., 2012) .

The design of experiments is the application of the scientific method to generate knowledge about a system or process, by means of properly planned tests. This methodology has been consolidated as a set of statistical and engineering techniques that allows a better understanding of complex cause-effect situations (Montgomery, 2017).

The objective of a factorial design is to study the effect of different factors on one or several responses, when the same interest is had on all factors (Lujan-Moreno et al., 2018). For example, one of the most important objectives of a factorial design is to determine the combination of levels of controllable factors that will improve process performance.

The factors can be qualitative or quantitative. In order to study the influence of each factor on the response variable, it is necessary to choose at least two test levels for each of them. With a full factorial design, all possible combinations between the levels of the factors to be tested are studied.

However, there are many experimental designs to study the different types of industrial processes. And it is necessary to know how to choose among these types of designs the one that is most appropriate for each situation (Lauro, C. H. et al., 2016) . There are five aspects that influence the selection of an experimental design, in the meaning that when these aspects change, they usually lead us to change the design;

1. The objective of the experiment,
2. The number of factors to be studied,
3. The number of levels tested on each factor,
4. The affects you are interested in investigating,
5. The cost of the experiment, the time and the desired precision.

According to their objective, experimental designs can be classified as follows (Pulido & Salazar, 2012):

- Designs to compare two or more treatments
  - Completely random designs (Montgomery, 2017)
  - Random complete block designs, (Cheng, 2016)
  - Latin Squares designs (Anderson, M. J., & Whitcomb, 2016)
  - Greco-Latin square designs (Montgomery, 2017), (MacCalman, et al., 2017)
- Designs to study the effect of several factors on one or more response variables
  - 2n factorial designs, (Kleijnen, 2015)
  - 3n factorial designs, (Malik & Pakzad, 2018)
  - 2n-p fractional factorial designs (Douaik, 2016)
- Designs for process optimization, these can be divided into designs for
  - First-order models

        - 2n factorial designs, (Kleijnen, 2015)
        - 2n-p fractional factorial designs, (Douaik, 2016), (Kleijnen, 2005)
        - Plackett burman design, (Douaik, 2016), (Kleijnen, 2005)
        - Simplex Design (Briones, F. Z. et al., 2016),
    - and designs for second-order models
        - Central composite design, (Malik & Pakzad, 2018), (Salvatori, P. E. et al., 2018), (Douaik, 2016).
        - Box-Behnken design, (Malik & Pakzad, 2018), (Douaik, 2016)
        - 3n factorial designs, (Malik & Pakzad, 2018)
        - 3n-p fractional factorial designs (Montgomery, 2017)
- Robust designs
    - Orthogonal Array Designs(Cheng, 2016), (Dersjö, T., & Olsson, 2012)
    - Design with internal and external arrangement (Heise, S. et al., 2018), (Briones, F. Z. et al., 2016)
- Mixture design
    - Axial design (Salvatori, P. E. et al.,2018)

Response Surface Methodology

Sometimes experiments do not give us the answers we are looking for or simply the level of improvement they present is not sufficient. In these cases it is necessary to carry out experiments sequentially so that the desired level of improvement is achieved (Khuri, A. I., & Cornell, 2018).

After a first experimental stage, it may be necessary to explore the experimental region in more detail, and it may even be necessary to move the experimental region in an appropriate direction. To do both, the response surface methodology is used (Myers, R. H. et al., 2016).

(Danmaliki, G. I. et al., 2017) define this methodology as a set of mathematical and statistical techniques for the development, optimization and improvement of industrial products and processes. Initiated from the design of experiments, this tool is used to determine the significant factors that affect the response variable. Its objective is to reduce the number of experimental runs while maximizing performance through the data generated.

This methodology involves three fundamental aspects: the design, the model and the optimization technique. The design and the model are thought at the same time, and depend on the type of behavior expected in the response. Specifically, the model can be first- or second-order (flat or curved); therefore, the type of design used and the optimization method are categorized, as the case may be, as first- or second-order.

The model generated facilitates the search for a better response to the process. The model goes on until an optimal process is identified or the experimental resources are consumed (Bezerra, M. A. et al., 2008). And finally, the aspect of optimization is made up of some mathematical techniques that serve to explore it, given an adjusted model, in order to obtain information about the optimum point.

The designs that can be used in the response surface methodology are classified according to the grade of the model to be used. These designs provide the treatments to generate data to fit a model that describes a response variable in an experimental region.

Some desirable properties in MSR designs are: That generates a satisfactory distribution of the experimental points on the experimental region. The design should require a minimum number of experimental runs, since in each test performed resources are spent that are always scarce. The design must allow other designs of a higher order to be built from it. The experiment should allow for the detection of mismatch. The design should provide a pure error variance estimator.

First-order designs: these models are used when the experimenter assumes that only the main effects of the factors are important. One criterion for choosing this type of design is that the variance of the response variable is minimal; the first-order designs most used are:

- 2n factorial designs, (Kleijnen, 2015)
- 2n-p fractional factorial designs (Douaik, 2016)· (Lujan-Moreno et al., 2018)
- Plackett burman design, (Douaik, 2016), (Kleijnen, 2005)
- Simplex Design (Briones, F. Z. et al., 2016)

The designs of second order: as its name indicates these allow to adjust models of second order that allow to study the linear effects and of interaction effects as well as the quadratic effects, it is for this reason that these designs are used when it is expected that the region to explore is complex or when it is thought that the optimal point already is inside the experimental region; the second order designs most used in the response surface methodology are:

- Box-Behnken design, (Malik & Pakzad, 2018), (Nissy, et al., 2018), (Latchubugata, et al., 2018) (Malik & Pakzad, 2018), (Douaik, 2016)
- Central Composite Design (Malik & Pakzad, 2018), (Yousefzadeh, S. et al., 2018)

Once the model is properly adjusted and validated, the surface described by the model is explored to find the combination of levels in the factors that result in an optimal value of the response, or to determine the optimal direction of movement in which it should be experienced in the future.

The optimization technique to use depends on the type of model adjusted and there are basically three methods, which are: Escalation ascending (or descending), the Canonical Analysis and the Analysis of mountain range. However, in the response surface methodology three stages are distinguished in the search for the optimum point, these three stages are:

- **Screening:** The optimization of a process begins with this stage, when it has many controllable factors (more than 8) that influence the response variable and that you do not have a clear idea of how each one of them influences; first of all it is necessary to run a design to identify the few factors that have the most influence.
- **First-order search:** This stage is applied when there are few controllable factors (less than 5), and it is known that these factors influence the response variable. In this stage, a first-order design is run that allows a preliminary characterization of the type of response surface and the detection of the presence of curvature. A complete or fractional factorial design with center repeats is usually used.
- **Second order search:** As soon as the presence of curvature is detected, a second order design is run to better characterize the surface and be able to model the curvature. With the adjusted model, the optimal operating conditions of the process are determined.

This methodology has been widely used in industry, in combination with the design of experiments has a wide range of applications, some examples are in the following topics: safe watter production (Yousefzadeh, S. et al., 2018), wastewater treatment (Bashir, M. J. et al., 2015); optimization of chemical reactions for mixing different products (Salvatori, P. E. et al., 2018), (Danmaliki, G. I. et al., 2017), (Elfghi, 2016); Extraction of Biosphenol A (Nissy et al., 2018); Biodiesel Production (Latchubugata et al., 2018), (Salamatinia, B. et al., 2011); or extraction of fatty oils (Mariod et al., 2010).

Simultaneous Perturbation Stochastic Approximation Algorithm

On the other hand, there are alternative techniques for the optimization of production systems. For example, the stochastic approximation procedure, presented by (Robbins, H., & Monro, 1951) is a method of linear search of the root of the unknown function $f\left(f:\Re \rightarrow \Re\right)$ that represents the expected value of a random variable. Later (Kiefer, J., & Wolfowitz, 1952) they modify it so that it can be used in the determination of the optimal of a function $f\left(x\right)$.

(Chin, 1997) classifies stochastic approximation procedures into two types: *Robbins-Monro type* and *Kiefer-Wolfowitz type*. The first are characterized by requiring direct observations of $f\left(x\right)$, in these procedures are: ascending steps, Newton-Raphson, perturbation analysis and probability rate). While the latter require estimators of $f\left(x\right)$, such as finite difference procedures, and simultaneous disturbance procedures)

The later ones are considered more useful, since they do not require a deep knowledge of the system to be optimized. In other words, they are applicable in situations in which the functional relationship between the response variable and the controllable values is unknown, a situation that occurs more frequently in practice (Olguín Tiznado, J. E. et al., 2011).

(Fu & Ho, 1988) and (Chin, 1997) point to the simultaneous perturbation stochastic algorithm as the most efficient, both theoretically and practically, since it has a higher convergence rate and requires a pair of observations in each iteration; this last one is of great Interest because the economic cost and simplicity of the experimental work depends directly on it.

An example of this type of technique is the Simultaneous Perturbation Stochastic Approximation (SPSA) Algorithm proposed by (J.C. Spall, 1992). In this technique, the relationship between controllable factors and the response variable is unknown. This algorithm is an optimization procedure easy to implement, since it does not require a deep knowledge of the system to be optimized (Olguín Tiznado, J. E. et al., 2011). Moreover, only two observations are required to start the iterative process that allows optimizing the evaluated response variables (Miranda & Del Castillo, 2010).

(Fu, M. C., & HILL, 1997) report that the Kiefer-Wolfowitz methods that use the simultaneous perturbation method have the following advantages over the Robbins- Monro type:

- Generality, is applicable in any system that can be simulated.
- Efficiency, requires fewer observations of the response variable to achieve the same or better results compared to any other method to estimate the vector gradient.
- Easy to use, it is as simple to apply as the finite difference method itself.

The simultaneous disturbance method presented by (J.C. Spall, 1992) calculates Y_k according to the following steps.

Step 1. Build a succession of vectors $\{\Delta_k\}\{\Delta_k \in \Re^d\}$, of mutually independent distributed identical random variables $\{\Delta_1,\ldots,\Delta_d\} = \Delta_k$, of which (Sadegh & Spall, 1997), establish that the optimal selection of the probability distribution for each of its components is the Bernoulli distribution with symmetrical values with respect to zero. With $|\Delta_k| < \delta$ almost anywhere and $E|\Delta_{ik}^{-1}| < \alpha_0$ almost anywhere, ((i= 1, 2, …,d) δ and α_0 represent positive real numbers.

Step 2. Consider the sequence $\{c_k\}$ of positive real numbers, which can be considered constant and equal to one for all k in the evaluation of the following equations:

$$Y_k^+ = f(X_k + c_k\Delta_k) + \varepsilon^+$$

$$Y_k^- = f(X_k - c_k\Delta_k) + \varepsilon^-$$

Where ε^+ ε^- represent noise measurements (experimental error) generated successively independent, respectively, for the observations Y_k^+ and Y_k^-. The estimation $Y(X_k)$ of the vector gradient $h(X_k)$ is defined by:

$$Y_k(X_K) = \begin{matrix} \dfrac{Y_k^+ - Y_k^-}{2C_k\Delta_1} \\ \cdot \\ \cdot \\ \cdot \\ \dfrac{Y_k^+ - Y_k^-}{2C_k\Delta_d} \end{matrix}$$

The convergence with probability 1 of the sequence $\{X_k\}$, which is generated with the procedure described in this section, when this procedure uses the simultaneous perturbation method to estimate the gradient, is based on the following assumptions:

1. The decreasing sequences of real numbers $\{a_k\}$ and $\{c_k\}$ converge to zero and also

$$\sum_{k=1}^{\infty} = \infty y \sum_{k=1}^{\infty} {a_k^2}/{c_k^2} < \infty;$$

2. For some positive real values $\alpha_0, \alpha_1, \alpha_2\ and\ \forall k$ it happens that $E\left(\varepsilon_k^{(\pm)2}\right) \le \alpha_0$

$$\tilde{Y} = \beta_0 + \sum_{i=1}^{k} \hat{\beta}_i X_i + \sum_{i=1}^{k} \hat{\beta}_{ii} x_i^2 + \sum_i \sum_j \hat{\beta}_{ij} X_i X_{j,i<j}$$

The Simultaneous Perturbation Stochastic Approximation (SPSA) Algorithm is a method of optimizing multi-variable stochastic systems. It is an iterative gradient free optimization algorithm designed for stochastic problems. It was initially proposed by (J.C. Spall, 1992) and (James C Spall, 1998), and was successfully applied to the optimization of a variety of systems in which precise system models are not available.

This algorithm has the following characteristics: it is especially efficient in large problems, this in terms of generating a good solution in a small number of measurements of the objective function (J.C. Spall, 2012). The gradient approximation uses only two values for the objective function per iteration, independently of the size of the problem, in order to greatly reduce the objective measurement frequency function for gradient estimation (Ding, et al., 2015). It only needs two observations for the estimation of the gradient in each iteration (Miranda, a K, 2009).

METHODOLOGY

This section presents the materials used in the article based on experimental simulation optimization work, which was developed with a laptop computer, with AMD E-450 processor with 1.65 GHz, 4GB RAM. Also, the software to carry out the statistical analysis of the data were Minitab 17®, to calculate the second order models, and MATLAB R2014®, to carry out the simulation of the Simultaneous Perturbation Stochastic Approximation (SPSA) Algorithm for the optimization of the response variables YIELD, EPA and DHA.

Next, the method used to obtain the necessary data for the experimental simulation design and analysis of this chapter will be described. In this research work we start from the analysis presented by (Gedi, et al., 2015) in their work entitled "Optimization of supercritical carbon dioxide (CO2) extraction of sardine (Sardinella lemuru Bleeker) oil using Response Surface Methodology (RSM)", in which a technique for the extraction of sardine oil by supercritical carbon dioxide (SC-CO2) and a few milliliters of ethanol is described, optimized by means of the response surface methodology (RSM).

In order to conduct the analysis proposed in this work, we start from the design of experiments used by (Gedi et al., 2015), the author proposes in his work a Central Composite Design, to evaluate each of the response variables, such as YIELD, and the ratios of eicosapentaenoic acid (EPA) and docosahexaenoic acid (DHA). The design used by the author is shown in table 1.

In his work, the author studies the influence of extraction pressure (200-400 bars) and temperature (40-70 °C) on total extraction of YIELD, and on the ratios of eicosapentaenoic acid (EPA) and docosahexaenoic acid (DHA). The results reported by the author with this extraction method are compared with those obtained with other extraction methods such as Soxhlet and the modified Kinsella method (MKM).

With this design the author calculates the second order regression models that are shown below:

$$\text{YIELD} = -34.08 + 0.151X_1 + 0.678X_2 - 0.000185X_1^2 - 0.00453X_2^2 - 0.000572X_1X_2$$

Table 1. Experimental Design proposed by (Gedi et al., 2015) for the optimization of the repestation variables YIEDL, EPA and DHA.

Run	Parameters		Yield (%)			EPA (%)			DHA (%)		
	X1	X2	y_0	y_1	y_0-y_1	y_0	y_1	y_0-y_1	y_0	y_1	y_0-y_1
1	400	40	6.04	6.64	-0.6	5.45	5.51	-0.06	20.23	20.69	0.05
2	300	76	5.32	6.09	-0.13	4.98	5.11	-0.13	15.70	15.69	0.01
3c	300	55	8.16	7.95	-0.9	5.39	5.48	-0.09	18.83	18.82	-0.44
4	441	55	5.76	5.42	0.02	5.54	5.51	0.02	19.12	18.86	0.25
5c	300	55	8.07	7.95	0.00	5.48	5.48	0.00	19.01	18.82	0.18
6c	300	55	8.14	7.95	0.09	5.57	5.48	0.09	18.89	18.82	0.06
7	400	70	5.43	5.18	0.04	5.61	5.56	0.04	16.50	16.94	-0.44
8c	300	55	7.62	7.95	0.01	5.49	5.48	0.01	19.06	18.82	0.23
9	300	33	6.15	5.73	0.01	5.86	5.84	0.01	20.35	20.25	0.09
10	158	55	2.39	3.08	-0.02	5.06	5.08	-0.02	17.38	17.53	-0.15
11c	300	55	7.78	7.95	0.10	5.58	5.48	0.10	18.79	18.82	-0.03
12	200	70	6.21	5.24	0.07	4.78	4.70	0.07	16.40	15.99	0.40
13	200	40	3.39	3.27	-0.03	5.74	5.77	-0.03	19.01	19.22	-0.21

$$\text{EPA} = 8.25 - 0.00346X_1 - 0.0730X_2 - 0.000009X_1^2 + 0.000187X_1X_2$$

$$\text{DHA} = 14.8 + 0.0235X_1 + 0.101X_2 - 0.000031X_1^2 - 0.00189X_2^2$$

For our research we will start from the central compound design and from the models reported by (Gedi et al., 2015) to recalculate linear regression models for the mean for each of the responses evaluated in this work, such as the YIELD, and the eicosapentaenoic acid (EPA) and docosahexaenoic acid (DHA) ratios. Using Minitab® Software, the following second order linear regression models were constructed.

$$\hat{Y}_{\mu\, YIELD} = -34.26 + 0.154X_1 + 0.6631X_2 - 0.000184X_1^2 - 0.00572X_1X_2 - 0.004413X_2^2$$

With correlation coefficient r^2 =97.76% for this model

$$\hat{Y}_{\mu\, EPA} = 7.972 - 0.0033X_1 - 0.06349X_2 - 0.000009X_1^2 + 0.000187X_1X_2 - 0.000087X_2^2$$

With correlation coefficient r^2= 98.77% for this model

$$\hat{Y}_{\mu\, DHA} = 12.796 + 0.03005X_1 + 0.1377X_2 - 0.000029X_1^2 - 0.000137X_1X_2 - 0.001879X_2^2$$

With correlation coefficient r^2= 99.66% for this model

Following this we will apply the Simultaneous Perturbation Stochastic Approximation (SPSA) Algorithm, following the steps defined by (James C Spall, 1988).

1. Initialization. Set the counter index K=1
2. To choose the initial values for the vector of the controllable factors X(k), in this work of investigation we made a variation in these values to evaluate each one of the different points of the central design compound proposed by (Gedi et al., 2015) as initial value to initiate the linear search with the Algorithm of Stochastic Approximations with Simultaneous Disturbance, the experiments that were analyzed were those that are shown below:
    a. Experiment 1 X1 =200 bars, X2= 40°C
    b. Experiment 2 X1 =200 bars, X2= 55°C
    c. Experiment 3 X1 =200 bars, X2= 70°C
    d. Experiment 4 X1 =300 bars, X2= 40°C
    e. Experiment 5 X1 =300 bars, X2= 55°C
    f. Experiment 6 X1 =300 bars, X2= 70°C
    g. Experiment 7 X1 =400 bars, X2= 30°C
    h. Experiment 8 X1 =400 bars, X2= 55°C
    i. Experiment 9 X1 =400 bars, X2= 70°C

Substituting the initial values of experiment 1, we would have the following values for each of the response variables YIELD, EPA and DHA:

$$\hat{Y}_{\mu\,YIELD} = -34.26 + 0.154(200) + 0.6631(40) - 0.000184(200^2)$$
$$-0.00572(200)(40) - 0.004413(40^2) = 3.3472$$

$$\hat{Y}_{\mu\,DHA} = 12.796 + 0.03005(200) + 0.1377(40) - 0.000029(200^2)$$
$$-0.000137(200)(40) - 0.001879(40^2) = 19.0516$$

In summary, for experiment 1 we obtain the values:

$$\hat{Y}_{\mu\,YIELD} = 3.3472$$

$$\hat{Y}_{\mu\,EPA} = 5.7692$$

$$\hat{Y}_{\mu\,DHA} = 19.0516$$

3. The next step consists on taking an assumed value from the initial gradient vector of values *a* and *c*, at this stage of the investigation we took the following values a = 0.16 and c = 1.

4. Define the non-negativity coefficients A, α and γ. A = 100, α = 0.602 and γ = 0.101 were used for this stage of the investigation.
5. In this research work we use the following succession measures
   a. $a_k = \frac{a}{k}$ reported by (Baltcheva et al., 2003) and (Bartkutė, V., & Sakalauskas, 2007).
   b. $c_k = \frac{c}{k^\gamma}$ reported by (James C Spall, 1998), (Baltcheva et al., 2003), (Reardon, et al., 2010) (Bartkutė, V., & Sakalauskas, 2007), (Bangerth et al., 2005) and (McClary et al., 2010)
6. The next step is to choose the vector of perturbation Δ_k, in our case: $\Delta_k^+ = 141.4\,bars$ and $\Delta_k^- = -21.5\,°C$
7. From the second-order model for the mean, we will evaluate the response variables reported in the article, such as total extraction (YIELD), eicosapentaenoic-language or edition- acid ratio (EPA) and docosahexaenoic –same- acid (DHA).
8. Following this, the values of the controllable factors X_k^+ are calculated from the initial value of the values of X_k, the succession measure c_k and the disturbance vector $\Delta_k^\pm$, with the formula $X_k^+ = X_k + c_k\left(\Delta_k^\pm\right)$.

So the new values for the controllable factors X_k^+ are $X_1^+ = 341.1\,bars$ and $X_2^+ = 18.5\,°C$. Values that will be substituted in the second order regression models of each response variable.

$$\hat{Y}^+_{\mu\,YIELD} = -34.26 + 0.154\left(341.1\right) + 0.6631\left(18.5\right) - 0.000184\left(341.1^2\right)$$
$$-0.00572\left(341.1\right)\left(18.5\right) - 0.004413\left(18.5^2\right) = 2.7849$$

$$\hat{Y}^+_{\mu\,EPA} = 7.972 - 0.0033\left(341.1\right) - 0.06349\left(18.5\right) - 0.000009\left(341.1^2\right)$$
$$+0.000187\left(341.1\right)\left(18.5\right) - 0.000087\left(18.5^2\right) = 5.7731$$

$$\hat{Y}^+_{\mu\,DHA} = 12.796 + 0.03005\left(341.1\right) + 0.1377\left(18.5\right) - 0.000029\left(341.1^2\right)$$
$$-0.000137\left(341.1\right)\left(18.5\right) - 0.001879\left(18.5^2\right) = 20.7141$$

9. After this we proceed to calculate the values of the controllable factors X_k^-, from the initial value of the values of X_k, the succession measure c_k and the disturbance vector $\Delta_k^\pm$, with the formula $X_k^- = X_k - c_k\left(\Delta_k^\pm\right)$. So the values for the controllable factors X_k^- are $X_1^- = 58.6\,bars$ and $X_2^- = 21.5\,°C$. These values will be substituted in the second order regression models of each response variable:

$$\hat{Y}^-_{\mu\,YIELD} = -34.26 + 0.154\left(58.6\right) + 0.6631\left(21.5\right) - 0.000184\left(58.6^2\right)$$
$$-0.00572\left(58.6\right)\left(21.5\right) - 0.004413\left(21.5^2\right) = -14.5823$$

$$\hat{Y}^{-}_{\mu\ EPA} = 7.972 - 0.0033(58.6) - 0.06349(21.5) - 0.000009(58.6^2)$$
$$+0.000187(58.6)(21.5) - 0.000087(21.5^2) = 6.5781$$

$$\hat{Y}^{-}_{\mu\ DHA} = 12.796 + 0.03005(58.6) + 0.1377(21.5) - 0.000029(58.6^2)$$
$$-0.000137(58.6)(21.5) - 0.001879(21.5^2) = 16.3767$$

10. With these values the gradient vectors are estimated $\frac{\partial YIELD(X_k)}{\partial X}$, $\frac{\partial EPA(X_k)}{\partial X}$ and $\frac{\partial DHA(X_k)}{\partial X}$ through the following equations:

a. $$\frac{\partial YIELD^{(+)}}{\partial X} = \begin{bmatrix} \frac{YIELD^{(+)} - YIELD^{(-)}}{2C_k\Delta_{k1}^{(\pm)}} \\ \bullet \\ \bullet \\ \bullet \\ \frac{YIELD^{(+)} - YIELD^{(-)}}{2C_k\Delta_{kd}^{(\pm)}} \end{bmatrix}$$

b. $$\frac{\partial EPA^{(+)}}{\partial X} = \begin{bmatrix} \frac{EPA^{(+)} - EPA^{(-)}}{2C_k\Delta_{k1}^{(\pm)}} \\ \bullet \\ \bullet \\ \bullet \\ \frac{EPA^{(+)} - EPA^{(-)}}{2C_k\Delta_{kd}^{(\pm)}} \end{bmatrix}$$

c. $$\frac{\partial DHA^{(+)}}{\partial X} = \begin{bmatrix} \frac{DHA^{(+)} - DHA^{(-)}}{2C_k\Delta_{k1}^{(\pm)}} \\ \bullet \\ \bullet \\ \bullet \\ \frac{DHA^{(+)} - DHA^{(-)}}{2C_k\Delta_{kd}^{(\pm)}} \end{bmatrix}$$

11. The estimated X_k value is updated to a new X_{k+1} value, this is calculated using the standard SPSA formulas as follows:

$$X_{k+1} = X_k - a_k\varphi_k^{\pm},$$

Where: $\varphi_k^{\pm}$ represents the gradient approximation and is estimated as follows, for the response variables being evaluated such as total extraction (YIELD), and on eicosapentaenoic acid (EPA), and docosahexaeonic acid (DHA) ratios.

$$\varphi_{k\,YIELD}^{\pm} = \frac{YIELD^{+} - YIELD^{-}}{2c_k \Delta_k^{\pm}}$$

$$\varphi_{k\,EPA}^{\pm} = \frac{EPA^{+} - EPA^{-}}{2c_k \Delta_k^{\pm}}$$

$$\varphi_{k\,DHA}^{\pm} = \frac{DHA^{+} - DHA^{-}}{2c_k \Delta_k^{\pm}}$$

Resulting the following values:

- For the response variable YIELD

$\varphi_{k\,YIELD}^{+} = 0.0614$,

$\varphi_{k\,YIELD}^{-} = -0.4039$,

- For the response variable EPA

$\varphi_{k\,EPA}^{+} = -0.0028$,

$\varphi_{k\,EPA}^{-} = 0.0187$,

- For the response variable DHA

$\varphi_{k\,DHA}^{+} = 0.0153$,

$\varphi_{k\,DHA}^{-} = -0.1009$

So the values for X_{k+1} are as follows:

- For the response variable YIELD

$X_{1+1} = 199.9902$

$X_{2+1} = 39.9888$

- For the response variable EPA

$X_{1+1} = 200.0005$

$X_{2+1} = 40.0000$

- For the response variable DHA

$X_{1+1} = 199.9975$

$X_{2+1} = 40.0000$

12. Consider $X_{k+1} = X_k$, increase k by k+1 and return to step 3.

The following are the results obtained from the different combinations of the different initial values of the controllable factors; which were carried out in the simulation processes carried out in each of the models analyzed: YIELD, EPA and DHA; reported by (Gedi et al., 2015).

These results are shown in the following tables which specify the experiment evaluated, the combination of the initial values of the controllable factors, the optimal values found for the controllable factors, the maximum value of the response variable (YIELD, EPA and DHA) and finally the number of iterations where the maximum value of the response variable was obtained.

Next, in Table 2, we show the maximum values of the response variable YIELD obtained in an iterative process of 100 iterations.

According to the information shown in table 2, we can note the nine experiments performed, in three of them the value of 7.2% reported by the author is exceeded as the maximum value obtained with the response surface methodology. The experiments mentioned are:

- Experiment 3, with initial values of 200 for X1 and 70 for X2 with a value of 8.052% in iteration number 6;
- Experiment 5, with initial values of 300 for X1 and 55 for X2 with a value of 7.98% in iteration number 100;
- Experiment 6, with initial values of 300 for X1 and 70 for X2 with a value of 7.21% in iteration number 100.
- Experiment 7, with initial values of 400 for X1 and 40 for X2 with a value of 8.045% in iteration number 100.

The values obtained in each experiment in the iterative simulation process of the response variable YIELD, from one to one hundred iterations are shown in Figure 1.

Table 2. Results of the response variable YIELD evaluated by SPSA, in an iterative process of 100 iterations reported by (Gedi et al., 2015).

Experiments	Initial Values of Controllable Factors		Response Variable	Optimum Values of Controllable Factors		Number of Iterations
	x_1^*	x_2^*	YIELD Max	x_1^*	x_2^*	Position
1	200	40	3.936	288.800	26.545	100
2	200	55	6.903	288.884	41.572	99
3	200	70	8.052	317.986	52.103	6
4	300	40	6.899	300.000	40.001	5
5	300	55	7.983	299.997	55.022	100
6	300	70	7.210	388.798	56.560	100
7	400	40	8.045	311.208	53.402	100
8	400	55	7.000	400.008	54.949	100
9	400	70	5.244	400.005	69.965	100

Figure 1. Graphical behavior of the response variable YIELD in the iterative process of 100 iterations.

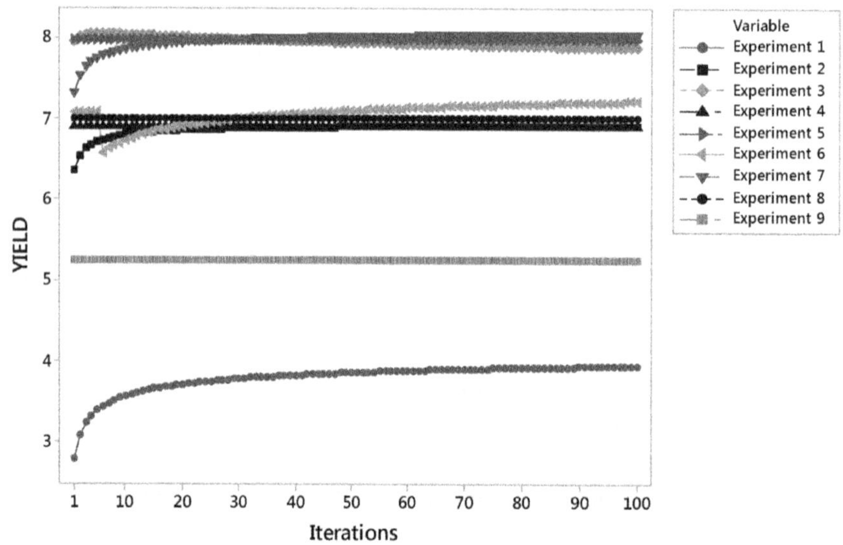

The following Table 3 shows the maximum values obtained for the response variable YIELD in an iterative process of 200 iterations

According to the results shown in table 3, it can be noted that one of the experiments is added to the list of those that achieve a result greater than the 7.20% reported by the author as maximum value. In this case the 4 experiments that exceed this value are:

- Experiment 3, with initial values of 200 for X1 and 70 for X2 with a value of 8.052% in iteration number 6;

Table 3. Results of the response variable YIELD evaluated by SPSA, in an iterative process of 200 iterations reported by (Gedi et al., 2015).

Experiments	Initial Values of Controllable Factors		Response Variable	Optimum Values of Controllable Factors		Number of Iterations
	x_1^*	x_2^*	YIELD Max	x_1^*	x_2^*	Position
1	200	40	3.999	282.794	27.468	200
2	200	55	6.903	288.884	41.572	99
3	200	70	8.052	317.986	52.103	6
4	300	40	6.899	300.000	40.001	5
5	300	55	7.983	299.996	55.029	200
6	300	70	7.299	382.791	57.491	200
7	400	40	8.057	317.216	52.469	200
8	400	55	7.000	400.010	54.937	200
9	400	70	5.245	400.006	69.959	200

- Experiment 5, with initial values of 300 for X2 and 55 for X2 with a maximum value of 7.983% in iteration number 200;
- Experiment 6, with initial values of 300 for X2 and 70 for X2 with a maximum value of 7.299% in iteration number 200;
- Experiment 7, with initial values of 400 for X1 and 40 for X2 with a value of 8.057% in iteration number 200.

The values obtained in each experiment in the iterative simulation process of the response variable YIELD, from one to two hundred iterations are shown in Figure 2.

Table 4 below shows the maximum values of the response variable YIELD, obtained in an iterative process of 500 iterations.

With the results shown in table 4, we can note that five of the nine experiments performed managed to obtain a value higher than the value of 7.2% reported by the author as the maximum value obtained with the response surface methodology.

The experiments that manage to improve this value are the following:

- Again, experiment 3 appears, with initial values of 200 for X1 and 70 for X2 with a value of 8.052% in iteration number 6;
- Experiment 5, with initial values of 300 for X2 and 55 for X2 with a maximum value of 7.983% in iteration number 500;
- Experiment 6, with initial values of 300 for X2 and 70 for X2 with a maximum value of 7.387% in iteration number 500;
- Experiment 7, with initial values of 400 for X1 and 40 for X2 with a value of 8.058% in iteration number 252;
- Experiment 8, with initial values of 400 for X2 and 55 for X2 with a maximum value of 7.418% in iteration number 500.

Figure 2. Graphical behavior of the response variable YIELD in the iterative process of 200 iterations.

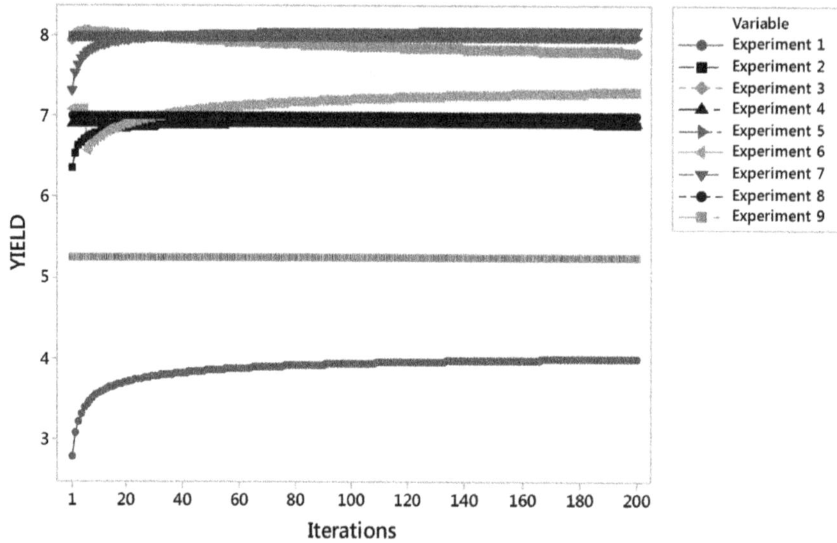

Table 4. Results of the response variable YIELD evaluated by SPSA, in an iterative process of 500 iterations reported by (Gedi et al., 2015).

Experiments	Initial Values of Controllable Factors		Response Variable	Optimum Values of Controllable Factors		Number of Iterations
	x_1^*	x_2^*	YIELD Max	x_1^*	x_2^*	Position
1	200	40	4.055	275.473	28.595	500
2	200	55	6.903	288.884	41.572	99
3	200	70	8.052	317.986	52.103	6
4	300	40	6.899	300.000	40.001	5
5	300	55	7.983	299.994	55.039	500
6	300	70	7.387	375.467	58.631	500
7	400	40	8.058	319.127	52.172	252
8	400	55	7.418	324.529	66.395	500
9	400	70	5.246	400.008	69.950	500

The values obtained in each experiment in the iterative simulation process of the response variable YIELD, from one to five hundred iterations are shown in Figure 3.

Table 5 shows the maximum values of the EPA response variable, obtained in an iterative process of 100 iterations.

With the information shown in table 5 we can affirm that of the nine experiments carried out, in three of them the value of 5.68% reported by the author is exceeded as the maximum value obtained with the response surface methodology. The experiments mentioned are:

Figure 3. Graphical behavior of the response variable YIELD in the iterative process of 200 iterations.

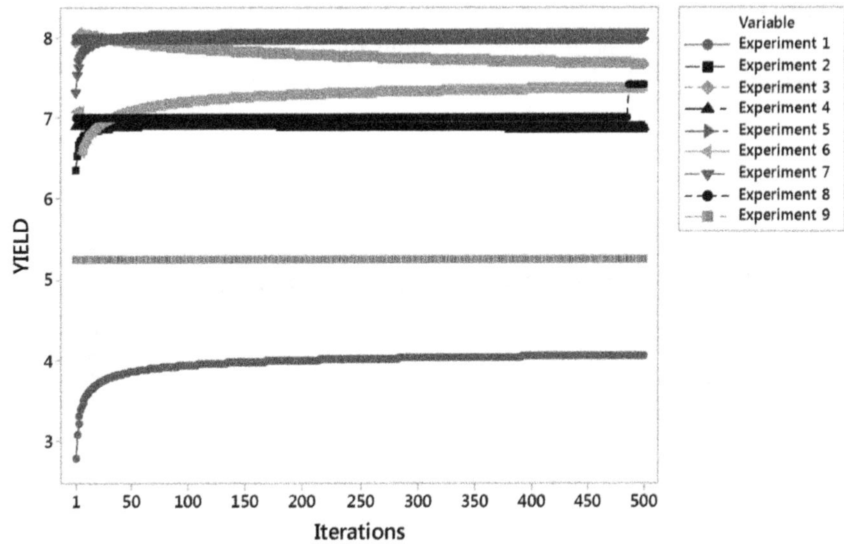

Table 5. Results of the response variable EPA evaluated by SPSA, in an iterative process of 100 iterations reported by (Gedi et al., 2015).

Experiments	Initial Values of Controllable Factors		Response Variable	Optimum Values of Controllable Factors		Number of Iterations
	x_1^*	x_2^*	EPA Max	x_1^*	x_2^*	Position
1	200	40	5.769	200.000	40.000	1
2	200	55	5.743	313.229	37.971	9
3	200	70	5.609	341.400	48.500	1
4	300	40	5.738	300.014	39.910	100
5	300	55	5.562	388.778	41.693	100
6	300	70	5.572	393.956	55.968	57
7	400	40	5.525	400.000	40.000	1
8	400	55	5.571	399.988	55.082	20
9	400	70	5.519	488.720	57.072	100

- Experiment 1 appears, with initial values of 200 for X1 and 40 for X2 with a value of 5.769% in iteration number 1;
- Experiment 2 appears, with initial values of 200 for X1 and 55 for X2 with a value of 5.743% in iteration number 9;
- Experiment 4 appears, with initial values of 300 for X1 and 40 for X2 with a value of 5.738% in iteration number 100.

The values obtained in each experiment in the iterative simulation process of the EPA response variable, from one to one hundred iterations are shown in Figure 4.

Table 6 below shows the maximum values obtained for the EPA response variable in an iterative process of 200 iterations.

Figure 4. Graphical behavior of the response variable EPA in the iterative process of 100 iterations.

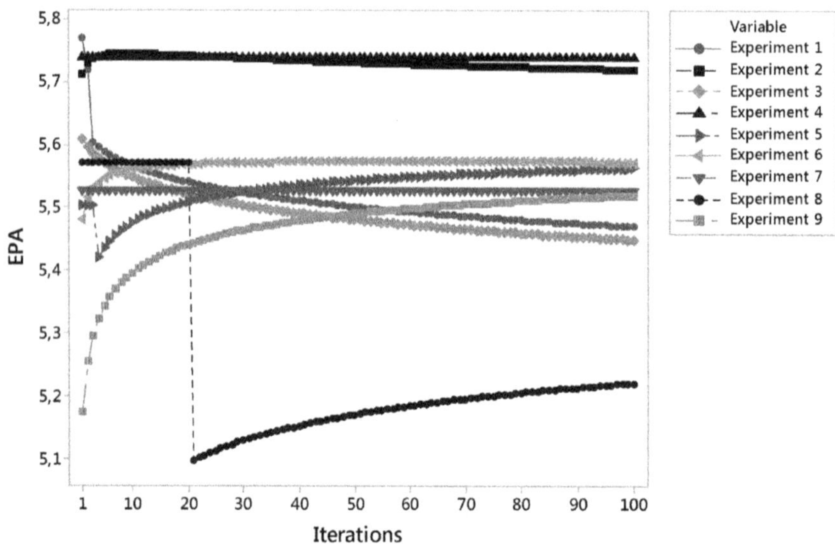

Table 6. Results of the response variable EPA evaluated by SPSA, in an iterative process of 200 iterations reported by (Gedi et al., 2015).

Experiments	Initial Values of Controllable Factors		Response Variable	Optimum Values of Controllable Factors		Number of Iterations
	x_1^*	x_2^*	EPA Max	x_1^*	x_2^*	Position
1	200	40	5.769	200.000	40.000	1
2	200	55	5.743	313.229	37.971	9
3	200	70	5.609	341.400	48.500	1
4	300	40	5.739	300.017	39.891	200
5	300	55	5.577	382.768	42.637	200
6	300	70	5.572	393.956	55.968	57
7	400	40	5.525	400.000	40.000	1
8	400	55	5.571	399.988	55.082	20
9	400	70	5.543	482.705	58.053	200

Based on the information shown in table 6, we can affirm that of the nine experiments performed, in three of them the value of 5.68% reported by the author as the maximum value obtained with the response surface methodology was exceeded. The experiments mentioned are:

- Experiment 1, with initial values of 200 for X1 and 40 for X2 with a value of 5.762% in iteration number 1;
- Experiment 2, with initial values of 200 for X1 and 55 for X2 with a value of 5.743% in iteration number 9;
- Experiment 4, with initial values of 300 for X1 and 40 for X2 with a value of 5.739% in iteration number 200.

The values obtained in each experiment in the iterative simulation process of the EPA response variable, from one to two hundred iterations are shown in Figure 5.

Table 7 below shows the maximum values obtained for the EPA response variable in an iterative process of 500 iterations.

With the results shown in table 7 we can affirm that of the nine experiments carried out, in three of them the value of 5.68% reported by the author is exceeded as the maximum value obtained with the response surface methodology. The experiments mentioned are:

- Experiment 1, with initial values of 200 for X1 and 40 for X2 with a value of 5.762% in iteration number 1;
- Experiment 2, with initial values of 200 for X1 and 55 for X2 with a value of 5.743% in iteration number 9;
- Experiment 4, with initial values of 300 for X1 and 40 for X2 with a value of 5.739% in iteration number 500.

Figure 5. Graphical behavior of the response variable EPA in the iterative process of 200 iterations.

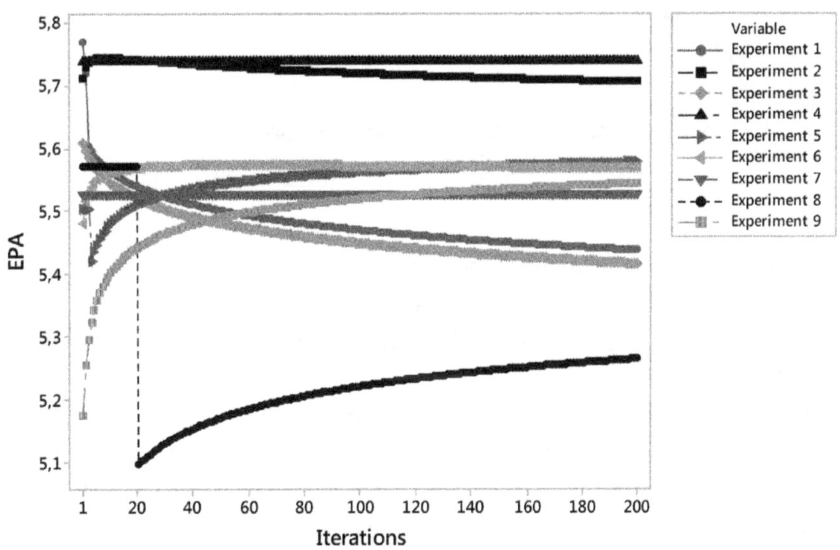

Table 7. Results of the response variable EPA evaluated by SPSA, in an iterative process of 500 iterations reported by (Gedi et al., 2015).

Experiments	Initial Values of Controllable Factors		Response Variable	Optimum Values of Controllable Factors		Number of Iterations
	x_1^*	x_2^*	EPA Max	x_1^*	x_2^*	Position
1	200	40	5.769	200.000	40.000	1
2	200	55	5.743	313.229	37.971	9
3	200	70	5.609	341.400	48.500	1
4	300	40	5.739	300.021	39.864	500
5	300	55	5.592	375.443	43.793	500
6	300	70	5.572	393.956	55.968	57
7	400	40	5.525	400.000	40.000	1
8	400	55	5.571	399.988	55.082	20
9	400	70	5.569	475.373	59.253	500

Figure 6. Graphical behavior of the response variable EPA in the iterative process of 500 iterations.

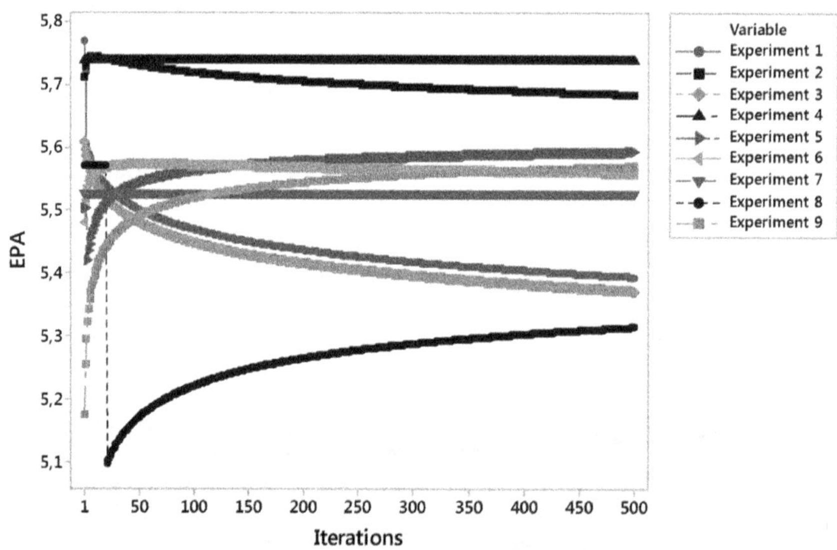

The values obtained in each experiment in the iterative simulation process of the EPA response variable, from one to five hundred iterations are shown in Figure 6.

Table 8 below shows the maximum values obtained for the response variable DHA in an iterative process of 100 iterations.

According to the information shown in table 8, we can note that none of the nine experiments performed manages to exceed the value of 20.09% reported by the author as the maximum value obtained with the response surface methodology. Although in the nine experiments very close values are found, the value that is closer to the maximum reported by the author is:

Table 8. Results of the response variable DHA evaluated by SPSA, in an iterative process of 100 iterations reported by (Gedi et al., 2015)

Experiments	Initial Values of Controllable Factors		Response Variable	Optimum Values of Controllable Factors		Number of Iterations
	x_1^*	x_2^*	DHA Max	x_1^*	x_2^*	Position
1	200	40	19,052	200,000	40,000	1
2	200	55	18,029	200,000	55,000	1
3	200	70	19,665	341,400	48,500	1
4	300	40	20,059	300,001	39,994	2
5	300	55	18,830	300,000	55,000	1
6	300	70	19,459	422,919	51,352	4
7	400	40	19,059	311,210	53,390	100
8	400	55	19,053	400,002	54,986	100
9	400	70	19,015	522,916	51,368	4

- Experiment 4, with initial values of 300 for X1 and 40 for X2 with a value of 20.059% in iteration number 2.

It is important to mention that for this model analyzed 200 and 500 iterations were analyzed, however there were no improvements to the results found in the first 100 iterations.

The values obtained in each experiment in the iterative simulation process of the response variable DHA, from one to one hundred iterations are shown in Figure 7.

From the results obtained in the three models analyzed YIELD, EPA and DHA; the best experiments were selected, looking for the maximum value of the response variable, in the least number of iterations in which this maximum value was found.

With this in mind, as a summary we can affirm that the best experiments were the following:

- For the response variable YIELD:
  - Experiment 3, with a value of 8.052 in iteration number 6
  - Experiment 4, with a value of 6.89 in iteration number 5
  - Experiment 7, with a value of 8.04 in iteration number 100
- For the response variable EPA:
  - Experiment 1, with a value of 5.76 in iteration number 1
  - Experiment 3, with a value of 5.60 in iteration number 1
  - Experiment 4, with a value of 5.73 in iteration 100
- For the response variable DHA:
  - Experiment 3, with a value of 19.66 in iteration number 1
  - Experiment 4, with a value of 20.05 in iteration number 2
  - Experiment 6, with a value of 19.45 in iteration number 4

Figure 7. Graphical behavior of the response variable DHA in the iterative process of 100 iterations.

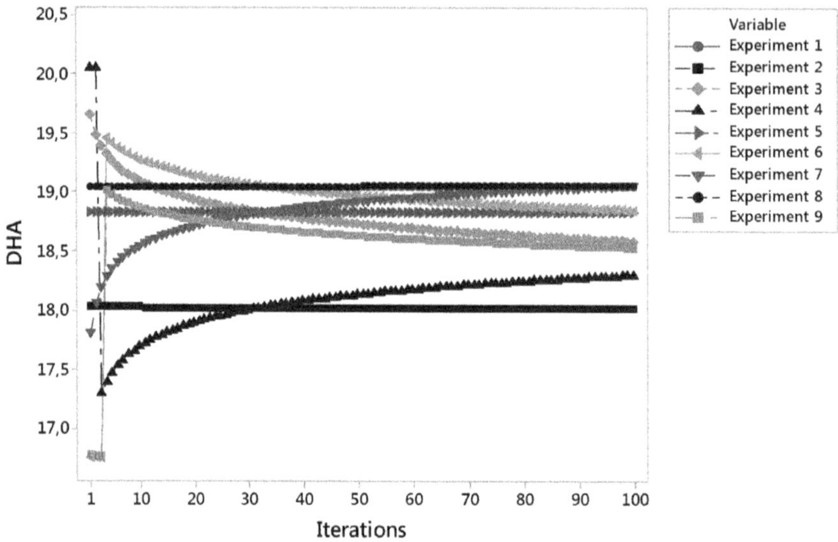

Note that experiments number 3 and 4, which appear in the three models analyzed, and appear in this particular study as the best experiments.

SOLUTIONS AND RECOMMENDATIONS

In this section, it is presented the results obtained from de nine experiments carry out with combination of the initial values for the controllable factors X1 and X2. In order to evaluate three models (YIELD, EPA and DHA) reported by (Gedi et al., 2015). This, with the purpose to determine which of the combinations in order to determine which of the combinations maximizes the value of the response variable in the least number of interactions.

Each of these experiments was evaluated by simulation process at one hundred, two hundred and five hundred iteration to evaluate the potential solution to these models, to be able to check if any of these experiments could be able to offer similar or better results for each of the response variables, compared with the results reported by the author.

It is important to highlight that for this particular case in which we are contrasting the Simultaneous Perturbation Stochastic Approximation (SPSA) algorithm against the response surface methodology used by (Gedi et al., 2015). Based on the work that has been done, we can conclude that the analysis carried out to use the Simultaneous Perturbation Stochastic Approximation (SPSA) algorithm is much simpler and more practical than the analysis required to implement an optimization tool such as the response surface methodology.

In the same way, the results obtained in these three analyzed models, allow us to conclude that at least in these three models the Algorithm of Stochastic Approaches with Simultaneous Disturbance is able to equal or even improve the optimal solutions obtained through classical optimization tools such as the response surface methodology.

In his work, the author (Gedi et al., 2015) mentions that the target values he seeks to overcome are the results of two other extraction methods which are as follows: method 1 Soxhlet extraction which presents as optimal values for the variables YIELD=5.1%, EPA=2.17% and DHA=6.46%; and method 2 modified Kinsella method (MKM) which presents as maximum values for the variables YIELD=6.83, EPA=5.43% and DHA=18.54%.

It is important to point out that, for this particular case, using the Stochastic Approach Algorithm with Simultaneous Disturbance we managed to equalize, even improve with some experiments, the optimal values reported by (Gedi et al., 2015). For the response variable YIELD with four of the nine experiments performed (experiments 3, 5, 6 and 7). For the EPA variable in three of the nine experiments performed (Experiments 1, 2 and 4). And for the DHA variable it is not possible to exceed the maximum value reported by the author, however it is possible to match this result.

REFERENCES

Anderson, M. J., & Whitcomb, P. J. (2016). *RSM simplified: optimizing processes using response surface methods for design of experiments*. Productivity Press. doi:10.1201/9781315382326

Baltcheva, I., Cristea, S., Felisa, J. V.-A., & De Prada, C. (2003). Simultaneous perturbation stochastic approximation for real-time optimization of model predictive control. In *Proceedings Of the 1st Industrial Simulation Conference,* (pp. 533-537).

Bangerth, W., Klie, H., Matossian, V., Parashar, M., & Wheeler, M. F. (2005). An autonomic reservoir framework for the stochastic optimization of well placement. *Cluster Computing*, *8*(4), 255–269. doi:10.100710586-005-4093-3

Bartkutė, V., & Sakalauskas, L. (2007). Simultaneous perturbation stochastic approximation of nonsmooth functions. *European Journal of Operational Research*, *181*(3), 1174–1188. doi:10.1016/j.ejor.2005.09.052

Bashir, M. J., Amr, S. S. A., Aziz, S. Q., Aun, N. C., & Sethupathi, S. (2015). Wastewater treatment processes optimization using response surface methodology (RSM) compared with conventional methods: Review and comparative study. *Middle East Journal of Scientific Research*, *23*(2), 244–252.

Bezerra, M. A., Santelli, R. E., Oliveira, E. P., Villar, L. S., & Escaleira, L. A. (2008). Response surface methodology (RSM) as a tool for optimization in analytical chemistry. *Talanta*, *76*(5), 965–977. doi:10.1016/j.talanta.2008.05.019 PMID:18761143

Briones, F. Z., Leal, J. S., & Pastrana, I. Y. M. (2016). Armentum: A hybrid direct search optimization methodology. *Journal of Industrial Engineering International*, *12*(4), 407–417. doi:10.100740092-016-0159-5

Cheng, C. S. (2016). *Theory of factorial design: Single-and multi-stratum experiments*. Chapman and Hall/CRC. doi:10.1201/b16253

Chin, D. C. (1997). Comparative study of stochastic algorithms for system optimization based on gradient approximations. *IEEE Transactions on Systems, Man, and Cybernetics. Part B, Cybernetics*, *27*(2), 244–249. doi:10.1109/3477.558808 PMID:18255862

Danmaliki, G. I., Saleh, T. A., & Shamsuddeen, A. A. (2017). Response surface methodology optimization of adsorptive desulfurization on nickel/activated carbon. *Chemical Engineering Journal, 313*, 993–1003. doi:10.1016/j.cej.2016.10.141

Dersjö, T., & Olsson, M. (2012). Efficient design of experiments for structural optimization using significance screening. *Structural and Multidisciplinary Optimization*, *45*(2), 185–196. doi:10.100700158-011-0677-0

Ding, S., Xia, N., Wang, P., Li, S., & Ou, Y. (2015). Optimization Algorithm Based on SPSA in Multi-channel Multi-radio Wireless Monitoring Network. *2015 International Conference on Cyber-Enabled Distributed Computing and Knowledge Discovery*, 517–524. 10.1109/CyberC.2015.84

Douaik, A. (2016). Design of experiments for screening and optimizing factors influencing extraction of bioactive compounds from saffron. *In V International Symposium on Saffron Biology and Technology: Advances in Biology, Technologies, Uses and Market 1184*, 151–158.

Elfghi, F. M. (2016). A hybrid statistical approach for modeling and optimization of RON: A comparative study and combined application of response surface methodology (RSM) and artificial neural network (ANN) based on design of experiment (DOE). *Chemical Engineering Research & Design*, *113*, 264–272. doi:10.1016/j.cherd.2016.05.023

Fu, M. C., & Hill, S. D. (1997). Optimization of discrete event systems via simultaneous perturbation stochastic approximation. *IIE Transactions*, *29*(3), 233–243. doi:10.1080/07408179708966330

Fu, M. C., & Ho, Y.-C. (1988). Using perturbation analysis for gradient estimation, averaging and updating in a stochastic approximation algorithm. *Winter Simulation Conference Proceedings. IEEE.*, 509–517. 10.1145/318123.318244

Gedi, M. A., Bakar, J., & Mariod, A. A. (2015). Optimization of supercritical carbon dioxide (CO 2) extraction of sardine (Sardinella lemuru Bleeker) oil using response surface methodology (RSM). *Grasas y Aceites*, *66*(June).

Heise, S., Virtanen, S., & Boccaccini, A. R. (2018). Taguchi Design of Experiments approach to determine process parameter for the electrophoretic deposition of chitosan/bioactive glass on Mg alloy substrates. *ECS Transactions*, *82*(1), 81–87. doi:10.1149/08201.0081ecst

Khuri, A. I., & Cornell, J. A. (2018). *Response surfaces: designs and analyses*. Routledge. doi:10.1201/9780203740774

Kiefer, J., & Wolfowitz, J. (1952). Stochastic estimation of the maximum of a regression function. *Annals of Mathematical Statistics*, *23*(3), 462–466. doi:10.1214/aoms/1177729392

Kleijnen, J. P. (2005). An overview of the design and analysis of simulation experiments for sensitivity analysis. *European Journal of Operational Research*, *164*(2), 287–300. doi:10.1016/j.ejor.2004.02.005

Kleijnen, J. P. (2015). Design and analysis of simulation experiments. *In International Workshop on Simulation*. pp. 3–22. Cham, Switzerland: Springer.

Latchubugata, C. S., Kondapaneni, R. V., Patluri, K. K., Virendra, U., & Vedantam, S. (2018). Kinetics and optimization studies using Response Surface Methodology in biodiesel production using heterogeneous catalyst. *Chemical Engineering Research & Design*, *135*, 129–139. doi:10.1016/j.cherd.2018.05.022

Lauro, C. H., Pereira, R. B., Brandão, L. C., & Davim, J. P. (2016). Design of Experiments Statistical and artificial intelligence analysis for the improvement of machining processes: A review. In In Design of Experiments in Production Engineering (pp. 89–107). doi:10.1007/978-3-319-23838-8_3

Lujan-Moreno, G. A., Howard, P. R., Rojas, O. G., & Montgomery, D. C. (2018). Design of experiments and response surface methodology to tune machine learning hyperparameters, with a random forest case-study. *Expert Systems with Applications*, *109*, 195–205. doi:10.1016/j.eswa.2018.05.024

MacCalman, A. D., Vieira, H., & Lucas, T. (2017). Second-order nearly orthogonal Latin hypercubes for exploring stochastic simulations. *Journal of Simulation*, *11*(2), 137–150. doi:10.1057/jos.2016.8

Malik, D., & Pakzad, L. (2018). Experimental investigation on an aerated mixing vessel through electrical resistance tomography (ERT) and response surface methodology (RSM). *Chemical Engineering Research & Design*, *129*, 327–343. doi:10.1016/j.cherd.2017.11.002

Mariod, A. A., Abdelwahab, S. I., Gedi, M. A., & Solati, Z. (2010). Supercritical carbon dioxide extraction of sorghum bug (Agonoscelis pubescens) oil using Response surface methodology. *JAOCS. Journal of the American Oil Chemists' Society*, *87*(8), 849–856. doi:10.100711746-010-1565-2

McClary, D. W., Syrotiuk, V. R., & Kulahci, M. (2010). Steepest-ascent constrained simultaneous perturbation for multiobjective optimization. *ACM Transactions on Modeling and Computer Simulation*, *21*(1), 1–22. doi:10.1145/1870085.1870087

Miranda, A. K., & Del Castillo, E. (2010). Robust parameter design optimization of simulation experiments using stochastic perturbation methods. *Journal of the Operational Research Society*, *62*(1), 198–205. doi:10.1057/jors.2009.163

Miranda, A. K. D. C. E. (2009). *Robust Parameter Design Optimization of Simulation Experiments Using Stochastic Perturbation Methods*. Academic Press.

Montgomery, D. C. (2017). *Design and analysis of experiments (9th)*. Hoboken, NJ: Wiley.

Myers, R. H., Montgomery, D. C., & Anderson-Cook, C. M. (2016). *Response surface methodology: process and product optimization using designed experiments*. John Wiley.

Nissy, S. M., Surendra, B., Raju, K. D., & Vangalapati, M. (2018). Optimization for the extraction of Bisphenol A; Response surface Methodology. *Materials Today: Proceedings*, *5*, 17914–17923. doi:10.1016/j.matpr.2018.06.120

Olguín Tiznado, J. E., García Martínez, R., Camargo Wilson, C., & López Barreras, J. A. (2011). Stochastic approximation algorithm for industrial process optimisation. *Ingenieria e Investigacion*, *31*(3), 100–111.

Pulido, H. G., De la Vara Salazar, R., González, P. G., Martínez, C. T., & Pérez, M. D. C. T. (2012). *Analysis and design of experiments*. New York, NY: McGraw-Hill. In Spanish

Reardon, B. E., Lloyd, J. M., & Perel, R. Y. (2010). *Tuning Missile Guidance and Control Algorithms Using Simultaneous Perturbation Stochastic Approximation.*, *29*(1), 85–100.

Robbins, H., & Monro, S. (1951). A stochastic approximation method. *Annals of Mathematical Statistics*, *22*(3), 400–407. doi:10.1214/aoms/1177729586

Sadegh, P., & Spall, J. C. (1997). Optimal random perturbations for stochastic approximation using a simultaneous perturbation gradient approximation. *Proceedings of the 1997 American Control Conference (Cat. No.97CH36041)*, 6, 3582–3586. 10.1109/ACC.1997.609490

Salamatinia, B., Bhatia, S., & Abdullah, A. Z. (2011). A response surface methodology based modeling of temperature variation in an ultrasonic-assisted biodiesel production process. *World Applied Sciences Journal*, *12*(9), 1549–1557.

Salvatori, P. E., Sánchez, G. R., Lombardi, A. G., Nicocia, E. J., & Boschetti, C. E. (2018). Strategy for the optimization of a rubber compound with soybean oil using design of experiments. *Matéria (Rio de Janeiro)*, *23*(2).

Spall, James C. (1988). *A stochastic approximation algorithm for large-dimensional system in the Kiefer-Wolfowitz setting.* (December), 1544–1548.

Spall, J. C. (1992). Multivariate stochastic approximation using a simultaneous perturbation gradient approximation. *IEEE Transactions on Automatic Control*, *37*(3), 332–341. doi:10.1109/9.119632

Spall, James C. (1998). *An Overview of the Simultaneous Perturbation Method for Efficient Optimization, 19*(4).

Spall, J. C. (2012). Stochastic optimization. In Handbook of computational statistics (pp. 173-201). Berlin, Germany: Springer. doi:10.1007/978-3-642-21551-3_7

Yousefzadeh, S., Matin, A. R., Ahmadi, E., Sabeti, Z., Alimohammadi, M., Aslani, H., & Nabizadeh, R. (2018). Response surface methodology as a tool for modeling and optimization of Bacillus subtilis spores inactivation by UV/nano-Fe0 process for safe water production. *Food and Chemical Toxicology*, *114*, 334–345. doi:10.1016/j.fct.2018.02.045 PMID:29481893

Chapter 15
Design of Experiments in Engineering Education:
Opportunities and Challenges

Isabel M. João
https://orcid.org/0000-0002-8690-1584
Instituto Superior de Engenharia de Lisboa, Instituto Politécnico de Lisboa, Portugal & CEG-IST, Instituto Superior Técnico, Universidade de Lisboa, Portugal

João M. Silva
https://orcid.org/0000-0003-1244-6483
Instituto Superior de Engenharia de Lisboa, Instituto Politécnico de Lisboa, Portugal & Instituto Superior Técnico, Universidade de Lisboa, Portugal

ABSTRACT

The chapter will start with an overview of today´s challenges of engineering education. DOE can be very effective for solving problems in view of the new pedagogical challenges in engineering education. The chapter reviews the progress of DOE in engineering teaching and learning for problem solving and for product/process optimization with focus on engineering education in this new millennium. The goal is to identify the main engineering areas accounting for the use of statistical experimental design in engineering education as well as the main teaching/learning strategies and the combination of other tools used to support the use of DOE in engineering education. The main contribution will be to bring up ideas from studies of DOE in teaching/learning engineering environments to better understand the deficit of utilization of such type of approaches in academic projects/experiments despite the common utilization of DOE in statistics and quality literature.

DOI: 10.4018/978-1-7998-1518-1.ch015

INTRODUCTION

Engineering students must develop skills with the aim of designing new and improved products and operating maintaining and optimizing industrial processes that are economically viable, valuable to society and with reduced negative environmental impacts. Due to the pressure to market that is nowadays an important issue the engineers must be able to develop better products in half the time. They need to understand the design and operation of processes, how changes in operations and in the environment will affect outcomes, and the actions needed to improve process performance and increase sustainability. The knowledge about the products and processes in engineering and related scientific fields is derived from experimentation mainly in situations where no scientific theory or principles are directly applicable. The Design of Experiments (DOE) methodology can become tremendously important in such circumstances to develop/improve new products/processes in an efficient way. DOE can be very effective for solving general or specific problems and is able to have into account the new pedagogical challenges in engineering education because a structured methodology will have the advantage of helping students to avoid the error of leaping to problem solve without a clear understanding of the problem.

This chapter starts by presenting an overview of the today´s challenges concerning engineering education. The speed of changes in engineering practices nowadays also require huge efforts in engineering education practice reason why this change demand a pedagogical paradigm where teaching and learning is student centred.

An overview of DOE is also presented in this chapter with a brief historical analysis of the important developments in DOE being restricted to that of a group of techniques in design that boost the exploration of a region of design variables in one or more response variables. This chapter also presents a literature review regarding the utilization of DOE in engineering teaching and learning for general problem solving as well as for the optimization of product/process with special focus on engineering education. The main objective of the literature review is to identify the main scientific engineering areas that contribute to the use of statistical experimental design in engineering education as well as the main teaching /learning strategies and the combination of DOE with other tools (e.g. simulation, virtual laboratories) used to support the use of DOE in engineering education.

The main contribution of the chapter will be to bring up ideas from those studies devoted to the use of DOE in teaching and learning engineering environments to better understand the deficit of utilization of such type of approaches in academic projects/experiments regardless of the common utilization of DOE in statistics and quality literature. Despite the statistical experimental design finds applications in several engineering areas (e.g. chemical industry, food industry, pharmaceutical research, environmental engineering) limited material have been published concerning the utilization of DOE in engineering education and the challenges for educators and students alike. It is crucial to understand the use of DOE in engineering education and to understand how to improve its use in industrial contexts and in academic and research projects. This work is intended to encourage further discussion between educators, practitioners, and experts in order to find ways to promote the usage of DOE methodology by future engineering graduates and researchers.

TODAY'S CHALLENGES IN ENGINEERING EDUCATION

The changing nature of industry and the use of new materials, new chemical and biological processes and the huge growth in information technology among others is part of the practice of engineering profession that was defined by UNESCO (2010) as *the field or discipline, practice, profession and art that relates to the development, acquisition and application of technical, scientific and mathematical knowledge about the understanding, design, development, invention, innovation and use of materials, machines, structures, systems and processes for specific purposes.*

Despite the invaluable contributions of engineers during the 20th century to health, quality of life, and progress there is much room to improve and the 21th century will be demarcated by some enormous challenges now facing humanity. This will require engineers that play a central role in sustainable development creating wealth and overcoming global problems such as the depletion of natural resources, the environmental damage, the climate change, the fast population growth and the devastation of the ecosystems. This requires a shift in engineering education towards programs that fit in student centred learning with a curriculum focused on the demanding challenges (i.e. societal, environmental, economic) of the 21th century (Graham, 2018). One important way to improve the academic quality of engineering programs and the relevance of the job market is the accreditation of engineering educational programs (Augusti, Freeston, Heitmann & Martin, 2008). Engineering educational program accreditation is a growing practice in Europe and in Portugal for example the accreditation of engineering programs came first (i.e. 1994) to the development of the quality assurance (QA) system for the higher education in the context of the Bologna Process that started with the Sorbonne and Bologna declarations and was planned to create a transparent system of comparable degrees in the European Higher Education Area (EHEA). In 2004 the European Commission (DG-Education and Culture) issued a call for projects proposals contributing to the realization of the EHEA (Bologna Process) and the EURopean ACredited Engineer (EUR-ACE) project (2004/2006) was launched with that purpose. The EUR-ACE Standards, define and require learning outcomes, i.e., what must be learned rather than how it is taught (Augusti, 2009). This preference for utilizing standards based on outcomes (i.e. what is achieved) is very advantageous as it can accommodate innovation in teaching/learning methods and practices and encourages the sharing of good practices. Also, in the EUA the focus is more on what is achieved rather than how it is achieved. In the EUA the Engineer's Council for Professional Development (ECPD) was originally founded in 1932 and later in 1980 became the Accreditation Board for Engineering and Technology (ABET). ABET adopted since 1997 the Engineering Criteria 2000 (EC2000) that also focuses on the outcomes toward what students learn and the attributes that engineers must possess to be excellent professionals in the fields of critical importance to society.

Also, the work of many other organizations has strong positive impact in the quality of engineering programs and engineering education. That is the case of organizations like the European Federation of National Engineering Associations (FEANI) that unites national engineering associations from 33 EHEA countries and as founding member of the World Federation of Engineering (WFEO) collaborates with many organizations related to engineering, technology and engineering education (FEANI, n.d.). Another example is the European Society for Engineering Education (SEFI) that is very much in support of the implementation of the Bologna process all over Europe (SEFI, 2012). The SEFI position paper about the Bologna process and the education for engineers recognize that the societal, economic and technological changes taking place in Europe and in the World constitute crucial factors to foster the excellence in education (i.e. promoting multidisciplinary in teaching and learning, emphasizing engagement, and stimulating entrepreneurship and innovation). Engineering students should be trained in line with the complexities and the evolution of the work environment and should be encouraged by developing more flexible curricula to be explored by the students always by putting students at the centre of the learning process.

Nowadays companies require engineers with good communication skills, able to work in multidisciplinary environments with strong problem-solving skills keeping up with the rapid technological progress (FEANI, 2010). Engineering students should be able to learn by themselves and need to discover what they need to know and where to go to find it (i.e. without counting on someone else to tell everything they need to know to solve problems) and understand the impact of the engineering solutions in an economic, environmental, and societal context.

One effective way to have students learn how to learn is to have students involved in experiments which usually require significant faculty time. The ABET as well as other accrediting bodies such as the EUR-ACE, all agree that design is an essential element of graduate outcomes in an engineering program. For example, the general criterion 3b. of ABET related to students' outcomes is defined as "an ability to design and conduct experiments, as well as to analyse and interpret data". The students must have the ability to design and conduct experiments as well as analyse and interpret data, the ability to design systems and processes that meet the desired needs, among others. Despite the ability to perform experiments and the ability to analyse and interpret data has been addressed in traditional engineering courses, much room is left for design of experiments (DOE).

With the speed of changes in engineering practice we observe that these trends also require huge efforts in engineering education practice reason why it is also required a new pedagogical paradigm. The methodology of DOE can be very effective for solving general or specific problems and is able to have into account the new pedagogical challenges in engineering education. A structured methodology will have the advantage of helping students to avoid the error of trying to evaluate solutions without a clear understanding of the problems.

AN OVERVIEW OF DESIGN OF EXPERIMENTS

The development of DOE goes back to 1920s´ and is mainly due to the work of Sir Ronald Fisher in the Rothamsted Agricultural Field Research Station in London. In 1935 he wrote the wholly new book, *The Design of Experiments.* where he discusses in extensive detail the logical principles of experimentation and the many complexities of factorial design and confounding effects. The book was even referred by Yates (1964) as one of the best expositions of the logical principles of DOE. The main objective of a cautiously planned designed experiment is to realize which set of input variables (i.e. factors) in a process affects the output functional performance and determine the best levels of these factors to obtain acceptable output functional performance in processes and products (Antony, 2003). DOE is a powerful and well-organized collection of methods suitable to meet the needs of future engineers dealing with modern technological advances and with the complexities associated with ever demanding products and processes. The constraints of time, cost and resources makes difficult the investigation of numerous factors that may be affecting complex processes using trial and error methods reason why the use of a structured experimental methodology is necessary. The application of screening factor analysis. with factorial design and some specific fractional factorial designs by Box and Hunter (Box & Hunter, 1961a, 1961b) to identify in an efficient way the vital factors that affect complex processes, is the first step to direct the experimental work and meet the increasing demand for better quality with lower cost. The DOE techniques offer powerful tools to achieve those objectives and are much more efficient than the one factor at a time (OFAT) procedure that involve changing one factor at time, keeping all the rest constant, to study the effect of the factor on the response variable of the product or process. OFAT is easy to understand, however it does not allow to identify and understand the presence of interactions between the factors and sometimes the interaction effects are more important than the effects of individual factors (Montgomery, 2012). After the screening procedure with the determination of a first order polynomial model the use of the sequential optimizing process is the key to improvement. Response Surface Methodology (RSM) is very useful to perform optimization and covers a collection of techniques developed in the 1950s´ that began with the work of Box and Wilson (1951). Plackett and Burman (1946) developed the design of multifactorial experiments. Second order models like the Central Composite Designs (CCD) are a class of very well know models and a class of special three level designs by Box and Behnken (1960) and also the work of Hill and Hunter (1966) are very important designs in the literature review work of RSM that is worth to mention. The 1980s´brought important quality requirements and a new urgency for RSM in the industry. The Japanese Genichi Taguchi proposed a methodology that differs from RSM in many ways (Taguchi, 1986). Taguchi planned the use of an orthogonal array (OA), often a Plackett-Burman design, and benefits from a main effect model to determine the levels of the factors that most likely conduct to product/process improvement. The Taguchi methods have its focus on improvement rather than optimization and generally ignore the interaction between the factors (Taguchi & Wu, 1979). Bharti, Khan and Singh (2010) review the Taguchi methods and multivariate loss function methodology and reveal that there are successful industrial applications of Taguchi robust design techniques for making products/processes insensible to any uncontrollable factors.

DOE and Taguchi Methods are very well recognized for optimizing and improving product quality and process performance, but research has shown that the application of such techniques by the engineering community in industry is limited (Antony, & Kaye, 1995). The work of engineers in the industry goes through experimentation, testing and analysis of data as essential part of their job. Nevertheless, there is still a noteworthy gap between their background in statistics and the effective application of DOE and Taguchi methods in industry. Tanco, Viles, Jesus Álvarez and Ilzarbe (2010) made a wide literature review to understand the main reasons why the DOE is not widely used by engineers in Europe. The survey conducted in Europe among practitioners, academics and consultants was related to the main barriers that hinder the widespread of DOE among engineers and the main reasons identified are related to the low commitment of managers and with the poor background of engineers in statistics. The management commitment is central in any quality management system according to ISO 9001 in order to improve quality. However, many managers are unaware of the benefits that DOE techniques can bring to product and process improvement and engineers still demonstrate some gaps in the knowledge and use of such techniques despite the important role of the DOE and Taguchi Methods in the industry. The difficulty that sometimes engineers have to spread on this type of techniques, combined with the deficit in communicating the benefits to management, restricts the application of DOE in many manufacturing and service industries. This is in accordance and reinforces the view of Antony and Capon (1998) that many years before stated that *there is a cognitive gap in the knowledge of statistics required by engineers in using DOE as a problem- solving tool* and even more when referring that many engineers usually make the remark that they are able to do the statistical exercises in the text book but do not feel comfortable when applying the concepts and principles of DOE in their work environment.

In this new millennium a lot of evolution has already occurred, and the new generations of students have different learning styles and different characteristics (e.g. their preference for experiential learning, their digital literacy, the need for interactivity) that requires different learning environments where students actively participate in the learning process. It is therefore important to identify and review the main teaching and learning strategies and the combination of tools used to support the utilization of DOE in engineering education.

DESIGN OF EXPERIMENTS IN ENGINEERING EDUCATION: A LITERATURE REVIEW

The methodology of DOE can be very effective for solving general or specific problems and is able to have into account the new pedagogical challenges in engineering education. Sometimes teachers receive questions from their students regarding the utilization of the statistical tools that they teach in the market labour (i.e. are the tools really used? Are they useful?). Other concerns of the students are related to their ability when graduated to transform their theoretical knowledge into practice in their workplace. Bergquist and Albing (2006) performed a survey to Sweden alumni to understand how and to what extent the statistical methods (i.e. statistical process control, DOE and capability analysis) were used by organizations hiring the alumni in order to search for answers related to the reasons that motivated organizations to implement statistical methods and to seek differences in use related to organizational types. They concluded that implementation techniques should be emphasized in the curriculum and that different types of courses should be offered (i.e. practical, hands on courses for engineers, managers and others working in organizations).

This section presents a literature review about the use of DOE in engineering teaching and learning for problem solving and for product/process optimization with focus on engineering education. The main goal is to identify the main engineering areas that contribute to the use of statistical experimental design in engineering education as well as the main teaching/learning strategies and the combination of other tools that can be used to support the use of DOE in engineering education.

It was central to the grounding of the review to establish the goal because the goal established the focus of the research and helped to clarify the criteria for study inclusion in the review. The location of the studies was carried on with search strings in electronic databases to find publications relevant to the scope of the review. The bibliometric analysis was carried out on the widely popular Web of Science (WoS) that allows searching in several databases with cross-disciplinary research, enabling in-depth search of subjects which include: science, social science, arts, humanities (i.e. supporting 256 disciplines). All subject areas were taken into consideration for the data search and selection. Google Scholar was also used to validate the previous searches to guarantee that all the appropriate studies that were handled within the searching criteria were included. The search criteria included peer reviewed articles and proceedings of conferences, excluding documents that did not comply with these criteria. The literature search was made founded on various search strings selected by research topics tracking papers from 2000 till February 2019. The search strings were selected to include terms related to statistical design of experiments (e.g. Design of experiments, factorial design, fractional factorial design, Taguchi methods, robust design, RSM) and combined with terms related to teaching and learning strategies and engineering education (e.g. active learning, problem based learning, teaching, learning, learning processes, problem solving, engineering, engineer, statistics education). This led to the definition of a research focus excluding studies when it was recognized that they did not belong to both areas or that they did not represent any relation. Even with this search and selection it was still necessary to perform a manual check for all the articles that were selected according to the search strings to remove the articles that clearly did not address the topic under research. The manually check was performed based on the abstract of each journal article or conference article. A huge number of studies were discarded because they were out of scope, including those that were focused on factorial and fractional factorial designs used in learning/teaching environments, but not related to teaching/learning activities of DOE tools and methods. Also, a lot of studies related to statistical techniques used in industrial/engineering environments were rejected in the filtering process because they were not related with the main objective of the investigation. A considerable number of research works related to the ability to conduct experiments and analyse and interpret data were also discarded because they were not related to statistical design of experiments, but with inquiry-based learning that despite forming the foundation for the design of any experiment is not the goal of this literature review. The search and filtering process finally resulted in a net list of 45 research works.

DESCRIPTIVE ANALYSIS

The starting point represents the beginning of the new millennium where new challenges in engineering education took place and a lot of evolution already occurred with new generation of students requiring different teaching/learning styles. The number of publications across the period studied (2000 – Feb 2019), the number and distribution of publications by journal and the number of publications by conference are summarized to quantitatively describe the features of the data.

Distribution of Publications Across the Period

In order to put our research into context it is necessary to follow the growth of publications in the field.

The number of publications across the period studied (2000-Feb. 2019) is depicted in Figure 1. The total number of publications that simultaneously addresses the topic of DOE and engineering education is quite limited with a total number of 45 and the evolution over time shows that the number of publications had a slight increase all over the years. The number of publications was limited to 9 papers in the period 2000-2004 and a total of 9 papers were also found in the period 2005-2009. A larger number of publications (i.e. 14 papers) were identified in the period between 2010-2014. In the last period from 2015 to February 2019 a little decrease in the number of publications occurred (i.e. 13 papers). We can observe that the number of published works across the period studied is globally low, but even so with a small increase from 2010 onwards. the number of recent research studies focusing on DOE in engineering education has not noted a significant growth in recent years and so far, as we know there have been no systematic literature reviews of the DOE in engineering education research that covers the same period of time.

Despite the high number of publications related to the use of DOE in industrial applications the studies related to DOE in engineering education are in low number. This may be explained by the recent rise in engineering education research (EER), as pointed out by de Graaff (2017) that evidenced the gaining of the importance of EER during the last decade and if the research in the area of engineering education is a rather recent topic, it is not surprising the limited number of works that exclusively focus on the use of DOE in engineering education.

Figure 1. Distribution of publications across the period studied

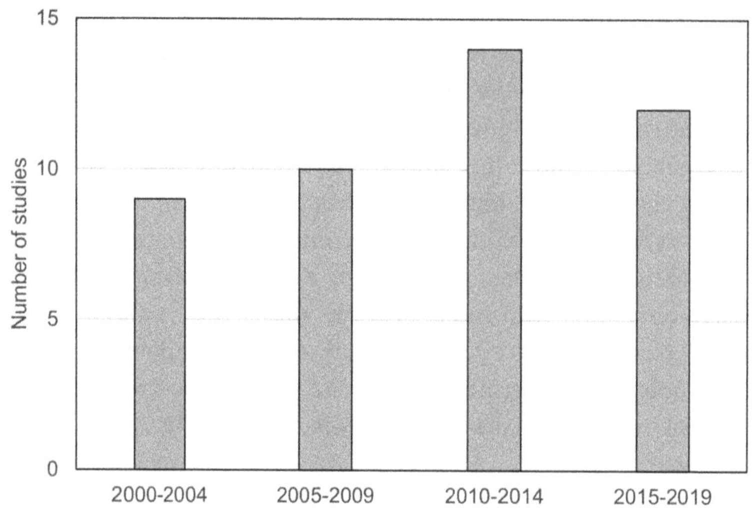

Distribution of Publications in Journals/Conferences

The 46 research works selected were also classified according to the journals/conferences in which the studies have been published. A total of 27 papers was published in specialized journals that cover topics typically related to areas of quality management, statistics or areas of education or engineering and technology (Figure 2).

Figure 2. Distribution of publications in specialized journals

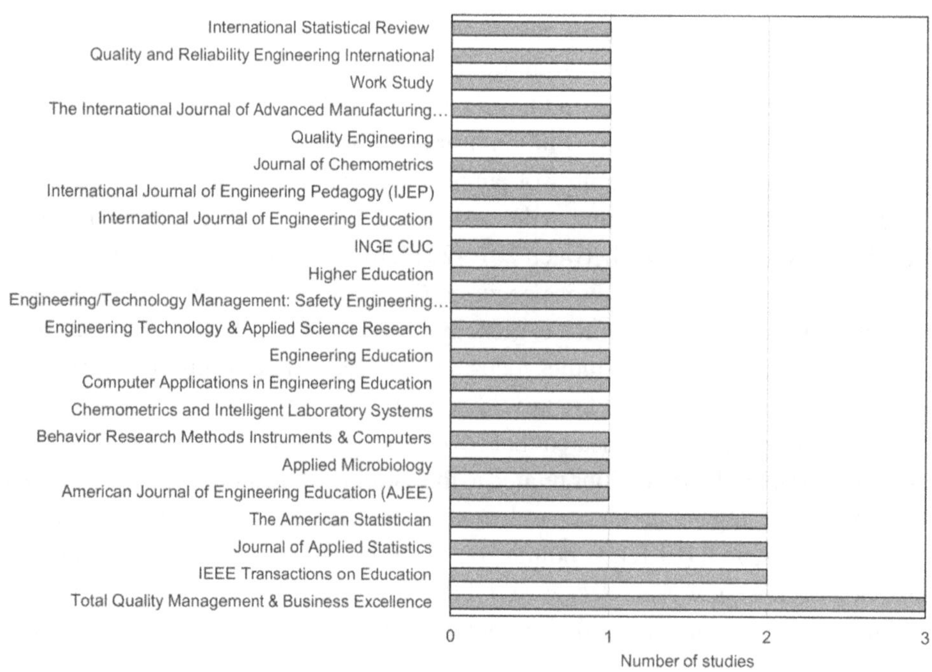

The Total Quality Management & Business Excellence is the journal with the highest number of published works about the topic of DOE in engineering education. The journal encourages thought and research in all subjects relating to total quality management, providing up to date research across the whole field including quality systems, tools and techniques. IEEE transactions on Education appears next in the list with a total of two Annual conference published works. This Journal publishes scholarly contributions to education in electrical and electronics engineering, computer engineering, computer science and other fields within the scope of education in engineering. Also, the Journal of Applied Statistics and the American Statistician appear in second place in the list and provide a forum for communication and debate for the application of statistical principles and statistical methodology across a wide range of disciplines. With a lower number of publications there are listed several journals focused on the topic of engineering education and in areas relating to engineering and technology.

A total of 19 works are proceedings presented at international conferences as EDULEARN, the IEEE International Systems Conference, The Annual Frontiers in Education Conference, the International Conference on Higher Education Advances (HEAd), the International Conference of the Portuguese Society for Engineering Education (CISPEE) and the ASEE and exposition just no name a few.

THEMATIC FINDINGS

DOE in engineering education evolved in the last decades and Table 1 provides and overview of the selected studies that represent examples of the nature of the research with the studies classified according to the scientific area and techniques that have been used. Table 1 provides an overview of the selected studies that somehow adopt the DOE perspective in engineering education. The papers included here represent those that we see as exemplars of the nature of the research within this stream especially if the themes of DOE are analysed through the engineering education perspective. The main objective of the classification is to identify the main engineering areas that account for the use of DOE in engineering education along with the main techniques and combination with other tools.

Applications in Scientific Area

Early at the beginning of this new millennium Montgomery (2001) made a historical perspective on industrial statistics and some of the findings include the need of modernizing education in SPC, DOE, and reliability engineering. In line with this need to modernize DOE teaching, it is important to analyse the works that have been selected in order to understand what has been done in recent years to educate students more broadly in DOE in engineering The selected papers were classified by scientific area to identify and analyse the type of studies developed and the weight of these studies by scientific area.

Almost 22% of the research works selected in this literature review refer to the scientific area of chemical engineering which is not surprising due to the growing demand of chemical engineers' skills to plan experiments and analyse experimental data. Also, the usefulness of the DOE techniques in chemistry and chemical engineering research was described by Liang, Fang and Xu (2001) and Xu, Xu, Li and Fang (2018). Some publications also combine the DOE in chemical engineering with the simulation as for example the development of a virtual chemical vapor deposition (CVD) reactor based on a numerical simulation where students learn and then actually apply DOE (Koretsky, Amatore, Barnes, & Kimura, 2006, 2008; Koretsky, Kimura, Barnes et al., 2006). Another example of DOE with simulation software

Table 1. Selected publications of DOE underpinned by engineering education perspective

Scientific Area	Techniques	Study
Chemical Engineering	Factorial and fractional factorial designs	(João, & Silva, 2007)
	Factorial and fractional factorial designs; Simulation tools	(João, & Silva, 2017)
	DOE	(Tanco, Viles, Jesus Álvarez, & Ilzarbe, 2010)
	DOE; simulation tools	(Koretsky, Amatore, Barnes, & Kimura, 2006)
	DOE; simulation tools	(Koretsky, Kimura, Barnes et al., 2006)
	DOE, Simulation tools, Taguchi methods	(Koretsky, Amatore, Barnes, & Kimura, 2008)
	Taguchi methods	(Allada & Jose, 2003)
	Random and mixed effects factorial designs. Two level factorial designs, Fractional factorial designs, Robust design, Evolutionary operation with two-level design, Polynomial models, RSM	(Fernandez, 2010)
	Factorial and fractional factorial design; Orthogonal design, D-optimal design; Uniform experimental design	(Liang, Fang, & Xu, 2001)
	DOE; Fractional factorial design with model uncertainty; Robust design; Uniform design	(Xu, Xu, Li, & Fang, 2018)
Electronic Engineering	DOE	(Brady & Allen, 2002)
	Design for Six Sigma; Statistical engineering	(Campean, Grove, & Henshall, 2005)
	DOE; Full factorial design; Simulation tools	(Salunke & Kittur, 2017)
Engineering	DOE; Taguchi methods; Genetic Algorithms.	(Woll & Burkhard, 2005)
	Factorial designs; RSM; Full quadratic design	(Annis, 2005)
	DOE; Capability analysis; Statistical process control (SPC)	(Bergquist & Albing, 2006)
	Taguchi methods; Robust design	(Platanitis & Pop-Iliev, 2010)
	Two level factorial designs; Taguchi methods	(Bharti, Khan, Singh, et al., 2010)
	Robust design (S/N)	(Squires & Cloutier, 2011)
	DOE; Regression; Monte Carlo Method	(Zhan, Fink, & Fang, 2011)
	DOE	(Wang & Li, 2014)
	Design for Six Sigma; Minitab statistical package	(Elbadawi, Aichouni, & Messaoudene, 2016)
	Regression analysis	(Pineda Becerril, García, Aguilar, & León, 2017)
	DOE; Web based tools	(Boon, Di Bucchianico, Rijpkema, & van Berkum, 2010)
	DOE; Factorial experiments	(Antony, 2002)
Food	DOE	(Tanco Viles, Jesus Álvarez, & Ilzarbe, 2010)
	Fractional factorial designs	(Bevilacqua, Corbo, & Sinigaglia, 2010)

continues on following page

Table 1. Continued

Scientific Area	Techniques	Study
Industrial and manufacturing engineering	DOE	(Wang & Li, 2014)
	DOE	(Tanco, Viles, Ilzarbe, & Álvarez, 2007)
	Taguchi methods	(Antony & Jiju Antony, 2001)
	Taguchi methods; Simulation based tools	(Allwood, Cox, & Latif, 2001)
	Taguchi method; Hybrid genetic algorithm	(Yildiz, 2013)
	Random and mixed effects factorial designs; Two level factorial designs; Fractional factorial designs; Robust design; Evolutionary operation with two-level design; Polynomial models; RSM	(Fernandez, 2010)
Mechanical engineering and automotive industry	DOE	(Tanco, Viles, Jesus Álvarez, & Ilzarbe, 2010)
	Evolutionary operation with two-level design;	(Lu, 2011)
	DOE	(Butler & Moses, 2005)
	DOE	(Burke, De Jonge, Avola, & Forte, 2017)
Quality and Environmental Engineering	Fractional factorial design	(João, & Silva, 2016)
	DOE	(João, & Silva, 2018)
Others	RSM; DOE; Web based tools	(Ketelaere, Bebber, Makar, & Sousa, 2014)
	SPC; DOE; Robust parameter design	(Montgomery, 2001)
	Complete and incomplete factorial designs with between-subjects, within-subjects, and quasi experimental factors, as well as mixed designs.	(Reips & Neuhaus, 2002)
	RSM	(Budé, Imbos, van de Wiel, & Berger, 2011)
	Taguchi Methods	(Ree, Park, & Yoo, 2014)
	DOE	(Reina Neira, de la Hoz, Felizzola Jimenez, & Hualpa Zuniga, 2016)
	Regression analysis; Bootstrapping, Simulation	(Reis & Kenett, 2017)
	DOE; SPC, Capability analysis	(Lundkvist, Bergquist, & Vanhatalo, 2018)
	Statistical thinking	(De Federico, 2018)
	RSM; Web based tools	(Darius, Schrevens, van der Knaap, et al., 2003)
	RSM; Virtual experiments	(Darius, Portier, & Schrevens, 2007)

is the use of Aspen HYSYS by chemical engineering students to model distillation systems (João, & Silva, 2017). Other examples are related to hands-on approach on laboratory sessions (João, & Silva, 2007) or hands-on approach to projects where the students are required to prepare the experiments and perform in the laboratory of chemistry (Fernandez, 2010).

The food industry has been a prime user of RSM since the early 1970s and this is reflected in a study of the content of journal articles and in industrial usage as well (Myers, Khuri, & Carter, 1989). However, the number of studies regarding DOE and RSM in food science education is limited. The study of Bevilacqua, Corbo and Sinigaglia (2010) presents a step-by-step description of the theory of DOE focusing on the 2^k design, CCD design and mixture approach with details on how researchers can use it for microbiological purposes. Despite the importance that the DOE has had over the years in the food sector, few studies have an exclusive focus on DOE teaching / learning in food engineering.

We combined the studies related to Industrial engineering and manufacturing engineering in the same class and a total of 11% studies were selected. The number of studies from this class focusing on DOE in engineering education is quite limited. This is in accordance with the work of Viles, Ilzarbe, and Álvarez (2007) that designed a survey to detect the need of DOE and the degree of implementation of this approach within manufacturing industries and the survey confirmed that DOE is hardly used. From the five studies selected in Industrial and manufacturing engineering related to engineering education three of them were based on Taguchi methods. Allwood, Cox and Latif (2001) developed a simulation-based tool to assist learning of the Taguchi method which was tested with a group of final year undergraduate students and most students felt that the simulation helped them to understand the Taguchi method and enhanced what they had been taught in the lectures. Antony and Jiju Antony (2001) presented a simple experiment which can be used in the classroom to teach engineers the basics of the Taguchi method and the experiment could support rapid understanding of the results with simple analytical and graphical tools. Also Fernandez (2010) reported the usefulness of a collaborative–problem based on the teaching-learning process of the Statistical Experimental Design course for Industrial Engineering students where the students developed the projects making use of tools as random and mixed effects factorial designs, two level factorial designs, fractional factorial designs, robust design, evolutionary operation with two-level designs and RSM. Yildiz (2013) presented a novel hybrid optimization approach based on teaching–learning based optimization algorithm and Taguchi's method of case studies application and the new optimization approach can solve optimization problems in the research and manufacturing area. Wang and Li (2014) made a recent effort at a research university in the U.S. to integrate renewable energy topics into the traditional DOE course to help industrial and systems engineering students update their knowledge base and foster environmental responsibility and sustainability awareness in their future careers.

Some selected studies are from mechanical and automotive engineering (i.e. around 9%). Tanco, Viles, Jesus Álvarez and Ilzarbe (2010) made a survey of different scientific, engineering areas where they considered automotive and machinery and in their exploratory analysis concluded that the main barriers that hinder the widespread use of DOE are related to low managerial commitment and engineers' general weakness in statistics. Lu (2011) made use of Design-Expert software and integrated it into an advanced engineering experimentation course that deals with the design, execution and analysis of experiments and concluded that the learning outcomes were better attained by the students that have used this software for designing and analysing their experiments. Butler and Moses (2005) examined student laboratory performance under two organizational models. In the old laboratory scheme, experiments were organized around a specific topic area (mechanics, thermal science, materials) and lab exercises were selected based only on their applicability to the current lab topic and the class level of the students. The new lab scheme uses essentially the same experiments in a more structured approach with an emphasis on designing experiments providing a better transition for students in their attempt to reconcile junior-level experiences with senior-level expectations. Burke, De Jonge, Avola and Forte (2017) proposed a virtual engine laboratory application for use in automotive engineering education where the laboratory software was built as a flexible MATLAB tool that can easily be transferred for applications in other disciplines promoting the link between teaching and research.

João and Silva (2016) made use of design of experiments in an active learning environment in a new Master course in Quality and Environmental Engineering and in this learning environment the students worked on projects designed to use fractional factorial designs in order to improve a system. The active learning approach required that the students developed their own projects and had to decide when and how to do the experiments. Taking the responsibility of their activities was very enriching forcing the students to think about the things they did and questioning some of their own decisions. Also, João and Silva (2018) designed a curricular unit of DOE to accomplish the learning outcomes and described how an active learning project is used as a teaching, learning methodology to help the students to attain the learning outcomes defined in the unit.

Salunke and Kittur (2017) made a work in electrical engineering providing an exposure to students on experimentation with factorial design and simulation environment as a part of the structured enquiry problem in the course and showed that the activity helped the students to analyse the results for the proposed problem and evaluated the competencies addressed for the activity (i.e. problem analysis, ability to design experiments, individual and team work and simulation). The study from Brady and Allen (2002) describe a case-based instructional approach illustrated with an exercise based on an actual case study of increasing yields for manufacturing electronic components. The case based instructional approach was designed to reinforce lecture material and motivate students to apply what they have learned. Students seem to identify with the case study much better than with the traditional lectures. Campean, Grove and Henshall (2005) developed an electronics case study to match the statistical engineering teaching materials suitable for use by Ford Motor Company's Technical Education Program but which could also be promoted by the UK's Royal Statistical Society within the university sector.

Some of the studies do not specify the engineering area and were generically aggregated by engineering with a total of near 27% of the selected studies. Antony (2002) presents the catapult experiment performed during a training program of DOE and the results showed rapid and easy understanding of engineers with limited statistical competency. Annis (2005) made use of Box's paper helicopter to teach experimental design due to its simplicity and because it is inexpensive and provides real data for multifactorial experiment emphasizing the value of applying problem-specific scientific judgement and at the same time thorough statistical techniques. Woll and Burkhard (2005) used different approaches for DOE from an educational program and explained how DOE can be used for student education in an effective way. Bergquist and Albing (2006) performed a survey to examining how and to what extent the methods, Statistical Process Control, Capability Analysis and DOE are used by organizations hiring the alumni and concluded that implementation techniques must be emphasized in the curriculum and that practical, hands-on courses for engineers should be offered at universities. Bharti, Khan and Singh (2010) made a Literature review of Taguchi methods and its successful implementation in several industries and research areas. Platanitis and Pop-Iliev (2010) introduced an assignment to supplement the need at the first year level of students to practice robust design techniques and DOE methodologies. The assignment was well received by the students looking favourably at the opportunity to apply design concepts to a realistic design problem. Boon, Di Bucchianico, Rijpkema and van Berkum (2010) presented Statlab an interactive web based teaching tool for DOE. With Statlab the engineering students were introduced to experimental design through virtual examples. The tool forces students to reflect about applying details since it hides options that students do not ask for. Squires and Cloutier (2011) provided an example of applying a robust design approach (i.e. making use of S/N) for two online course design factors and the framework is applied to the challenge of improving online course design by investigating the impact of the pedagogy used to deliver course lectures and discuss course content on student satisfaction and learning. Elbadawi, Aichouni and Messaoudene (2016) employed a hands-on project based on Lean Six Sigma experiments for engineering education and found the hands-on project to be extremely valuable to students´ learning. Pineda Becerril, García, Aguilar and León (2017) developed an e-learning course, adapting the subject regression and linear correlation in the subjects of Statistics that are imparted in the careers of Engineering offering the student a greater flexibility with respect to the conventional method in the classroom.

In the category named others, we have placed all the studies that are based on the teaching of industrial statistics with reference to the teaching practices and pedagogical aspects of DOE teaching with almost 18% of the studies. Darius, Schrevens, van der Knaap et al. (2003) described a few accessible web-based tools used in several statistics courses, used for introductory statistics courses as well as to more specialized DOE courses. Reips and Neuhaus (2002) developed WEXTOR that can be used to plan Experimental Design and Web-based Experimentation in order to enable learning by doing in (Web) experimental design. Also, Darius, Portier and Schrevens (2007) developed some virtual experiments based on software environments that mimic the real conditions with data generated by an underlying realistic stochastic model that as soon as the data are collected, they can be transferred to a standard statistical package and the user can train his/her design skills by relating the quality of the statistical results obtained to the data collection strategy used. Budé, Imbos, van de Wiel and Berger (2011) investigated and compared across course lengths the conceptual understanding of topics in statistical courses and concluded that distributed practice enhances short term conceptual understanding of statistics. Ketelaere, Bebber, Makar and Sousa (2014) developed a web-based application based on a real life example to apply response surface methodology in a simple way as by inserting a design into the application the corresponding responses are generated and can be used for analysis and follow-up experimentation. Ree et al. (2014) introduced a methodology to improve lecture quality by using the Taguchi method in education settings and discussed a case study for it. Reina Neira, de la Hoz, Felizzola Jimenez and Hualpa Zuniga (2016) in their work explored the use of Problem Based Learning (PBL) in the area of DOE, based on the principles of the model and implementation of its methodology. Reis and Kenett (2017) mapped a wide range of simulation-based tools used in teaching statistical methods and concluded that simulators offer opportunities for hands-on experience in the classroom, bridging the gap between theory and practice. De Federico (2018) did a study where he concluded that learning with dynamic animations collaborates with the insertion in the mind of the students of the statistical thinking as the main axis on which the other specific knowledge of the career will be based. Lundkvist, Bergquist and Vanhatalo (2018) performed a study to assess the use of statistical process control (SPC), process capability analysis, and DOE over time and concluded that SPC is the most prevalent of the three methods while DOE is least used. The greatest barriers identified for increasing the use of statistical methods were insufficient resources (i.e. time and money), low commitment of managers, inadequate statistical knowledge, and lack of methods to guide the user through experimentations.

Statistical Methods and Combination With Simulation Tools

The selected studies of Taguchi methods account for 18% of the selected research works. This is not surprising as Dr. Genichi Taguchi´s method for quality improvement is a widely used engineering technique in the world and recognized by most engineers since the Western awareness of Taguchi methods that arose largely as a natural consequence of the importing of successful quality control practices from Japan in the last part of the twentieth century. Many explicit references also appear to factorial design and fractional factorial design as well as some references to RSM and evolutionary operation. Most of the studies do not explicitly refer to the method used, but only to the use of DOE. For a better understanding of the of the statistical methods aggregated by scientific area and selected studies, see Table 1.

Some studies combine the statistical experimental design techniques with simulation as for example Allwood, Cox and Latif (2001) that developed a computerized learning tool as an iterative process to assist learning of the Taguchi method where the students found the simulation very useful to better understand the Taguchi experimental design methods. Some authors developed virtual experiments and web-based tools that mimic the real conditions allowing the students to train their DOE skills (Darius, Portier, & Schrevens, 2007; Darius, Schrevens, van der Knaap et al., 2003; Darius, van der Knaap, Schrevens et al., 2003). Also, Boon, Di Bucchianico, Rijpkema and van Berkum (2010) developed a web-based teaching tool that can be a stimulating learning environment, forcing students to think about the things that they are doing. Some virtual experiments were also developed specifically for chemical engineering students as for example the works of Koretsky and co-authors (Koretsky, Amatore, Barnes, & Kimura, 2006, 2008; Koretsky, Kimura, Barnes et al., 2006) that made the development of a virtual chemical vapor deposition (CVD) reactor based on a numerical simulation where students learned how to apply DOE. The development of an integrated capstone experience proved to be very enriching, giving the students a hands-on learning experience and provided the opportunities for students' engagement in DOE and engineering science. Also, Ketelaere, Bebber, Makar and Sousa (2014) developed a web application to mimic a real life problem allowing the users to interact with the tool and develop their own designs. Reis and Kenett (2017) mapped several wide range simulation tools with the main objective of helping educators to access and integrate options related to the incorporation of simulators as a teaching aid in an educational and pedagogical roadmap.

CONCLUSION

Most courses available in engineering statistics often tend to concentrate on the theoretical statistical issues and the more mathematical aspects of the subject without a focus on the engineering education. This is in accordance with Antony and Capon (1998) that at the end of the twentieth century already stated that *statistical education for engineers at university level is generally inadequate" and that "many engineers would deem statistics as useless in their late careers in industries.*

Weissman and Anderson (2015) made a review of recent publications of DOE selected publications from Organic Process Research & Development and considered DOE a powerful approach to optimizing chemical processes concluding that the number of publications per year had a huge increase in the last years. Yet, although its use has increased considerably in the industry, it is evident that the studies dedicated to its teaching/learning in academia are still scarce as we can observe in this literature review work. According to the research carried out we could conclude that although there is not a very high number of studies dedicated to the teaching / learning of DOE the scientific area of chemical engineering is the one that presents a greater number of selected studies dedicated to DOE in engineering education. Despite the number of studies of DOE in chemical engineering education it is our opinion that DOE needs a more systematic approach in teaching and learning in engineering education and this is in accordance with the position of Laird (2002) the editor of Organic Process Research & Development that expressed the same opinion related to the DOE teaching in academia and the need to be more often used in industry.

Currently industry require engineers able to work in complex settings with strong problem-solving skills, keeping up with the rapid technological progress reason why educators should prepare students to be able to learn by themselves. DOE is one effective way to have students involved in experiments because unlike other statistical techniques that are focused on extracting information after data has been collected, DOE is pro-active and requires an active learning approach that forces the engineering students to actively think about the things that they are doing. The students should actively think how to set up an experimental design and room should be left for students making mistakes in order to learn from them. Such a learning environment will make teaching DOE more like the work an engineer will have to perform in industry. By a student centred approach the students identify the design problem, organize the ideas and develop some representations of the problem (i.e. analyse), and also choose the working conditions, the course of action, testing designs (i.e. evaluate) and design and optimize solutions (i.e. create). Some of the selected studies combine virtual experiments that are carried out in a software environments with DOE in order to mimic a real situation of interest allowing the creation of accessible and rich experiences that have the potential to provide a learning environment well beyond those of traditional textbook exercises. Some selected examples of DOE performed through simulation also have the advantage of providing the execution of the experiences with less resources of time and money that would be necessary for the accomplishment of the physical experiences and real projects.

Most of the selected publications refer to the importance of the students to decide for themselves how to plan the experiments, their execution and analyses referring the advantages of the active learning environments in the acquisition of skills in DOE. The spirit of student-centred learning is also fully reflected in the Engineering Criteria of the Accreditation Board for Engineering and Technology (ABET), whose the approach focuses on learning outcome not on teaching input Felder and Brent (2003). The rapid changes taking place in the industry today require a pedagogical paradigm in which teaching is student-cantered and based on active learning methodologies. The selected studies present DOE teaching and learning approaches that can serve as a starting point for future work and development of teaching practices to be used to engage students in DOE education and promote the usage of DOE methodology by future engineering graduates and researchers.

REFERENCES

Allada, V., & Jose, V. (2003). Case-Based Problem Solving Exercises for Engineering Education. *Engineering/Technology Management* [ASME.]. *Safety Engineering and Risk Analysis, Technology and Society, Engineering Business Management, and Homeland Security, 2003*, 137–142.

Allwood, J. M., Cox, B. M., & Latif, S. S. (2001). The structured development of simulation-based learning tools with an example for the Taguchi method. *IEEE Transactions on Education, 44*(4), 347–354. doi:10.1109/13.965783

Annis, D. H. (2005). Rethinking the Paper Helicopter. *The American Statistician, 59*(4), 320–326. doi:10.1198/000313005X70777

Antony, J. (2002). Training for design of experiments using a catapult. *Quality and Reliability Engineering International, 18*(1), 29–35. doi:10.1002/qre.444

Antony, J. (2003). *Design of experiments for engineers and scientists*. Butterworth-Heinemann.

Antony, J., & Capon, N. (1998). Teaching experimental design techniques to industrial engineers. *International Journal of Engineering*, *14*, 335–343.

Antony, J., & Jiju Antony, F. (2001). Teaching the Taguchi method to industrial engineers. *Work Study*, *50*(4), 141–149. doi:10.1108/00438020110391873

Antony, J., & Kaye, M. (1995). Experimental Quality. *Journal of Manufacturing Engineer*, *74*(4), 178–181. doi:10.1049/me:19950409

Augusti, G. (2009). EUR-ACE: The European Accreditation system of Engineering Education and its global context. *Engineering Education Quality Assurance: A Global Perspective*, 41–49.

Augusti, G., Freeston, I., Heitmann, G., & Martin, R.-P. (2008). Accreditation and QA of engineering education in Europe: Setting up a pan-European system. In A. Beso, L. Bollaert, B. Curvale, H. T. Jensen, L. Harvey, E. Helle, … A. Sursock (Eds.), *Implementing and using quality assurance: Strategy and practice a selection of papers from the 2nd European Quality Assurance Forum* (pp. 42–47). European University Association.

Bergquist, B., & Albing, M. (2006). Statistical methods – Does anyone really use them? *Total Quality Management & Business Excellence*, *17*(8), 961–972. doi:10.1080/14783360600747762

Bevilacqua, A., Corbo, M. R., & Sinigaglia, M. (2010). Design of experiments : A powerful tool in food microbiology. *Applied Microbiology*, 1419–1429.

Bharti, P. K., Khan, M. I., & Singh, H. (2010). Taguchi's off line method and multivariate loss function approach for quality management and optimization of process parameters-A review. In S. Paruya, S. Kar, & S. Roy, (Eds.), *International Conference on Modeling, Optimization, and Computing* (p. 486+). 10.1063/1.3516353

Bharti, P. K., Khan, M. I., Singh, H., Paruya, S., Kar, S., & Roy, S. (2010). Taguchi's off line method and multivariate loss function approach for quality management and optimization of process parameters -A review. *AIP Conference Proceedings*, *1298*, 486–491. doi:10.1063/1.3516353

Boon, M. A. A., Di Bucchianico, A., Rijpkema, J. J. M., & van Berkum, E. E. M. (2010). Statlab: An interactive teaching tool for DOE. In A. D. Fitt, J. Norbury, H. Ockendon, & R. E. Wilson (Eds.), *Progress In Industrial Mathematics at ECMI 2008* (pp. 663–668). doi:10.1007/978-3-642-12110-4_107

Box, G. E. P., & Behnken, D. W. (1960). Some new three level designs for the study of quantitative variables. *Technometrics*, *2*(4), 455–475. doi:10.1080/00401706.1960.10489912

Box, G. E. P., & Hunter, J. S. (1961a). The 2^{k-p} fractional factorial designs. Part I. *Technometrics*, *3*, 311–351.

Box, G. E. P., & Hunter, J. S. (1961b). The 2^{k-p} fractional factorial designs Part II. *Technometrics*, *3*, 449–458.

Box, G. E. P., & Wilson, K. B. (1951). On the experimental attainment of optimum conditions. *Journal of the Royal Statistical Society. Series B. Methodological*, *13*(1), 1–45. doi:10.1111/j.2517-6161.1951.tb00067.x

Brady, J. E., & Allen, T. T. (2002). Case study based instruction of DOE and SPC. *The American Statistician, 56*(4), 312–315. doi:10.1198/000313002614

Budé, L., Imbos, T., van de Wiel, M. W., & Berger, M. P. (2011). The effect of distributed practice on students' conceptual understanding of statistics. *Higher Education, 62*(1), 69–79. doi:10.100710734-010-9366-y

Burke, R. D., De Jonge, N., Avola, C., & Forte, B. (2017). A virtual engine laboratory for teaching powertrain engineering. *Computer Applications in Engineering Education, 25*(6), 948–960. doi:10.1002/cae.21847

Butler, A. J., & Moses, W. M. (2005). Introducing experimental design in mechanical engineering laboratories. In *Proceedings 2005 American Society for Engineering Education Annual Conference and Exposition*, 10.827.1-10. American Society for Engineering Education.

Campean, I. F., Grove, D., & Henshall, E. (2005). A course in statistical engineering. *International Journal of Engineering Education, 21*, 502–511.

Darius, P., Schrevens, E., van der Knaap, H., Portier, K., Massonnet, G., Lievens, L., ... Thas, O. (2003). *Using web-based tools for teaching statistical concepts and experimentation skills*. IASE/ISI Satelite.

Darius, P. L., Portier, K. M., & Schrevens, E. (2007). Virtual experiments and their use in teaching experimental design. *International Statistical Review / Revue Internationale de Statistique, 75*, 281–294.

Darius, P. L., van der Knaap, H., Schrevens, E., Portier, K., Massonnet, G., Lievens, L., & Duchateau, L. (2003). Virtual experiments and their role in teaching design of experiments. In *Ingénierie Des Systèmes D'information*. ISI.

De Federico, S. (2018). Consolidation of statistical thinking in engineering students through animation and simulation of the statistic methods and results of stochastic processes analysis. *EDULEARN18 Proceedings, 1*, 8442–8451.

de Graaff, E. (2017). Ten years in engineering education research: Looking back ahead. *European Journal of Engineering Education, 42*(6), 587–590. doi:10.1080/03043797.2017.1408942

Elbadawi, I., Aichouni, M., & Messaoudene, N. A. (2016). Developing an innovative and creative hands-on lean six sigma manufacturing experiments for engineering education. *Engineering Technology & Applied Science Research, 6*, 1297–1302.

ESG. (2015). *Standards and Guidelines for Quality Assurance in the European Higher Education Area (ESG)*. ESG.

FEANI. (2010). *Position paper on Educational Policy*. Document approved by the FEANI General Assembly on 1 October 2010.

FEANI. (n.d.). FEANI. Retrieved from https://www.feani.org/feani/what-feani

Felder, R. M., & Brent, R. (2003). Designing and teaching courses to satisfy the ABET engineering criteria. *Journal of Engineering Education, 92*(1), 7–25. doi:10.1002/j.2168-9830.2003.tb00734.x

Fernandez, C. G. (2010). Introducing statistical experimental designs to chemical and industrial engineering using the collaborative and problem based learning approach. *Edulearn10: International Conference on Education and New Learning Technologies*, 1883–1888.

Graham, R. (2018). The global state of the art in engineering education. Massachusetts Institute of Technology (MIT).

Hill, W. J., & Hunter, W. G. (1966). A review of response surface methodology: A literature survey. *Technometrics*, *8*(4), 571–590. doi:10.2307/1266632

João, I. M., & Silva, J. M. (2007). Design of experiments in chemical engineering education. Application to liquid-liquid extraction. *Proceedings 56th session of the International Statistical Institute*, 657–658. International Statistical Institute.

João, I. M., & Silva, J. M. (2016). Student engagement with statistical design of experiments by active learning projects. *Proceedings of the 2nd International Conference on Higher Education Advances*, 203–210. Valencia: Universitat Politècnica València. 10.4995/HEAD16.2016.2650

João, I. M., & Silva, J. M. (2017). Designing solutions by a student centred approach: Integration of chemical process simulation with statistical tools to improve distillation systems. [IJEP]. *International Journal of Engineering Pedagogy*, *7*(3), 4. doi:10.3991/ijep.v7i3.6795

João, I. M., & Silva, J. M. (2018). Designing and teaching a curricular unit to accomplish the outcomes related learning objectives. In *Proceedings 2018 3rd International Conference of the Portuguese Society for Engineering Education (CISPEE)*, 1–7. IEEE.

Ketelaere, D., Bebber, V., Makar, I. K., & Sousa, B. De. (2014). The garden sprinkler: An interactive web-based application for teaching design of experiments. *ICOTS9*, 1–4.

Koretsky, M. D., Amatore, D., Barnes, C., & Kimura, S. (2006). The virtual CVD learning platform. *Proceedings. Frontiers in Education. 36th Annual Conference*, 25–30. IEEE.

Koretsky, M. D., Amatore, D., Barnes, C., & Kimura, S. (2008). Enhancement of student learning in experimental design using a virtual laboratory. *IEEE Transactions on Education*, *51*(1), 76–85. doi:10.1109/TE.2007.906894

Koretsky, M. D., Kimura, S., Barnes, C., Amatore, D., & Meyers-Graham, D. (2006). Experiential learning of design of experiments using a virtual CVD reactor. *ASEE Annual Conference and Exposition, Conference Proceedings*.

Laird, T. (2002). Design of experiments (DoE). *Organic Process Research & Development*, *6*(4), 337. doi:10.1021/op0255581

Liang, Y., Fang, K., & Xu, Q. (2001). Uniform design and its applications in chemistry and chemical engineering. *Chemometrics and Intelligent Laboratory Systems*, *58*(1), 43–57. doi:10.1016/S0169-7439(01)00139-3

Lu, Y. C. (2011). Enhancing the students' abilities in experimental designs through Design-Expert. In *Proceedings 2011 ASEE Annual Conference & Exposition*. ASEE.

Lundkvist, P., Bergquist, B., & Vanhatalo, E. (2018). Statistical methods – still ignored? The testimony of Swedish alumni. *Total Quality Management & Business Excellence, 0*, 1–18. doi:10.1080/14783363.2018.1426449

Montgomery, D. C. (2001). Opportunities and challenges for industrial statisticians. *Journal of Applied Statistics, 28*(3-4), 427–439. doi:10.1080/02664760120034153

Montgomery, D. C. (2012). *Design and Analysis of Experiments (8th ed.)*. John Wiley & Sons.

Myers, R. H., Khuri, A. I., & Carter, W. H. (1989). Response surface methodology: 1966-1988. *Technometrics, 31*, 137–157.

Pineda Becerril, M., García, O., Aguilar, A., & León, F. (2017). Teaching the regression and simple linear correlation theme with e-learning support. In L. Gómez Chova, A. L. Martínez, & I. C. Torres (Eds.), *9th International Conference on Education and New Learning Technologies - EDULEARN17* (pp. 6324–6330). 10.21125/edulearn.2017.2034

Plackett, R. L., & Burman, J. P. (1946). The design of optimum multifactorial experiments. *Biometrika, 33*(4), 305–325. doi:10.1093/biomet/33.4.305

Platanitis, G., & Pop-Iliev, R. (2010). Early introduction of robust design into the engineering curriculum. *Proceedings of the ASME International Design Engineering Technical Conferences and Computers and Information in Engineering Conference, DETC2010, VOL6*, 599–606. ASME. 10.1115/DETC2010-28221

Ree, S., Park, Y. H., & Yoo, H. (2014). A study on education quality using the Taguchi method. *Total Quality Management & Business Excellence, 25*(7-8), 935–943. doi:10.1080/14783363.2014.906114

Reina Neira, M. A., de la Hoz, L. A., Felizzola Jimenez, H. A., & Hualpa Zuniga, A. M. (2016). Problem-based learning for teaching of design and analysis of experiments. *INGE CUC, 12*, 86–96. doi:10.17981/ingecuc.12.2.2016.09

Reips, U.-D., & Neuhaus, C. (2002). WEXTOR: A Web-based tool for generating and visualizing experimental designs and procedures. *Behavior Research Methods, Instruments, & Computers, 34*(2), 234–240. doi:10.3758/BF03195449

Reis, M., & Kenett, R. S. (2017). A structured overview on the use of computational simulators for teaching statistical methods. *Quality Engineering, 29*(4), 730–744. doi:10.1080/08982112.2016.1272122

Salunke, M., & Kittur, J. (2017). An experiential learning in electrical machines laboratory course. In *Proceedings 2017 5th IEEE International Conference on MOOCs, Innovation and Technology in Education (MITE)*, 15–20. IEEE.

SEFI. (2012). *Engineering Education and the Bologna Process.* Position Paper, a Joint communication of SEFI and BEST in view of the 8th Ministerial Conference in Bucharest, 26-27 April 2012. SEFI aisbl, Brussels.

Squires, A., & Cloutier, R. (2011). Applying a robust design approach to improve online systems engineering education. In *Proceedings 2011 IEEE International Systems Conference (SYSCON 2011)*, 371–378. 10.1109/SYSCON.2011.5929103

Taguchi, G. (1986). *Introduction to quality engineering: designing quality into products and processes*. Tokyo: Asian Productivity Organization.

Taguchi, G., & Wu, Y. (1979). *Introduction to Off-Line Quality Control*. Nagoya: Central Japan Quality Control Association.

Tanco, M., Viles, E., Ilzarbe, L., & Álvarez, M. J. (2007). Manufacturing industries need design of experiments (DoE). *World Congress on Engineering*, *20*, 1108–1113.

Tanco, M., Viles, E., Jesus Álvarez, M., & Ilzarbe, L. (2010). Why is not design of experiments widely used by engineers in Europe? *Journal of Applied Statistics*, *37*(12), 1961–1977. doi:10.1080/02664760903207308

UNESCO. (2010). *Engineering: issues, challenges and opportunities for development*. UNESCO Publishing.

Wang, Y., & Li, L. (2014). Inclusion of renewable energy topics in the design of experiments course for industrial and systems engineering students. *121st ASEE Annual Conference and Exposition, Conference Proceedings*. Indianapolis, IN. ASEE.

Weissman, S. A., & Anderson, N. G. (2015). Design of experiments (DoE) and process optimization. A review of recent publications. *Organic Process Research & Development*, *19*(11), 1605–1633. doi:10.1021/op500169m

Woll, R., & Burkhard, C. (2005). Full factorial design, Taguchi design or genetic algorithms — Teaching different approaches to design of experiments. In *Innovations in Classification, Data Science, and Information Systems* (pp. 567–574). Berlin, Germany: Springer-Verlag. doi:10.1007/3-540-26981-9_65

Xu, Q.-S., Xu, Y.-D., Li, L., & Fang, K.-T. (2018). Uniform experimental design in chemometrics. *Journal of Chemometrics*, *32*(11), e3020. doi:10.1002/cem.3020

Yates, F. (1964). Sir Ronald Fisher and the design of experiments. *Biometrics*, *20*(2), 307–321. doi:10.2307/2528399

Yildiz, A. R. (2013). Optimization of multi-pass turning operations using hybrid teaching learning-based approach. *International Journal of Advanced Manufacturing Technology*, *66*(9-12), 1319–1326. doi:10.100700170-012-4410-y

Zhan, W., Fink, R., & Fang, A. (2011). Application of statistics in engineering technology programs. [AJEE]. *American Journal of Engineering Education*, *1*, 65–78.

Glossary

Absorbance: Is a measure of the quantity of light absorbed by a sample. is calculated based on either the amount of light reflected or scattered by a sample or by the amount transmitted through a sample. If all light passes through a sample, none was absorbed, so the absorbance would be zero and the transmission would be 100%. On the other hand, if no light passes through a sample, the absorbance is infinite and the percent transmission is zero.

Active Petroleum: Considered as a number of wells and oil facilities of a region, it is generally part of a paleontological sub-basin of reserves of this natural resource.

Amperage: The amount of energy applied to the materials at the time of electroplating. The unit of measurement is amps.

ANOVA: Analysis Of Variance.

ASEA: Agency responsible for conducting, monitoring and enforcing the environmental regulations of the energy sector in Mexico

Biodiesel: Diesel obtained from triglycerides.

Bioethanol: Term used for ethanol obtained from fermentation of biological sugars to be used as a fuel.

Biofuel: Fuel obtained from biological sources.

Biorefinery: Factory which transforms the biomass into fuel.

Box-Behnken: is an independent quadratic design in that it does not contain an embedded factorial or fractional factorial design. In this design, the treatment combinations are at the midpoints of edges of the process space and at the center. These designs are rotatable (or near rotatable) and require 3 levels of each factor. The designs have limited capability for orthogonal blocking compared to the central composite designs.

Carry-Over: Effect that persist from one experimental condition to another. Whenever subjects perform in more than one condition (as they do in within-subject designs) there is a possibility of carryover effects.

Clinical Assay: Medical research conducted on people who voluntarily participate in these studies and who help discover better ways to treat, prevent, diagnose and understand diseases that affect humans.

Confusing Factor: Variable that influences both the dependent variable and independent variable, causing a mistaken association.

Copper: Cu is chemical element, a reddish, extremely ductile metal of Group 11 (Ib) of the periodic table that is an unusually good conductor of electricity and heat. Copper is found in the free metallic state in nature.

Cross-Over: Observational epidemiological design to assess whether a given intermittent or unusual exposure may have triggered an immediate short-term, acute event.

Cultivar: Is a plant or group of plants that have been selected from a naturally occurring species and bred to enhance or maintain a particular set of desirable characteristics. These plants almost always originate from human cultivation, propagated through cutting or grafting, and often cannot be grown from seeds from the original plant.

Current: Also called electrical intensity, is the load flow per unit of time that a material travels. It is due to a movement of the electrons inside the material.

Design on Experiments: Set of active techniques that manipulate a process to induce it to provide the information required to improve it through changes in its variables and their interaction or sequence of execution.

Diffusion: A movement of a solvent from an area of high concentration to an area of low concentration.

DoE: Design of Experiments

Drug Delivery: Is the process of the system to achieve a drug on the target site.

Edaphology: Science that studies the classification of soil according to the study of the properties of the horizons of a profile

Effect of a Factor: It is the change observed in the response variable due to a change in level in the factor.

Experiment: It is the moment of scientific research in which theories and hypotheses are put into practice to observe the results of them. The experiment, a word from Latin that means 'put to the test', is precisely the mechanism that is developed to check, verify or correct the postulates of the hypotheses that have been created.

Factorial Design: Is one involving two or more factors in a single experiment. The number of levels of each factor and the number of factors classify such designs; is often used by scientists wishing to understand the effect of two or more independent variables upon a single dependent variable.

Flatness: The surface is considered flat when it is totally comprised between two parallel planes, separated from one another by tolerances.

Flow Injection Analysis: FIA is based on the injection of a liquid sample into a moving, non- segmented continuous carrier stream of a suitable liquid. The injected sample forms a zone, which is then transported toward a detector that continuously records the changes in absorbance, electrode potential, or other physical parameter resulting from the passage of the sample material through the flow cell.

Frecuency: It is the magnitude that measures the number of repetitions per unit of time of any phenomenon or periodic event.

Input Factors: Independent variables of the studied system; they are introduced as the first approach in the optimization process.

Interaction Effect: Two factors interact significantly on the response variable when the effect of one depends on the level in which the other is.

Lipid: A class of organic compounds as fatty acid or their derivates, soluble in organic solvents.

Manganese: Mn is chemical element, one of the silvery white, hard, brittle metals of Group 7 (VIIb) of the periodic table. It was recognized as an element in 1774 by the Swedish chemist Carl Wilhelm Scheele while working with the mineral pyrolusite and was isolated the same year by his associate, Johan Gottlieb Gahn. Although it is rarely used in pure form, manganese is essential to steelmaking.

Nutrition: food intake in relation to the dietary needs of the human body.

Optimization: Is a process to get target particle size, finding best concentrations of reagents with a percent of confidence.

OQ: Operational Qualification.

Orthogonality: Forms of comparison that ensures that all the parameters can be estimated independently.

Output Factors: Dependent variables or response of the studied system.

Parameters: Characteristics that, through their numerical value, describe a set of elements or individuals.

Particle Size: The hydrodynamic diameter of a solid particle.

Permeation: The activity of the molecule to cross de stratum cornea and the epidermis.

Plackett-Burman: Used to identify the most important factors early in the experimentation phase when complete knowledge about the system is usually unavailable. Developed in 1946 by statisticians Robin L. Plackett and J.P. Burman, it is an efficient screening method to identify the active factors using as few experimental runs as possible.

Pressure: It is a scalar physical quantity that measures the force in a perpendicular direction per unit of surface, and serves to characterize how a determined force is applied on a surface.

Process: It is a unique combination of machines, tools, methods, materials and human being.

Radiofrecuency: Also called radio frequency spectrum or RF, it is applied to the least energetic portion of the electromagnetic spectrum, located between about 3 Hz and about 300 Ghz. The Hertz is the unit of measurement of the frequency of the waves and corresponds to one cycle per second.

Randomization: It is the process, in experimental studies, by which the subjects are randomly assigned to the treatment and control groups.

Regression Model: Statistical method to determine the correlations between the factors.

Representative Sample: It is a part of a population, properly selected, that preserves the key aspects of the population.

Revalidation: It is the process validation of a device for a second time due to the product specification, the equipment and the parameters of the processor the materials have changed.

Sequence Effects: potential confounding influences in experiments where subjects are exposed to multiple conditions.

Solid Lipid Nanoparticles: Particles with nanometric size elaborated with lipid and stabilized with surfactant.

Surfactant: Is a class of water-soluble polymer used to make stable emulsions oil/water.

Topical: Site on the skin to applicate a drug formulation as a enter drug point on the body.

Variation: Action and effect of varying. Modification, change or transformation.

Variety: means a plant grouping, within a single botanical taxon of the lowest known rank, defined by the reproducible expression of its distinguishing and other genetic characteristics.

VIS Spectrophotometry: Refers to absorption spectroscopy in the visible spectral region; is used to determine the absorption or transmission of VIS light (400 to 820 nm) by a sample. It can also be used to measure concentrations of absorbing materials based on developed calibration curves of the material.

Voltage: Electrical voltage or potential differential is a physical quantity that quantifies the electrical potential difference between two points.

Wine: An alcoholic drink that is usually made from grapes, but can also be made from other fruits or flowers. It is made by fermenting the fruit with water and sugar.

Compilation of References

Abdel-Ghani, N. T., Hegazy, A. K., El-Chaghaby, G. A., & Lima, E. C. (2009). Factorial experimental design for biosorption of iron and zinc using *Typha domingensis*. Phytomass. *Desalination, 249*(1), 343–347. doi:10.1016/j.desal.2009.02.065

Abdullah, B., Syed Muhammad, S. A. F., Shokravi, Z., Ismail, S., Kassim, K. A., Mahmood, A. N., & Aziz, M. M. A. (2019). Fourth generation biofuel: A review on risks and mitigation strategies. *Renewable & Sustainable Energy Reviews, 107*, 37–50. doi:10.1016/j.rser.2019.02.018

Abidi, N., Cabrales, L., & Haigler, C. H. (2014). Changes in the cell wall and cellulose content of developing cotton fibers investigated by FTIR spectroscopy. *Carbohydrate Polymers, 100*, 9–16. doi:10.1016/j.carbpol.2013.01.074 PMID:24188832

Abramczyk, H., Placek, I., Brożek-Płuska, B., Kurczewski, K., Morawiec, Z., & Tazbir, M. (2008). Human breast tissue cancer diagnosis by Raman spectroscopy. *Spectroscopy (Springfield, Or.), 22*(2-3), 113–121. PMID:20678135

Adams, R. H., Álvarez-Ovando, A. L., & Castañón, N. (2015). Efecto de la concentración de hidrocarburos sobre la producción del pasto (*Brachiaria humidicola*) en Texistepec, Veracruz. *Phyton (Buenos Aires), 84*(1), 222–232.

Adams, R. H., Kanga-Leyva, K., Guzmán-Osorio, F. J., & Escalante-Espinosa, E. (2011). Comparison of moisture management methods for the bioremediation of hydrocarbon contaminated soil. *African Journal of Biotechnology, 10*(3), 394–404.

Adams, R. H., Zavala-Cruz, J., & Morales-García, F. (2008). Concentración residual de hidrocarburos en suelo del trópico. II: Afectación a la fertilidad y su recuperación. *Interciencia, 33*(7), 483–489.

Adeleke, O. A., Latiff, A. A. A., Saphira, M. R., Daud, Z., Ismail, N., Ahsan, A., & Hijab, M. (2018). Locally derived activated carbon from domestic, agricultural and industrial wastes for the treatment of palm oil mill effluent. In *Nanotechnology in water and wastewater treatment* (pp. 35–62). Elsevier; doi:10.1016/b978-0-12-813902-8.00002-2

Adewale, P., Vithanage, L. N., & Christopher, L. (2017). Optimization of enzyme-catalyzed biodiesel production from crude tall oil using Taguchi method. *Energy Conversion and Management, 154*, 81–91. doi:10.1016/j.enconman.2017.10.045

Adhikari, P., Pal, P., Das, A. K., Ray, S., Bhattacharjee, A., & Mazumder, B. (2017). Nano lipid-drug conjugate: An integrated review. *International Journal of Pharmaceutics, 529*(1–2), 629–641. doi:10.1016/j.ijpharm.2017.07.039 PMID:28723407

Adhikary, S. N. (2014). A case study in increasing productivity by DOE in manufacturing sector. *International journal of Advancement in Engineering Technology, Management, and Applied Science, 1*(7).

Agrawal, U., Gupta, M., & Vyas, S. P. (2015). Capsaicin delivery into the skin with lipidic nanoparticles for the treatment of psoriasis. *Artificial Cells, Nanomedicine, and Biotechnology, 43*(1), 33–39. doi:10.3109/21691401.2013.832683 PMID:24040836

Ahmadi, M., Vahabzadeh, F., Bonakdarpour, B., Mofarrah, E., & Mehranian, M. (2005). Application of the central composite design and response surface methodology to the advanced treatment of olive oil processing wastewater using Fenton's peroxidation. *Journal of Hazardous Materials*, *123*(1), 187–195. doi:10.1016/j.jhazmat.2005.03.042 PMID:15993298

Ail, S. S., & Dasappa, S. (2016). Biomass to liquid transportation fuel via Fischer Tropsch synthesis – Technology review and current scenario. *Renewable & Sustainable Energy Reviews*, *58*, 267–286. doi:10.1016/j.rser.2015.12.143

Ajala, E. O., Aberuagba, F., Olaniyan, A. M., Ajala, M. A., & Sunmonu, M. O. (2017). Optimization of a two stage process for biodiesel production from shea butter using response surface methodology. *Egyptian Journal of Petroleum*, *26*(4), 943–955. doi:10.1016/j.ejpe.2016.11.005

Akinoso R., Igbeka J., Olayanju T. (2006). Process optimization of oil expression from sesame seed (Sesamum indicum Linn). *Agricultural Engineering International,* (8), 1-7

Akkarawatkhoosith, N., Kaewchada, A., & Jaree, A. (2019). Simultaneous development of biodiesel synthesis and fuel quality via continuous supercritical process with reactive co-solvent. *Fuel*, *237*, 117–125. doi:10.1016/j.fuel.2018.09.077

Akyuz, S., Akyuz, T., Celik, O., & Atak, Ç. (2013). FTIR and EDXRF investigations of salt tolerant soybean mutants. *Journal of Molecular Structure*, *1044*, 67–71. doi:10.1016/j.molstruc.2013.01.025

Akyuz, S., Akyuz, T., Celik, O., & Atak, C. (2018). FTIR spectroscopy of protein isolates of salt-tolerant soybean mutants. *Journal of Applied Spectroscopy*, *84*(6), 1019–1023. doi:10.100710812-018-0580-1

Albano, J. M. R., Morais Ribeiro, L., Muñiz Couto, V., Barbosa Messias, M., Rodrigues da Silva, G. H., Breitkreitz, M. C., ... Pickholz, M. (2019). Rational desing of polymer-lipid nanoparticles for docetaxel delivery. *Colloids and Surfaces. B, Biointerfaces*, *175*(1), 56–64. doi:10.1016/j.colsurfb.2018.11.077 PMID:30517905

Alexander, K.Clarkson, P. J. (2000). Good design practice for medical devices and equipment. *Journal of Medical Engineering & Technology*, *24*(1), 5–13. doi:10.1080/030919000293987 PMID:10849882

Ali, M. E. M., Abdelsalam, H., Ammar, N. S., & Ibrahim, H. S. (2018). Response surface methodology for optimization of the adsorption capability of ball-milled pomegranate peel for different pollutants. *Journal of Molecular Liquids*, *250*, 433–445. doi:10.1016/j.molliq.2017.12.025

Al-Kassas, R., Wen, J., Cheng, A. E. M., Kim, A. M. J., Liu, S. S. M., & Yu, J. (2016). Transdermal delivery of propranolol hydrochloride through chitosan nanoparticles dispersed in mucoadhesive gel. *Carbohydrate Polymers*, *153*, 176–186. doi:10.1016/j.carbpol.2016.06.096 PMID:27561485

Allada, V., & Jose, V. (2003). Case-Based Problem Solving Exercises for Engineering Education. *Engineering/Technology Management* [ASME.]. *Safety Engineering and Risk Analysis, Technology and Society, Engineering Business Management, and Homeland Security*, *2003*, 137–142.

Alleleyn, A. M. E., van Avesaat, M., Ripken, D., Bleiel, S. B., Keszthelyi, D., Wilms, E., ... Masclee, A. A. M. (2018). The effect of encapsulated nutrients mixture on food intake and satiety: A double blind randomized cross-over proof of concept study. *Nutrients*, *10*(11), 1787. doi:10.3390/nu10111787 PMID:30453597

Allwood, J. M., Cox, B. M., & Latif, S. S. (2001). The structured development of simulation-based learning tools with an example for the Taguchi method. *IEEE Transactions on Education*, *44*(4), 347–354. doi:10.1109/13.965783

Alvarez-Ordóñez, A., & Prieto, M. (2012). Technical and methodological aspects of fourier transform infrared spectroscopy in food microbiology research. In *Fourier Transform Infrared Spectroscopy in Food Microbiology* (pp. 1–18). Boston, MA: Springer. doi:10.1007/978-1-4614-3813-7_1

Alvino, L., van der Lubbe, R. H., & Constantinides, E. (2018). Investigating individual preferences and brain activity in a wine tasting experience: a neuromarketing approach. *2018 Global Marketing Conference at Tokyo*, 559–564. 10.15444/GMC2018.05.03.04

Amer, M. S. (Ed.). (2009). *Raman spectroscopy for soft matter applications*. Hoboken, NJ: Wiley. doi:10.1002/9780470475997

Amini, M., & Younesi, H. (2009). Biosorption of Cd (II), Ni (II) and Pb (II) from aqueous solution by dried biomass of *Aspergillus niger*: Application of response surface methodology to the optimization of process parameters. *Clean-Soil, Air, Water (Basel)*, *37*(10), 776–786.

Amini, M., Younesi, H., Bahramifar, N., Lorestani, A. A. Z., Ghorbani, F., Daneshi, A., & Sharifzadeh, M. (2008). Application of response surface methodology for optimization of lead biosorption in an aqueous solution by Aspergillus niger. *Journal of Hazardous Materials*, *154*(1-3), 694–702. doi:10.1016/j.jhazmat.2007.10.114 PMID:18068898

Anand, P., & Bley, K. (2011). Topical capsaicin for pain management: Therapeutic potential and mechanisms of action of the new high-concentration capsaicin 8% patch. *British Journal of Anaesthesia*, *107*(4), 490–502. doi:10.1093/bja/aer260 PMID:21852280

Anderson, M. J. (2005). Trimming the fat out of experimental methods. *OE Magazine*, *5*(6), 32–33.

Anderson, M. J., & Whitcomb, P. J. (2015). *DOE simplified: Practical tools for effective experimentation (3rd Ed.)*. Boca Raton, FL: CRC Press Taylor & Francis Group.

Anderson, M. J., Whitcomb, P. J., & Whitcomb, P. J. (2016). *DOE simplified*. Productivity Press; doi:10.1201/9781315382326

Anderson, R. J., Bendell, D. J., & Groundwater, P. W. (2004). *Organic spectroscopic analysis* (Vol. 22). Royal Society of Chemistry.

Angerosa, F., Mostallino, R., Basti, C., & Vito, R. (2001). Influence of malaxation temperature and time on the quality of virgin olive oils. *Food Chemistry*, *72*(1), 19–28. doi:10.1016/S0308-8146(00)00194-1

Annis, D. H. (2005). Rethinking the Paper Helicopter. *The American Statistician*, *59*(4), 320–326. doi:10.1198/000313005X70777

Antony, *J.* (2014). Fractional factorial designs. Elsevier; doi:10.1016/B978-0-08-099417-8.00007-9

Antony, J. (2014b). 4 - A systematic methodology for design of experiments. Design of experiments for engineers and scientists (2nd ed., pp. 33–50). Elsevier; doi:10.1016/B978-0-08-099417-8.00004-3

Antony, J., (2003a). Design of experiments for engineers and scientists. *Design of Experiments for Engineers and Scientists*. doi:10.1016/B978-0-7506-4709-0.X5000-5

Antony, J. (2002). Training for design of experiments using a catapult. *Quality and Reliability Engineering International*, *18*(1), 29–35. doi:10.1002/qre.444

Antony, J. (2003b). Fundamentals of design of experiments. In *Design of experiments for engineers and scientists* (pp. 6–16). Butterworth-Heinemann; doi:10.1016/B978-075064709-0/50003-X

Antony, J. (2014). *Full factorial designs*, doi:10.1016/B978-0-08-099417-8.00006-7

AntonyJ, . (2014a). *1 - Introduction to industrial experimentation. Design of experiments for engineers and scientists* (2nd ed., pp. 1–6). Elsevier; doi:10.1016/B978-0-08-099417-8.00001-8

Antony, J. (2014a). 6 - Full factorial designs. In S. E. Antony (Ed.), *J. B. T.-D. of E. for E. and S* (pp. 63–85). Oxford, UK: Elsevier;

Antony, J. (2014b). *Design of experiments for engineers and scientists* (2nd ed.). London, UK: Elsevier.

Antony, J., & Capon, N. (1998). Teaching experimental design techniques to industrial engineers. *International Journal of Engineering*, *14*, 335–343.

Antony, J., & Jiju Antony, F. (2001). Teaching the Taguchi method to industrial engineers. *Work Study*, *50*(4), 141–149. doi:10.1108/00438020110391873

Antony, J., & Kaye, M. (1995). Experimental Quality. *Journal of Manufacturing Engineer*, *74*(4), 178–181. doi:10.1049/me:19950409

Antony, J., & Kaye, M. (2012). *Experimental quality: a strategic approach to achieve and improve quality. New York*: Springer Science Business Media.

Arana, L., Salado, C., Vega, S., Aizpurua-Olaizola, O., Arada, I., de la, & Suarez, T. (2015). Solid lipid nanoparticles for delivery of Calendula officinalis extract. *Colloids and Surfaces. B, Biointerfaces*, *135*, 18–26. doi:10.1016/j.colsurfb.2015.07.020 PMID:26231862

Arguijo-Portillo, N. V., Guerrero-Peña, A., Domínguez-Rodríguez, V. I., Carrillo-Ávila, E., & Zavala-Cruz, J. (2019). Modelos de calibración para la cuantificación espectrofotométrica de hidrocarburos totales del petróleo en suelo. *Revista Internacional de Contaminación Ambiental*, *35*(2), 469–479. doi:10.20937/RICA.2019.35.02.17

Armstrong, N. A. (2006). Pharmaceutical Experimental Design and Interpretation (Second Edition). Boca Raton, FL: Taylor & Francis Group.

Arvidsson, M., & Gremyr, I. (2008). Principles of robust design methodology. *Quality and Reliability Engineering International*, *9*(3), 185–202.

Ashton, L., Lau, K., Winder, C. L., & Goodacre, R. (2011). Raman spectroscopy: Lighting up the future of microbial identification. *Future Microbiology*, *6*(9), 991–997. doi:10.2217/fmb.11.89 PMID:21958140

Aslan, N. (2008). Application of response surface methodology and central composite rotatable design for modeling and optimization of a multi-gravity separator for chromite concentration. *Powder Technology*, *185*(1), 80–86. doi:10.1016/j.powtec.2007.10.002

Astakhov, V. P. (2012). *Design of experiment methods in manufacturing: basics and practical applications*. Berlin, Germany: Springer.

Augusti, G. (2009). EUR-ACE: The European Accreditation system of Engineering Education and its global context. *Engineering Education Quality Assurance: A Global Perspective*, 41–49.

Augusti, G., Freeston, I., Heitmann, G., & Martin, R.-P. (2008). Accreditation and QA of engineering education in Europe: Setting up a pan-European system. In A. Beso, L. Bollaert, B. Curvale, H. T. Jensen, L. Harvey, E. Helle, … A. Sursock (Eds.), *Implementing and using quality assurance: Strategy and practice a selection of papers from the 2nd European Quality Assurance Forum* (pp. 42–47). European University Association.

Auner, G. W., Koya, S. K., Huang, C., Broadbent, B., Trexler, M., Auner, Z., ... Brusatori, M. A. (2018). Applications of Raman spectroscopy in cancer diagnosis. *Cancer and Metastasis Reviews*, *37*(4), 691–717. doi:10.100710555-018-9770-9 PMID:30569241

Ayerza, R. (1995). Oil content and fatty acid composition of chia (Salvia hispanica L.) from five northwestern locations in Argentina. *Journal of the American Oil Chemists' Society*, *72*(9), 1079–1081. doi:10.1007/BF02660727

Ayerza, R., & Coates, W. (2001). *Chia seeds: new source of omega-3 fatty acids, natural antioxidants, and dietetic fiber, Southwest center for natural products research & commercialization* (p. 3). Tucson, AZ: Office of Arid Lands Studies.

Ayerza, R., & Coates, W. (2004). Composition of chia (*Salvia hispanica*) grown in six tropical and subtropical ecosystems of South America. *Tropical Science*, *44*(3), 131–135. doi:10.1002/ts.154

Azapagic, A. (2015). Special issue: Sustainability issues in the food–energy–water nexus. *Sustainable Production and Consumption*, *2*, 1–2. doi:10.1016/j.spc.2015.08.002

Badui, S. (2006). *Químca de los alimentos* (p. 285). México: Pearson Educación.

Bağcıoğlu, M., Fricker, M., Johler, S., & Ehling-Schulz, M. (2019). Detection and identification of Bacillus cereus, Bacillus cytotoxicus, Bacillus thuringiensis, Bacillus mycoides and Bacillus weihenstephanensis via machine learning based FTIR Spectroscopy. *Frontiers in Microbiology*, *10*, 902. doi:10.3389/fmicb.2019.00902 PMID:31105681

Bahar, S., & Zakerian, R. (2012). Speciation of Fe(II) and Fe(III) by using dispersive liquid-liquid microextraction and flame atomic absorption spectrometry. *Journal of the Brazilian Chemical Society*, *23*(5), 944–950. doi:10.1590/S0103-50532012000500021

Bal, D. K., & Bhasarkar, J. B. (2019). Mechanistic investigation of Sono–Phosphotungstic acid/phase transfer agent assisted oxidative desulfurization of liquid fuel. *Asia-Pacific Journal of Chemical Engineering*, *14*(1). doi:10.1002/apj.2271

Baltcheva, I., Cristea, S., Felisa, J. V.-A., & De Prada, C. (2003). Simultaneous perturbation stochastic approximation for real-time optimization of model predictive control. In *Proceedings Of the 1st Industrial Simulation Conference*, (pp. 533-537).

Bangerth, W., Klie, H., Matossian, V., Parashar, M., & Wheeler, M. F. (2005). An autonomic reservoir framework for the stochastic optimization of well placement. *Cluster Computing*, *8*(4), 255–269. doi:10.100710586-005-4093-3

Barakat, M. A. (2011). New trends in removing heavy metals fromindustrial wastewater. *Arabian Journal of Chemistry*, *4*(4), 361–377. doi:10.1016/j.arabjc.2010.07.019

Barba-Macías, E., Rangel-Mendoza, J., & Ramos-Reyes, R. (2014). Clasificación de los humedales de Tabasco mediante sistemas de información geográfica. *Ecosistemas Recursos Agropecuarios*, *22*(2).

Barrado, E., Vega, M., Pardo, R., Grande, P., & Del Valle, J. L. (1996). Optimisation of a purification method for metal-containing wastewater by use of a Taguchi experimental design. *Water Research*, *30*(10), 2309–2314. doi:10.1016/0043-1354(96)00119-4

Barrientos del Monte, F., & Añorve Añorve, D. (2014). México 2013: Acuerdos, reformas y descontento. *Revista de ciencia política (Santiago)*, *34*(1), 221–247. doi:10.4067/S0718-090X2014000100011

Barth, A. (2000). The infrared absorption of amino acid side chains. *Progress in Biophysics and Molecular Biology*, *74*(3-5), 141–173. doi:10.1016/S0079-6107(00)00021-3 PMID:11226511

Barth, A. (2007). Infrared spectroscopy of proteins. *Biochimica et Biophysica Acta (BBA)-. Bioenergetics*, *1767*(9), 1073–1101. doi:10.1016/j.bbabio.2007.06.004

Bart, J. C. J. (2006). *Plastics additives: advanced industrial analysis*. IOS Press.

Bartkutė, V., & Sakalauskas, L. (2007). Simultaneous perturbation stochastic approximation of nonsmooth functions. *European Journal of Operational Research*, *181*(3), 1174–1188. doi:10.1016/j.ejor.2005.09.052

Baş, D., & Boyacı, I. H. (2007). Modeling and optimization I: Usability of response surface methodology. *Journal of Food Engineering*, *78*(3), 836–845. doi:10.1016/j.jfoodeng.2005.11.024

Bashir, M. J., Amr, S. S. A., Aziz, S. Q., Aun, N. C., & Sethupathi, S. (2015). Wastewater treatment processes optimization using response surface methodology (RSM) compared with conventional methods: Review and comparative study. *Middle East Journal of Scientific Research*, *23*(2), 244–252.

Becker, K., Al Laham, N., Fegeler, W., Proctor, R. A., Peters, G., & von Eiff, C. (2006). Fourier-transform infrared spectroscopic analysis is a powerful tool for studying the dynamic changes in Staphylococcus aureus small-colony variants. *Journal of Clinical Microbiology*, *44*(9), 3274–3278. doi:10.1128/JCM.00847-06 PMID:16954260

Bel'skaya, L. V. (2019). Use of IR spectroscopy in cancer diagnosis. A review. *Journal of Applied Spectroscopy*, *86*(2), 187–205. doi:10.100710812-019-00800-w

Bendicho, S., Barbosa-Cánovas, G. V., & Martín, O. (2002). Milk processing by high intensity pulsed electric fields. *Trends in Food Science & Technology*, *13*(6–7), 195–204. doi:10.1016/S0924-2244(02)00132-2

Bergquist, B., & Albing, M. (2006). Statistical methods – Does anyone really use them? *Total Quality Management & Business Excellence*, *17*(8), 961–972. doi:10.1080/14783360600747762

Bermúdez, Y. G., Rico, I. L. R., Guibal, E., de Hoces, M. C., & Martín-Lara, M. A. (2012). Biosorption of hexavalent chromium from aqueous solution by *Sargassum muticum* brown alga. Application of statistical design for process optimization. *Chemical Engineering Journal*, *183*, 68–76. doi:10.1016/j.cej.2011.12.022

Berrios, M., Gutiérrez, M. C., Martín, M. A., & Martín, A. (2009). Application of the factorial design of experiments to biodiesel production from lard. *Fuel Processing Technology*, *90*(12), 1447–1451. doi:10.1016/j.fuproc.2009.06.026

Bevilacqua, A., Corbo, M. R., & Sinigaglia, M. (2010). Design of experiments : A powerful tool in food microbiology. *Applied Microbiology*, 1419–1429.

Bezerra, M. A., Santelli, R. E., Oliveira, E. P., Villar, L. S., & Escaleira, L. A. (2008). Response surface methodology (RSM) as a tool for optimization in analytical chemistry. *Talanta*, *76*(5), 965–977. doi:10.1016/j.talanta.2008.05.019 PMID:18761143

Bharti, P. K., Khan, M. I., & Singh, H. (2010). Taguchi's off line method and multivariate loss function approach for quality management and optimization of process parameters-A review. In S. Paruya, S. Kar, & S. Roy, (Eds.), *International Conference on Modeling, Optimization, and Computing* (p. 486+). 10.1063/1.3516353

Bhote, K. R. (2000). *World class quality. Using design of experiments to make it happen* (2nd ed.). New York: Amacom.

Biglar, M., Moghaddam, G., Sadeghi, N., Reza Oveisi, M., Jannat, B., Kaboli, A., ... Hajimahmoodi, M. (2012). Profiling of major fatty acids in different raw and roasted sesame sedes cultivars. *African Journal of Biotechnology*, *11*(24), 6619–6623.

Bondi, R. W. Jr, Igne, B., Drennen, J. K. III, & Anderson, C. A. (2012). Effect of experimental design on the prediction performance of calibration models based on near-infrared spectroscopy for pharmaceutical applications. *Applied Spectroscopy*, *66*(12), 1442–1453. doi:10.1366/12-06689 PMID:23231907

Booker, B. W., & Lyth, D. M. (2005). Quality engineering from 1988 through 2005: Lessons from the past and trends for the future. *Quality Engineering*, *18*(1), 1–4. doi:10.1080/08982110500439452

Boon, M. A. A., Di Bucchianico, A., Rijpkema, J. J. M., & van Berkum, E. E. M. (2010). Statlab: An interactive teaching tool for DOE. In A. D. Fitt, J. Norbury, H. Ockendon, & R. E. Wilson (Eds.), *Progress In Industrial Mathematics at ECMI 2008* (pp. 663–668). doi:10.1007/978-3-642-12110-4_107

Borah, M. J., Das, A., Das, V., Bhuyan, N., & Deka, D. (2019). Transesterification of waste cooking oil for biodiesel production catalyzed by Zn substituted waste egg shell derived CaO nanocatalyst. *Fuel*, *242*, 345–354. doi:10.1016/j.fuel.2019.01.060

Bouazizi, S., Jamoussi, B., & Bousta, D. (2016). Application of response surface methodology for optimization of heavy metals biosorption on natural gum of Acacia Nilotica. *International Journal of Engineering Research & Technology*, *5*, 2278–0181.

Box & Wilson. (1951). On the experimental attainment of optimum conditions. *Journal of the Royal Statistical Society. Series B. Methodological*, (13), 1–45.

Box, G. E. P., & Behnken, D. W. (1960). Some new three level designs for the study of quantitative variables. *Technometrics*, *2*(4), 455–475. doi:10.1080/00401706.1960.10489912

Box, G. E. P., & Hunter, J. S. (1961a). The 2^{k-p} fractional factorial designs. Part I. *Technometrics*, *3*, 311–351.

Box, G. E. P., & Hunter, J. S. (1961b). The 2^{k-p} fractional factorial designs Part II. *Technometrics*, *3*, 449–458.

Box, G. E. P., & Liu, P. Y. T. (1999). Statistics as a catalyst to learning by scientific method part 1. *Journal of Quality Technology*, *31*(1), 1–15. doi:10.1080/00224065.1999.11979889

Brady, J. E., & Allen, T. T. (2002). Case study based instruction of DOE and SPC. *The American Statistician*, *56*(4), 312–315. doi:10.1198/000313002614

Brennan, L., & Owende, P. (2010). Biofuels from microalgae—A review of technologies for production, processing, and extractions of biofuels and co-products. *Renewable & Sustainable Energy Reviews*, *14*(2), 557–577. doi:10.1016/j.rser.2009.10.009

Brereton, R. G. (2007). *Applied chemometrics for scientists*. John Wiley & Sons. doi:10.1002/9780470057780

Briones, F. Z., Leal, J. S., & Pastrana, I. Y. M. (2016). Armentum: A hybrid direct search optimization methodology. *Journal of Industrial Engineering International*, *12*(4), 407–417. doi:10.100740092-016-0159-5

Buasri, A., Worawanitchaphong, P., Trongyong, S., & Loryuenyong, V. (2014). Utilization of scallop waste shell for biodiesel production from palm oil–optimization using Taguchi method. *APCBEE Procedia*, *8*, 216–221. doi:10.1016/j.apcbee.2014.03.030

Budé, L., Imbos, T., van de Wiel, M. W., & Berger, M. P. (2011). The effect of distributed practice on students' conceptual understanding of statistics. *Higher Education*, *62*(1), 69–79. doi:10.100710734-010-9366-y

Bureau, S., Cozzolino, D., & Clark, C. J. (2019). Contributions of Fourier-transform mid infrared (FT-MIR) spectroscopy to the study of fruit and vegetables: A review. *Postharvest Biology and Technology*, *148*, 1–14. doi:10.1016/j.postharvbio.2018.10.003

Burke, R. D., De Jonge, N., Avola, C., & Forte, B. (2017). A virtual engine laboratory for teaching powertrain engineering. *Computer Applications in Engineering Education*, *25*(6), 948–960. doi:10.1002/cae.21847

Butler, A. J., & Moses, W. M. (2005). Introducing experimental design in mechanical engineering laboratories. In *Proceedings 2005 American Society for Engineering Education Annual Conference and Exposition*, 10.827.1-10. American Society for Engineering Education.

Cacho, J., Castells, J., & Barroso, C. (2002). *Analytica Chimica Acta*, *458*, 197–202. doi:10.1016/S0003-2670(01)01499-4

Cahill, J. P. (2003). Ethnobotany of chia, *Salvia hispanica* L. (Lamiaceae). *Economic Botany*, *57*(4), 604–618. doi:10.1663/0013-0001(2003)057[0604:EOCSHL]2.0.CO;2

Calder, P. C. (2008). Polyunsaturated fatty acids, inflammatory processes and inflammatory bowel diseases. *Molecular Nutrition & Food Research, 52*(8), 885–897. doi:10.1002/mnfr.200700289 PMID:18504706

Campean, I. F., Grove, D., & Henshall, E. (2005). A course in statistical engineering. *International Journal of Engineering Education, 21*, 502–511.

Candioti, L. V., De Zan, M. M., Cámara, M. S., & Goicoechea, H. C., (2014). Experimental design and multiple response optimization. Using the desirability function in analytical methods development. doi:10.1016/j.talanta.2014.01.034

Cañizares, M. P. (2002). Análisis por inyección en flujo: Herramienta clave para la automatización analítica. *Revista de la Sociedad Química de México, 46*(2), 167–174.

Cao, Y., Liu, Z., Cheng, G., Jing, X., & Xu, H. (2010). Exploring single and multi-metal biosorption by immobilized spent Tricholoma lobayense using multi-step response surface methodology. *Chemical Engineering Journal, 164*(1), 183–195. doi:10.1016/j.cej.2010.08.055

Carbone, C., Tomasello, B., Ruozi, B., Renis, M., & Puglisi, G. (2012). Preparation and optimization of PIT solid lipid nanoparticles via statistical factorial design. *European Journal of Medicinal Chemistry, 49*, 110–117. doi:10.1016/j.ejmech.2012.01.001 PMID:22244589

Careghini, A., Mastorgio, A. F., Saponaro, S., & Sezenna, E. (2015). Bisphenol A, nonylphenols, benzophenones, and benzotriazoles in soils, groundwater, surface water, sediments, and food: A review. *Environmental Science and Pollution Research International, 22*(8), 5711–5741. doi:10.100711356-014-3974-5 PMID:25548011

Carlos, C., Maretto, D. A., Poppi, R. J., Sato, M. I. Z., & Ottoboni, L. M. (2011). Fourier transform infrared microspectroscopy as a bacterial source tracking tool to discriminate fecal E. coli strains. *Microchemical Journal, 99*(1), 15–19. doi:10.1016/j.microc.2011.03.002

Carrillo, E. G., Haro, M. P., Díaz, G. C., & Cañizares, M. P. (2014). Plackett-Burman factorial design for the optimization of a spectrophotometric flow injection method for phenol determination in tap and bottled water using 4-aminoantipyrine. (S. Q. México, Ed.) J. Mex. Chem. Soc., 58(2), 99-105.

Carrillo, E. G., Cañizares, M. P., Ames, A. I., & Sepúlveda, R. G. (2016). Aplicación de Diseño Factorial Plackett-Burman en la Optimización de Parámetros para Determinar Cobre. *Aportaciones Científicas en Química Analítica, 2016*, 464–469.

Castellanos, M. L., Isaza, R. J., & Torres, J. M. (2015). Evaluación de los hidrocarburos totales de petróleo (TPH) sobre suelos urbanos en Maicao, Colombia. *Revista Colombiana de Quimica, 44*(3), 11–17. doi:10.15446/rev.colomb.quim.v44n3.55605

Castorena-Cortés, G., Roldán-Carrillo, T., Zapata-Peñasco, I., Reyes-Avila, J., Quej-Ake, L., Marin-Cruz, J., & Olguín-Lora, P. (2009). Microcosm assays and Taguchi experimental design for treatment of oil sludge containing high concentration of hydrocarbons. *Bioresource Technology, 100*(23), 5671–5677. doi:10.1016/j.biortech.2009.06.050 PMID:19635663

Castro, A., Acevedo-Berruecos, D., Urbieta, J., Iturbe, R., & Delgado-Rodríguez, O. (2013). Caracterización geoquímica y geoeléctrica de un sitio contaminado por hidrocarburos en el estado de Puebla, México. *Boletín de la Sociedad Geológica Mexicana, 65*(2), 405–418. doi:10.18268/BSGM2013v65n2a20

Cavalli, R., Caputo, O., & Gasco, M. R. (2000). Preparation and characterization of solid lipid nanospheres containing paclitaxel. *European Journal of Pharmaceutical Sciences, 10*(4), 305–309. doi:10.1016/S0928-0987(00)00081-6 PMID:10838020

Cazón, P., Vázquez, M., & Velazquez, G. (2018). Cellulose-glycerol-polyvinyl alcohol composite films for food packaging: Evaluation of water adsorption, mechanical properties, light-barrier properties and transparency. *Carbohydrate Polymers, 195*, 432–443. doi:10.1016/j.carbpol.2018.04.120 PMID:29804996

Celik, A. E., & Karatepe, Y. (2007). Evaluating and forecasting banking crisis through neural network models: An application for Turkish banking sector. *Expert Systems with Applications, 33*(4), 809–815. doi:10.1016/j.eswa.2006.07.005

Cepo, D. V., Pelajic, M., Vrcek, I. V., & Krivohlavek, A. (2018). Differences in the levels of pesticides, metals, sulphites and ochratoxin A between organically and conventionally produced wines. *Food Chemistry*, (246), 349–403. PMID:29291864

Challagulla, V., Nayar, S., Walsh, K., & Fabbro, L. (2017). Advances in techniques for assessment of microalgal lipids. *Critical Reviews in Biotechnology, 37*(5), 566–578. doi:10.1080/07388551.2016.1206058 PMID:27417693

Charles-Rodríguez, A. V., Nevárez-Moorillón, G. V., Zhang, Q. H., & Ortega-Rivas, E. (2007). Comparison of thermal processing and pulsed electric fields treatment in pasteurization of apple juice. *Food and Bioproducts Processing, 85*(2), 93–97. doi:10.1205/fbp06045

Chauvet, R., Lagarde, F., Charrier, T., Assaf, A., Thouand, G., & Daniel, P. (2017). Microbiological identification by surface-enhanced raman spectroscopy. *Applied Spectroscopy Reviews, 52*(2), 123–144. doi:10.1080/05704928.2016.1209760

Chávez Torres, M. M. (2009). *Estudio de la composición de ácidos grasos insaturados (oleico, linoleico y linolénico) en aceites de oliva extra virgen de venta en el mercado mexicano, por cromatografía de gases. Tesis de Licenciatura en Ingeniería Química*. México: FES Cuautitlán, Universidad Nacional Autónoma de México.

Chen, H., Liu, J., Chang, X., Chen, D., Xue, Y., & Liu, P., … Han, S. (2017). A review on the pretreatment of lignocellulose for high-value chemicals. *Fuel Processing Technology, 160*, 196–206.

Cheng, C. S. (2016). *Theory of factorial design: Single-and multi-stratum experiments*. Chapman and Hall/CRC. doi:10.1201/b16253

Chen, W., Ma, L., Zhou, P., Zhu, Y., Wang, X., Luo, X., ... Yu, L. (2015). A novel feedstock for biodiesel production: The application of palmitic acid from Schizochytrium. *Energy, 86*, 128–138. doi:10.1016/j.energy.2015.03.110

Chin, D. C. (1997). Comparative study of stochastic algorithms for system optimization based on gradient approximations. *IEEE Transactions on Systems, Man, and Cybernetics. Part B, Cybernetics, 27*(2), 244–249. doi:10.1109/3477.558808 PMID:18255862

Chisanga, M., Muhamadali, H., Ellis, D. I., & Goodacre, R. (2018). Surface-enhanced raman scattering (SERS) in microbiology: Illumination and enhancement of the microbial world. *Applied Spectroscopy, 72*(7), 987–1000. doi:10.1177/0003702818764672 PMID:29569946

Chobot, V., Hadacek, F., Bachmann, G., Weckwerth, W., & Kubicova, L. (2018). Antioxidant properties and the formation of iron coordination complexes of 8-hydroxyquinoline. *International Journal of Molecular Sciences, 19*(12), 3917. doi:10.3390/ijms19123917 PMID:30544490

Christy, A. A., Ozaki, Y., & Gregoriou, V. G. (2001). *Modern Fourier transform infrared spectroscopy* (Vol. 2001). Amsterdam, The Netherlands: Elsevier.

Chryssolouris, G. (2006). *Manufacturing systems: theory and practice* (2nd ed.). New York: Springer-Verlag.

Ciurczak, E. W., & Igne, B. (2015). Pharmaceutical and medical applications of near-infrared spectroscopy. Boca Raton, FL: CRC Press.

Clark, A., Wilkes, E., & Scollary, G. (2015). Chemistry of copper in white wine: A review. *Australian Journal of Grape and Wine Research*, *21*(3), 339–350. doi:10.1111/ajgw.12159

Cocke, J., & Maaß, S. (2017). Cross linking between the baffling effect and phase inversion during liquid–liquid monomer mixing. *Macromolecular Reaction Engineering*, *11*(4), 1–17. doi:10.1002/mren.201700015

Coleman, D. E., & Montgomery, D. C. (1993). A systematic approach to planning for a designed industrial experiment. *Technometrics*, *35*(1), 1–12. doi:10.1080/00401706.1993.10484984

Conover, W. J., & Iman, R. L. (1981). Rank transformation as a bridge between parametric and nonparametric statistics. *The American Statistician*, *35*(3), 124–129.

Cooper, L. B. (2006). Obtención de un concentrado de ácidos grasos poliinsaturados a partir del aceite de chía (Salvia hispánica). Tesis de Licenciatura. Facultad de Química. UNAM. México. D.F.

Cornell, J. A., & Gorman, J. W. (1984). Fractional design plans for process variables in mixture experiments. *Journal of Quality Technology*, *16*(1), 20–38. doi:10.1080/00224065.1984.11978883

Correa, D. F., Beyer, H. L., Possingham, H. P., Thomas-Hall, S. R., & Schenk, P. M. (2017). Biodiversity impacts of bioenergy production: Microalgae vs. first generation biofuels. *Renewable & Sustainable Energy Reviews*, *74*, 1131–1146. doi:10.1016/j.rser.2017.02.068

Costagli, G., & Betti, M. (2015). Avocado oil extraction processes: Method for cold-pressed high-quality edible oil production versus traditional production. *Journal of Agricultural Engineering*, *46*(3), 115–122. doi:10.4081/jae.2015.467

Costa, N. R. P., Pires, A. R., & Ribeiro, C. O. (2006). Guidelines to help practitioners of design of experiments. *The TQM Magazine*, *18*(4), 386–399. doi:10.1108/09544780610671057

Coste, A., Sousa, P., & Malfeito-Ferreira, M. (2018). Wine tasting based on emotional responses: An expedite approach to distinguish between warm and cool climate dry red wine styles. *Food Research International*, *106*, 11–21. doi:10.1016/j.foodres.2017.12.039 PMID:29579904

Cox, D. R., & Reid, N. (2000). *The theory of the design of experiments* (1st ed.). New York: Chapman and Hall/CRC. doi:10.1201/9781420035834

Cozzolino, D., Cynkar, W. U., Shah, N., & Smith, P. (2011). Multivariate data analysis applied to spectroscopy: Potential application to juice and fruit quality. *Food Research International*, *44*(7), 1888–1896. doi:10.1016/j.foodres.2011.01.041

Crompton, T. R. (2006). *Polymer reference book*. iSmithers Rapra Publishing.

Cronje, K. J., Chetty, K., Carsky, M., Sahu, J. N., & Meikap, B. C. (2011). Optimization of chromium (VI) sorption potential using developed activated carbon from sugarcane bagasse with chemical activation by zinc chloride. *Desalination*, *275*(1-3), 276–284. doi:10.1016/j.desal.2011.03.019

Cui, S., Zhang, S., & Yue, S. (2018). Raman spectroscopy and imaging for cancer diagnosis. *Journal of Healthcare Engineering*, 2018. PMID:29977484

Czajkowska-Kośnik, A., Szekalska, M., & Winnicka, K. (2019). Nanostructured lipid carriers: A potential use for skin drug delivery systems. *Pharmacological Reports*, *71*(1), 156–166. doi:10.1016/j.pharep.2018.10.008 PMID:30550996

Czitrom, V. (1999). One factor at a time versus designed experiments. *The American Statistician*.

Czitrom, V. (1999). One-factor-at-a-time versus designed experiments. *The American Statistician*, *53*(2). Retrieved from https://www.jstor.org/stable/2685731

Damartzis, T., & Zabaniotou, A. (2011). Thermochemical conversion of biomass to second generation biofuels through integrated process design—A review. *Renewable & Sustainable Energy Reviews, 15*(1), 366–378. doi:10.1016/j.rser.2010.08.003

Daniel, C. (1973). One at a time plans. *Journal of the American Statistical Association, 68*(342), 353–360. doi:10.1080/01621459.1973.10482433

Danmaliki, G. I., Saleh, T. A., & Shamsuddeen, A. A. (2017). Response surface methodology optimization of adsorptive desulfurization on nickel/activated carbon. *Chemical Engineering Journal, 313*, 993–1003. doi:10.1016/j.cej.2016.10.141

Daoud, S., Bou-maroun, E., Dujourdy, L., Waschatko, G., Billecke, N., & Cayot, P. (2019). Fast and direct analysis of oxidation levels of oil-in-water emulsions using ATR-FTIR. *Food Chemistry, 293*, 307–314. doi:10.1016/j.foodchem.2019.05.005 PMID:31151616

Darius, P. L., Portier, K. M., & Schrevens, E. (2007). Virtual experiments and their use in teaching experimental design. *International Statistical Review / Revue Internationale de Statistique, 75*, 281–294.

Darius, P. L., van der Knaap, H., Schrevens, E., Portier, K., Massonnet, G., Lievens, L., & Duchateau, L. (2003). Virtual experiments and their role in teaching design of experiments. In *Ingénierie Des Systèmes D'information*. ISI.

Darius, P., Schrevens, E., van der Knaap, H., Portier, K., Massonnet, G., Lievens, L., ... Thas, O. (2003). *Using web-based tools for teaching statistical concepts and experimentation skills*. IASE/ISI Satelite.

Das, A. K., & Dewanjee, S. (2018a). Optimization of extraction using mathematical models and computation. In *Computational phytochemistry* (pp. 75–106). Elsevier; doi:10.1016/B978-0-12-812364-5.00003-1

Das, S., Ng, W. K., Kanaujia, P., Kim, S., & Tan, R. B. H. (2011). Formulation design, preparation and physicochemical characterizations of solid lipid nanoparticles containing a hydrophobic drug: Effects of process variables. *Colloids and Surfaces. B, Biointerfaces, 88*(1), 483–489. doi:10.1016/j.colsurfb.2011.07.036 PMID:21831615

Davim, J. P. (2012). *Statistical and computational techniques in manufacturing*. Statistical and Computational Techniques in Manufacturing; doi:10.1007/978-3-642-25859-6

Davim, J. P. (2015). *Design of experimental in production engineering*. New York.

Davis, R., & Mauer, L. J. (2010). Fourier transform infrared (FT-IR) spectroscopy: a rapid tool for detection and analysis of foodborne pathogenic bacteria. *Current research, technology, and education topics in applied microbiology and microbial biotechnology, 2*, 1582-1594.

De Bruyne, S., Speeckaert, M. M., & Delanghe, J. R. (2018). Applications of mid-infrared spectroscopy in the clinical laboratory setting. *Critical Reviews in Clinical Laboratory Sciences, 55*(1), 1–20. doi:10.1080/10408363.2017.1414142 PMID:29239240

De Federico, S. (2018). Consolidation of statistical thinking in engineering students through animation and simulation of the statistic methods and results of stochastic processes analysis. *EDULEARN18 Proceedings, 1*, 8442–8451.

De Freitas, A. P. B. R., de Freitas, L. V., Loures, C. C. A., dos Santos, M. A. R., Ricardo, G. D., Marins, F. A. S., & Silva, M. B. (2013). Taguchi method applied to environmental engineering. In *Design of experiments-applications*. IntechOpen. doi:10.5772/56531

de Graaff, E. (2017). Ten years in engineering education research: Looking back ahead. *European Journal of Engineering Education, 42*(6), 587–590. doi:10.1080/03043797.2017.1408942

De Meester, F., & Watson, R. R. (2008) Wild-type food in health promotion and disease prevention. In R. Ayerza (Ed.), Chapter 14, Chia as a new source of ω-3 fatty acids. Towa, NJ: Human Press Inc.

De Oliveira, R. R., das Neves, L. S., & de Lima, K. M. G. (2012). Experimental design, near-infrared spectroscopy, and multivariate calibration: an advanced project in a chemometrics course. *Journal of Chemical Education*, *89*(12), 1566–1571. doi:10.1021/ed200765j

De Plano, L. M., Fazio, E., Rizzo, M. G., Franco, D., Carnazza, S., Trusso, S., ... Guglielmino, S. P. (2019). Phage-based assay for rapid detection of bacterial pathogens in blood by Raman spectroscopy. *Journal of Immunological Methods*, *465*, 45–52. doi:10.1016/j.jim.2018.12.004 PMID:30552870

Dean, A., Voss, D., & Draguljić, D. (2017). Principles and techniques. In *Design and analysis of experiments* (pp. 1–5). Cham, Switzerland: Springer International Publishing; doi:10.1007/978-3-319-52250-0_1

Dean, J. R. (2003). *Methods for environmental trace analysis* (Vol. 12). John Wiley & Sons. doi:10.1002/0470863250

Decker, E. A., Chen, B., Panya, A., & Elias, R. J. (2010). Understanding mechanisms of oxidation and antioxidant activity. In E. A. Decker, B. Chen, A. Panya, & R. J. Elias (Eds.), *Oxidation in foods and beverages and antioxidant applications* (pp. 225–243). London, UK: Woodhead Publishing Series in Food Science, Technology, and Nutrition; doi:10.1533/9780857090447.2.225

Deiab, I. M., & Elbestawi, M. A. (2005). Experimental determination of the friction coefficient on the workpiece-fixture contact surface in workholding applications. *International Journal of Machine Tools & Manufacture*, *45*(6), 705–712. doi:10.1016/j.ijmachtools.2004.09.021

Deidda, R., Sacre, P. Y., Clavaud, M., Coïc, L., Avohou, H., Hubert, P., & Ziemons, E. (2019). Vibrational spectroscopy in analysis of pharmaceuticals: Critical review of innovative portable and handheld NIR and Raman spectrophotometers. *TrAC Trends in Analytical Chemistry*.

Del Castillo, E. (2007). *Process optimization: A statistical approach. International Series in Operations Research & Management Science*. Springer. doi:10.1007/978-0-387-71435-6

DeMast, J. A. (2004). A methodological comparison of three strategies for quality improvement. *International Journal of Quality & Reliability Management*, *21*(2), 198–212. doi:10.1108/02656710410516989

Deming, W. E. (1982). Out of the crisis. Cambridge, MA: MIT Center for Advanced Engineering Study.

Deming, S. N., & Morgan, S. L. (1993). *Experimental design: A chemometric approach.* Amsterdam, The Netherlands: Elsevier.

Depciuch, J., Stanek-Widera, A., Lange, D., Biskup-Fruzynska, M., Stanek-Tarkowska, J., Czarny, W., & Cebulski, J. (2018). Spectroscopic analysis of normal and neoplastic (WI-FTC) thyroid tissue. *Spectrochimica Acta. Part A: Molecular and Biomolecular Spectroscopy*, *204*, 18–24. doi:10.1016/j.saa.2018.06.010 PMID:29902767

Derman, S. (2015). Caffeic Acid Phenethyl Ester Loaded PLGA nanoparticles: Effect of various process parameters on reaction yield, encapsulation efficiency, and particle size. *Journal of Nanomaterials*, *2015*, 1–12. doi:10.1155/2015/341848

Dersjö, T., & Olsson, M. (2012). Efficient design of experiments for structural optimization using significance screening. *Structural and Multidisciplinary Optimization*, *45*(2), 185–196. doi:10.100700158-011-0677-0

Deshpande, R. P., Chinnan, M. S., & McWatters, K. H. (2008). Optimization of a chocolate-flavored, peanut–soy beverage using response surface methodology (RSM) as applied to consumer acceptability data. *Lebensmittel-Wissenschaft + Technologie*, *41*(8), 1485–1492. doi:10.1016/j.lwt.2007.08.013

Design Expert 7 software, Stat Ease, Inc. (Minneapolis, MN). Retrieved from https://www.statease.com

Devahastin, S. (2010). Physicochemical aspects of food engineering and processing. doi:10.1201/9781420082425

Devpura, S., Barton, K. N., Brown, S. L., Palyvoda, O., Kalkanis, S., Naik, V. M., ... Chetty, I. J. (2014). Vision 20/20: The role of Raman spectroscopy in early stage cancer detection and feasibility for application in radiation therapy response assessment. *Medical Physics*, *41*(5). doi:10.1118/1.4870981 PMID:24784365

Dhawane, S. H., Bora, A. P., Kumar, T., & Halder, G. (2017). Parametric optimization of biodiesel synthesis from rubber seed oil using iron doped carbon catalyst by Taguchi approach. *Renewable Energy*, *105*, 616–624. doi:10.1016/j.renene.2016.12.096

Dhawane, S. H., Karmakar, B., Ghosh, S., & Halder, G. (2018). Parametric optimisation of biodiesel synthesis from waste cooking oil via Taguchi approach. *Journal of Environmental Chemical Engineering*, *6*(4), 3971–3980. doi:10.1016/j.jece.2018.05.053

Dhawane, S. H., Kumar, T., & Halder, G. (2015). Central composite design approach towards optimization of flamboyant pods derived steam activated carbon for its use as heterogeneous catalyst in transesterification of Hevea brasiliensis oil. *Energy Conversion and Management*, *100*, 277–287. doi:10.1016/j.enconman.2015.04.083

Dhawane, S. H., Kumar, T., & Halder, G. (2016). Biodiesel synthesis from Hevea brasiliensis oil employing carbon supported heterogeneous catalyst: Optimization by Taguchi method. *Renewable Energy*, *89*, 506–514. doi:10.1016/j.renene.2015.12.027

Dhawan, S., Kapil, R., & Singh, B. (2011). Formulation development and systematic optimization of solid lipid nanoparticles of quercetin for improved brain delivery. *The Journal of Pharmacy and Pharmacology*, *63*(3), 342–351. doi:10.1111/j.2042-7158.2010.01225.x PMID:21749381

Díaz-Ramírez, I., Escalante-Espinosa, E., Schroeder, R. A., Fócil-Monterrubio, R., & Ramírez-Saad, H. (2013). Hydrocarbon biodegradation potential of native and exogenous microbial inocula in Mexican tropical soils. In R. Chamy, & F. Rosenkranz (Eds.), In biodegradation of hazardous and special products (pp. 155-178). InTech. doi:10.5772/56233

Din, F., Mustapha, O., Kim, D. W., Rashid, R., Park, J. H., Choi, J. Y., ... Choi, H.-G. (2015). Novel dual-reverse thermosensitive solid lipid nanoparticle-loaded hydrogel for rectal administration of flurbiprofen with improved bioavailability and reduced initial burst effect. *European Journal of Pharmaceutics and Biopharmaceutics*, *94*, 64–72. doi:10.1016/j.ejpb.2015.04.019 PMID:25979136

Ding, S., Xia, N., Wang, P., Li, S., & Ou, Y. (2015). Optimization Algorithm Based on SPSA in Multi-channel Multi-radio Wireless Monitoring Network. *2015 International Conference on Cyber-Enabled Distributed Computing and Knowledge Discovery*, 517–524. 10.1109/CyberC.2015.84

Ding, Y., Pyo, S. M., & Muller, R. H. (2017). smartLipids ® as third solid lipid nanoparticle generation – stabilization of retinol for dermal application. *Original Articles*, *72*, 728–735. doi:10.1691/ph.2017.7016 PMID:29441957

Dixon, D., Eatock, J., Meenan, B., & Morgan, M. (2006). Application of Design of Experiment (DOE). Techniques to process validation in medical device manufacture. Newtownabbey, Irlanda del Norte, Irlanda: Journal of Validation Technology.

Dixon, D., Eatock, J., Meenan, B. J., & Morgan, M. (2006). Application of design of experiment (DOE) techniques to process validation in medical device manufacture. *Journal of Validation Technology*, *12*(2), 92–100.

DOF. Diario oficial de la Federación. (2002). *Norma oficial Mexicana NOM-021-SEMARNAT-2000. Que establece las especificaciones de fertilidad, salinidad y clasificación de suelos, estudio, muestreo y análisis*. Retrieved from http://biblioteca.semarnat.gob.mx/janium/Documentos/Ciga/libros2009/DO2280n.pdf

DOF. Diario oficial de la Federación. (2012). *NOM-138-SEMARNAT/SSA1-2012. Límites máximos permisibles de hidrocarburos en suelos y lineamientos para el muestreo en la caracterización y especificaciones para la remediación* Retrieved from http://www.dof.gob.mx/nota_detalle.php?codigo=5313544&fecha=10/09/2013

Dominguez, T., Aurell, J., Gullett, B., Eninger, R., & Yamamoto, D. (2018). Characterizing emissions from open burning of military food waste and ration packaging compositions. *Journal of Material Cycles and Waste Management*, *20*(2), 902–913. doi:10.100710163-017-0652-y

Douaik, A. (2016). Design of experiments for screening and optimizing factors influencing extraction of bioactive compounds from saffron. *In V International Symposium on Saffron Biology and Technology: Advances in Biology, Technologies, Uses and Market 1184*, 151–158.

Duraipandian, S., Petersen, J. C., & Lassen, M. (2019). Authenticity and concentration analysis of extra virgin olive oil using spontaneous Raman spectroscopy and multivariate data analysis. *Applied Sciences*, *9*(12), 2433. doi:10.3390/app9122433

Durakovic, B. (2017). Design of experiments application, concepts, examples: state of the art. *Periodicals of engineering and natural sciences, 5(3).* pp. 421–439. Available at http://pen.ius.edu.ba

Durmaz, U., & Ozel, M. B. (2019). An experimental study on extraction of sugar from carob using with Taguchi method. *Sakarya University Journal of Science*, *23*(5), 916–923. doi:10.16984aufenbilder.541940

Dutcosky, S. D., Grossmann, M. V. E., Silva, R. S., & Welsch, A. K. (2006). Combined sensory optimization of a prebiotic cereal product using multicomponent mixture experiments. *Food Chemistry*, *98*(4), 630–638. doi:10.1016/j.foodchem.2005.06.029

Dutta, K., Daverey, A., & Lin, J.-G. (2014). Evolution retrospective for alternative fuels: First to fourth generation. *Renewable Energy*, *69*, 114–122. doi:10.1016/j.renene.2014.02.044

Dziuba, B. (2013). Identification of propionibacteria to the species level using Fourier transform infrared spectroscopy and artificial neural networks. *Polish Journal of Veterinary Sciences*, *16*(2), 351–357. doi:10.2478/pjvs-2013-0047 PMID:23971204

Ebrahimi, H. A., Javadzadeh, Y., Hamidi, M., & Barzegar Jalali, M. (2016). Development and characterization of a novel lipohydrogel nanocarrier: Repaglinide as a lipophilic model drug. *The Journal of Pharmacy and Pharmacology*, *68*(4), 450–458. doi:10.1111/jphp.12537 PMID:27114047

Eder, K. (1995). Gas chromatographic analysis of fatty acid methyl esters. *Journal of Chromatography. B, Biomedical Applications*, *671*(1-2), 113–131. doi:10.1016/0378-4347(95)00142-6 PMID:8520689

Editorial, N. (2019). It's time to talk about ditching statistical significance. *Nature*, *567*(7748), 283–283. doi:10.1038/d41586-019-00874-8 PMID:30894740

Ekambaram, P., & Abdul, H. S. A. (2011). Formulation and evaluation of solid lipid nanoparticles of ramipril. *Journal of Young Pharmacists: JYP*, *3*(3), 216–220. doi:10.4103/0975-1483.83765 PMID:21897661

Ekins, S. (2008). *Pharmaceutical applications of Raman spectroscopy*. John Wiley & Sons.

El Assasy, A. E.-H. I., Younes, N. F., & Makhlouf, A. I. A. (2019). Enhanced oral absorption of Amisulpride via a nanostructured lipid carrier-based capsules: development, optimization applying the desirability function approach and in vivo pharmacokinetic study. *AAPS PharmSciTech*, *20*(2), 82. doi:10.120812249-018-1283-x PMID:30652198

Elbadawi, I., Aichouni, M., & Messaoudene, N. A. (2016). Developing an innovative and creative hands-on lean six sigma manufacturing experiments for engineering education. *Engineering Technology & Applied Science Research*, *6*, 1297–1302.

Eldem, T., Speiser, P., & Hincal, A. (1991). *Optimization of spray-dried and -congealed lipid micropellets and characterization of their surface morphology by scanning electron microscopy. Pharmaceutical Research.* An Official Journal of the American Association of Pharmaceutical Scientists; doi:10.1023/A:1015874121860

Elfghi, F. M. (2016). A hybrid statistical approach for modeling and optimization of RON: A comparative study and combined application of response surface methodology (RSM) and artificial neural network (ANN) based on design of experiment (DOE). *Chemical Engineering Research & Design*, *113*, 264–272. doi:10.1016/j.cherd.2016.05.023

Elhalil, A., Tounsadi, H., Elmoubarki, R., Mahjoubi, F. Z., Farnane, M., Sadiq, M., ... Barka, N. (2016). Factorial experimental design for the optimization of catalytic degradation of malachite green dye in aqueous solution by Fenton process. *Water Resources and Industry*, *15*, 41–48. doi:10.1016/j.wri.2016.07.002

Elmi, F., Movaghar, A. F., Elmi, M. M., Alinezhad, H., & Nikbakhsh, N. (2017). Application of FT-IR spectroscopy on breast cancer serum analysis. *Spectrochimica Acta. Part A: Molecular and Biomolecular Spectroscopy*, *187*, 87–91. doi:10.1016/j.saa.2017.06.021 PMID:28666157

Elston, R. C., Olson, J. M., & Palmer, L. (Eds.). (2002). *Biostatistical genetics and genetic epidemiology*. John Wiley & Sons.

Endut, A., Abdullah, S. H. Y. S., Hanapi, N. H. M., Hamid, S. H. A., Lananan, F., & Kamarudin, M. K. A., … Khatoon, H. (2017). Optimization of biodiesel production by solid acid catalyst derived from coconut shell via response surface methodology. *International Biodeterioration & Biodegradation, 124*, 250–257.

ESG. (2015). *Standards and Guidelines for Quality Assurance in the European Higher Education Area (ESG)*. ESG.

European Food Safety Authority (EFSA). (2011). Scientific and technical guidance for the preparation and presentation of an application for authorization of a health claim. *EFSA Journal*, *9*(5), 2170. doi:10.2903/j.efsa.2011.2170

Ezekannagha, C. B., Ude, C. N., & Onukwuli, O. D. (2017). Optimization of the methanolysis of lard oil in the production of biodiesel with response surface methodology. *Egyptian Journal of Petroleum*, *26*(4), 1001–1011. doi:10.1016/j.ejpe.2016.12.004

Fakayode, S. O., Baker, G. A., Bwambok, D. K., Bhawawet, N., Elzey, B., Siraj, N., ... Warner, I. M. (2019). Molecular (Raman, NIR, and FTIR) spectroscopy and multivariate analysis in consumable products analysis. *Applied Spectroscopy Reviews*, 1–77. doi:10.1080/05704928.2019.1631176

Fan, T., & Buckley, J. S. (2002). Rapid and accurate SARA analysis of medium gravity crude oils. *Energy & Fuels*, *16*(6), 1571–1575. doi:10.1021/ef0201228

Farooq, M. A., Nóvoa, H., Araújo, A., & Tavares, S. M. O. (2016). An innovative approach for planning and execution of pre-experimental runs for Design of Experiments. *European Research on Management and Business Economics*, *22*(3), 155–161. doi:10.1016/j.iedee.2014.12.003

FEANI. (2010). *Position paper on Educational Policy*. Document approved by the FEANI General Assembly on 1 October 2010.

FEANI. (n.d.). FEANI. Retrieved from https://www.feani.org/feani/what-feani

Felder, R. M., & Brent, R. (2003). Designing and teaching courses to satisfy the ABET engineering criteria. *Journal of Engineering Education*, *92*(1), 7–25. doi:10.1002/j.2168-9830.2003.tb00734.x

Fernandez, C. G. (2010). Introducing statistical experimental designs to chemical and industrial engineering using the collaborative and problem based learning approach. *Edulearn10: International Conference on Education and New Learning Technologies*, 1883–1888.

Fernández-López, J. A., Angosto, J. M., Roca, M. J., & Miñarro, M. D. (2019). Taguchi design-based enhancement of heavy metals bioremoval by agroindustrial waste biomass from artichoke. *The Science of the Total Environment*, *653*, 55–63. doi:10.1016/j.scitotenv.2018.10.343 PMID:30404069

Ferraro, J. R., & Nakamoto, K. (1994). *Introductory Raman spectroscopy*. Academic Press, Inc.

Ferreira, S. L., Souza, A. S., Brandao, G. C., Ferreira, H. S., dos Santos, W. N., Pimentel, M. F., & Vale, M. G. R. (2008). Direct determination of iron and manganese in wine using the reference element technique and fast sequential multi-element flame atomic absortion spectrometry. *Science Direct*, (74), 699-702.

Ferreira, M., Silva, E., Barreiros, L., Segundo, M. A., Costa Lima, S. A., & Reis, S. (2016). Methotrexate loaded lipid nanoparticles for topical management of skin-related diseases: Design, characterization and skin permeation potential. *International Journal of Pharmaceutics*, *512*(1), 14–21. doi:10.1016/j.ijpharm.2016.08.008 PMID:27530292

FEUM. C. P. (2011). Farmacopea de los Estados Unidos Mexicanos (Décima edición ed., Vol. I). México: Secretaría de Salud.

Fisher, R. A. (1935). *The design of experiments*. New York: John Wiley.

Fisher, R. A. (1936). The use of multiple measurements in taxonomic problems. *Annals of Eugenics*, *7*(2), 179–188. doi:10.1111/j.1469-1809.1936.tb02137.x

Flock, M. R., Harris, W. S., & Kris-Etherton, P. M. (2013). Long-chain omega-3 fatty acids: Time to establish a dietary reference intake. *Nutr. Rev. 71*, 692-707.

Fordellone, M., Bellincontro, A., & Mencarelli, F. (2018). Partial least squares discriminant analysis: A dimensionality reduction method to classify hyperspectral data. *arXiv preprint arXiv:1806.09347*.

Frankel, E. N. (2005). *Lipid oxidation* (2nd ed.). Bridgewater, UK: The Oily Press. doi:10.1533/9780857097927

Fraser, D. A. S. (1956). *Nonparametric methods in statistics*. Hoboken, NJ: John Wiley & Sons. doi:10.1037/13191-000

Freddi, A., & Salmon, M. (2019). Introduction to the Taguchi Method. In *Design principles and methodologies: from conceptualization to first prototyping with examples and case studies* (pp. 159–180). Cham, Switzerland: Springer International Publishing; doi:10.1007/978-3-319-95342-7_7

Frey, D. D., & Wang, H. (2006). Adaptive one-factor at a time experimentation and expected value of improvement. *Technometrics*, *48*(3), 418–431. doi:10.1198/004017006000000075

Frost, J., Ludeman, L., Hillaby, K., Gornall, R., Lloyd, G., Kendall, C., ... Stone, N. (2017). Raman spectroscopy and multivariate analysis for the noninvasive diagnosis of clinically inconclusive vulval lichen sclerosus. *Analyst (London)*, *142*(8), 1200–1206. doi:10.1039/C6AN02009G PMID:27827479

Frost, S. C., Blackman, J. W., Ebeler, S. E., & Heymann, H. (2018). Analysis of temporal dominance of sensation data using correspondence analysis on Merlot wine with differing maceration and cap management regimes. *Food Quality and Preference*, *64*, 245–252. doi:10.1016/j.foodqual.2016.11.011

Fu, F., & Wang, Q. (2011). Removal of heavy metal ions from wastewaters: A review. *Journal of Environmental Management, 92*(3), 407–418. doi:10.1016/j.jenvman.2010.11.011 PMID:21138785

Fujioka, N., Morimoto, Y., Arai, T., & Kikuchi, M. (2004). Discrimination between normal and malignant human gastric tissues by Fourier transform infrared spectroscopy. *Cancer Detection and Prevention, 28*(1), 32–36. doi:10.1016/j.cdp.2003.11.004 PMID:15041075

Fukuda, I. M., Fidelis, C. F., Moreira, C., Morais, A., & Rebello, F. (2018). Design of Experiments (DoE) applied to pharmaceutical and analytical Quality by Design (QbD). *Brazilian Journal of Pharmaceutical Sciences*, (54): 1–16.

Fu, M. C., & Hill, S. D. (1997). Optimization of discrete event systems via simultaneous perturbation stochastic approximation. *IIE Transactions, 29*(3), 233–243. doi:10.1080/07408179708966330

Fu, M. C., & Ho, Y.-C. (1988). Using perturbation analysis for gradient estimation, averaging and updating in a stochastic approximation algorithm. *Winter Simulation Conference Proceedings. IEEE.*, 509–517. 10.1145/318123.318244

Funkenbush, P. D. (2005). *Practical guide to designed experiments. A unified modular approach.* Marcel Dekker.

Furimsky, E. (2013). Hydroprocessing challenges in biofuels production. *Catalysis Today, 217*(0), 13–56. doi:10.1016/j.cattod.2012.11.008

Gajjar, K., Heppenstall, L. D., Pang, W., Ashton, K. M., Trevisan, J., Patel, I. I., ... Martin, F. L. (2013). Diagnostic segregation of human brain tumours using Fourier-transform infrared and/or Raman spectroscopy coupled with discriminant analysis. *Analytical Methods, 5*(1), 89–102. doi:10.1039/C2AY25544H PMID:24098310

Galvain, F., Barolo, M., Macchietto, S., & Bezzo, F. (2010). Identification of physiological models of type 1 diabetes mellitus by model-based design of experiments. Process Systems Engineering, 7, 545-581. https://doi.org doi:10.1002/9783527631209.ch76

Gaona, J. A. S., Manrique, J. P. B., & Majé, Y. M. (2010). Predicción de la estabilidad de los asfaltenos mediante la utilización del análisis SARA para petróleos puros. *Ingeniería y Región, 7*, 103–110. doi:10.25054/22161325.805

Garapati, V. K., & Mishra, S. (2012). Hydrocarbon degradation using fungal isolate: Nutrients optimized by combined grey relational analysis. *Int. J. Eng. Res. Appl, 2*(2), 390–399.

Garba, Z. N., Bello, I., Galadima, A., & Lawal, A. Y. (2016). Optimization of adsorption conditions using central composite design for the removal of copper (II) and lead (II) by defatted papaya seed. *Karbala International Journal of Modern Science, 2*(1), 20–28. doi:10.1016/j.kijoms.2015.12.002

García, D., Viloria Matos, A., Belén, D., & Moreno-Álvarez, M. (2003). Características fisicoquímicas y composición de ácidos grasos del aceite crudo extraído de residuos de mora (*Rubus glaucus* Benth). *Grasas y Aceites, 54*(3), 259–263.

García-López, E., Zavala-Cruz, J., & Palma-López, D. J. (2006). Caracterización de las comunidades vegetales en un área afectada por derrames de hidrocarburos. *Terra Latinoamericana, 24*(1), 17–26.

García, M. A. H., Bautista, C. M. M., Olán, C. M., & Schoreder, R. H. A. (2016). Extracción de Hidrocarburo Pesado en Suelo Acrisol. *Journal of Basic Sciences, 2*(5).

García, M. A., Soberón, E., Cortés, M., Rodríguez, R., Herrera, J. L., & Alcántara, A. (2002). *Guía de Validación de Métodos Analíticos. C.D.Mx.* México: Colegio Nacional de Químicos Farmacéuticos Biólogos.

García-Torres, R., Ríos-Leal, E., Martínez-Toledo, Á, Ramos-Morales, F. R., Cruz-Sánchez, J. S., & Cuevas-Díaz, M. D. C. (2011). Uso de cachaza y bagazo de caña de azúcar en la remoción de hidrocarburos en suelo contaminado. *Revista Internacional de Contaminación Ambiental, 27*(1), 31–39.

Gardi, C., Angelini, M., Barceló, S., Comerma, J., Cruz Gaistardo, C., Encina Rojas, A., . . . Vargas, R. (Eds.). (2014). Atlas de suelos de América Latina y el Caribe, Comisión Europea Oficina de Publicaciones de la Unión Europea, L-2995 Luxembourg.

Gardiner, D. J., & Graves, P. R. (1989). *Practical Raman spectroscopy*. Springer. doi:10.1007/978-3-642-74040-4

Garg, U. K., Kaur, M. P., Garg, V. K., & Sud, D. (2008). Removal of nickel (II) from aqueous solution by adsorption on agricultural waste biomass using a response surface methodological approach. *Bioresource Technology*, *99*(5), 1325–1331. doi:10.1016/j.biortech.2007.02.011 PMID:17383868

Gasper, R., Dewelle, J., Kiss, R., Mijatovic, T., & Goormaghtigh, E. (2009). IR spectroscopy as a new tool for evidencing antitumor drug signatures. *Biochimica et Biophysica Acta (BBA)-. Biomembranes*, *1788*(6), 1263–1270. doi:10.1016/j.bbamem.2009.02.016 PMID:19250921

Gaur, P. K., Mishra, S., & Purohit, S. (2013). Solid lipid nanoparticles of guggul lipid as drug carrier for transdermal drug delivery. *BioMed Research International*, *2013*, 1–10. doi:10.1155/2013/750690 PMID:24058913

Gedi, M. A., Bakar, J., & Mariod, A. A. (2015). Optimization of supercritical carbon dioxide (CO 2) extraction of sardine (Sardinella lemuru Bleeker) oil using response surface methodology (RSM). *Grasas y Aceites*, *66*(June).

Geier, A. B., Rozin, P., & Doros, G. (2006). Unit bias: A new heuristic that helps explain the effect of portion size on food intake. *Psychological Science*, *17*(6), 578–582. doi:10.1111/j.1467-9280.2006.01738.x PMID:16771803

Gemperline, P. (2006). Practical guide to chemometrics. Boca Raton, FL: CRC Press. doi:10.1201/9781420018301

Geneus, V., Li, C., Weerahandi, S., Whalen, E., & Ray Yu, C. (2018). Interferences on treatment and carryover effects in two-period two-sequence crossover designs. *Current Trends in Biomedical Engineering & Biosciences*, *11*(5). doi:10.19080/CTBEB.2018.11.555821

Ghadiri, M., Fatemi, S., Vatanara, A., Doroud, D., Najafabadi, A. R., Darabi, M., & Rahimi, A. A. (2012). Loading hydrophilic drug in solid lipid media as nanoparticles: Statistical modeling of entrapment efficiency and particle size. *International Journal of Pharmaceutics*, *424*(1–2), 128–137. doi:10.1016/j.ijpharm.2011.12.037 PMID:22227603

Ghatak, K. L. (2010). *Techniques and methods in biology*. PHI Learning Pvt. Ltd.

Ghill, J. (2003). Ethnobotany of chia, *Salvia hispánica* L. (*Lamiaceae*). *Economic Botany*, *54*(4), 604–618.

Ghosh, D., Dasgupta, D., Agrawal, D., Kaul, S., Adhikari, D. K., Kurmi, A. K., ... Negi, M. S. (2015). Fuels and chemicals from lignocellulosic biomass: an integrated biorefinery approach. *Energy & Fuels*, *29*(5), 3149–3157. doi:10.1021/acs.energyfuels.5b00144

Girshick, M. A. (1939). On the sampling theory of roots of determinantal equations. *Annals of Mathematical Statistics*, *10*(3), 203–224. doi:10.1214/aoms/1177732180

Giunta, A., Wojtkiewicz, S., & Eldred, M. (2013). Overview of modern design of experiments methods for computational simulations. In *41st Aerospace Sciences Meeting and Exhibit*, Reston, VA. American Institute of Aeronautics and Astronautics. doi:10.2514/6.2003-649

Gleason, P. M., Harris, J., Sheean, P. M., Boushey, C. J., & Bruemmer, B. (2010). Publishing nutrition research: Validity, reliability, and diagnostic test assessment in nutrition-related research. *Journal of the American Dietetic Association*, *110*(3), 409–419. doi:10.1016/j.jada.2009.11.022 PMID:20184991

Glyk, A., Solle, D., Scheper, T., & Beutel, S., (2015, Dec. 15). Optimization of PEG-salt aqueous two-phase systems by design of experiments. *Chemometrics and Intelligent Laboratory Systems*. Elsevier. doi:10.1016/j.chemolab.2015.09.014

Go, A. W., Conag, A. T., & Bertumen, M. M. N. (2019). Taguchi method to improve the production of sugar-rich hydrolysate from non-delipidated spent coffee grounds, and subsequent recovery of lipids and bioactive compounds. *Biofuels, 10*(2), 193–205. doi:10.1080/17597269.2017.1309855

Godolphin, J. D. (2019). Two-level factorial and fractional factorial replicates in blocks of size two. *Computational Statistics & Data Analysis, 133*(1), 120–137. doi:10.1016/j.csda.2018.09.006

Goh, T. N. (1993). Taguchi methods: Some technical, cultural, and pedagogical perspectives. *Quality and Reliability Engineering International, 9*(3), 185–202. doi:10.1002/qre.4680090306

Golskowski, D., Friede, T., & Kieser, M. (2014). Blinded sample re-estimation in crossover bioequivalence trials. *Pharmaceutical Statistics, 13*(3), 157–162. doi:10.1002/pst.1617 PMID:24715672

Gomez, F., & Sartaj, M. (2014). Optimization of field scale biopiles for bioremediation of petroleum hydrocarbon contaminated soil at low temperature conditions by response surface methodology (RSM). *International Biodeterioration & Biodegradation, 89*, 103–109. doi:10.1016/j.ibiod.2014.01.010

Gomez-Miguel, V., & Sotes, V. (2014). *El Manganeso y la Viticultura: Una revisión*. Madrid, España: Ministerio de Agricultura, Alimentación y Medio Ambiente.

Gómez, R. S. G., Pandiyan, T., Iris, V. A., Figueroa, M. R., & de Bazúa, C. D. (2004). Identificación y cuantificación de hidrocarburos aromáticos policíclicos en el suelo usando métodos espectroscópicos. *Tecnología, Ciencia. La Educación, 19*(2), 69–82.

Gonzalez-Fernandez, I., Iglesias-Otero, M. A., Esteki, M., Moldes, O. A., Mejuto, J. C., & Simal-Gandara, J. (2019). A critical review on the use of artificial neural networks in olive oil production, characterization and authentication. *Critical Reviews in Food Science and Nutrition, 59*(12), 1913–1926. doi:10.1080/10408398.2018.1433628 PMID:29381389

Goormaghtigh, E. (2016). Infrared imaging in histopathology: Is a unified approach possible? *Biomedical Spectroscopy and Imaging, 5*(4), 325–346. doi:10.3233/BSI-160151

Goormaghtigh, E., Gasper, R., Bénard, A., Goldsztein, A., & Raussens, V. (2009). Protein secondary structure content in solution, films and tissues: Redundancy and complementarity of the information content in circular dichroism, transmission and ATR FTIR spectra. *Biochimica et Biophysica Acta (BBA)-. Proteins and Proteomics, 1794*(9), 1332–1343. doi:10.1016/j.bbapap.2009.06.007 PMID:19540367

Gopinadh, R., Ayyanna, C., Ramakrishna, Ch., Narayana, K. V., Kandisa, R. V., & Jagadhi, R. (2015). Optimization of chemical parameters for the production of citric acid using Box-Behnken design. *Journal of Bioprocessing & Biotechniques*, (5): 1–6.

Goupy-Creightom. (2007). Introduction to design of experiments, with JMP examples (3rd ed.), SAS Press Series.

Goupy, J. L. (1993). *Methods for experimental design*. Amsterdam, The Netherlands: Elsevier.

Gow-Chin, Y., & Shyi-Liang, S. (1989). Oxidative Stability of sesame oil prepared from sesame seed with different roasting temperatures. *Food Chemistry, 31*(3), 215–224. doi:10.1016/0308-8146(89)90059-9

Graham, R. (2018). The global state of the art in engineering education. Massachusetts Institute of Technology (MIT).

Granato, D., Putnik, P., Kovačević, D. B., Santos, J. S., Calado, V., Rocha, R. S., ... Pomerantsev, A. (2018). Trends in chemometrics: Food authentication, microbiology, and effects of processing. *Comprehensive Reviews in Food Science and Food Safety, 17*(3), 663–677. doi:10.1111/1541-4337.12341

Gremlich, H. U., & Yan, B. (2000). Infrared and Raman spectroscopy of biological materials. Boca Raton, FL: CRC Press. doi:10.1201/b16936

Griffith, C. R. (2006). The principles of experimental method. In *Principles of systematic psychology* (pp. 436–470). Elsevier/Butterworth-Heinemann; doi:10.1037/11148-012

Griffiths, C. A., Howarth, J., De Almeida-Rowbotham, G., Rees, A., & Kerton, R. (2016). A design of experiments approach for the optimisation of energy and waste during the production of parts manufactured by 3D printing. *Journal of Cleaner Production, 139*, 74–85. doi:10.1016/j.jclepro.2016.07.182

Griffiths, P. R., & De Haseth, J. A. (2007). *Fourier transform infrared spectrometry* (Vol. 171). John Wiley & Sons. doi:10.1002/047010631X

Griggs, D., Stafford-Smith, M., Gaffney, O., Rockström, J., Öhman, M. C., Shyamsundar, P., ... Noble, I. (2013). Policy: Sustainable development goals for people and planet. *Nature, 495*(7441), 305–307. doi:10.1038/495305a PMID:23518546

Grompone, M. A., & Irigaray, B. (2013) Aceites de oliva: de la planta al consumidor. 2. Montevideo: INIA; Hemisferio Sur. 123-140.

Guijarro-Aldaco, A., Hernandez-Montoya, V., Bonilla-Petriciolet, A., Montes-Morán, M. A., & Mendoza-Castillo, D. I. (2011). Improving the adsorption of heavy metals from water using commercial carbons modified with egg shell wastes. *Industrial & Engineering Chemistry Research, 50*(15), 9354–9362. doi:10.1021/ie2006627

Guimaraes, L. L., Moreira, L. P., Lourenço, B. F., Toma, W., Zangaro, R. A., Pacheco, M. T. T., & Silveira, L. (2018). Multivariate method based on Raman spectroscopy for quantification of dipyrone in oral solutions. *Journal of Spectroscopy*, 2018.

Gupta, J., Agarwal, M., & Dalai, A. K. (2016). Optimization of biodiesel production from mixture of edible and nonedible vegetable oils. *Biocatalysis and Agricultural Biotechnology, 8*, 112–120. doi:10.1016/j.bcab.2016.08.014

Gupta, T. B., & Lataye, D. H. (2018). Adsorption of indigo carmine and methylene blue dye: Taguchi's design of experiment to optimize removal efficiency. *Sadhana, 43*(10), 170. doi:10.100712046-018-0931-x

Gutiérres Pulido, H., & De la Vara Salazar, R. (2012). Análisis y diseño de experimentos. *Journal of Chemical Information and Modeling*. doi:10.1017/CBO9781107415324.004

Gutiérrez, H., & Salazar, R. (2012). Análisis y Diseños de Experimentos (3rd. ed.). D. F., México: McGraw-Hill.

Gutiérrez, A. J., Rubio, C., Moreno, I. M., González, A. G., Gonzalez-Weller, D., Bencharki, N., ... Revert, C. (2017). Estimation of dietary intake and target hazard quotients for metals by consumption of wines from the Canary Islands. *Food and Chemical Toxicology, 108*, 10–18. doi:10.1016/j.fct.2017.07.033 PMID:28733233

Gutierrez, H., & De la Vara, R. (2008). Análisis y diseño de experimentos. In *Análisis y diseño de experimentos (pág. 195)*. D.F., México: McGraw-Hill Interamericana.

Gutiérrez, M. D. C., & Zavala, J. (2002). Rasgos hidromórficos de suelos tropicales contaminados con hidrocarburos. *Terra Latinoamericana, 20*(2).

Guzmán-Osorio, F. J., & Adams, R. H. (2015). Mitigation of water repellency in the treatment of contaminated muds using the chemical–biological stabilization process. *International Journal of Environmental Science and Technology, 12*(6), 2071–2078. doi:10.100713762-014-0606-z

Hadiani, M. R., Darani, K. K., Rahimifard, N., & Younesi, H. (2018). Biosorption of low concentration levels of Lead (II) and Cadmium (II) from aqueous solution by Saccharomyces cerevisiae: Response surface methodology. *Biocatalysis and Agricultural Biotechnology*, *15*, 25–34. doi:10.1016/j.bcab.2018.05.001

Hadiani, M. R., Khosravi-Darani, K., & Rahimifard, N. (2019). Optimization of As (III) and As (V) removal by *Saccharomyces cerevisiae* biomass for biosorption of critical levels in the food and water resources. *Journal of Environmental Chemical Engineering*, *7*(2). doi:10.1016/j.jece.2019.102949

Hamilton Beach 31199 User Manual, Glen Allen, VA, PN 840156300, p. 5.

Hamze, H., Akia, M., & Yazdani, F. (2015). Optimization of biodiesel production from the waste cooking oil using response surface methodology. *Process Safety and Environmental Protection*, *94*, 1–10. doi:10.1016/j.psep.2014.12.005

Hanrahan, G., Montes, R., & Gomez, F. A. (2008). Chemometric experimental design based optimization techniques in capillary electrophoresis: A critical review of modern applications. *Analytical and Bioanalytical Chemistry*, *390*(1), 169–179. doi:10.100700216-007-1619-y PMID:17909758

Hansen, E., & Miró, M. (2007). How flow-injection analysis (FIA) over the past 25 years has changed our way of performing chemical analyses. *Trends in Analytical Chemistry*, *26*(1), 18–26. doi:10.1016/j.trac.2006.07.010

Hansen, S. H., Pedersen-Bjergaard, S., & Rasmussen, K. (2012). *Introduction to pharmaceutical chemical analysis*. John Wiley & Sons.

Hanson, C. (2013). *Recent advances in liquid-liquid extraction*. Elsevier.

Hanvey, J. S., Lewis, P. J., Lavers, J. L., Crosbie, N. D., Pozo, K., & Clarke, B. O. (2017). A review of analytical techniques for quantifying microplastics in sediments. *Analytical Methods*, *9*(9), 1369–1383. doi:10.1039/C6AY02707E

Hao, J., Fang, X., Zhou, Y., Wang, J., Guo, F., Li, F., & Peng, X. (2011). Development and optimization of solid lipid nanoparticle formulation for ophthalmic delivery of chloramphenicol using a Box-Behnken design. *International Journal of Nanomedicine*, *6*, 683–692. doi:10.2147/IJN.S17386 PMID:21556343

Hao, J., Wang, F., Wang, X., Zhang, D., Bi, Y., Gao, Y., ... Zhang, Q. (2012). Development and optimization of baicalin-loaded solid lipid nanoparticles prepared by coacervation method using central composite design. *European Journal of Pharmaceutical Sciences*, *47*(2), 497–505. doi:10.1016/j.ejps.2012.07.006 PMID:22820033

Harris, J. E., & Raynor, H. A. (2017). Crossover designs in nutrition and dietetics research. *Journal of the Academy of Nutrition and Dietetics*, *117*(7), 1023–1030. doi:10.1016/j.jand.2017.03.017 PMID:28479137

Harrison, J. P., & Berry, D. (2017). Vibrational spectroscopy for imaging single microbial cells in complex biological samples. *Frontiers in Microbiology*, *8*, 675. doi:10.3389/fmicb.2017.00675 PMID:28450860

Hasan, S. H., Srivastava, P., & Talat, M. (2010). Biosorption of lead using immobilized Aeromonas hydrophila biomass in up flow column system: Factorial design for process optimization. *Journal of Hazardous Materials*, *177*(1-3), 312–322. doi:10.1016/j.jhazmat.2009.12.034 PMID:20036460

Hayes, A. W. (2008). Principles and methods of toxicology. Boca Raton, FL: CRC Press.

Hayman, M., & Kam, P. C. A. (2008). Capsaicin: A review of its pharmacology and clinical applications. *Current Anaesthesia and Critical Care*, *19*(5–6), 338–343. doi:10.1016/j.cacc.2008.07.003

He, H., Sun, D. W., Pu, H., Chen, L., & Lin, L. (2019). Applications of Raman spectroscopic techniques for quality and safety evaluation of milk: A review of recent developments. *Critical Reviews in Food Science and Nutrition*, *59*(5), 770–793. doi:10.1080/10408398.2018.1528436 PMID:30614242

Heiberger, R. M., & Holland, B. (2015). Design of experiments-factorial designs. In *Statistical analysis and data display: an intermediate course with examples in R* (pp. 427–478). New York, NY: Springer New York; doi:10.1007/978-1-4939-2122-5_13

Heise, S., Virtanen, S., & Boccaccini, A. R. (2018). Taguchi Design of Experiments approach to determine process parameter for the electrophoretic deposition of chitosan/bioactive glass on Mg alloy substrates. *ECS Transactions*, *82*(1), 81–87. doi:10.1149/08201.0081ecst

Helm, D., Labischinski, H., Schallehn, G., & Naumann, D. (1991). Classification and identification of bacteria by Fourier-transform infrared spectroscopy. *Microbiology*, *137*(1), 69–79. PMID:1710644

Hernández Sampieri, R. (2014). *Metodología de la investigación. 6th Edición*. McGraw-Hill.

Hernández, M. A., Santillán, S., Sabelkin, V., & Parra, M. L. (2003). Caso de aplicación del método Taguchi en el diseño de herramientas de muestreo de Hidrocarburos. *Ingeniería Mecánica, Tecnología y Desarrollo*, *1*(3), 91–101.

Hibbert, D. B. (2012). Experimental design in chromatography: A tutorial review. *Journal of Chromatography. B, Analytical Technologies in the Biomedical and Life Sciences*, *910*, 2–13. doi:10.1016/j.jchromb.2012.01.020 PMID:22333438

Hill, W. J., & Hunter, W. G. (1966). A review of response surface methodology: A literature survey. *Technometrics*, *8*(4), 571–590. doi:10.2307/1266632

Ho, C. S., Jean, N., Hogan, C. A., Blackmon, L., Jeffrey, S. S., Holodniy, M., . . . Dionne, J. (2019). Rapid identification of pathogenic bacteria using Raman spectroscopy and deep learning. *arXiv preprint arXiv:1901.07666.*

Hollas, J. M. (2004). *Modern spectroscopy*. John Wiley & Sons.

Horwitz, W. (1980). Official methods of analysis of the association of official analytical chemists. AOAC. Washington.

Hotelling, H. (1933). Analysis of a complex of statistical variables into principal components. *Journal of Educational Psychology*, *24*(6), 417–441. doi:10.1037/h0071325

Hotelling, H. (1936). Simplified calculation of principal components. *Psychometrika*, *1*(1), 27–35. doi:10.1007/BF02287921

Hsia, C., Plack, R., & Nagel, C. (1975). Influence of must processing on iron and copper contents of experimental wines. *American Journal of Enology and Viticulture*, *26*(2), 57–61.

Hu, L., Jia, Y., & Ding, W. (2010). Preparation and characterization of solid lipid nanoparticles loaded with epirubicin for pulmonary delivery. *Die Pharmazie*, *65*(8), 585–587. doi:10.1691/ph.2010.0023 PMID:20824958

Hu, Q., Sommerfeld, M., Jarvis, E., Ghirardi, M., Posewitz, M., Seibert, M., & Darzins, A. (2008). Microalgal triacylglycerols as feedstocks for biofuel production: Perspectives and advances. *The Plant Journal*, *54*(4), 621–639. doi:10.1111/j.1365-313X.2008.03492.x PMID:18476868

Ibelings, M. S., Maquelin, K., Endtz, H. P., Bruining, H. A., & Puppels, G. J. (2005). Rapid identification of Candida spp. in peritonitis patients by Raman spectroscopy. *Clinical Microbiology and Infection*, *11*(5), 353–358. doi:10.1111/j.1469-0691.2005.01103.x PMID:15819860

IEA. (2011). Biofuels can provide up to 27% of world transportation fuel by 2050, IEA report says - IEA 'roadmap' shows how biofuel production can be expanded in a sustainable way, and identifies needed technologies and policy actions. Retrieved from https://www.iea.org/newsroom/news/2011/april/biofuels-can-provide-up-to-27-of-world-transportation-fuel-by-2050-iea-report-.html

IEA. (2019a). Biofuels for transport: Tracking clean energy progress. Retrieved from https://www.iea.org/tcep/transport/biofuels/#

IEA. (2019b). Key world energy statistics. Retrieved from https://www.iea.org/statistics/kwes/consumption/

Ilzarbe, L., Álvarez, M. J., Viles, E., & Tanco, M. (2008). Practical applications of Design of Experiments in the field of engineering. A bibliographical review. *Quality and Reliability Engineering International*, *24*(4), 417–428. doi:10.1002/qre.909

Infante, C., & Morales García, F. A. (2012). Evaluación de la toxicidad en desechos y suelos petrolizados empleando semillas de Lactuca sativa L. *Interciencia*, *37*(10).

Islas-Flores, C. A., Buenrostro-Gonzalez, E., & Lira-Galeana, C. (2005). Comparisons between open column chromatography and HPLC SARA fractionations in petroleum. *Energy & Fuels*, *19*(5), 2080–2088. doi:10.1021/ef050148+

Islas, G., Rodriguez, J. A., Perez-Silva, I., Miranda, J. M., & Ibarra, I. S. (2018). Solid-phase extraction and large-volume sample stacking-capillary electrophoresis for determination of tetracycline residues in milk. *Journal of Analytical Methods in Chemistry*. PMID:29675286

Ixtaina, V. Y., Nolasco, S. M., & Tomás, M. C. (2012). Oxidative stability of Chia (Salvia hispanica L.) Seed oil: Effect of antioxidants and storage conditions. *Journal of the American Oil Chemists' Society*, *89*(6), 1077–1090. doi:10.100711746-011-1990-x

Jacyna, J., Kordalewska, M., & Markuszewski, M. (2019). Design of experiments in metabolomics-related studies: An overview. *Journal of Pharmaceutical and Biomedical Analysis*, *164*, 598–606. doi:10.1016/j.jpba.2018.11.027 PMID:30469109

Jain, M., Garg, V. K., & Kadirvelu, K. (2011). Investigation of Cr (VI) adsorption onto chemically treated *Helianthus annuus*: Optimization using response surface methodology. *Bioresource Technology*, *102*(2), 600–605. doi:10.1016/j.biortech.2010.08.001 PMID:20739179

Jarvis, R. M., & Goodacre, R. (2004). Discrimination of bacteria using surface-enhanced Raman spectroscopy. *Analytical Chemistry*, *76*(1), 40–47. doi:10.1021/ac034689c PMID:14697030

Jayashree, P. K., Sharma, S. S., Shetty, R., Mahato, A., & Gowrishankar, M. C. (2018). Optimization of TIG welding parameters for 6061Al alloy using Taguchi's design of experiments. *Materials Today: Proceedings*, *5*(11), 23648–23655.

Jeevanandam, J., Chan, Y. S., & Danquah, M. K. (2016). Nano-formulations of drugs: Recent developments, impact and challenges. *Biochimie*, *128–129*, 99–112. doi:10.1016/j.biochi.2016.07.008 PMID:27436182

Jiang, S., Zhu, R., He, X., Wang, J., Wang, M., Quian, Y., & Wang, S. (2017). Enhanced photocytotoxicity of curcumin delivered by solid lipid nanoparticles. *International Journal of Nanomedicine*, *12*, 167–178. doi:10.2147/IJN.S123107 PMID:28053531

Jing, P., Ruan, S.-Y., Dong, Y., Zhang, X.-G., Yue, J., Kan, J.-Q., ... Yu, L. L. (2011). Optimization of purification conditions of radish (Raphanus sativus L.) anthocyanin-rich extracts using chitosan. *Lebensmittel-Wissenschaft + Technologie*, *44*(10), 2097–2103. doi:10.1016/j.lwt.2011.06.003

Joachim, K. (2000). *Experimental design - a handbook and dictionary for medical and behavioral research. Techniques in the behavioral and neural sciences* (Vol. 14). Elsevier; doi:10.1016/S0921-0709(00)80012-7

João, I. M., & Silva, J. M. (2007). Design of experiments in chemical engineering education. Application to liquid-liquid extraction. *Proceedings 56th session of the International Statistical Institute*, 657–658. International Statistical Institute.

João, I. M., & Silva, J. M. (2016). Student engagement with statistical design of experiments by active learning projects. *Proceedings of the 2nd International Conference on Higher Education Advances*, 203–210. Valencia: Universitat Politècnica València. 10.4995/HEAD16.2016.2650

João, I. M., & Silva, J. M. (2018). Designing and teaching a curricular unit to accomplish the outcomes related learning objectives. In *Proceedings 2018 3rd International Conference of the Portuguese Society for Engineering Education (CISPEE)*, 1–7. IEEE.

João, I. M., & Silva, J. M. (2017). Designing solutions by a student centred approach: Integration of chemical process simulation with statistical tools to improve distillation systems. [IJEP]. *International Journal of Engineering Pedagogy, 7*(3), 4. doi:10.3991/ijep.v7i3.6795

Jonkman, J., Bloemhof, J. M., van der Vorst, J. G. A. J., & van der Padt, A. (2017). Selecting food process designs from a supply chain perspective. *Journal of Food Engineering, 195*, 52–60. doi:10.1016/j.jfoodeng.2016.09.015

Joshi, G., Pandey, J. K., Rana, S., & Rawat, D. S. (2017). Challenges and opportunities for the application of biofuel. *Renewable & Sustainable Energy Reviews, 79*, 850–866. doi:10.1016/j.rser.2017.05.185

Joshi, M. D., & Müller, R. H. (2009). Lipid nanoparticles for parenteral delivery of actives. *European Journal of Pharmaceutics and Biopharmaceutics, 71*(2), 161–172. doi:10.1016/j.ejpb.2008.09.003 PMID:18824097

Jribi, S., Molnàr, H., Antal, O. T., Adànyi, N., Kheriji, O., Naàr, Z., & Debbabi, H. (2019). Zinc fortification as a tool for improving sprout hygienic and nutritional quality: A factorial design approach. *Journal of the Science of Food and Agriculture, 0*(0). doi:10.1002/jsfa.9765 PMID:31032927

Jung, M. Y., Kang, J. H., Choi, Y. S., Lee, D. Y., Lee, J. Y., & Park, J. S. (2019). Analytical features of microwave plasma-atomic emission spectrometry (MP-AES) for the quantitation of manganese (Mn) in wild grape (Vitis coignetiae) red wines: Comparison with inductively coupled plasma-optica emission spectrometry (ICP-OES) l. *Food Chemistry, 274*, 20–25. doi:10.1016/j.foodchem.2018.08.114 PMID:30372927

Kademi, H. I., Ulusoy, B. H., & Hecer, C. (2019). Applications of miniaturized and portable near infrared spectroscopy (NIRS) for inspection and control of meat and meat products. *Food Reviews International, 35*(3), 201–220. doi:10.1080/87559129.2018.1514624

Kandpal, L. M., & Cho, B. K. (2014). A review of the applications of spectroscopy for the detection of microbial contaminations and defects in agro foods. *Journal of Biosystems Engineering, 39*(3), 215–226. doi:10.5307/JBE.2014.39.3.215

Karabas, H. (2013). Biodiesel production from crude acorn (Quercus frainetto L.) kernel oil: An optimisation process using the Taguchi method. *Renewable Energy, 53*, 384–388. doi:10.1016/j.renene.2012.12.002

Karmakar, B., Dhawane, S. H., & Halder, G. (2018). Optimization of biodiesel production from castor oil by Taguchi design. *Journal of Environmental Chemical Engineering, 6*(2), 2684–2695. doi:10.1016/j.jece.2018.04.019

Karnitz Jr., O., Gurgel, L. V. A., De Melo, J. C. P., Botaro, V. R., Melo, T. M. S., de Freitas Gil, R. P., & Gil, L. F. (2007). Adsorption of heavy metal ion from aqueous single metal solution by chemically modified sugarcane bagasse. *Bioresource Technology, 98*(6), 1291–1297. doi:10.1016/j.biortech.2006.05.013 PMID:16843656

Kaufmann, K. C., Favero, F. D. F., de Vasconcelos, M. A. M., Godoy, H. T., Sampaio, K. A., & Barbin, D. F. (2019). Portable NIR spectrometer for prediction of palm oil acidity. *Journal of Food Science, 84*(3), 406–411. doi:10.1111/1750-3841.14467 PMID:30758058

Kavkler, K., Gunde-Cimerman, N., Zalar, P., & Demšar, A. (2011). FTIR spectroscopy of biodegraded historical textiles. *Polymer Degradation & Stability, 96*(4), 574–580. doi:10.1016/j.polymdegradstab.2010.12.016

Kaznowska, E., Depciuch, J., Łach, K., Kołodziej, M., Koziorowska, A., Vongsvivut, J., ... Cebulski, J. (2018). The classification of lung cancers and their degree of malignancy by FTIR, PCA-LDA analysis, and a physics-based computational model. *Talanta, 186*, 337–345. doi:10.1016/j.talanta.2018.04.083 PMID:29784370

Kelidari, H. R., Saeedi, M., Akbari, J., Morteza-Semnani, K., Gill, P., Valizadeh, H., & Nokhodchi, A. (2015). Formulation optimization and in vitro skin penetration of spironolactone loaded solid lipid nanoparticles. *Colloids and Surfaces. B, Biointerfaces*, *128*, 473–479. doi:10.1016/j.colsurfb.2015.02.046 PMID:25797482

Keller, M. D., Kanter, E. M., & Mahadevan-Jansen, A. (2006). Raman spectroscopy for cancer diagnosis. *Spectroscopy-Springfield then eugene then duluth-*, *21*(11), 33.

Kelsall, R. W., Hamley, I. W., & Geoghegan, M. (Eds.). (2005). *Nanoscale science and technology* (p. 108). Chichester, UK: Wiley. doi:10.1002/0470020873

Kenkel, J. (2014). Analytical chemistry for technicians. Boca Raton, FL: CRC Press.

Ketelaere, D., Bebber, V., Makar, I. K., & Sousa, B. De. (2014). The garden sprinkler: An interactive web-based application for teaching design of experiments. *ICOTS9*, 1–4.

Ketty, G. S. L., Khairunnisak, M., Faridah, I., Roosnoor, F. H., Izwan, I., Norakmar, I., ... Saipul, B. A. R. (2017). Preliminary study on classification of raw bovine's milk using atr-ftir coupled with PCA from peninsular Malaysia. *Malaysian Journal of Veterinary Research*, *8*(1), 95–101.

Khalifa, E. B., Rzig, B., Chakroun, R., Nouagui, H., & Hamrouni, B. (2019). Application of response surface methodology for chromium removal by adsorption on low-cost biosorbent. *Chemometrics and Intelligent Laboratory Systems*, *189*, 18–26. doi:10.1016/j.chemolab.2019.03.014

Khan, A. B., & Kathi, S. (2014). Evaluation of heavy metal and total petroleum hydrocarbon contamination of roadside surface soil. *International Journal of Environmental Science and Technology*, *11*(8), 2259–2270. doi:10.100713762-014-0626-8

Khanmohammadi, M., & Garmarudi, A. B. (2011). Infrared spectroscopy provides a green analytical chemistry tool for direct diagnosis of cancer. *TrAC Trends in Analytical Chemistry*, *30*(6), 864–874. doi:10.1016/j.trac.2011.02.009

Kharia, A. A., & Singhai, A. K. (2013). Screening of most effective variables for development of gastroretentive mucoadhesive nanoparticles by Taguchi design. *ISRN Nanomaterials*, 2013.

Khayati, G., & Barati, M. (2017). Bioremediation of petroleum hydrocarbon contaminated soil: optimization strategy using taguchi design of experimental (DOE) methodology. *Environmental Processes*, *4*(2), 451–461. doi:10.100740710-017-0244-9

Khorasani, A. M., Yazdi, M. R. S., & Safizadeh, M. S. (2011). Tool life prediction in face milling machining of 7075 Al by using artificial neural networks (ANN) and Taguchi Design of Experiment (DOE). *IACSIT International Journal of Engineering and Technology*, *3*(1), 30–35. doi:10.7763/IJET.2011.V3.196

Khuri, A. I., & Cornell, J. A. (2018). *Response surfaces: designs and analyses*. Routledge. doi:10.1201/9780203740774

Kiefer, J. (2015). Recent advances in the characterization of gaseous and liquid fuels by vibrational spectroscopy. *Energies*, *8*(4), 3165–3197. doi:10.3390/en8043165

Kiefer, J., & Wolfowitz, J. (1952). Stochastic estimation of the maximum of a regression function. *Annals of Mathematical Statistics*, *23*(3), 462–466. doi:10.1214/aoms/1177729392

Kılıç, M., Uzun, B. B., Pütün, E., & Pütün, A. E. (2013). Optimization of biodiesel production from castor oil using factorial design. *Fuel Processing Technology*, *111*, 105–110. doi:10.1016/j.fuproc.2012.05.032

Kim, J. H., Ko, J. A., Kim, J. T., Cha, D. S., Cho, J. H., Park, H. J., & Shin, G. H. (2014). Preparation of a capsaicin-loaded nanoemulsion for improving skin penetration. *Journal of Agricultural and Food Chemistry*, *62*(3), 725–732. doi:10.1021/jf404220n PMID:24417234

Kim, S., Diab, R., Joubert, O., Canilho, N., & Pasc, A. (2016). Core–shell microcapsules of solid lipid nanoparticles and mesoporous silica for enhanced oral delivery of curcumin. *Colloids and Surfaces. B, Biointerfaces*, *140*, 161–168. doi:10.1016/j.colsurfb.2015.12.040 PMID:26752213

Kira, R. S. (1991). Pearson's composition and analysis of foods. Ed. Longman Scientific & Technical. USA. 420-445.

Kleijnen, J. P. (2015). Design and analysis of simulation experiments. *In International Workshop on Simulation*. pp. 3–22. Cham, Switzerland: Springer.

Kleijnen, J. P. (2005). An overview of the design and analysis of simulation experiments for sensitivity analysis. *European Journal of Operational Research*, *164*(2), 287–300. doi:10.1016/j.ejor.2004.02.005

Klein, D., Breuch, R., von der Mark, S., Wickleder, C., & Kaul, P. (2019). Detection of spoilage associated bacteria using Raman-microspectroscopy combined with multivariate statistical analysis. *Talanta*, *196*, 325–328. doi:10.1016/j.talanta.2018.12.094 PMID:30683371

Knapp, G. (2003). How DoE saved a company 50 million dollars in raw materials costs. *DEO FAQ Alert, 3*(2).

Kobilinsky, A., Monod, H., & Bailey, R. A. (2017). Automatic generation of generalised regular factorial designs. *Computational Statistics & Data Analysis*, *113*, 311–329. doi:10.1016/j.csda.2016.09.003

Kocbek, P., Baumgartner, S., & Kristl, J. (2006). Preparation and evaluation of nanosuspensions for enhancing the dissolution of poorly soluble drugs. *International Journal of Pharmaceutics*, *312*(1–2), 179–186. doi:10.1016/j.ijpharm.2006.01.008 PMID:16469459

Kochan, K., Perez-Guaita, D., Pissang, J., Jiang, J. H., Peleg, A. Y., McNaughton, D., ... Wood, B. R. (2018). In vivo atomic force microscopy–infrared spectroscopy of bacteria. *Journal of the Royal Society, Interface*, *15*(140). doi:10.1098/rsif.2018.0115 PMID:29593091

Kontoudakis, N., Smith, M., Smith, P. A., Wilkes, E. N., & Clark, A. C. (2018). The determination of total Cu in white wine by BCA colorimetric analysis. *National Wine and Grape Industry Centre & Charles Sturt University*, (1), 1-4.

Kontoudakis, N., Guo, A., Scollary, G. R., & Clark, A. C. (2017). The impact of aging wine in high and low oxygen conditions on the fractionation of Cu and Fe in Chardonnay wine. *Food Chemistry*, *229*, 319–328. doi:10.1016/j.foodchem.2017.02.065 PMID:28372180

Kontoudakis, N., Schmidtke, L. M., Bekker, M., Smith, M., Smith, P. A., Scollary, G. R., ... Clark, A. C. (2019). Analytical strategies for the measurement of different forms of Cu and Fe in wine: Comparison between approaches in relation to wine composition. *Food Chemistry*, *274*, 89–99. doi:10.1016/j.foodchem.2018.08.084 PMID:30373024

Koretsky, M. D., Amatore, D., Barnes, C., & Kimura, S. (2006). The virtual CVD learning platform. *Proceedings. Frontiers in Education. 36th Annual Conference*, 25–30. IEEE.

Koretsky, M. D., Amatore, D., Barnes, C., & Kimura, S. (2008). Enhancement of student learning in experimental design using a virtual laboratory. *IEEE Transactions on Education*, *51*(1), 76–85. doi:10.1109/TE.2007.906894

Koretsky, M. D., Kimura, S., Barnes, C., Amatore, D., & Meyers-Graham, D. (2006). Experiential learning of design of experiments using a virtual CVD reactor. *ASEE Annual Conference and Exposition, Conference Proceedings*.

Kotanen, C. N., Martinez, L., Alvarez, R., & Simecek, J. W. (2016). Surface enhanced Raman scattering spectroscopy for detection and identification of microbial pathogens isolated from human serum. *Sensing and bio-sensing research, 8*, 20-26.

Kris-Etherton, P. M., Harris, W. S., & Appel, L. J. (2002). Fish consumption, fish oil, omega-3 fatty acids, and cardiovascular disease. *Circulation, 106*(21), 2747–2757. doi:10.1161/01.CIR.0000038493.65177.94 PMID:12438303

Kubow, S. (1992). Routes of formation and toxic consequences of lipid oxidation-products in foods. *Free Radical Biology & Medicine, 12*(1), 63–81. doi:10.1016/0891-5849(92)90059-P PMID:1537572

Kuehl, R. O. (2001). Diseño de Experimentos Principios Estadísticos de Diseño y Análisis de Investigación. (Second Edition), Arizona: Thomson Learning.

Kuehl, R. O. (2001). *Diseño de experimentos: principios estadísticos para el diseño y análisis de investigaciones.* Thomson Learning.

Kul, M., & Çetinkaya, Ü. (2010). Application of the Taguchi orthogonal array design to Cr (VI) solvent extraction. *Mining, Metallurgy, & Exploration, 27*(2), 102–109. doi:10.1007/BF03402386

Kumari, A., Kaur, J., & Bhattacharyya, S. (2018). Application of fourier transform-infrared spectroscopy as a tool for early cancer detection. *American Journal of Biomedical Sciences, 10*(3).

Kumar, R. S., Sureshkumar, K., & Velraj, R. (2015). Optimization of biodiesel production from Manilkara zapota (L.) seed oil using Taguchi method. *Fuel, 140*, 90–96. doi:10.1016/j.fuel.2014.09.103

Kumar, S. V., Pai, K. V., Narayanaswamy, R., & Sripathy, M. (2013). Experimental optimization for cu removal from aqueous solution using neem leaves based on Taguchi method. *International Journal of Science, Environmental Technology, 2*(1), 103–114.

Kumar, S., Srinivasan, A., & Nikolajeff, F. (2018). Role of infrared spectroscopy and imaging in cancer diagnosis. *Current Medicinal Chemistry, 25*(9), 1055–1072. doi:10.2174/0929867324666170523121314 PMID:28545365

Kwon, H. Y., Lee, J. Y., Choi, S. W., Jang, Y., & Kim, J. H. (2001). Preparation of PLGA nanoparticles containing estrogen by emulsification-diffusion method. *Colloids and Surfaces. A, Physicochemical and Engineering Aspects, 182*(1–3), 123–130. doi:10.1016/S0927-7757(00)00825-6

Laird, T. (2002). Design of experiments (DoE). *Organic Process Research & Development, 6*(4), 337. doi:10.1021/op0255581

Lander, R., Manger, W., Scouloudis, M., Ku, A., Davis, C., & Lee, A. (2000). Gaulin homogenization: A mechanistic study. *Biotechnology Progress, 16*(1), 80–85. doi:10.1021/bp990135c PMID:10662494

Langlet, B., Bach, M. T., Odegi, D., Fagerberg, P., & Ioakimidis, I. (2018). The effect of food unit sizes and meal serving occasions on eating behavior characteristics: Within person randomized crossover studies on healthy women. *Nutrients, 10*(7), 880. doi:10.3390/nu10070880 PMID:29986529

Larkin, P. (2011). *Infrared and Raman spectroscopy: principles and spectral interpretation.* Elsevier.

Lasa-Saracibar, B., Estella-Hermoso De Mendoza, A., Guada, M., Dios-Vieitez, C., & Blanco-Prieto, M. J. (2012). Lipid nanoparticles for cancer therapy: State of the art and future prospects. *Expert Opinion on Drug Delivery, 9*(10), 1245–1261. doi:10.1517/17425247.2012.717928 PMID:22950878

Lasch, P., Stämmler, M., Zhang, M., Baranska, M., Bosch, A., & Majzner, K. (2018). FT-IR hyperspectral imaging and artificial neural network analysis for identification of pathogenic bacteria. *Analytical Chemistry, 90*(15), 8896–8904. doi:10.1021/acs.analchem.8b01024 PMID:29944341

Latchubugata, C. S., Kondapaneni, R. V., Patluri, K. K., Virendra, U., & Vedantam, S. (2018). Kinetics and optimization studies using Response Surface Methodology in biodiesel production using heterogeneous catalyst. *Chemical Engineering Research & Design, 135*, 129–139. doi:10.1016/j.cherd.2018.05.022

Lauro, C. H., Pereira, R. B., Brandão, L. C., & Davim, J. P. (2016). Design of Experiments Statistical and artificial intelligence analysis for the improvement of machining processes: A review. In In Design of Experiments in Production Engineering (pp. 89–107). doi:10.1007/978-3-319-23838-8_3

Lee Seung, W., Jeung Min, K., Park Min, H., & Lee Soo, Y. (2010). Effects of roasting conditions of sesame seeds on the oxidative stability of pressed oil during thermal oxidation. *Food Chemistry, 118*(3), 681–685. doi:10.1016/j.foodchem.2009.05.040

Lee, L. C., Liong, C. Y., & Jemain, A. A. (2017). A contemporary review on Data Preprocessing (DP) practice strategy in ATR-FTIR spectrum. *Chemometrics and Intelligent Laboratory Systems, 163*, 64–75. doi:10.1016/j.chemolab.2017.02.008

Lee, L. C., Liong, C. Y., & Jemain, A. A. (2018). Partial least squares-discriminant analysis (PLS-DA) for classification of high-dimensional (HD) data: A review of contemporary practice strategies and knowledge gaps. *Analyst (London), 143*(15), 3526–3539. doi:10.1039/C8AN00599K PMID:29947623

Lemos-Senna, E., Wouessidjewe, D., Lesieur, S., & Duche, D. (1998). Preparation of amphiphilic cyclodextrin nanospheres using the emulsification solvent evaporation method. Influence of the surfactant on preparation and hydrophobic drug loading, *170*, 119–128.

Lennartsson, P. R., Erlandsson, P., & Taherzadeh, M. J. (2014). Integration of the first and second generation bioethanol processes and the importance of by-products. *Bioresource Technology, 165*, 3–8. doi:10.1016/j.biortech.2014.01.127 PMID:24582951

Leonard, G., Lafrenaye, S., & Goffaux, P. (2012). Randomized placebo-controlled cross-over designs in clinical trials: A gold standard to be reassessed. *Current Medical Research and Opinion, 28*(2), 245–248. doi:10.1185/03007995.2011.653558 PMID:22208674

Lessa, E. F., Medina, A. L., Ribeiro, A. S., & Fajardo, A. R. (2017). Removal of multi-metals from water using reusable pectin/cellulose microfibers composite beads. *Arabian Journal of Chemistry*. doi:10.1016/j.arabjc.2017.07.011

Levy, D. M., Abraham, R. R., & Tomlinson, D. R. (1991). Topical capsaicin in the treatment of painful diabetic neuropathy. *The New England Journal of Medicine, 324*(11), 776–777. doi:10.1056/NEJM199103143241116 PMID:1997848

Lewis, I. R., & Edwards, H. (2001). Handbook of Raman spectroscopy: from the research laboratory to the process line. Boca Raton, FL: CRC Press. doi:10.1201/9781420029253

Liang, Y., Fang, K., & Xu, Q. (2001). Uniform design and its applications in chemistry and chemical engineering. *Chemometrics and Intelligent Laboratory Systems, 58*(1), 43–57. doi:10.1016/S0169-7439(01)00139-3

Lim, S., & Cohenford, M. (2014, September). FT-IR spectra analysis toward cancer detection. In *Proceedings 2014 25th International Workshop on Database and Expert Systems Applications* (pp. 77-81). IEEE. 10.1109/DEXA.2014.31

Li, R., Dhankhar, D., Chen, J., Krishnamoorthi, A., Cesario, T. C., & Rentzepis, P. M. (2019). Identification of live and dead bacteria: A Raman spectroscopic study. *IEEE Access: Practical Innovations, Open Solutions, 7*, 23549–23559. doi:10.1109/ACCESS.2019.2899006

Li, S., Li, L., Zeng, Q., Zhang, Y., Guo, Z., Liu, Z., ... Liu, S. (2015). Characterization and noninvasive diagnosis of bladder cancer with serum surface enhanced Raman spectroscopy and genetic algorithms. *Scientific Reports*, *5*(1), 9582. doi:10.1038rep09582 PMID:25947114

Li, T., Tsung, Y., Hawkins, B. S., & Dickersin, K. (2015). Design, analysis and reporting of crossover trials for inclusion in a meta-analysis. *PLoS One*. doi:10.1371/journal.pone.0133023 PMID:26284684

Liu, W., Luo, Y., Teng, Y., Li, Z., & Ma, L. Q. (2010). Bioremediation of oily sludge-contaminated soil by stimulating indigenous microbes. *Environmental Geochemistry and Health*, *32*(1), 23–29. doi:10.100710653-009-9262-5 PMID:19363671

Liu, W., Sun, Z., Chen, J., & Jing, C. (2016). Raman spectroscopy in colorectal cancer diagnostics: Comparison of PCA-LDA and PLS-DA models. *Journal of Spectroscopy*, 2016.

Liu, Y. H., Zhan, H. W., Heng, L. J., & Ma, W.-X. (2013). Spectrophotometric determination of manganese in wine by oxidizing acid fuchsin. *Asian Journal of Chemistry*, *25*(4), 2270–2272. doi:10.14233/ajchem.2013.13824

Li, X., Yang, T., & Lin, J. (2012). Spectral analysis of human saliva for detection of lung cancer using surface-enhanced Raman spectroscopy. *Journal of Biomedical Optics*, *17*(3). doi:10.1117/1.JBO.17.3.037003 PMID:22502575

Li, Y., Huang, W., Pan, J., Ye, Q., Lin, S., Feng, S., ... Chen, R. (2015). Rapid detection of nasopharyngeal cancer using Raman spectroscopy and multivariate statistical analysis. *Molecular and Clinical Oncology*, *3*(2), 375–380. doi:10.3892/mco.2014.473 PMID:25798270

Li, Z., Li, C., Lin, D., Huang, Z., Pan, J., Chen, G., ... Chen, R. (2014). Surface-enhanced Raman spectroscopy for differentiation between benign and malignant thyroid tissues. *Laser Physics Letters*, *11*(4). doi:10.1088/1612-2011/11/4/045602

López, A., Aisa, J., Martinez, A., & Mercado, D. (2016). Injection moulding parameters influence on weight quality of complex parts by means of DOE application: Case study. *Measurement*, *90*, 349–356. doi:10.1016/j.measurement.2016.04.072

Lorenz, B., Wichmann, C., Stöckel, S., Rösch, P., & Popp, J. (2017). Cultivation-free Raman spectroscopic investigations of bacteria. *Trends in Microbiology*, *25*(5), 413–424. doi:10.1016/j.tim.2017.01.002 PMID:28188076

Lui, K. J. (2016). *Crossover designs: testing, estimation, and sample size*. Hoboken, N.J.: John Wiley & Sons; doi:10.1002/9781119114710

Lui, K. J. (2018). Sample size determination for testing equality in frequency data under an incomplete block crossover design. *Journal of Applied Statistics*, *45*(8), 1517–1529. doi:10.1080/02664763.2017.1380788

Lujan-Moreno, G. A., Howard, P. R., Rojas, O. G., & Montgomery, D. C. (2018). Design of experiments and response surface methodology to tune machine learning hyperparameters, with a random forest case-study. *Expert Systems with Applications*, *109*, 195–205. doi:10.1016/j.eswa.2018.05.024

Luna-Pizarro, P., Lopes-Almeida, E., Sammán, N. C., & Chang, Y. K. (2013). Evaluation of whole chia (*Salvia hispanica* L.) flour and hydrogenated vegetable fat in pound cake. *Lebensmittel-Wissenschaft + Technologie*, *54*(1), 73–79. doi:10.1016/j.lwt.2013.04.017

Lundkvist, P., Bergquist, B., & Vanhatalo, E. (2018). Statistical methods – still ignored? The testimony of Swedish alumni. *Total Quality Management & Business Excellence*, *0*, 1–18. doi:10.1080/14783363.2018.1426449

Lu, Y. C. (2011). Enhancing the students' abilities in experimental designs through Design-Expert. In *Proceedings 2011 ASEE Annual Conference & Exposition*. ASEE.

Lye, L. M. (2005). *Tools and toys for teaching design of experiments methodology. 33rd Annual General Conference of the Canadian Society for Civil Engineering*, Toronto, Canada.

Lynk, T. P., Sit, C. S., & Brosseau, C. L. (2018). Electrochemical surface-enhanced Raman spectroscopy as a platform for bacterial detection and identification. *Analytical Chemistry*, *90*(21), 12639–12646. doi:10.1021/acs.analchem.8b02806 PMID:30350616

MacCalman, A. D., Vieira, H., & Lucas, T. (2017). Second-order nearly orthogonal Latin hypercubes for exploring stochastic simulations. *Journal of Simulation*, *11*(2), 137–150. doi:10.1057/jos.2016.8

Mackanos, M. A., & Contag, C. H. (2010). Fiber-optic probes enable cancer detection with FTIR spectroscopy. *Trends in Biotechnology*, *28*(6), 317–323. doi:10.1016/j.tibtech.2010.04.001 PMID:20452071

Macnamara, J. (1969). *The Irish Journal of Education / Iris Eireannach an Oideachais, 3*(1), 71-73. Retrieved from http://www.jstor.org/stable/30076687

Madhavi, S. K., Sreeramulu, D., & Venkatesh, M. (2017). Evaluation of optimum turning process of process parameters using DOE and PCA Taguchi method. *Materials today: Proceedings*, *4*(2), 1937–1946.

Mahant, S., Rao, R., & Nanda, S. (2018). *Nanostructured lipid carriers: Revolutionizing skin care and topical therapeutics. Design of nanostructures for versatile therapeutic applications*. Elsevier; doi:10.1016/B978-0-12-813667-6.00003-6

Mahdi Jafari, S., Masoudi, S., & Bahrami, A. (2019). A Taguchi approach production of spray-dried whey powder enriched with nanoencapsulated vitamin D3. *Drying Technology*, 1–13.

Maity, S. K. (2015). Opportunities, recent trends and challenges of integrated biorefinery: Part I. *Renewable & Sustainable Energy Reviews*, *43*, 1427–1445. doi:10.1016/j.rser.2014.11.092

Malam, Y., Loizidou, M., & Seifalian, A. M. (2009). Liposomes and nanoparticles: Nanosized vehicles for drug delivery in cancer. *Trends in Pharmacological Sciences*, *30*(11), 592–599. doi:10.1016/j.tips.2009.08.004 PMID:19837467

Maldonado-Chávez, E., Rivera-Cruz, M. C., Izquierdo-Reyes, F., & Palma-López, D. J. (2010). Efectos de rizósfera, microorganismos y fertilización en la biorremediación y fitorremediación de suelos con petróleos crudo nuevo e intemperizado. *Universidad y ciencia, 26*(2), 121-136.

Malik, D., & Pakzad, L. (2018). Experimental investigation on an aerated mixing vessel through electrical resistance tomography (ERT) and response surface methodology (RSM). *Chemical Engineering Research & Design*, *129*, 327–343. doi:10.1016/j.cherd.2017.11.002

Mandal, V., Mohan, Y., & Hemalatha, S. (2008). Microwave assisted extraction of curcumin by sample–solvent dual heating mechanism using Taguchi L9 orthogonal design. *Journal of Pharmaceutical and Biomedical Analysis*, *46*(2), 322–327. doi:10.1016/j.jpba.2007.10.020 PMID:18309573

Mandenius, C.-F. (2016, March 29). Challenges for bioreactor design and operation. *Bioreactors*.

Mante, A., Heider, M., Zlomke, C., & Mäder, K. (2016). PLGA nanoparticles for peroral delivery: How important is pancreatic digestion and can we control it? *European Journal of Pharmaceutics and Biopharmaceutics*, *108*, 32–40. doi:10.1016/j.ejpb.2016.08.009 PMID:27553262

Marín-García, D. C., Adams, R. H., & Hernández-Barajas, R. (2016). Effect of crude petroleum on water repellency in a clayey alluvial soil. *International Journal of Environmental Science and Technology*, *13*(1), 55–64. doi:10.100713762-015-0838-6

Mariod, A. A., Abdelwahab, S. I., Gedi, M. A., & Solati, Z. (2010). Supercritical carbon dioxide extraction of sorghum bug (Agonoscelis pubescens) oil using Response surface methodology. *JAOCS. Journal of the American Oil Chemists' Society*, *87*(8), 849–856. doi:10.100711746-010-1565-2

Marston, L., & Sedgwich, P. (2014). Randomised controlled trials: Missing data. *British Medical Journal, 349*(jul18 2), g4656. doi:10.1136/bmj.g4656 PMID:25038075

Martin, T., Elisabeth, V., & Lourdes, P. (2009). Diferentes Enfoques de Diseño de Experimentos. Memoria de trabajos de difusión cientifica y técnica, 19 - 37.

Martínez, A. M. L., Marcos, J. M. F., Manzanares, M. V., Urbina, E. M. C., Iliana, M., & Hernández, J. A. (2018). Cuantificación de etanol en gasolina por microextracción líquido-líquido y valoración potenciométrica. *Revista de Ciencia y Tecnología, 34*(1).

Martínez-Chávez, C. G., Morales-Bautista, C. M., & Alor-Chávez, M. D. J. (2017). Extracción de hidrocarburo pesado en suelo arenoso. *Avances en Ciencias e Ingeniería, 8*(1).

Martínez-Lara, J. M., & Páez Melo, M. I. (2017). Diseño de experimentos aplicado en la optimización del método de extracción QuEChERS para la determinación de plaguicidas organoclorados y organofosforados en suelos. *Revista Internacional de Contaminación Ambiental, 33*(4), 559–573. doi:10.20937/RICA.2017.33.04.02

Martins, Q. S., Aguirre, C. A., & Farias, J. L. B. (2019). Approach by Raman and infrared spectroscopy in three vegetable oils from the Brazilian Amazon. *Revista Mexicana de Física, 65*(4 Jul-Aug), 328-332.

Masetto, A., Flores-Cotera, L. B., Díaz, C., Langley, E., & Sanchez, S. (2001). Application of a complete factorial design for the production of zeaxanthin by Flavobacterium sp. *Journal of Bioscience and Bioengineering, 92*(1), 55–58. doi:10.1016/S1389-1723(01)80199-7 PMID:16233058

Massart, D. L., Vandeginste, B. G. M., Buydens, L. M. C., De Jong, S., Lewi, P. J., & Smeyers-Verbeke. (1997). In Handbook of chemometrics and qualimetrics: part A. Amsterdam, The Netherlands: Elsevier.

Massart, D. L., Vandegiste, B. G. M., Buydens, L. M. C., Lewi, S., de Jong, P. J., & Smeyers-Verbeke, J. (1997). Handbook of chemometrics and qualimetrics. Amsterdam, The Netherlands: Elsevier.

Mathimani, T., & Pugazhendhi, A. (2019). Utilization of algae for biofuel, bio-products and bio-remediation. *Biocatalysis and Agricultural Biotechnology, 17*, 326–330. doi:10.1016/j.bcab.2018.12.007

Ma, X., & Ouyang, F. (2013). Adsorption properties of biomass-based activated carbon prepared with spent coffee grounds and pomelo skin by phosphoric acid activation. *Applied Surface Science, 268*, 566–570. doi:10.1016/j.apsusc.2013.01.009

Mažeikiene, A., Rimeika, M., Valentukevičiene, M., Oškinis, V., Paškauskaite, N., & Brannvall, E. (2005). Removal of petroleum products from water using natural sorbent zeolite. *Journal of Environmental Engineering and Landscape Management, 13*(4), 187–191. doi:10.3846/16486897.2005.9636870

McClary, D. W., Syrotiuk, V. R., & Kulahci, M. (2010). Steepest-ascent constrained simultaneous perturbation for multiobjective optimization. *ACM Transactions on Modeling and Computer Simulation, 21*(1), 1–22. doi:10.1145/1870085.1870087

McCrickerd, K., & Forde, C. G. (2017). Consistency of eating rate, oral processing behaviors and energy intake across meals. *Nutrients, 9*(8), 891. doi:10.3390/nu9080891 PMID:28817066

Medina, P. D., & López, A. M. (2011). Análisis crítico de Diseño factorial 2k sobre Casos Aplicados. *Sciences et Techniques (Paris)*, 101–106.

Mehlig, J. (1939). Colorimetric determination of manganese with periodate. *Industrial & Engineering Chemistry, 11*(5), 274–278.

Mehnert, W., & Mäder, K. (2012). Solid lipid nanoparticles: Production, characterization and applications. *Advanced Drug Delivery Reviews, 64*(SUPPL.), 83–101. doi:10.1016/j.addr.2012.09.021 PMID:11311991

Meisel, S., Stöckel, S., Elschner, M., Melzer, F., Rösch, P., & Popp, J. (2012). Raman spectroscopy as a potential tool for detection of Brucella spp. in milk. *Applied and Environmental Microbiology, 78*(16), 5575–5583. doi:10.1128/AEM.00637-12 PMID:22660699

Meksiarun, P., Ishigaki, M., Huck-Pezzei, V. A., Huck, C. W., Wongravee, K., Sato, H., & Ozaki, Y. (2017). Comparison of multivariate analysis methods for extracting the paraffin component from the paraffin-embedded cancer tissue spectra for Raman imaging. *Scientific Reports, 7*(1), 44890. doi:10.1038rep44890 PMID:28327648

Melendez, L. V., Lache, A., Orrego-Ruiz, J. A., Pachon, Z., & Mejía-Ospino, E. (2012). Prediction of the SARA analysis of Colombian crude oils using ATR–FTIR spectroscopy and chemometric methods. *Journal of Petroleum Science Engineering, 90*, 56–60. doi:10.1016/j.petrol.2012.04.016

Mendoza, C. (1542). *Edition of Francisco del Paso y Troncoso* (p. 1825). Museo Nacional de Arqueología, Historia y Etnología.

Meneghetti, S. M. P., Meneghetti, M. R., Serra, T. M., Barbosa, D. C., & Wolf, C. R. (2007). Biodiesel production from vegetable oil mixtures: Cottonseed, soybean, and castor oils. *Energy & Fuels, 21*(6), 3746–3747. doi:10.1021/ef070039q

Merchand, M. A. (2015). Estado y reforma energética en México. *Problemas del Desarrollo, 46*(183), 117–139. doi:10.1016/j.rpd.2015.10.006

Merkl, N., Schultze-Kraft, R., & Infante, C. (2005). Assessment of tropical grasses and legumes for phytoremediation of petroleum-contaminated soils. *Water, Air, and Soil Pollution, 165*(1-4), 195–209. doi:10.100711270-005-4979-y

Miller, J. N., & Miller, J. C. (2010) *Statistics and chemometrics for analytical chemistry*, Pearson Education Limited, Exxes.

Mills, E. J., An-Wen, C., Wu, P., Vail, A., Guyatt, G. H., & Altman, D. G. (2009). Design, analysis, and presentation of crossover trials. *Trials, 10*(1), 27. doi:10.1186/1745-6215-10-27 PMID:19405975

Miranda, A. K. D. C. E. (2009). *Robust Parameter Design Optimization of Simulation Experiments Using Stochastic Perturbation Methods*. Academic Press.

Miranda, A. K., & Del Castillo, E. (2010). Robust parameter design optimization of simulation experiments using stochastic perturbation methods. *Journal of the Operational Research Society, 62*(1), 198–205. doi:10.1057/jors.2009.163

Miranda-Ackerman, M. A., Azzaro-Pantel, C., Montastruc, L., Domenech, S., & Lasserre, A. A. A. (2013). Optimal design of a green supply chain: application to the agro-food industry supplier selection problem. Coll. Récents Progrès En Génie Des Procédés, N° 104 Les Sciences Du Génie Des Procédés Pour Une Industrie Durable XIVe Congrès SFGP 2013 8 Au 10 Octobre 2013, Lyon, 104. Société Française de Génie des Procédés.

Miranda-Ackerman, M. A., Azzaro-Pantel, C., & Aguilar-Lasserre, A. A. (2017). A green supply chain network design framework for the processed food industry: Application to the orange juice agrofood cluster. *Computers & Industrial Engineering, 109*, 369–389. doi:10.1016/j.cie.2017.04.031

Miranda, J. M., Anton, X., Redondo-Valbuena, C., Roca-Saavedra, P., Rodriguez, J. A., Lamas, A., ... Cepeda, A. (2015). Egg and egg-derived foods: Effects of human health and use as functional foods. *Nutrients, 7*(1), 706–729. doi:10.3390/nu7010706 PMID:25608941

Mitić, M., Tošić, S., Pavlović, A., Mašković, P., Kostić, D., Mitićc, J., & Stevanović, V. (2019). Optimization of the extraction process of minerals from *Salvia officinalis* L. using factorial design methodology. *Microchemical Journal, 145*, 1224–1230. doi:10.1016/j.microc.2018.12.047

Mitu, R. K. (2012). *Process validation SOP template*. ROFRI MED.

Mobili, P., Londero, A., De Antoni, G., Gomez-Zavaglia, A., Araujo-Andrade, C., Avila-Donoso, H., ... Frausto-Reyes, C. (2010). Multivariate analysis of Raman spectra applied to microbiology: Discrimination of microorganisms at the species level. *Revista Mexicana de Física, 56*(5), 378–385.

Moise, G. (2015). Method of analysis for the determination of heavy metals from wine. *Scientific Papers. Series Management, Economic, Engineering in Agriculture, and Rural Development, 15*(3), 201–204.

Moisoiu, V., Socaciu, A., Stefancu, A., Iancu, S. D., Boros, I., Alecsa, C. D., ... Socaciu, C. (2019). Breast cancer diagnosis by Surface-Enhanced Raman Scattering (SERS) of urine. *Applied Sciences, 9*(4), 806. doi:10.3390/app9040806

Molina, A. G., & Diego, S. M. M. (2016). Avanzada científica. Avanzada Científica, 19(1), 2016 (Ejemplar dedicado a: ENERO-FEBRERO-MARZO-ABRIL), pp. 1-11. Centro de Información y Gestión Tecnológica.

Montgomery, D. C. (2002). Changing roles for the industrial statisticians. *Quality and Reliability Engineering International, 18*(5).

Montgomery, D. C. (2001). Opportunities and challenges for industrial statisticians. *Journal of Applied Statistics, 28*(3-4), 427–439. doi:10.1080/02664760120034153

Montgomery, D. C. (2005). *Design and analysis of experiment* (5th ed.). John Wiley & Sons.

Montgomery, D. C. (2005). *Design and Analysis of Experiments*. Phoenix, AZ: John Wiley & Sons.

Montgomery, D. C. (2005a). Generation III Six Sigma. *Quality and Reliability Engineering International, 21*(6), 3–4. doi:10.1002/qre.751

Montgomery, D. C. (2005b). *Design and analysis of experiments*. John Wiley & Sons.

Montgomery, D. C. (2009). *Introduction to statistical quality control* (6th ed.). Jefferson City: John Wiley & Sons.

Montgomery, D. C. (2009). *Statistical quality control*. New York: Wiley.

Montgomery, D. C. (2012). *Design and Analysis of Experiments (8th ed.)*. John Wiley & Sons.

Montgomery, D. C. (2017). *Design and analysis of experiments (9th)*. Hoboken, NJ: Wiley.

Montgomery, D. C. (2017). *Montgomery: design and analysis of experiments. John Wiley & Sons* (8th ed.)

Morales-Bautista, C. M., Adams, R. H., & Guzmán-Osorio, F. D. Marín-García. (2013). Dilution-extrapolation hydrometer method for easy determination of API gravity of heavily weathered hydrocarbons in petroleum contaminated soil. Energy and Environmental Research, 3(1), 115–124. doi:10.5539/eer.v3n1p115 doi:10.5539/eer.v3n1p115

Morales-Bautista, C. M., Adams, R. H., Hernández-Barajas, J. R., Lobato-García, C. E., & Torres-Torres, J. G. (2016). Characterization of fresh and weathered petroleum for potential impacts to soil fertility. *International Journal of Environmental Science and Technology, 13*(11), 2689–2696. doi:10.100713762-016-1097-x

Morales-Bautista, C., Méndez-Olán, C., Hernández-Jiménez, G., & Adams-Schroeder, R. (2018). Normalization of calibration curves and their application in the determination of total oil hydrocarbons (HTP). *Revista Técnica de la Facultad de Ingeniería. Universidad del Zulia, 41*(3), 116–124.

Morales, S. A. (2007). *Elaboración de un alimento funcional tipo palanqueta*. D.F., México: Tesis de Licenciatura. Escuela Nacional de Ciencias Biológicas. IPN.

Moreno, L. A., Joyanes, M., Mesana, M. I., González-Gross, M., Gil, C. M., Gutiérrez, A., ... Marcos, A. (2006). Harmonization of anthropometric measurements for a multicenter nutrition survey in Spanish adolescents. *Nutrition (Burbank, Los Angeles County, Calif.), 19*(6), 481–486. doi:10.1016/S0899-9007(03)00040-6 PMID:12781845

Moss, D. (2011). *Biomedical applications of Synchrotron Infrared Microspectroscopy: A practical approach*. Royal Society of Chemistry.

Mostaço-Guidolin, L. B., & Bachmann, L. (2011). Application of FTIR spectroscopy for identification of blood and leukemia biomarkers: A review over the past 15 years. *Applied Spectroscopy Reviews*, *46*(5), 388–404. doi:10.1080/05704928.2011.565534

Mostafaei, M., Javadikia, H., & Naderloo, L. (2016). Modeling the effects of ultrasound power and reactor dimension on the biodiesel production yield: Comparison of prediction abilities between response surface methodology (RSM) and adaptive neuro-fuzzy inference system (ANFIS). *Energy*, *115*, 626–636. doi:10.1016/j.energy.2016.09.028

Mukherjee, S., Ray, S., & Thakur, R. S. (2009). Solid lipid nanoparticles: A modern formulation approach in drug delivery system. *Indian Journal of Pharmaceutical Sciences*, *71*(4), 349–358. doi:10.4103/0250-474X.57282 PMID:20502539

Müller, R. H., Mder, K., & Gohla, S. (2000). Solid lipid nanoparticles (SLN) for controlled drug delivery - A review of the state of the art. *European Journal of Pharmaceutics and Biopharmaceutics*, *50*(1), 161–177. doi:10.1016/S0939-6411(00)00087-4 PMID:10840199

Muthukumar, M., & Mohan, D. (2005). Studies on furan polymer concrete. *Journal of Polymer Research*, *12*(3), 231–241. doi:10.100710965-004-3206-7

Myers, R. H., Khuri, A. I., & Carter, W. H. (1989). Response surface methodology: 1966-1988. *Technometrics*, *31*, 137–157.

Myers, R. H., Montgomery, C. D., & Anderson-Cook, C. M. (2016). *Response surface methodology: process and product optimization using designed experiments* (4th ed.). Wiley.

Nair, V. N., Abraham, B., MacKay, J., Box, G., Kacker, R. N., Lorenzen, T. J., ... Jeff Wu, C. F. (1992). Taguchi's parameter design: A panel discussion. *Technometrics*, *31*(2), 127–161. doi:10.1080/00401706.1992.10484904

Nakamura, Y., Mochida, A., Choyke, P. L., & Kobayashi, H. (2016). Nanodrug delivery: Is the enhanced permeability and retention effect sufficient for curing cancer? *Bioconjugate Chemistry*, *27*(10), 2225–2238. doi:10.1021/acs.bioconjchem.6b00437 PMID:27547843

Namany, S., Al-Ansari, T., & Govindan, R. (2019). Sustainable energy, water and food nexus systems: A focused review of decision-making tools for efficient resource management and governance. *Journal of Cleaner Production*, *225*, 610–626. doi:10.1016/j.jclepro.2019.03.304

Narenderan, S. T., Meyyanathan, S. N., & Karri, V. V. S. R. (2019). Experimental design in pesticide extraction methods: A review. *Food Chemistry*, *289*, 384–395. doi:10.1016/j.foodchem.2019.03.045 PMID:30955627

Nataraj, M. (2005). Optimizing diesel engine parameters for low emissions using Taguchi Method: Variation risk analysis approach –part 1. *Indian Journal of Engineering and Materials Sciences*, *12*(3), 169–181.

Naumann, D. (2000). Infrared spectroscopy in microbiology. Encyclopedia of analytical chemistry.

Navarro, P., Etxebarria, N., & Arana, G. (2009). Development of a focused ultrasonic-assisted extraction of polycyclic aromatic hydrocarbons in marine sediment and mussel samples. *Analytica Chimica Acta*, *648*(2), 178–182. doi:10.1016/j.aca.2009.06.062 PMID:19646582

Naveena, B. J., Altaf, M., Bhadriah, K., & Reddy, G. (2005). Selection of medium components by Plackett–Burman design for production of L(+) lactic acid by Lactobacillus amylophilus GV6 in SSF using wheat bran. *Science Direct*, (96), 485-490.

Nesakumar, N., Baskar, C., Kesavan, S., Rayappan, J. B. B., & Alwarappan, S. (2018). Analysis of moisture content in beetroot using fourier transform infrared spectroscopy and by principal component analysis. *Scientific Reports*, *8*(1), 7996. doi:10.103841598-018-26243-5 PMID:29789563

Neugebauer, U., Roesch, P., & Popp, J. (2015). Raman spectroscopy towards clinical application: Drug monitoring and pathogen identification. *International Journal of Antimicrobial Agents*, *46*, S35–S39. doi:10.1016/j.ijantimicag.2015.10.014 PMID:26612228

Nguyen, V. H. (2015). *Scanning angle Raman spectroscopy in polymer thin film characterization (No. IS-T 3192). Ames Laboratory*. Ames, IA: AMES. doi:10.2172/1409182

Niad, M., Zaree, S., & Tahanzadeh, N. (2016). Response surface methodology for optimization of Cd (II) biosorption by *Cystoseria myricaas*. *Journal of Biomedicine and Health*, *1*(2), 6264.

Niedz, R. P., & Evens, T. J. (2016). Design of experiments (DOE)—history, concepts, and relevance to in vitro culture. *In Vitro Cellular & Developmental Biology. Plant*, *52*(6), 547–562. doi:10.100711627-016-9786-1

Nissy, S. M., Surendra, B., Raju, K. D., & Vangalapati, M. (2018). Optimization for the extraction of Bisphenol A; Response surface Methodology. *Materials Today: Proceedings*, *5*, 17914–17923. doi:10.1016/j.matpr.2018.06.120

NIST/SEMATECH. (2012). What is design of experiments (DOE)? Engineering Statistics Handbook. e-Handbook of Statistical Methods. Retrieved from https://www.itl.nist.gov/div898/handbook/pmd/section3/pmd31.htm

NIST/SEMATECH. (2013). *How do you select an experimental design?* Retrieved from https://www.itl.nist.gov/div898/handbook/pri/section3/pri33.htm#DesignSelection

NMX-F-592-SCFI-2017. (2017). *Aceites y grasas vegetales o animales-aceite de semilla de chia-especificaciones*. Secretaría de Economía, Subsecretaría de Competitividad y Normatividad, Dirección General de Normas.

NOM-199-SCFI-2017. (2017). Bebidas alcohólicas-Denominación, especificaciones fisicoquímicas, información comercial y métodos de prueba. *Norma Oficial Mexicana, Diario Oficial de la Federación*, 1-37.

Noorbakhsh-Soltani, S. M., Zerafat, M. M., & Sabbaghi, S. (2018). A comparative study of gelatin and starch-based nano-composite films modified by nano-cellulose and chitosan for food packaging applications. *Carbohydrate Polymers*, *189*, 48–55. doi:10.1016/j.carbpol.2018.02.012 PMID:29580425

Nyquist, R. A. (2001). *Interpreting infrared, Raman, and nuclear magnetic resonance spectra*. Academic Press.

O'Donnell, C. P., Fagan, C., & Cullen, P. J. (Eds.). (2014). *Process analytical technology for the food industry*. Springer. doi:10.1007/978-1-4939-0311-5

Ofer, R., Yerachmiel, A., & Shmuel, Y. (2003). Marine macroalgae as biosorbents for cadmium and nickel in water. *Water Environment Research*, *75*(3), 246–253. doi:10.2175/106143003X141033 PMID:12837031

OI Analytical, a xylem brand, Analysis of Thiophene in Benzene using ASTM Method D7011 and a Pulsed Flame Photometric Detector (PFPD), Publication 43600916.

Ojeda, J. J., & Dittrich, M. (2012). Fourier transform infrared spectroscopy for molecular analysis of microbial cells. In *Microbial Systems Biology* (pp. 187–211). Totowa, NJ: Humana Press. doi:10.1007/978-1-61779-827-6_8

Ojeda-Morales, M. E., Hernández-Rivera, M. Á., Martínez-Vázquez, J. G., Díaz-Flores, L. L., & Rivera-Cruz, M. D. C. (2012). Tres dosis inoculantes con Proteus sp en la biodegradación de petróleo crudo. *Revista Internacional de Contaminación Ambiental*, *28*(4), 323–332.

Okoro, O. V., Sun, Z., & Birch, J. (2019). Thermal depolymerization of biogas digestate as a viable digestate processing and resource recovery strategy. In *Advances in eco-fuels for a sustainable environment* (pp. 277–308). Woodhead Publishing; doi:10.1016/B978-0-08-102728-8.00010-3

Olguín Tiznado, J. E., García Martínez, R., Camargo Wilson, C., & López Barreras, J. A. (2011). Stochastic approximation algorithm for industrial process optimisation. *Ingenieria e Investigacion*, *31*(3), 100–111.

Omar, W. N. N. W., & Amin, N. A. S. (2011). Optimization of heterogeneous biodiesel production from waste cooking palm oil via response surface methodology. *Biomass and Bioenergy*, *35*(3), 1329–1338. doi:10.1016/j.biombioe.2010.12.049

Ong, S. K., Yuan, M. L., & Nee, A. Y. C. (2008). Augmented reality applications in manufacturing: a survey. *International Journal of Production Research*, *46*(10), 2707–2742. doi:10.1080/00207540601064773

ONUDI. (2015). Organización de las Naciones Unidas para el Desarrollo Industrial. Informe sobre el Desarrollo Industrial 2016. El rol de la tecnología y la innovación en el desarrollo industrial inclusivo y sostenible. Resumen. Viena. Número de identificación de ONUDI: 447. Retrieved from https://www.unido.org/sites/default/files/2015-12/EBOOK_IDR2016_OVERVIEW_SPANISH_0.pdf

Onukwuli, D. O., Emembolu, L. N., Ude, C. N., Aliozo, S. O., & Menkiti, M. C. (2017). Optimization of biodiesel production from refined cotton seed oil and its characterization. *Egyptian Journal of Petroleum*, *26*(1), 103–110. doi:10.1016/j.ejpe.2016.02.001

Orellana, S., Johansen, A. M., & Gazis, C. (2019). Geographic classification of U.S. Washington State wines using elemental and water isotope composition. *Food Chemistry*, *X*(100007), 1–8. PMID:31432007

Ortega, M., Pérez-Rodrigo, C., & López-Sobaler, A. M. (2015). Dietary assessment methods: Dietary reports. *Nutrición Hospitalaria*, *31*, 38–45. doi:10.3305/nh.2015.31.sup3.8749 PMID:25719769

Ortiz-Salinas, R., Cram, S., & Sommer, I. (2012). Hidrocarburos aromáticos policíclicos (HAPs) en suelos de la llanura aluvial baja del estado de Tabasco, México. *Universidad y ciencia*, *28*(2), 131-144.

Otto, M. (2017). *CHEMOMETRICS: Statistics and computer application in analytical chemistry*. Weinheim, Germany: Wiley-VCH.

Ou, S., Li, C., & Wei-Xing, M. (2013). Spectrophotometric determination of manganese in cooking wine samples by oxidation of chromotrope 2B. *Food Science China*, *34*(12), 156–158.

Ou, Z., Zhang, M., & Qin, H. (2019). Tripling of fractional factorial designs. *Journal of Statistical Planning and Inference*, *199*, 151–159. doi:10.1016/j.jspi.2018.06.002

Ozaki, Y., McClure, W. F., & Christy, A. A. (Eds.). (2006). *Near-infrared spectroscopy in food science and technology*. John Wiley & Sons. doi:10.1002/0470047704

Ozbay, N., & Yargic, A. S. (2018). Statistical analysis of Cu(II) and Co(II) sorption by apple pulp carbon using factorial design approach. *Journal of Industrial and Engineering Chemistry*, *57*, 275–283. doi:10.1016/j.jiec.2017.08.033

Padilla, D. (2008). *Aplicación del Diseño de Experimentos a los ensayos realizados en la Unidad de Negocio PROPAGACION, El Quinche*. Escuela Superior Politécnica de Chimborazo, Facultad de Ciencias, Escuela de Física y Matemática.

Pahlow, S., Meisel, S., Cialla-May, D., Weber, K., Rösch, P., & Popp, J. (2015). Isolation and identification of bacteria by means of Raman spectroscopy. *Advanced Drug Delivery Reviews*, *89*, 105–120. doi:10.1016/j.addr.2015.04.006 PMID:25895619

Pahlow, S., Weber, K., Popp, J., Bayden, R. W., Kochan, K., Rüther, A., ... Gardner, B. (2018). Application of vibrational spectroscopy and imaging to point-of-care medicine: A review. *Applied Spectroscopy*, *72*(101), 52–84. PMID:30265133

Palma-Cruz, F. D. J., Pérez-Vargas, J., Casado, N. A. R., Guzmán, O. G., & Calva-Calva, G. (2016). Phytoremediation potential and ecological and phenological changes of native pioneer plants from weathered oil spill-impacted sites at tropical wetlands. *Environmental Science and Pollution Research International*, *23*(16), 16359–16371. doi:10.100711356-016-6675-4 PMID:27164872

Palma-López, D. J., Cisneros, D. J., Moreno, C. E., & Rincón-Ramírez, J. A. (2007). *Suelos de Tabasco: su uso y manejo sustentable* (p. 195). Villahermosa, México: Colegio de Postgraduados-ISPROTAB-FUPROTAB.

Palmeros Parada, M., Osseweijer, P., & Posada Duque, J. A. (2017). Sustainable biorefineries, an analysis of practices for incorporating sustainability in biorefinery design. *Industrial Crops and Products*, *106*, 105–123. doi:10.1016/j.indcrop.2016.08.052

Pande, P., Mewman, R., & Cavanagh, R. (2000). *The six sigma way*. McGraw-Hill.

Pander, A., Hatta, A., & Furuta, H. (2016). Optimization of catalyst formation conditions for synthesis of carbon nanotubes using Taguchi method. *Applied Surface Science*, *371*, 425–435. doi:10.1016/j.apsusc.2016.02.216

Pandey, M. P., & Kim, C. S. (2011). Lignin depolymerization and conversion: a review of thermochemical methods. *Chemical Engineering & Technology*, *34*(1), 29–41. doi:10.1002/ceat.201000270

Pandiarajan, S., Kumaran, S. S., Kumaraswamidhas, L. A., & Saravanan, R. (2016). Interfacial microstructure and optimization of friction welding by Taguchi and ANOVA method on SA 213 tube to SA 387 tube plate without backing block using an external tool. *Journal of Alloys and Compounds*, *654*, 534–545. doi:10.1016/j.jallcom.2015.09.152

Paris Agreement, U. N. (2016). *The Paris Agreement*. Retrieved from https://unfccc.int/process-and-meetings/the-paris-agreement/the-paris-agreement

Pavel, L. V., & Gavrilescu, M. (2008). Overview of ex situ decontamination techniques for soil cleanup. [EEMJ]. *Environmental Engineering and Management Journal*, *7*(6), 815–834. doi:10.30638/eemj.2008.109

Pavia, D. L., Lampman, G. M., Kriz, G. S., & Engel, R. G. (2018). *A microscale approach to organic laboratory techniques*. Cengage Learning.

Pearson, K. (1901). LIII. On lines and planes of closest fit to systems of points in space. *The London, Edinburgh, and Dublin Philosophical Magazine and Journal of Science*, *2*(11), 559–572. doi:10.1080/14786440109462720

Pérez-Alonso, M., Castro, K., & Madariaga, J. M. (2006). Vibrational spectroscopic techniques for the analysis of artefacts with historical, artistic and archaeological value. *Current Analytical Chemistry*, *2*(1), 89–100. doi:10.2174/157341106775197385

Pérez-Hernández, I., Ochoa-Gaona, S., Adams, R. H., Rivera-Cruz, M. C., Pérez-Hernández, V., Jarquín-Sánchez, A., & Martínez-Zurimendi, P. (2017). Growth of four tropical tree species in petroleum-contaminated soil and effects of crude oil contamination. *Environmental Science and Pollution Research International*, *24*(2), 1769–1783. doi:10.100711356-016-7877-5 PMID:27796985

Pérez-Hernández, I., Ochoa-Gaona, S., Schroeder, R. A., Rivera-Cruz, M. C., & Geissen, V. (2013). Tolerance of four tropical tree species to heavy petroleum contamination. *Water, Air, and Soil Pollution*, *224*(8), 1637. doi:10.100711270-013-1637-7

Pignatello, J. (1991). Top ten triumphs and tragedies of Genechi Taguchi. *Quality Engineering*, *4*(2), 211–225. doi:10.1080/08982119108918907

Pindado, O., Pérez, R. M., & García, S. (2014). *Análisis del TPH y las fracciones de hidrocarburos alifáticos y aromáticos en matrices de interés medioambiental*. Editorial CIEMAT.

Pineau, N., Moser, M., Rawyler, F., Lepage, M., Antille, N., & Rytz, A. (2019). Design of experiment with sensory data: A pragmatic data analysis approach. *Journal of Sensory Studies*, *34*(2). doi:10.1111/joss.12489

Pineda Becerril, M., García, O., Aguilar, A., & León, F. (2017). Teaching the regression and simple linear correlation theme with e-learning support. In L. Gómez Chova, A. L. Martínez, & I. C. Torres (Eds.), *9th International Conference on Education and New Learning Technologies - EDULEARN17* (pp. 6324–6330). 10.21125/edulearn.2017.2034

Pino, V., Ayala, J. H., Afonso, A. M., & González, V. (2001). Ultrasonic micellar extraction of polycyclic aromatic hydrocarbons from marine sediments. *Talanta*, *54*(1), 15–23. doi:10.1016/S0039-9140(00)00622-6 PMID:18968221

Plackett, R. L., & Burman, J. P. (1946). The design of optimum multifactorial experiments. *Biometrika*, *33*(4), 305–325. doi:10.1093/biomet/33.4.305

Platanitis, G., & Pop-Iliev, R. (2010). Early introduction of robust design into the engineering curriculum. *Proceedings of the ASME International Design Engineering Technical Conferences and Computers and Information in Engineering Conference, DETC 2010, VOL 6*, 599–606. ASME. 10.1115/DETC2010-28221

Pohl, P. (2007). What do metals tell us about wine? *Trends in Analytical Chemistry*, *26*(9), 941–949. doi:10.1016/j.trac.2007.07.005

Polewczyk, A., Marchut-Mikolajczyk, O., & Smigielski, K. B. (2015). Diesel fuel extraction from soil. *Acta Scientiarum Polonorum. Formatio Circumiectus*, *14*(2), 165–173. doi:10.15576/ASP.FC/2015.14.2.165

Pons-Jiménez, M., Guerrero-Peña, A., Zavala-Cruz, J., & Alarcón, A. (2011). Extracción de hidrocarburos y compuestos derivados del petróleo en suelos con características físicas y químicas diferentes. *Universidad y ciencia*, *27*(1), 1-15.

Portalagrario, B. (2014, Dec. 10) *Comercialización de aceite de chía (Salvia hispanica) como ingrediente alimentario.* Retrieved from http://www.besana.es/es/web/201412/comercializacion-aceite-chia

Prasad, V., & Grady, C. (2014). The misguided ethics of crossover trials. *Contemporary Clinical Trials*, *37*(2), 167–169. doi:10.1016/j.cct.2013.12.003 PMID:24365533

Preetha, B., & Viruthagiri, T. (2007). Application of response surface methodology for the biosorption of copper using *Rhizopus arrhizus*. *Journal of Hazardous Materials*, *143*(1-2), 506–510. doi:10.1016/j.jhazmat.2006.09.077 PMID:17084526

Priyanka, K., & Sathali, A. A. H. (2012). Preparation and evaluation of montelukast sodium loaded solid lipid nanoparticles. *Journal of Young Pharmacists: JYP*, *4*(3), 129–137. doi:10.4103/0975-1483.100016 PMID:23112531

Prow, T. W., Grice, J. E., Lin, L. L., Faye, R., Butler, M., Becker, W., ... Roberts, M. S. (2011). Nanoparticles and microparticles for skin drug delivery. *Advanced Drug Delivery Reviews*, *63*(6), 470–491. doi:10.1016/j.addr.2011.01.012 PMID:21315122

Pulido, H. (2004). *Analisis y diseño de experimentos.* McGraw-Hill, Interamericana.

Pulido, H. G., De la Vara Salazar, R., González, P. G., Martínez, C. T., & Pérez, M. D. C. T. (2008). *Análisis y diseño de experimentos*. McGraw-Hill.

Pulido, H. G., De la Vara Salazar, R., González, P. G., Martínez, C. T., & Pérez, M. D. C. T. (2012). *Analysis and design of experiments*. New York, NY: McGraw-Hill. In Spanish

Pundir, R., Chary, G. H. V. C., & Dastidar, M. G. (2018). Application of Taguchi method for optimizing the process parameters for the removal of copper and nickel by growing *Aspergillus sp. Water Resources and Industry*, *20*, 83–92. doi:10.1016/j.wri.2016.05.001

Pyrzynska, K. (2004). Analytical methods for the determination of trace metals in wine. *Critical Reviews in Analytical Chemistry*, *34*(2), 69–83. doi:10.1080/10408340490475858

Quesada, G. M., & Del Castillo, E. (2004). A dual response approach to the multivariate robust Parameter Design problem. *Technometrics*, *46*(2), 176–187. doi:10.1198/004017004000000220

Quintanar-Guerrero, D., Tamayo-Esquivel, D., Ganem-Quintanar, A., Allémann, E., & Doelker, E. (2005). Adaptation and optimization of the emulsification-diffusion technique to prepare lipidic nanospheres. *European Journal of Pharmaceutical Sciences*, *26*(2), 211–218. doi:10.1016/j.ejps.2005.06.001 PMID:16046105

Quintelas, C., Ferreira, E. C., Lopes, J. A., & Sousa, C. (2018). An overview of the evolution of infrared spectroscopy applied to bacterial typing. *Biotechnology Journal*, *13*(1). doi:10.1002/biot.201700449 PMID:29090857

Rahimi, M., Aghel, B., Alitabar, M., Sepahvand, A., & Ghasempour, H. R. (2014). Optimization of biodiesel production from soybean oil in a microreactor. *Energy Conversion and Management*, *79*, 599–605. doi:10.1016/j.enconman.2013.12.065

Rahman, Z., Zidan, A. S., & Khan, M. A. (2010). Non-destructive methods of characterization of risperidone solid lipid nanoparticles. *European Journal of Pharmaceutics and Biopharmaceutics*, *76*(1), 127–137. doi:10.1016/j.ejpb.2010.05.003 PMID:20470882

Ramirez, E., Recasens, F., Fernández, M., & Larrayoz, M. A. (2004). Sunflower oil hydrogenation on Pd/C in SC propane in a continuous recycle reactor. *AIChE Journal. American Institute of Chemical Engineers*, *50*(7), 1545–1555. doi:10.1002/aic.10142

Ramirez-Elias, M. G., & González, F. J. (2018). Raman Spectroscopy for in vivo medical diagnosis. In *Raman Spectroscopy*. IntechOpen. doi:10.5772/intechopen.72933

Ranga, S., Jaimini, M., Sharma, S. K., Chauhan, B. S., & Kumar, A. (2014). A review on design of experiments (DOE). *International Journal of Pharmaceutical and Chemical Sciences*, *3*(1), 216–224. Retrieved from www.ijpcsonline.com

Ratola, N., Lacorte, S., Barceló, D., & Alves, A. (2009). Microwave-assisted extraction and ultrasonic extraction to determine polycyclic aromatic hydrocarbons in needles and bark of Pinus pinaster Ait. and Pinus pinea L. by GC–MS. *Talanta*, *77*(3), 1120–1128. doi:10.1016/j.talanta.2008.08.010 PMID:19064100

Rauckyte, T., Żak, S., Pawlak, Z., & Oloyede, A. (2010). Determination of oil and grease, total petroleum hydrocarbons and volatile aromatic compounds in soil and sediment samples. *Journal of Environmental Engineering and Landscape Management*, *18*(3), 163–169. doi:10.3846/jeelm.2010.19

Ravanshad, R., Karimi Zadeh, A., Amani, A. M., Mousavi, S. M., Hashemi, S. A., Savar Dashtaki, A., ... & Zare, B. (2018). Application of nanoparticles in cancer detection by Raman scattering based techniques. *Nano Reviews & Experiments, 9*(1).

Raveendran, S., Parameswaran, B., Ummalyma, S. B., Abraham, A., Mathew, A. K., Madhavan, A., ... Pandey, A. (2018). Applications of microbial enzymes in food industry. *Food Technology and Biotechnology*, *56*(1), 16–30. doi:10.17113/ftb.56.01.18.5491 PMID:29795993

Rawat, M. K., Jain, A., & Singh, S. (2011). Studies on binary lipid matrix based solid lipid nanoparticles of repaglinide: in vitro and in vivo evaluation. *Journal of Pharmaceutical Sciences*, *100*(6), 2366–2378. doi:10.1002/jps.22435 PMID:21491449

Ray, A. (2004). Development of coating colour formation for duplex board based on Plackett Burman statistical design. *Cellulose Chemistry and Technology*, *38*(3-4), 265–275.

Raza, K., Singh, B., Singal, P., Wadhwa, S., & Katare, O. P. (2013). Systematically optimized biocompatible isotretinoin-loaded solid lipid nanoparticles (SLNs) for topical treatment of acne. *Colloids and Surfaces. B, Biointerfaces*, *105*(January), 67–74. doi:10.1016/j.colsurfb.2012.12.043 PMID:23357735

Reardon, B. E., Lloyd, J. M., & Perel, R. Y. (2010). *Tuning Missile Guidance and Control Algorithms Using Simultaneous Perturbation Stochastic Approximation.*, *29*(1), 85–100.

Ree, S., Park, Y. H., & Yoo, H. (2014). A study on education quality using the Taguchi method. *Total Quality Management & Business Excellence*, *25*(7-8), 935–943. doi:10.1080/14783363.2014.906114

Reich, N. G., & Milstone, A. M. (2014). Improving efficiency in cluster-randomized study design and implementation: Taking advantage of a crossover. *Open Access Journal of Clinical Trials*, *14*(6), 11–15. doi:10.2147/oajct.S56730

Reich, N. G., Myers, J. A., Obeng, D., Milstone, A. M., & Perl, T. M. (2012). Empirical power and sample size calculations for cluster-randomized and cluster-randomized crossover studies. *PLoS One*, *7*(4), e35564. doi:10.1371/journal.pone.0035564 PMID:22558168

Reina Neira, M. A., de la Hoz, L. A., Felizzola Jimenez, H. A., & Hualpa Zuniga, A. M. (2016). Problem-based learning for teaching of design and analysis of experiments. *INGE CUC*, *12*, 86–96. doi:10.17981/ingecuc.12.2.2016.09

Reips, U.-D., & Neuhaus, C. (2002). WEXTOR: A Web-based tool for generating and visualizing experimental designs and procedures. *Behavior Research Methods, Instruments, & Computers*, *34*(2), 234–240. doi:10.3758/BF03195449

Reis, M., & Kenett, R. S. (2017). A structured overview on the use of computational simulators for teaching statistical methods. *Quality Engineering*, *29*(4), 730–744. doi:10.1080/08982112.2016.1272122

Reyes, E., Tecante, A., & Valdivia, M. (2008). Dietary fibre content and antioxidant activity of phenolic compounds present in Mexican chia (Salvia hispánica L.) seeds. *Food Chemistry*, *107*(2), 656–663. doi:10.1016/j.foodchem.2007.08.062

Ribereau-Gayon, P., Glories, Y., Maujean, A., & Dubourdieu, D. (2006). *Handbook of enology, the chemistry of wine stabilization and treatments (* (2nd ed., Vol. 2). Bordeaux, France: John Wiley & Sons. doi:10.1002/0470010398

Ricker, K. (2007). *FDA letter describing regulatory status of Salvia hispanica (chia)*. US Dept. of Health and Human Services, Food and Drug Administration, Public Health Services.

Ridge, E., & Kudenko, D. (2010). Tuning an algorithm using design of experiments. In *Experimental methods for the analysis of optimization algorithms* (pp. 265–286). Springer. doi:10.1007/978-3-642-02538-9_11

Ridzuan, N., Adam, F., & Yaacob, Z. (2016). Screening of factor influencing wax deposition using full factorial experimental design. *Petroleum Science and Technology*, *34*(1), 84–90. doi:10.1080/10916466.2015.1122625

Riveroll-Larios, J., Escalante-Espinosa, E., Fócil-Monterrubio, R. L., & Díaz-Ramírez, I. J. (2015). Biological activity assessment in Mexican tropical soils with different hydrocarbon contamination histories. *Water, Air, and Soil Pollution*, *226*(10), 353. doi:10.100711270-015-2621-1 PMID:26478633

Robbins, H., & Monro, S. (1951). A stochastic approximation method. *Annals of Mathematical Statistics*, *22*(3), 400–407. doi:10.1214/aoms/1177729586

Robinson, T. J., Borror, C. M., & Myers, R. H. (2004). Robust parameter design: A review. *Quality and Reliability Engineering International*, *20*(1), 81–101. doi:10.1002/qre.602

Rodriguez-Saona, L. E., Khambaty, F. M., Fry, F. S., & Calvey, E. M. (2001). Rapid detection and identification of bacterial strains by Fourier transform near-infrared spectroscopy. *Journal of Agricultural and Food Chemistry*, *49*(2), 574–579. doi:10.1021/jf000776j PMID:11261995

Romero Aguilar, M. A., & Vargas-Martínez, M. G. (2016). *Identificación y clasificación de la chía (Salvia hispanica L.) silvestre y comercial mediante el perfil de proteínas obtenido por electroforesis capilar* (Undergraduate dissertation). Universidad Nacional Autónoma de México.

Romeu, J. L. (2006). Teaching Engineering Statistics to Practicing Engineers, ICOTS-7, In *7th International Conference on Teaching Statistics*, Salvador, Brazil.

Rousseva, M., Kontoudakis, N., Schmidtke, L., Scollary, G., & Clark, A. (2016). Impact of wine production on the fractionation of copper and iron in Chardonnay wine: Implications for oxygen consumption. *Food Chemistry*, *203*, 440–447. doi:10.1016/j.foodchem.2016.02.081 PMID:26948636

Roy, R. K. (2001). *Design of experiments using the Taguchi approach: 16 steps to product and process improvement*. Wiley.

Roy, R. K. (2010). *A primer on the Taguchi method*. USA: Society of Manufacturing Engineers.

Rozet, E., Lebrun, P., Hubert, P., Debrus, B., & Boulanger, B. (2013). Design spaces for analytical methods. *TrAC Trends in Analytical Chemistry*, *42*, 157–167. doi:10.1016/j.trac.2012.09.007

Ruiz-Álvarez, O., Arteaga-Ramírez, R., Vázquez-Peña, M. A., Ontiveros Capurata, R. E., & López-López, R. (2012). Balance hídrico y clasificación climática del estado de Tabasco, México. *Universidad y ciencia*, *28*(1), 1-14.

Rytz, A., Moser, M., Lepage, M., Mokdad, C., Perrot, M., Antille, N., & Pineau, N. (2017). Using fractional factorial designs with mixture constraints to improve nutritional value and sensory properties of processed food. *Food Quality and Preference*, *58*, 71–75. doi:10.1016/j.foodqual.2017.01.004

Sadegh, P., & Spall, J. C. (1997). Optimal random perturbations for stochastic approximation using a simultaneous perturbation gradient approximation. *Proceedings of the 1997 American Control Conference (Cat. No.97CH36041)*, 6, 3582–3586. 10.1109/ACC.1997.609490

Sadh, P. K., Duhan, S., & Duhan, J. S. (2018). Agro-industrial wastes and their utilization using solid state fermentation: A review. *Bioresources and Bioprocessing*, *5*(1), 1–15. doi:10.118640643-017-0187-z

Saeidpour, S., Lohan, S. B., Solik, A., Paul, V., Bodmeier, R., Zoubari, G., ... Teutloff, C. (2017). Drug distribution in nanostructured lipid particles. *European Journal of Pharmaceutics and Biopharmaceutics*, *110*, 19–23. doi:10.1016/j.ejpb.2016.10.008 PMID:27789357

Sahagún, B. (1579). *Historia general de las cosas de la Nueva España* (p. 1982). Santa Fe, NM: Reprinted by School of American Research.

Sakkas, V. A., Islam, M. A., Stalikas, C., & Albanis, T. A. (2010). Photocatalytic degradation using design of experiments: A review and example of the Congo red degradation. *Journal of Hazardous Materials*, *175*(1–3), 33–44. doi:10.1016/j.jhazmat.2009.10.050 PMID:19931983

Salamatinia, B., Bhatia, S., & Abdullah, A. Z. (2011). A response surface methodology based modeling of temperature variation in an ultrasonic-assisted biodiesel production process. *World Applied Sciences Journal*, *12*(9), 1549–1557.

Salazar, J. C., & Baena-Zapata, A. (2009). Dyna. *Dyna (Bilbao)*, *76*(159), 249–257.

Salunke, M., & Kittur, J. (2017). An experiential learning in electrical machines laboratory course. In *Proceedings 2017 5th IEEE International Conference on MOOCs, Innovation and Technology in Education (MITE)*, 15–20. IEEE.

Salvatori, P. E., Sánchez, G. R., Lombardi, A. G., Nicocia, E. J., & Boschetti, C. E. (2018). Strategy for the optimization of a rubber compound with soybean oil using design of experiments. *Matéria (Rio de Janeiro), 23*(2).

Sanchez, G. S. (2009). *Medición del ruido en la línea de transmisión eléctrica generado por electrodomésticos, para aplicaciones de PLC*. Universidad de las Américas Puebla.

Sandt, C., Madoulet, C., Kohler, A., Allouch, P., De Champs, C., Manfait, M., & Sockalingum, G. D. (2006). FT-IR microspectroscopy for early identification of some clinically relevant pathogens. *Journal of Applied Microbiology, 101*(4), 785–797. doi:10.1111/j.1365-2672.2006.02969.x PMID:16968290

Santos, I. P., Barroso, E. M., Schut, T. C. B., Caspers, P. J., van Lanschot, C. G., Choi, D. H., ... Hegt, V. N. (2017). Raman spectroscopy for cancer detection and cancer surgery guidance: Translation to the clinics. *Analyst (London), 142*(17), 3025–3047. doi:10.1039/C7AN00957G PMID:28726868

Santos, M. C., Nascimento, P. A. M., Guedes, W. N., Pereira Filho, E. R., Filletti, É. R., & Pereira, F. M. V. (2019). Chemometrics in analytical chemistry–an overview of applications from 2014 to 2018. *Eclética Química, 44*(2), 11–25. doi:10.26850/1678-4618eqj.v44.2.2019.p11-25

Santos, M. C., Nascimento, Y. M., Araújo, J. M., & Lima, K. M. (2017). ATR-FTIR spectroscopy coupled with multivariate analysis techniques for the identification of DENV-3 in different concentrations in blood and serum: A new approach. *RSC Advances, 7*(41), 25640–25649. doi:10.1039/C7RA03361C

Santos, W., Brandao, G., Portugal, L., David, J., & Ferreira, S. (2009). A photo-oxidation procedure using UV radiation/ H_2O_2 for decomposition of wine samples - Determination of iron and manganese content by flame atomic absorption spectrometry. *Spectrochimica Acta. Part B, Atomic Spectroscopy, 64*(6), 601–604. doi:10.1016/j.sab.2009.04.012

Saval, S. (2012). Aprovechamiento de residuos agroindustriales: Pasado, presente y futuro. *BioTecnología, 16*, 14–46.

Schmitt, J., & Flemming, H. C. (1998). FTIR-spectroscopy in microbial and material analysis. *International Biodeterioration & Biodegradation, 41*(1), 1–11. doi:10.1016/S0964-8305(98)80002-4

Schonning, A., Nayfeh, J., & Zarda, R. (2005). An integrated design and optimization environment for industrial large scaled systems. *Research in Engineering Design, 16*(1-2), 86–95. doi:10.100700163-005-0006-y

Schork, N. J., & Goetz, L. H. (2017). Single-subject studies in translational nutrition research. *Annual Review of Nutrition, 37*(1), 395–422. doi:10.1146/annurev-nutr-071816-064717 PMID:28715990

Schut, S., Zauner, S., Hampel, G., Konig, H., & Claus, H. (2011). Biosorption of copper by wine-relevant lactobacilli. *International Journal of Food Microbiology, 145*(1), 126–131. doi:10.1016/j.ijfoodmicro.2010.11.039 PMID:21195499

Schwartz, G., Eshel, G., & Ben-Dor, E. (2011). Reflectance spectroscopy as a tool for monitoring contaminated soils. In Soil contamination. InTechOpen.

Sedwick, P. (2014). Randomised controlled trials: "Within subject" versus "between subject" designs. *British Medical Journal, 349*(aug06 2), g6435. doi:10.1136/bmj.g6435 PMID:25954990

Sedwick, P. (2015). Bias in randomized controlled trials: Comparison of crossover group and parallel group designs. *British Medical Journal, 351*, h4283. doi:10.1136/bmj.h4283 PMID:26253465

SEFI. (2012). *Engineering Education and the Bologna Process*. Position Paper, a Joint communication of SEFI and BEST in view of the 8th Ministerial Conference in Bucharest, 26-27 April 2012. SEFI aisbl, Brussels.

Sergeant, S., Rahbar, E., & Chilton, F. (2016). Gamma-linolenic acid, Dihommo-gamma linolenic, Eicosanoids and Inflammatory Processes. *European Journal of Pharmacology, 785*, 77–86. doi:10.1016/j.ejphar.2016.04.020 PMID:27083549

Serrano Guzmán, M. F., Torrado Gómez, L. M., & Pérez Ruiz, D. D. (2013). Impact of oil spills on the mechanical properties of sandy soils. *Revista Científica General José María Córdova, 11*(12), 233–244.

Settle, F. A. (1997). *Handbook of instrumental techniques for analytical chemistry*. Prentice Hall PTR.

Shah, M., & Pathak, K. (2010). Development and statistical optimization of solid lipid nanoparticles of simvastatin by using 23 full-factorial design. *AAPS PharmSciTech, 11*(2), 489–496. doi:10.120812249-010-9414-z PMID:20309652

Shainin, D., & Shainin, P. (1988). Better than Taguchi orthogonal tables. *Quality and Reliability Engineering International, 4*(2), 143–149. doi:10.1002/qre.4680040209

Sharif Makhmal Zadeh, B., Niro, H., Rahim, F., & Esfahani, G. (2018). Ocular delivery system for propranolol hydrochloride based on nanostructured lipid carrier. *Scientia Pharmaceutica, 86*(2), 16. doi:10.3390cipharm86020016 PMID:29677103

Sharma, A., Jindal, M., Aggarwal, G., & Jain, S. (2010). Development of a novel method for fabrication of solid lipid nanoparticles: Using high shear homogenization and ultrasonication. *Research Journal of Pharmaceutical, Biological and Chemical Sciences, 1*(2), 265–274.

Sheean, P. M., Bruemmen, B., Gleason, P., Harris, J., Boushey, C., & Van Horn, L. (2011). Publishing nutrition research: A review of multivariate techniques- Part 1. *Journal of the American Dietetic Association, 111*(1), 103–110. doi:10.1016/j.jada.2010.10.010 PMID:21185971

Shi, C., Ahmad Khan, S., Wang, K., & Schneider, M. (2015). Improved delivery of the natural anticancer drug tetrandrine. *International Journal of Pharmaceutics, 479*(1), 41–51. doi:10.1016/j.ijpharm.2014.12.022 PMID:25510598

Shinzawa, H., Awa, K., Kanematsu, W., & Ozaki, Y. (2009). Multivariate data analysis for Raman spectroscopic imaging. *Journal of Raman Spectroscopy: An International Journal for Original Work in all Aspects of Raman Spectroscopy, Including Higher Order Processes, and also Brillouin and Rayleigh Scattering, 40*(12), 1720-1725.

Siafaka, P. I., Üstündağ Okur, N., Karavas, E., & Bikiaris, D. N. (2016). Surface modified multifunctional and stimuli responsive nanoparticles for drug targeting: Current status and uses. *International Journal of Molecular Sciences, 17*(1440). doi:10.3390/ijms17091440 PMID:27589733

Sil, S., Mukherjee, R., Kumar, N. S., Aravind, S., Kingston, J., & Singh, U. K. (2017). Detection and classification of bacteria using Raman spectroscopy combined with multivariate analysis. *Defence Life Science Journal, 2*(4), 435–441. doi:10.14429/dlsj.2.12275

Silva, V., (2018). *Statistical approaches with emphasis on design of experiments applied to chemical processes*. BoD--Books on Demand.

Silva, D. A. L., Filleti, R. A. P., Christoforo, A. L., Silva, E. J., & Ometto, A. R. (2015). Application of Life Cycle Assessment (LCA) and Design of Experiments (DOE) to the monitoring and control of a grinding process. *Procedia CIRP, 29*, 508–513. doi:10.1016/j.procir.2015.01.037

Silva, G. F., Camargo, F. L., & Ferreira, A. L. O. (2011). Application of response surface methodology for optimization of biodiesel production by transesterification of soybean oil with ethanol. *Fuel Processing Technology, 92*(3), 407–413. doi:10.1016/j.fuproc.2010.10.002

Silva, M. B., Carneiro, L. M., Silva, J. P. A., dos Santos Oliveira, I., Izário Filho, H. J., & de Oliveira Almeida, C. R. (2014). An application of the Taguchi method (robust design) to environmental engineering: Evaluating advanced oxidative processes in polyester-resin wastewater treatment. *American Journal of Analytical Chemistry, 5*(13), 828–837. doi:10.4236/ajac.2014.513092

Silveira-Coelho, M., & Salas-Mellado, M. (2015). Effects of substituting chia (*Salvia hispanica* L.) flour or seeds for wheat flour on the quality of the bread. *Lebensmittel-Wissenschaft + Technologie*, *60*(2), 729–736. doi:10.1016/j.lwt.2014.10.033

SINC Science News and Information Service. The Aztec and Maya Seed Arrives in Europe. (2013, Dec. 2) Retrieved from http://www.agenciasinc.es/en/News/The-Aztec-and-Maya-seed-arrives-in-Europe

Sindhu, R., Binod, P., & Pandey, A. (2016). A novel sono-assisted acid pretreatment of chili post harvest residue for bioethanol production. *Bioresource Technology*, *213*, 58–63. doi:10.1016/j.biortech.2016.02.079 PMID:26949055

Sindhu, R., Kuttiraja, M., Binod, P., Sukumaran, R. K., & Pandey, A. (2014). Bioethanol production from dilute acid pretreated Indian bamboo variety (Dendrocalamus sp.) by separate hydrolysis and fermentation. *Industrial Crops and Products*, *52*, 169–176. doi:10.1016/j.indcrop.2013.10.021

Singh, I., Swami, R., Jeengar, M. K., Khan, W., & Sistla, R. (2015). p-Aminophenyl-a-d-mannopyranoside engineered lipidic nanoparticles for effective delivery of docetaxel to brain. *Chemistry and Physics of Lipids*, *188*, 1–9. doi:10.1016/j.chemphyslip.2015.03.003 PMID:25819559

Siqueira, L. F., & Lima, K. M. (2016). A decade (2004–2014) of FTIR prostate cancer spectroscopy studies: An overview of recent advancements. *TrAC Trends in Analytical Chemistry*, *82*, 208–221. doi:10.1016/j.trac.2016.05.028

Siqueira, L. F., & Lima, K. M. (2016). MIR-biospectroscopy coupled with chemometrics in cancer studies. *Analyst (London)*, *141*(16), 4833–4847. doi:10.1039/C6AN01247G PMID:27433557

Sjöström, B., & Bergenståhl, B. (1992). Preparation of submicron drug particles in lecithin-stabilized o w emulsions: I. Model studies of the precipitation of cholesteryl acetate. *International Journal of Pharmaceutics*, *84*(2), 107–116. doi:10.1016/0378-5173(92)90051-3

Skorupskaite, V., Makareviciene, V., & Levisauskas, D. (2015). Optimization of mixotrophic cultivation of microalgae Chlorella sp. for biofuel production using response surface methodology. *Algal Research*, *7*, 45–50. doi:10.1016/j.algal.2014.12.001

Slade, R., & Bauen, A. (2013). Micro-algae cultivation for biofuels: Cost, energy balance, environmental impacts and future prospects. *Biomass and Bioenergy*, *53*, 29–38. doi:10.1016/j.biombioe.2012.12.019

Smilauer, P., & Leps, J. (2014). *Multivariate analysis of ecological data using CANOCO 5*. Cambridge University Press. doi:10.1017/CBO9781139627061

Smith, B. C. (2011). Fundamentals of Fourier transform infrared spectroscopy. Boca Raton, FL: CRC Press. doi:10.1201/b10777

Smith, E., & Dent, G. (2005). *Modern Raman spectroscopy: a practical approach*. Hoboken, NJ: John Wiley & Sons.

Sohrabi, M. R., Jamshidi, S., & Esmaeilifar, A. (2012). Cloud point extraction for determination of Diazinon: Optimization of the effective parameters using Taguchi method. *Chemometrics and Intelligent Laboratory Systems*, *110*(1), 49–54. doi:10.1016/j.chemolab.2011.09.009

Sorak, D., Herberholz, L., Iwascek, S., Altinpinar, S., Pfeifer, F., & Siesler, H. W. (2012). New developments and applications of handheld Raman, mid-infrared, and near-infrared spectrometers. *Applied Spectroscopy Reviews*, *47*(2), 83–115. doi:10.1080/05704928.2011.625748

Soto, L. A., Botello, A. V., Licea-Durán, S., Lizárraga-Partida, M. L., & Yáñez-Arancibia, A. (2014). The environmental legacy of the Ixtoc-I oil spill in Campeche Sound, southwestern Gulf of Mexico. *Frontiers of Materials Science*, *1*, 57.

Souza, M. F., Francisco, R. L., Sanchez, J. L., Guimarães-Inácio, A., Valderrama, P., Bona, E., ... Gonçalves, O. H. (2017). Fatty acids profile of chia oil-loaded Lipid microparticles. *Brazilian Journal of Chemical Engineering, 34*(3), 659–669. doi:10.1590/0104-6632.20170343s20150669

Spada, A., Conte, A., & Del Nobile, M. A. (2018). The influence of shelf life on food waste: A model-based approach by empirical market evidence. *Journal of Cleaner Production, 172*, 3410–3414. doi:10.1016/j.jclepro.2017.11.071

Spall, J. C. (2012). Stochastic optimization. In Handbook of computational statistics (pp. 173-201). Berlin, Germany: Springer. doi:10.1007/978-3-642-21551-3_7

Spall, James C. (1988). *A stochastic approximation algorithm for large-dimensional system in the Kiefer-Wolfowitz setting.* (December), 1544–1548.

Spall, James C. (1998). *An Overview of the Simultaneous Perturbation Method for Efficient Optimization, 19*(4).

Spall, J. C. (1992). Multivariate stochastic approximation using a simultaneous perturbation gradient approximation. *IEEE Transactions on Automatic Control, 37*(3), 332–341. doi:10.1109/9.119632

Spearman, C. (1904). General Intelligence, objectively determined and measured. *The American Journal of Psychology, 15*(2), 201–292. doi:10.2307/1412107

Spivey, N., Thompson, P., Shelton, C., & Kavan, A. (2015). The analysis of copper, iron, and manganese in wine with FAST flame atomic absorption. *Perkin Elmer*, 1-4.

Squires, A., & Cloutier, R. (2011). Applying a robust design approach to improve online systems engineering education. In *Proceedings 2011 IEEE International Systems Conference (SYSCON 2011)*, 371–378. 10.1109/SYSCON.2011.5929103

Srinophakun, P., Thanapimmetha, A., Rattanaphanyapan, K., Sahaya, T., & Saisriyoot, M. (2017). Feedstock production for third generation biofuels through cultivation of Arthrobacter AK19 under stress conditions. *Journal of Cleaner Production, 142*, 1259–1266. doi:10.1016/j.jclepro.2016.08.068

Srivastava, V. C., Mall, I. D., & Mishra, I. M. (2007). Multicomponent adsorption study of metal ions onto bagasse fly ash using Taguchi's design of experimental methodology. *Industrial & Engineering Chemistry Research, 46*(17), 5697–5706. doi:10.1021/ie0609822

Srivastava, V. C., Mall, I. D., & Mishra, I. M. (2008). Optimization of parameters for adsorption of metal ions onto rice husk ash using Taguchi's experimental design methodology. *Chemical Engineering Journal, 140*(1-3), 136–144. doi:10.1016/j.cej.2007.09.030

Starkey, M., Aughton, J., & Brewin, R. (1997). Extending process thinking: Design of experiments in sales and marketing. *The TQM Magazine, 9*(6), 434–439. doi:10.1108/09544789710187046

Statgraphics Centurion, V. I. Statgraphics Technologies, Inc., The Plains, Virginia. Retrieved from http://www.statgraphics.com

Stöckel, S., Kirchhoff, J., Neugebauer, U., Rösch, P., & Popp, J. (2016). The application of Raman spectroscopy for the detection and identification of microorganisms. *Journal of Raman Spectroscopy: JRS, 47*(1), 89–109. doi:10.1002/jrs.4844

Stuart, B. H. (2004). *Infrared spectroscopy: Fundamentals and applications.* John Wiley & Sons. doi:10.1002/0470011149

Sunde, K., Brekke, A., & Solberg, B. (2011). Environmental impacts and costs of woody Biomass-to-Liquid (BTL) production and use — A review. *Forest Policy and Economics, 13*(8), 591–602. doi:10.1016/j.forpol.2011.05.008

Supelco, Sigma-Aldrich Corporation subsidiary, St. Louis, MO. Retrieved from https://www.sigmaaldrich.com/catalog/product/supelco/crm18917?lang=es&region=MX

Su, W. H., & Sun, D. W. (2019). Mid-infrared (MIR) spectroscopy for quality analysis of liquid foods. *Food Engineering Reviews*, 1–17.

Tabata, M., & Morita, H. (1997). Spectrophotometric determination of a nanomolar amount of ascorbic acid using its catalytic effect on copper (II) porphyrin formation. *Talanta*, *44*(2), 151–157. doi:10.1016/S0039-9140(96)02010-3 PMID:18966728

Tadros, T. (2015). Viscoelastic properties of sterically stabilised emulsions and their stability. *Advances in Colloid and Interface Science*, *222*, 692–708. doi:10.1016/j.cis.2015.03.001 PMID:25900262

Taguchi, G. (1986). Introduction to quality engineering. New York: UNIPUB/Kraus International.

Taguchi, G. (1986). *Introduction to quality engineering: designing quality into products and processes*. Tokyo: Asian Productivity Organization.

Taguchi, G., Chowdhury, S., & Wu, Y. (2005). *Taguchi's quality engineering handbook*. Hoboken, NJ: Wiley.

Taguchi, G., & Wu, Y. (1979). *Introduction to Off-Line Quality Control*. Nagoya: Central Japan Quality Control Association.

Tanco, M., Viles, E., & Pozueta, L. (2009). Diferentes enfoques del diseño de experimentos (DOE), *7*, 29–37.

Tanco, M., Viles, E., Ilzarbe, L., & Álvarez, M. J. (2007). Manufacturing industries need design of experiments (DoE). *Proceedings of the World Congress on Engineering, Vol. II*, WCE, London, UK.

Tanco, M. (2006). *Is DoE really used in Industries? ENBIS 6, Wroclaw, Poland.*

Tanco, M., Ilzarbe, L., Viles, E., & Alvarez, M. J. (2008). Applying design of experiments to improve a laser welding process. *Proceedings of the Institution of Mechanical Engineers. Part B, Journal of Engineering Manufacture*, *222*(8), 1035–1042. doi:10.1243/09544054JEM1155

Tanco, M., Viles, E., Ilzarbe, L., & Álvarez, M. J. (2007). Manufacturing industries need design of experiments (DoE). *World Congress on Engineering*, *20*, 1108–1113.

Tanco, M., Viles, E., Ilzarbe, L., & Alvarez, M. J. (2008). Is design of experiments really used? A survey of Basque industries. *Journal of Engineering Design*, *19*(5), 5. doi:10.1080/09544820701749124

Tanco, M., Viles, E., Ilzarbe, L., & Álvarez, M. J. (2009). Barriers faced by engineers when applying design of experiments. *The TQM Journal*, *21*(6), 565–575. doi:10.1108/17542730910995846

Tanco, M., Viles, E., Jesus Álvarez, M., & Ilzarbe, L. (2010). Why is not design of experiments widely used by engineers in Europe? *Journal of Applied Statistics*, *37*(12), 1961–1977. doi:10.1080/02664760903207308

Tang, S. H., Tan, Y. J., Sapuan, S. M., Sulaiman, S., Ismail, N., & Samin, R. (2007). The use of Taguchi method in the design of plastic injection mould for reducing warpage. *Journal of Materials Processing Technology*, *182*(1-3), 418–426. doi:10.1016/j.jmatprotec.2006.08.025

Tan, Y. H., Abdullah, M. O., Nolasco-Hipolito, C., & Zauzi, N. S. A. (2017). Application of RSM and Taguchi methods for optimizing the transesterification of waste cooking oil catalyzed by solid ostrich and chicken-eggshell derived CaO. *Renewable Energy*, *114*, 437–447. doi:10.1016/j.renene.2017.07.024

Tas, T. (2008). *Katışıklı zeytinyağlarında katıştırılan yağ cinsi ve oranının ftır spektroskopisi ve kemometrik yöntemlerle belirlenmesi*. Yüksek Lisans Tezi.

Taveira, S. F., Araújo, L. M. P. de C., de Santana, D. C. A. S., Nomizo, A., de Freitas, L. A. P., & Lopez, R. F. V. (2012). Development of cationic solid lipid nanoparticles with factorial design-based studies for topical administration of doxorubicin. *Journal of Biomedical Nanotechnology*, *8*(2), 219–228. doi:10.1166/jbn.2012.1383 PMID:22515073

Tay, K. M., & Butler, C. (1999). Methodologies for experimental design: A survey, comparison and future predictions. *Quality Engineering*, *11*(3), 343–356. doi:10.1080/08982119908919250

Theophanides, T. (Ed.). (2012). *Infrared spectroscopy: Materials science, engineering, and technology*. Published by InTech. doi:10.5772/2055

Therdthai, N., & Zhou, W. (2009). Characterization of microwave vacuum drying and hot air drying of mint leaves (Mentha cordifolia Opiz ex Fresen). *Journal of Food Engineering*, *91*(3), 482–489. doi:10.1016/j.jfoodeng.2008.09.031

Thuan, V. T., Quynh, B. T. P., Nguyen, T. D., & Bach, L. G. (2017). Response surface methodology approach for optimization of Cu^{2+}, Ni^{2+} and Pb^{2+} adsorption using KOH-activated carbon from banana peel. *Surfaces and Interfaces*, *6*, 209–217. doi:10.1016/j.surfin.2016.10.007

Tiwary, A. K., Puri, M., Sharma, D., & Barrow, C. J. (2012). Optimisation of novel method for the extraction of steviosides from Stevia rebaudiana leaves. *Food Chemistry*, *132*(3), 1113–1120. doi:10.1016/j.foodchem.2011.11.063 PMID:29243590

Tobin, E. (2015). *A guide to process validation for medical devices*. Solo Validation Resources Limited.

Toms, D., Deardon, R., & Ungrin, M. (2017). Climbing the mountain: Experimental design for the efficient optimization of stem cell bioprocessing. *Journal of Biological Engineering*, *11*(1), 35. doi:10.118613036-017-0078-z PMID:29213303

Tong, K. W., Kwong, C. K., & Yu, K. M. (2004). Process optimisation of ranfer moulding for electronic packages using artificial neural networks and multiobjective optimisation techniques. *International Journal of Advanced Manufacturing Technology*, *24*(9-10), 675–685. doi:10.100700170-003-1739-2

Tonon, R. V., Brabet, C., & Hubinger, M. D. (2008). Influence of process conditions on the physicochemical properties of acai (S Euterpe oleraceae Mart.) powder produced by spray drying. *Journal of Food Engineering*, *88*(3), 411–418. doi:10.1016/j.jfoodeng.2008.02.029

Tounsadi, H., Khalidi, A., Abdennouri, M., & Barka, N. (2016a). Activated carbon from *Diplotaxis Harra* biomass: Optimization of preparation conditions and heavy metal removal. *Journal of the Taiwan Institute of Chemical Engineers*, *59*, 348–358. doi:10.1016/j.jtice.2015.08.014

Tounsadi, H., Khalidi, A., Farnane, M., Abdennouria, M., & Barka, N. (2016b). Experimental design for the optimization of preparation conditions of highly efficient activated carbon from *Glebionis coronaria* L. and heavy metals removal ability. *Process Safety and Environmental Protection*, *102*, 710–723. doi:10.1016/j.psep.2016.05.017

Tounsadi, H., Khalidi, A., Farnane, M., Machrouhi, A., Elhalil, A., & Barka, N. (2019). Efficient removal of heavy metals by koh activated diplotaxis harra biomass: Experimental design optimization. *Environmental Engineering and Management Journal*, *18*(3).

Tounsadi, H., Khalidi, A., Machrouhia, A., Farnane, M., & Elmoubarkia, R. (2016c). Highly efficient activated carbon from *Glebionis coronaria L.* biomass: Optimization of preparation conditions and heavy metals removal using experimental design approach. *Journal of Environmental Chemical Engineering*, *4*(4), 4549–4564. doi:10.1016/j.jece.2016.10.020

Tovar-Gómez, R., Moreno-Virgen, M., Moreno-Pérez, J., Bonilla-Petriciolet, A., Hernández-Montoya, V., & Durán-Valle, C. J. (2015). Analysis of synergistic and antagonistic adsorption of heavy metals and acid blue 25 on activated carbon from ternary systems. *Chemical Engineering Research & Design*, *93*, 755–772. doi:10.1016/j.cherd.2014.07.012

Trujillo-Narcía, A., Rivera-Cruz, M. D. C., Lagunes-Espinoza, L. D. C., Palma-López, D. J., Soto-Sánchez, S., & Ramírez-Valverde, G. (2012). Efecto de la restauración de un Fluvisol contaminado con petróleo crudo. *Revista Internacional de Contaminación Ambiental, 28*(4), 360–374.

Tsukahara, K., & Sawayama, S. (2005). Liquid fuel production using microalgae. *Journal of the Japan Petroleum Institute, 48*(5), 251–259. doi:10.1627/jpi.48.251

Tu, A. T. (1982). *Raman spectroscopy in biology. Principles and Applications.* John Wiley & Sons.

Tye, H. (2004). Application of statistical 'design of experiments' methods in drug discovery. *Drug Discovery Today, 9*(11), 485–491. doi:10.1016/S1359-6446(04)03086-7 PMID:15149624

UNESCO. (2010). *Engineering: issues, challenges and opportunities for development.* UNESCO Publishing.

United Nations. (2015). Sustainable development goals, knowledge platform. Goal N.12 of the 2030 Agenda for Sustainable Development aims to ensure sustainable consumption and production patterns. Retrieved from https://sustainabledevelopment.un.org/topics/sustainableconsumptionandproduction

Urbán-Morlán, Z., Ganem-Rondero, A., Melgoza-Contreras, L. M., Escobar-Chávez, J. J., Nava-Arzaluz, M. G., & Quintanar-Guerrero, D. (2010). Preparation and characterization of solid lipid nanoparticles containing cyclosporine by the emulsification-diffusion method. *International Journal of Nanomedicine, 5*(1), 611–620. doi:10.2147/IJN.S12125 PMID:20856836

Urpí, J. L. (2012). *Tecnología y margen de refino del petróleo.* Ediciones Díaz de Santos.

Vale, M. G. R., Silva, M. M., Damin, I. C., Sanches Filho, P. J., & Welz, B. (2008). Determination of volatile and non-volatile nickel and vanadium compounds in crude oil using electrothermal atomic absorption spectrometry after oil fractionation into saturates, aromatics, resins and asphaltenes. *Talanta, 74*(5), 1385–1391. doi:10.1016/j.talanta.2007.09.009 PMID:18371794

van Hal, J. W., Huijgen, W. J. J., & López-Contreras, A. M. (2014). Opportunities and challenges for seaweed in the biobased economy. *Trends in Biotechnology, 32*(5), 231–233. doi:10.1016/j.tibtech.2014.02.007 PMID:24767734

Vanaja, K., & Rani, R. H. S. (2007). *Design of experiments: Concept and applications of Plackett Burman design. Clinical Research and Regulatory Affairs.* Taylor & Francis; doi:10.1080/10601330701220520

Vandenabeele, P. (2013). *Practical Raman spectroscopy: an introduction.* Chichester, UK: Wiley. doi:10.1002/9781119961284

Vannini, L., Lanciotti, R., Gardini, F., & Guerzoni, M. E. (1996). Evaluation of Fourier-transform infrared spectroscopy for data capture in predictive microbiology. *World Journal of Microbiology & Biotechnology, 12*(1), 85–90. doi:10.1007/BF00327808 PMID:24415096

Vargas, M. G., Vander Heyden, Y., Maftouh, M., & Massart, D. L. (1999). Rapid development of the enantiomeric separation of beta-blockers by capillary electrophoresis using an experimental design approach. *Journal of Chromatography. A, 855*(2), 681–693. doi:10.1016/S0021-9673(99)00701-3 PMID:10519104

Vargas-Martínez, M. G., & Beltrán-Orozco, C. (2011) Capítulo 7: Pharmacological activity of the chia seeds (Salvia hispanica L.). In R. M. Perez Gutierrez (Ed.), Edible plants in traditional medicine: Special Review Books, Research Singpost, Trivandrum, India.

Varmuza, K., & Filzmoser, P. (2009). *Introduction to multivariate statistical analysis in chemometrics.* Taylor & Francis Group.

Vázquez-Luna, D. (2014). Chronic toxicity of weathered oil-contaminated soil. In M. C. Hernandez-Soriano (Ed)., Environmental risk assessment of soil contamination (pp. 87-104). InTech. doi:10.5772/57253

Veličković, A. V., Stamenković, O. S., Todorović, Z. B., & Veljković, V. B. (2013). Application of the full factorial design to optimization of base-catalyzed sunflower oil ethanolysis. *Fuel*, *104*, 433–442. doi:10.1016/j.fuel.2012.08.015

Verdugo, C., Luna, D., Posadillo, A., Sancho, E. D., Rodríguez, S., Bautista, F., ... Romero, A. A. (2011). Production of a new second generation biodiesel with a low-cost lipase derived from Thermomyces lanuginosus: Optimization by response surface methodology. *Catalysis Today*, *167*(1), 107–112. doi:10.1016/j.cattod.2010.12.028

Vicente, G., Coteron, A., Martinez, M., & Aracil, J. (1998). Application of the factorial design of experiments and response surface methodology to optimize biodiesel production. *Industrial Crops and Products*, *8*(1), 29–35. doi:10.1016/S0926-6690(97)10003-6

Vinodh, S., Bharathi, S. K., & Gopi, N. (2016). Parametric optimization of submerged arc welding using Taguchi Method BT - Design of experiments in production engineering. In J. P. Davim (Ed.), Design of experiments in production engineering (pp. 183–194). Cham, Switzerland: Springer; doi:10.1007/978-3-319-23838-8_8

Vitorino, C., Carvalho, F. A., Almeida, A. J., Sousa, J. J., & Pais, A. A. C. C. (2011). The size of solid lipid nanoparticles: An interpretation from experimental design. *Colloids and Surfaces. B, Biointerfaces*, *84*(1), 117–130. doi:10.1016/j.colsurfb.2010.12.024 PMID:21242064

Volk, H., & George, S. C. (2019). Using petroleum inclusions to trace petroleum systems–a review. *Organic Geochemistry*, *129*, 99–123. doi:10.1016/j.orggeochem.2019.01.012

Vystavna, Y., Rushenko, L., Diadin, D., Klymenko, O., & Klymenko, M. (2014). Trace metals in wine and vineyard environment in southern Ukraine. *Food Chemistry*, *146*, 339–344. doi:10.1016/j.foodchem.2013.09.091 PMID:24176352

Wadikar, D. D., Nanjappa, C., Premavalli, K. S., & Bawa, A. S. (2010). Development of ginger based ready-to-eat appetizers by response surface methodology. *Appetite*, *55*(1), 76–83. doi:10.1016/j.appet.2010.04.004 PMID:20417239

Wahla, A. Q., Iqbal, S., Anwar, S., Firdous, S., & Mueller, J. A. (2019). Optimizing the metribuzin degrading potential of a novel bacterial consortium based on Taguchi design of experiment. *Journal of Hazardous Materials*, *366*, 1–9. doi:10.1016/j.jhazmat.2018.11.054 PMID:30500692

Wang, C., Vickers, T. J., Schlenoff, J. B., & Mann, C. K. (1992). In situ monitoring of emulsion polymerization using fiber-optic Raman spectroscopy. *Applied Spectroscopy*, *46*(11), 1729–1731. doi:10.1366/0003702924926961

Wang, J. S., Shi, J. S., Xu, Y. Z., Duan, X. Y., Zhang, L., Wang, J., ... Wu, J. G. (2003). FT-IR spectroscopic analysis of normal and cancerous tissues of esophagus. *World Journal of Gastroenterology*, *9*(9), 1897. doi:10.3748/wjg.v9.i9.1897 PMID:12970871

Wang, J., & Buckley, J. S. (2003). Asphaltene stability in crude oil and aromatic solvents the influence of oil composition. *Energy & Fuels*, *17*(6), 1445–1451. doi:10.1021/ef030030y

Wang, S., Chen, F., Wu, J., Wang, Z., Liao, X., & Hu, X. (2007). Optimization of pectin extraction assisted by microwave from apple pomace using response surface methodology. *Journal of Food Engineering*, *78*(2), 693–700. doi:10.1016/j.jfoodeng.2005.11.008

Wang, X. R., Gao, S. Q., Niu, X. Q., Li, L. J., Ying, X. Y., Hu, Z. J., & Gao, J. Q. (2017). Capsaicin-loaded nanolipoidal carriers for topical application: Design, characterization, and in vitro/in vivo evaluation. *International Journal of Nanomedicine*, *12*, 3881–3898. doi:10.2147/IJN.S131901 PMID:28579775

Wang, Y., Ding, Y., Wei, W., Cao, Y., Davidson, D. F., & Hanson, R. K. (2019). On estimating physical and chemical properties of hydrocarbon fuels using mid-infrared FTIR spectra and regularized linear models. *Fuel*, *255*, 115715. doi:10.1016/j.fuel.2019.115715

Wang, Y., Feng, J., Lin, Q., Lyu, X., Wang, X., & Wang, G. (2013). Effects of crude oil contamination on soil physical and chemical properties in Momoge wetland of China. *Chinese Geographical Science*, *23*(6), 708–715. doi:10.100711769-013-0641-6

Wang, Y., & Li, L. (2014). Inclusion of renewable energy topics in the design of experiments course for industrial and systems engineering students. *121st ASEE Annual Conference and Exposition, Conference Proceedings*. Indianapolis, IN. ASEE.

Wang, Y., Yang, Q., Bao, G., Xun, Y., & Zhang, L. (2011). Optimization design and experiment of fruit and vegetable picking manipulator. Nongye Jixie Xuebao=. *Nongye Jixie Xuebao*, *42*(7), 191–195.

Wasserstein, R. L., Schirm, A. L., & Lazar, N. A. (2019). Moving to a world beyond "p < 0.05." *The American Statistician, 73*(supl), 1–19. doi:10.1080/00031305.2019.1583913

Wehrung, D., Geldenhuys, W. J., & Oyewumi, M. O. (2012). Effects of gelucire content on stability, macrophage interaction and blood circulation of nanoparticles engineered from nanoemulsions. *Colloids and Surfaces. B, Biointerfaces*, *94*, 259–265. doi:10.1016/j.colsurfb.2012.02.005 PMID:22386863

Weissman, S. A., & Anderson, N. G. (2014). Design of experiments (DoE) and process optimization. A review of recent publications. *Organic Process Research & Development*, *19*(11), 1605–1633. doi:10.1021/op500169m

Weiss, R., Palatinszky, M., Wagner, M., Niessner, R., Elsner, M., Seidel, M., & Ivleva, N. P. (2019). Surface-enhanced Raman spectroscopy of microorganisms: Limitations and applicability on the single-cell level. *Analyst (London)*, *144*(3), 943–953. doi:10.1039/C8AN02177E PMID:30574650

Wenning, M., & Scherer, S. (2013). Identification of microorganisms by FTIR spectroscopy: Perspectives and limitations of the method. *Applied Microbiology and Biotechnology*, *97*(16), 7111–7120. doi:10.100700253-013-5087-3 PMID:23860713

Wiedemair, V., De Biasio, M., Leitner, R., Balthasar, D., & Huck, C. W. (2018). Application of design of experiment for detection of meat fraud with a portable near-infrared spectrometer. *Current Analytical Chemistry*, *14*(1), 58–67. doi:10.2174/1573411013666170207121113

Wijnhoven, H. A. H., van der Meij, B. S., & Viser, M. (2015). Variety within a cooked meal increases meal energy intake in older women with a poor appetite. *Appetite*, *95*, 571–576. doi:10.1016/j.appet.2015.08.029 PMID:26321418

Wilks, S. S. (1932). Certain generalizations in the analysis of variance. *Biometrika*, *24*(3-4), 471–494. doi:10.1093/biomet/24.3-4.471

Wingbrant, H., Svenningstorp, H., Salomonsson, P., Tengström, P., Lundström, I., & Lloyd Spetz, A. (2003). Using a MISiCFET device as a cold start sensor. *Sensors and Actuators. B, Chemical*, *93*(1-3), 295–303. doi:10.1016/S0925-4005(03)00227-2

Wissing, S., Kayser, O., & Müller, R. (2004). Solid lipid nanoparticles for parenteral drug delivery. *Advanced Drug Delivery Reviews*, *56*(9), 1257–1272. doi:10.1016/j.addr.2003.12.002 PMID:15109768

Woll, R., & Burkhard, C. (2005). Full factorial design, Taguchi design or genetic algorithms — Teaching different approaches to design of experiments. In *Innovations in Classification, Data Science, and Information Systems* (pp. 567–574). Berlin, Germany: Springer-Verlag. doi:10.1007/3-540-26981-9_65

Wong, H. L., Bendayan, R., Rauth, A. M., & Wu, X. Y. (2004). Development of solid lipid nanoparticles containing ionically complexed chemotherapeutic drugs and chemosensitizers. *Journal of Pharmaceutical Sciences*, *93*(8), 1993–2008. doi:10.1002/jps.20100 PMID:15236449

Wong, W. H., Lee, W. X., Ramanan, R. N., Tee, L. H., Kong, K. W., Galanakis, C. M., ... Prasad, K. N. (2015). Two level half factorial design for the extraction of phenolics, flavonoids, and antioxidants recovery from palm kernel by-product. *Industrial Crops and Products*, *63*, 238–248. doi:10.1016/j.indcrop.2014.09.049

Woods, D. C., & Lewis, S. M. (2017). Design of experiments for screening. Handbook of uncertainty quantification, 1143–1185.

World Bank. (2017). The World Bank Group, IBRID-IDA. Manufacturing, value added (% of GDP). World Bank national accounts data, and OECD National Accounts data files. License: CC BY-4.0. Retrieved from https://data.worldbank.org/indicator/NV.IND.MANF.ZS

Wu, C. H. (2016). Randomized response model versus mixture model in the crossover design for clinical trials. *Communication in Statistics-Theory and Methods, 45*(15), 4611-4627. doi:. doi:10.1080/03610926.2014.927490

Wu, C. F. J. (2000). *Experiments*. Wiley-Interscience.

Wu, Y., Pang, H., Liu, Y., Wang, X., Yu, S., Fu, D., ... Wang, X. (2019). Environmental remediation of heavy metal ions by novel-nanomaterials: A review. *Environmental Pollution*, *246*, 608–620. doi:10.1016/j.envpol.2018.12.076 PMID:30605816

Xu, Q.-S., Xu, Y.-D., Li, L., & Fang, K.-T. (2018). Uniform experimental design in chemometrics. *Journal of Chemometrics*, *32*(11), e3020. doi:10.1002/cem.3020

Yadav, L. D. S. (2005). *Organic spectroscopy*. Springer Science & Business Media. doi:10.1007/978-1-4020-2575-4

Yadav, M., Singh, A., Balan, V., Pareek, N., & Vivekanand, V. (2019). Biological treatment of lignocellulosic biomass by Chaetomium globosporum: Process derivation and improved biogas production. *International Journal of Biological Macromolecules*, *128*, 176–183. doi:10.1016/j.ijbiomac.2019.01.118 PMID:30684578

Yang, S.-T., & Yu, M. (2013, July 12). Integrated biorefinery for sustainable production of fuels, chemicals, and polymers. *Bioprocessing Technologies in Biorefinery for Sustainable Production of Fuels, Chemicals, and Polymers*.

Yang, B., Yao, L., & Pan, T. (2017). Near-infrared spectroscopy combined with partial least squares discriminant analysis applied to identification of liquor brands. *Engineering*, *9*(2), 181–189. doi:10.4236/eng.2017.92009

Yang, J., Xu, J., Zhang, X., Wu, C., Lin, T., & Ying, Y. (2019). Deep learning for vibrational spectral analysis: Recent progress and a practical guide. *Analytica Chimica Acta*, *1081*, 6–17. doi:10.1016/j.aca.2019.06.012 PMID:31446965

Yang, S. C., & Zhu, J. B. (2002). Preparation and characterization of Camptothecin solid lipid nanoparticles. *Drug Development and Industrial Pharmacy*, *28*(3), 265–274. doi:10.1081/DDC-120002842 PMID:12026219

Yan, H., Zhang, J., Gao, J., Huang, Y., Xiong, Y., & Min, S. (2018). Towards improvement in prediction of iodine value in edible oil system based on chemometric analysis of portable vibrational spectroscopic data. *Scientific Reports*, *8*(1), 14729. doi:10.103841598-018-33022-9 PMID:30283065

Yan, J. H., Guan, B. J., Gao, H. Y., & Peng, X. E. (2018). Omega-3 polyunsaturated fatty acid supplementation and non-alcoholic fatty liver disease. *Medicine*, *97*(37). doi:10.1097/MD.0000000000012271 PMID:30212963

Yates, F. (1964). Sir Ronald Fisher and the design of experiments. *Biometrics*, *20*(2), 307–321. doi:10.2307/2528399

Yildiz, A. R. (2013). Optimization of multi-pass turning operations using hybrid teaching learning-based approach. *International Journal of Advanced Manufacturing Technology*, *66*(9-12), 1319–1326. doi:10.100700170-012-4410-y

Yondo, R., Andrés, E., & Valero, E. (2018). A review on design of experiments and surrogate models in aircraft real-time and many-query aerodynamic analyses. *Progress in Aerospace Sciences*, *96*, 23–61. doi:10.1016/j.paerosci.2017.11.003

Young, R. J., & Eichhorn, S. J. (2009). Raman applications in synthetic and natural polymer fibers and their composites. Raman spectroscopy for soft matter applications, 63-94.

Young-Chul, L., Se-Wook, O., Jaehyn, C., & In-Hwan, K. (2004). Chemical composition and oxidative stability of safflower oil prepared from safflower seed roasted with different temperaturas. *Food Chemistry*, *84*(1), 1–6. doi:10.1016/S0308-8146(03)00158-4

You, P., Yuan, R., & Chen, C. (2017). Design and evaluation of lidocaine- and prilocaine-coloaded nanoparticulate drug delivery systems for topical anesthetic analgesic therapy: A comparison between solid lipid nanoparticles and nanostructured lipid carriers. *Drug Design, Development, and Therapy*, *11*, 2743–2752. doi:10.2147/DDDT.S141031 PMID:29075099

Yousefzadeh, S., Matin, A. R., Ahmadi, E., Sabeti, Z., Alimohammadi, M., Aslani, H., & Nabizadeh, R. (2018). Response surface methodology as a tool for modeling and optimization of Bacillus subtilis spores inactivation by UV/nano-Fe0 process for safe water production. *Food and Chemical Toxicology*, *114*, 334–345. doi:10.1016/j.fct.2018.02.045 PMID:29481893

You, Z. (2017). Application of infrared Raman spectroscopy in analysis of food agricultural products. *Chemical Engineering Transactions*, *59*, 763–768.

Yu, D., Wang, L., & Wu, M. (2018). Simultaneous removal of dye and heavy metal by banana peels derived hierarchically porous carbons. *Journal of the Taiwan Institute of Chemical Engineers*, *93*, 543–553. doi:10.1016/j.jtice.2018.08.038

Yu, P., Low, M. Y., & Zhou, W. (2018). Design of experiments and regression modelling in food flavour and sensory analysis: A review. *Trends in Food Science & Technology*, *71*, 202–215. doi:10.1016/j.tifs.2017.11.013

Zamora, A., Ramos, J., & Arias, M. (2012). Efecto de la contaminación por hidrocarburos sobre algunas propiedades químicas y microbiológicas de un suelo de sabana. *Bioagro-*, *24*(1), 5–12.

Zapata Gómez, A., & Sarache Castro, W. A. (2014). Mejoramiento de la calidad del café soluble utilizando el método Taguchi. *Ingeniare. Revista Chilena de Ingeniería*, *22*(1), 116–124. doi:10.4067/S0718-33052014000100011

Zavala Cruz, J., Salgado García, S., Marín Aguilar, Á., Palma López, D. J., Castelán Estrada, M., & Ramos Reyes, R. (2014). Transecto de suelos en terrazas con plantaciones de cítricos en Tabasco. *Ecosistemas Recursos Agropecuarios*, *1*(2), 123–137.

Zavala-Cruz, J., Gavi-Reyes, F., Adams-Schroeder, R. H., Ferrera-Cerrato, R., Palma-López, D. J., Vaquera-Huerta, H., & Domínguez-Ezquivel, J. M. (2005). Derrames de petróleo en suelos y adaptación de pastos tropicales en el Activo Cinco Presidentes, Tabasco, México. *Terra Latino Americana*, *23*(3), 293–302.

Zeng, H. Y., Cai, L. H., Cai, X. L., Wang, Y. J., & Li, Y. Q. (2011). Structure characterization of protein fractions from lotus (Nelumbo nucifera) seed. *Journal of Molecular Structure*, *1001*(1-3), 139–144. doi:10.1016/j.molstruc.2011.06.031

Zhang, C. X., Long, W. Q., Ye, Y. B., Lu, M. S., Zhang, N. Q., Xu, M., ... Su, Y. X. (2018). Effects of chocolate-based products intake on blood glucose, insulin and ghrelin levels on satiety in young people: A cross-over experimental study. *International Journal of Food Sciences and Nutrition*, *69*(7), 882–891. doi:10.1080/09637486.2018.1426737 PMID:29457516

Zhang, J., Fan, Y., & Smith, E. (2009). Experimental design for the optimization of lipid nanoparticles. *Journal of Pharmaceutical Sciences*, *98*(5), 1813–1819. doi:10.1002/jps.21549 PMID:18781636

Zhang, K., Hao, C., Man, B., Zhang, C., Yang, C., Liu, M., ... Chen, C. (2018). Diagnosis of liver cancer based on tissue slice surface enhanced Raman spectroscopy and multivariate analysis. *Vibrational Spectroscopy*, *98*, 82–87. doi:10.1016/j.vibspec.2018.07.010

Zhan, W., Fink, R., & Fang, A. (2011). Application of statistics in engineering technology programs. [AJEE]. *American Journal of Engineering Education*, *1*, 65–78.

Zhou, G., Fu, L., & Li, X. (2015). Optimisation of ultrasound-assisted extraction conditions for maximal recovery of active monacolins and removal of toxic citrinin from red yeast rice by a full factorial design coupled with response surface methodology. *Food Chemistry*, *170*, 186–192. doi:10.1016/j.foodchem.2014.08.080 PMID:25306334

Zhou, J., Wu, X., Chen, Z., You, J., & Xiong, S. (2019). Evaluation of freshness in freshwater fish based on near infrared reflectance spectroscopy and chemometrics. *LWT*, *106*, 145–150. doi:10.1016/j.lwt.2019.01.056

Zhu, Y., Wang, M., Zhang, J., Peng, W., Firempong, C. K., Deng, W., ... Zhang, W. (2015). Improved oral bioavailability of capsaicin via liposomal nanoformulation: Preparation, in vitro drug release and pharmacokinetics in rats. *Archives of Pharmacal Research*, *38*(4), 512–521. doi:10.100712272-014-0481-7 PMID:25231341

Zinoubi, K., Majdoub, H., Barhoumi, H., Boufi, S., & Jaffrezic-Renault, N. (2017). Determination of trace heavy metal ions by anodic stripping voltammetry using nanofibrillated cellulose modified electrode. *Journal of Electroanalytical Chemistry*, *799*, 70–77. doi:10.1016/j.jelechem.2017.05.039

Zolgharnein, J., Asanjarani, N., & Shariatmanesh, T. (2013). Taguchi L16 orthogonal array optimization for Cd(II) removal using *Carpinus betulus* tree leaves: Adsorption characterization. *International Biodeterioration & Biodegradation*, *85*, 66–77. doi:10.1016/j.ibiod.2013.06.010

About the Contributors

Karina Cecilia Arredondo-Soto studied Industrial Engineer at the Tijuana Institute of Technology where she continued studying a Master Degree in Industrial Engineering too; with a minor in Manufacturing. She has a Ph.D. in Industrial Engineering at the Ciudad Juarez Institute of Technology. She is a full-time teacher at the Autonomous University of Baja California since 2012 and an invited Professor at the Tijuana Institute of Technology since 2008. Her interest area is work studies to improve productivity in organizations through the collaborative work with the industrial regional sector.

Jose Antonio Rodríguez Ávila is an Assistant Professor at Chemistry Academic Area from Universidad Autónoma del Estado de Hidalgo (Mexico) since 2007. He received his Ph. D. degree in Applied Analytical Chemistry from the University of Valladolid (Spain) in 2005 and then he worked as assistant research in a private industry. His research interests include development of sample preparation procedures combined with, flow, chromatography and electrochemical techniques, which are applied in the analysis of emerging pollutants in environmental and food samples. He has been awarded by different national and international societies for his contribution in food analysis.

Eugenia Gabriela Carrillo-Cedillo Education 1990 B.S., Industrial Chemistry. Universidad Autonoma de Baja California (UABC). Mexico. 1997 M.S., Chemical Sciences. UABC, Mexico. 2008 Ph.D., Sciences. UABC, Mexico Academic and Professional Positions Held 1990-1999 Laboratory Technician Faculty of Chemical Sciences UABC. 1999-present Professor. Faculty of Chemical Sciences and Engineering UABC. 2013-2015 President of the Academy of Natural and Exact Sciences, UABC. 2016-2018 Member of the Board of Representatives of the Mexican Association of Analytical Chemistry. Teaching Undergraduate and Graduate Courses on Analytical, Flow Chemistry, Design of Experiments, Water Analysis, Statistical Quality Control, Probability and Statistics. Research Publications Indexed in JCR: 8, Book Chapters: 4 and Books: 1. Thesis Advised Undergraduate Thesis: 4, Master Degree Thesis: 5 and Doctorate Degree thesis: 1. Awards National Research System Member (Candidate): 2017-2020 and Professor with Desirable Profile (SEP): 2006-2020.

José Manuel Cornejo-Bravo received a BS in Chemistry, and a MS in Biopharmaceutics from the Autonomus University of Baja California. He received his Ph.D. in Pharmaceutical Chemistry from the University of California San Francisco. He is a full-time Professor-Researcher at the Faculty of Chemical Sciences and Engineering, Autonomous University of Baja California since 1992. His research interest are in the field of drug delivery and pharmacotherapy.

* * *

Carlos Manuel Franco Abuin is associate professor in the Univerdad de Santiago de Compostela. He has published 130 JCR articles, as well as several books, book chapters and other scientific documents. He is actually member of the scientific commitee of the Spanish Agency for Food Safety and Nutrition.

Sevim Akyuz is a professor of physics, currently working at the Physics Department of Istanbul Kultur University (Istanbul-Turkey). Her research interest is on experimental and theoretical molecular spectroscopy and mainly focused on the study of a relationship between the structure and spectral properties of macromolecules using combined theoretical and experimental methods. She has 182 scientific papers as SCI record and total number of citations according to Web of Knowledge is 1335. Her H-index according to Web of Knowledge is 22.

Jesús Ivan Alaniz-Muñoz is Masters In Industrial engineering from Tecnológico Nacional de México/ Instituto Tecnológico de Tijuana. Profesional experience as production manager, process engineering, logistics and operations resource planning.

Ali Tugrul Albayrak is Research Assistant Dr. at the Chemical Engineering Department of Istanbul University - Cerrahpasa (Turkey). His research is focused on the elucidation of the structure of ionic liquids using experimental methods. He has three scientific papers as SCI record.

Enrique Barrado is Professor in the Analytical Chemistry Department (Universidad de Valladolid) Experience on application of Design of experiment in Instrumental techniques (separation and electroanalytical).

Juan Andrés López Barreras is Industrial Engineer, Instituto Tecnológico de Huatabampo. Master of Science in Industrial Engineering, Instituto Tecnológico de Tijuana. Doctor in global development, at Universidad Autónoma de Baja California. Professor and researcher, Universidad Autónoma de Baja California.

Sefa Celik is an Associate Professor at the Physics Department of Istanbul University (Turkey). His research is focused on the elucidation of the structure of biomacromolecules using combined experimental (molecular spectroscopy) and theoretical methods (ab initio, molecular dynamics), particularly on the molecular docking studies for shedding light on new drug design. He has 21 scientific papers as SCI record.

Paola Cervantes-Covarrubias Bachelor degree as Biological and Pharmaceutical Chemist. Honorable mention as Master Chemical Sciences. Currently working as Subject Professor and Ph. D. Sciences student. Faculty of Chemistry Science and Engineer, Autonomous University of Baja California, developing projects with solid lipid nanoparticles of different drugs to deliver on the skin.

Karla Frida Madrigal Estrada is a Research Professor at Universidad Autonoma de Baja California. She has a Master science degree in Manufacturing Systems.

T. Alexandra Ferreira is M. Sc. degree in Analytical Chemistry.

Carlos Manuel Franco Abuin is associate professor in the Univerdad de Santiago de Compostela. He has published 130 JCR articles, as well as several books, book chapters and other scientific documents. He is actually member of the scientific commitee of the Spanish Agency for Food Safety and Nutrition.

Alexandre Lamas Freire is graduated in Veterinary Medicine, he received his doctorate in 2018 as a Doctor of Veterinary Medicine. He is a postdoctoral researcher at the University of Santiago de Compostela and has published 25 JCR articles.

Carlos Andrés Galán-Vidal received his Ph.D. in Analytical Chemistry in 1996 from the Universtitat Autònoma de Barcelona, currently he is working at the Analytical Chemistry and Physical Chemistry of Solutions and Surfaces Group of the Universidad Autónoma del Estado de Hidalgo. His main research interests concern the development of analytical methodologies and design of electrochemical sensors.

Germán Adán Sánchez Gallegos. German Adan Sanchez-Gallegos. Education (2012) B.S., Industrial Chemistry. Autonomous University of Baja California (UABC). Mexico. (2016) M.S., Chemical Sciences. UABC, Mexico. He is currently studying a Ph.D. Sciences and interested in the analysis of water and food of the regiôn of Baja California. He is a paramedic of the Red Cross in Rosarito beaches.

Juan Carlos Castillo Garcia is Industrial Engineer from the Autonomous University of Baja California, Ensenada, Mexico. Master of Science in Industrial Engineering from the Autonomous University of Baja California, Ensenada, Mexico. Current PhD student at Universidad Autonoma de Baja California, Ensenada, Mexico.

Alejandra García-Lechuga Subject Teacher and Doctoral Student at the Faculty of Chemical Sciences and Engineering, Industrial Engineering Program. She teaches among other courses Planning and Design of Industrial Installations. Here current research work focuses on Product-Service System design strategies for emerging market conditions.

Gabriela Islas Guerrero is actually collaborating at Universidad Politécnica de Francisco I. Madero. She has worked in separation methods and actually the design of experiments.

Evelin Gutiérrez is actually collaborating at Universidad Politécnica de Metropolitana del Estado de Hidalgo. She has worked in corrosion topics and actually the design of experiments.

Guadalupe Hernández-Escobedo is Profile PROMEP; PhD from the University of Leeds, UK; more than 20 years of experience in teaching, research and application of Industrial Engineering at Tecnológico Nacional de México/Instituto Tecnológico de Tijuana.

Israel S. Ibarra currently works at the Department of Analytical Chemistry, Universidad Autonoma del Estado de Hidalgo has worked on the development of techniques of separation and synthesis process in the analysis of emerging contaminants through Chemometric tools, Design of experiments and Validation tests.

Gabriela Islas is actually collaborating at Universidad Politécnica de Francisco I. Madero. She has worked in separation methods and actually the design of experiments.

Alejandro Jiménez-Zaragoza studied Industrial Engineer at the Universidad Autónoma de Baja California and he is currently studying a Master Degree in Manufacturing Processes and optimization.

Isabel Maria João is a Professor at the Polytechnic Institute of Lisbon, School of Engineering (ISEL), at the Chemical Engineering Department since 2000. She has 9 years of Industrial experience in a Plastic industry where she was the chief of the laboratory of I&D and quality control and improvement. Isabel received her PhD degree in Industrial Engineering and Management from IST/UTL and a Master degree in Management and Quality of Materials from FCT/UNL and she also holds a graduation in Chemical Engineering from IST/UTL. She is a researcher at the Centre for Management Studies of IST (CEG-IST) and her areas of interest include Innovative Product and Process Design, Quality Management and Decision Analysis and Information Systems.

Alfredo Guevara Lara was born in México City. He gained his B. Sc. and M. Sc. degrees in Chemical Engineering from Autónoma Metropolitana-Iztapalapa university and completed his Ph.D. under the direction of professor Michel Vrinat at the institute des Recherches Sur la Catalyse et l'environnement de Lyon, France in 2003. He is currently in the Research in Chemical Area at the Universidad Autónoma del Estado de Hidalgo, México. His research centers on the synthesis and use of heterogeneous catalysis with current focus on morphological nanostructured control of metallic catalysts by applications in fuel and biofuel hydrotreatment themes.

Sugey López-Martínez is Professor-Researcher in environmental remediation.

Rafael García Martínez B.Sc. in Mathematics, Universidad Autónoma de Nuevo León. Maestro Master and Doctor of Science in Industrial Engineering, Instituto Tecnológico de Ciudad Juárez. Professor- researcher at the Instituto Tecnológico del Hermosillo. Sonora. México.

María Gabriela Vargas Martínez is a professor with 27 years of experience at the National Autonomous University of Mexico. She obtained her PhD in Analytical Chemistry from the University of Masaryk, Czech Republic. She works in the development of analytical methods using capillary electrophoresis, liquid chromatography and gas chromatography applied in fundamental studies and in the determination of several substances present in food, biological fluids, medicines and environmental samples. Other areas of incursion are chemometrics and the use of other techniques such as absorption and atomic emission, all for the solution of problems in the field of analytical chemistry.

Candelario Méndez-Olán is Professor-Researcher in statistics applied to environmental assessments.

Marco A. Miranda-Ackerman is Research Professor at the Faculty of Chemical Sciences and Engineering, Industrial Engineering Program. He teaches among other courses Design of Experiments at an undergraduate level and Quantitative Methods in geography at a graduate level. Candidate Member of the National Researcher System of Mexico. Research Member of the "Network to Reduce and Value Food Loss and Waste: Towards Sustainable Food Systems". Current research interest is in Sustainable Food Supply Chain Design and Management.

José Manuel Miranda is actually associate professor at the Universidad de Santiago de Compostela (Spain). He has extensive extra-university experience as a health inspector and has published more than 100 JCR articles and several books.

Carlos Mario Morales-Bautista. Professor-Researcher in the area of Chemistry at the Universidad Juárez Autónoma de Tabasco since 2013. Has 6 years of experience in environmental assessment in soils contaminated with hydrocarbons. Has a PhD in Science in Ecology and Tropical Systems Management and received her Master's Degree in Engineering and Environmental Protection as well as a Bachelor's Degree in Chemistry. He is a researcher recognized by the SNI and leader in his Research Group in Applied Chemistry whose line of application of knowledge is analytical geochemistry and his areas of interest include management, diagnosis and evaluation in ecology's systems.

Mydory Nakasima-López is a full-time research professor in the industrial engineering education program at the Faculty of Chemical Sciences and Engineering, She has a PhD in engineering and her line of applied knowledge generation is in renewable energies and sustainability.

Sukey Nakasima-López is a subject teacher in the computer engineering and industrial engineering programs for the Faculty of Chemical Sciences and Engineering, and she is currently studying the last year of the doctorate in science and engineering focused on data science.

Marcia E. Ojeda-Morales is Professor-Researcher in environmental evaluation.

María del Carmen Beltrán Orozco is researcher at the National School of Biological Sciences of the National Polytechnic Institute, Mexico.

Ayşen E. Özel is a Professor at the Physics Department of Istanbul University (Turkey). She is a head of Atom and Molecular Physics division. Her research is focused on the molecular spectroscopy and on the molecular modeling. She has 34 scientific papers as SCI record.

Selene Pascual-Bustamante is Food Engineer from the National Autonomous University of Mexico (UNAM). Master in Food Sciences from the National Polytechnic Institute. Professor of the Department of Engineering and Technology at the FESC (UNAM).

Alicia del Carmen Mondragon Portocarrero holds a doctorate from the University of Santiago de Compostela. She has worked as a full professor at the Autonomous University of the State of Hidalgo (Mexico), and currently works as head of R+D+i in the biotechnology company Deinal Soluciones Agroalimentarias.

Arturo Realyvásquez-Vargas has studied Ergonomics since 2008. At the beginins, he contributed to this sicence developing an expert system to measure ergonomic compatibility of advanced manufacturing technology. Most recently, he developed the Macroergonomic Compatibility Questionnaire (MCQ), he proved that macroergonomic compatibility has positive effects on work systems performance, and he proposed the macroergonomic compatiblity index (MCI). He has published and edited books in the most recognised publishers. In addition, he has pusblised several articles in JCR journals. Currently, he works as full time professor in Instituto Tecnológico de Tijuana. He offers the classes of Ergonomics, Operations Research, and Research Fundamentals and he is member of the National System of Researchers of the National Council of Science and Technology (CONACYT).

Alberto Cepeda Saez is full professor in the Universidade de Santiago de Compostela (Spain). He is head of the food inspection and control laboratory research group. He has published 185 JRC articles and his h index is 32, being cited for more than 3.500 times.

Erika Beltrán Salomón is Master of Science in Industrial Administration Research professor of the Industrial Engineering program by the Faculty of Chemical Sciences and Engineering.

Germán Adán Sánchez-Gallegos Education (2012) B.S., Industrial Chemistry. Autonomous University of Baja California (UABC). Mexico. (2016) M.S., Chemical Sciences. UABC, Mexico. He is currently studying a Ph.D. Sciences and interested in the analysis of water and food of the regi6n of Baja California. He is a paramedic of the Red Cross in Rosarito beaches.

Aracely Serrano-Medina B.S. Biopharmaceutical Chemistry, MSc. Chemical Sciences, Ph.D. Sciences.

João Miguel Silva is a Coordinator Professor at the Polytechnic Institute of Lisbon, School of Engineering (ISEL) and the Head of the Chemical Engineering Department since 2010. He is member of the research center CQE – Centro de Química Estrutural in the Heterogeneous Catalysis and Catalytic Processes (CATHPRO) group at IST. João received his PhD degree in Chemical Engineering in 1994 from IST/UTL and is co-author of over 50 articles in peer-review journals. His areas of interest include Innovative Product and Process Design, Process Simulation and Heterogeneous Catalysis, specifically with zeolites.

Jesús Andrés Tavizón-Pozos received his MS and PhD degrees in Universidad Autónoma Metropolitana - Iztapalapa in Mexico City in 2011 and 2017 respectively. In 2011 he was hired as Chemistry teacher in the Human Medicine program in the Universidad Autónoma de Zacatecas. In 2017, he was hired in the Universidad Autónoma Metropolitana - Iztapalapa in the Chemical Engineering program. He has studied catalytic systems for hydrodesulfurization and hydrodeoxygenation for biofuels. Currently he works for CONACYT in the program of Cátedras. He was assigned to develop new catalytic systems for biofuel productions in the Universidad Autónoma del Estado de Hidalgo.

Jesús Everardo Olguín Tiznado. Industrial Engineer from the Huatabampo Technological Institute, Sonora, Mexico. Master of Science in Industrial Engineering from the Hermosillo Technological Institute, Sonora, Mexico. Doctor of Science in Industrial Engineering from the Autonomous University of

Baja California, Ensenada Campus. Professor - Researcher of the Faculty of Engineering, Architecture and Design at the Autonomous University of Baja California, Ensenada Campus. He has developed in the area of occupational and environmental health, and in the modeling of optimization of industrial processes. He has collaborated in different academic and administrative functions for more than 14 years in the Autonomous University of Baja California. He has participated in research projects in occupational health, ergonomics, and optimization of industrial processes through the modeling of stochastic approximation algorithms, among others. He has participated in agreements with Industry and Public Institutions; he promotes science and education in previous educational systems.

María del Pilar Haro Vázquez Education 1986 B.S. Industrial Chemistry, Autonomus University of Baja California (UABC) Mexico, 1994 M.S. Chemical Science Tijuana Technologic Institute (ITT), Mexico; 2008 Ph.D Sciences UABC, Mexico, Full time profesor in analytical área. Severas Research publication indexes in JCR and other book publications.

Ayla Vea-Barragan Bachelor's degree in Biology and Pharmaceutical Chemistry later I study a Master Degree in chemist science, development a buccal device for the dentist. I am a postgraduate student of Ph.D. in Chemist Science with the work in mucoadhesive systems and new construction of methods for drugs with modified in the release. I am half time professor in the Autonomous University of the State of Baja California for pharmaceuticals students.

Claudia Camargo Wilson is Industrial Engineering, Instituto Tecnológico de Los Mochis. Master of Science in Industrial Engineering, Instituto Tecnológico de Hermosillo.Doctor of Science in Industrial Engineering in the Autonomous University of Baja California, Professor-researcher, in Universidad Autónoma de Baja California. Campus Ensenada México.

Index

A

B

C

D

E

F

G

H

I

K

L

M

N

O

P

Q

R

S

T

V

W

www.ingramcontent.com/pod-product-compliance
Lightning Source LLC
LaVergne TN
LVHW062110300926
849161LV00010B/2416